Marine Mam  Conservation

Techniques in Ecology and Conservation Series

Series Editor: William J. Sutherland

Bird Ecology and Conservation: A Handbook of Techniques
William J. Sutherland, Ian Newton, and Rhys E. Green

Conservation Education and Outreach Techniques
Susan K. Jacobson, Mallory D. McDuff, and Martha C. Monroe

Forest Ecology and Conservation: A Handbook of Techniques
Adrian C. Newton

Habitat Management for Conservation: A Handbook of Techniques
Malcolm Ausden

Conservation and Sustainable Use: A Handbook of Techniques
E.J. Milner-Gulland and J. Marcus Rowcliffe

Invasive Species Management: A Handbook of Principles and Techniques
Mick N. Clout and Peter A. Williams

Amphibian Ecology and Conservation: A Handbook of Techniques
C. Kenneth Dodd, Jr.

Insect Conservation: Approaches and Methods
Michael J. Samways, Melodie A. McGeoch, and Tim R. New

Marine Mammal Ecology and Conservation: A Handbook of Techniques
Ian L. Boyd, W. Don Bowen, and Sara J. Iverson

Remote Sensing for Ecology and Conservation: A Handbook of Techniques
Ned Horning, Julie A. Robinson, Eleanor J. Sterling, Woody Turner, and Sacha Spector

Marine Mammal Ecology and Conservation

A Handbook of Techniques

Edited by

Ian L. Boyd
W. Don Bowen

and

Sara J. Iverson

OXFORD
UNIVERSITY PRESS

Great Clarendon Street, Oxford OX2 6DP

Oxford University Press is a department of the University of Oxford.
It furthers the University's objective of excellence in research, scholarship,
and education by publishing worldwide in

Oxford New York

Auckland Cape Town Dar es Salaam Hong Kong Karachi
Kuala Lumpur Madrid Melbourne Mexico City Nairobi
New Delhi Shanghai Taipei Toronto

With offices in

Argentina Austria Brazil Chile Czech Republic France Greece
Guatemala Hungary Italy Japan Poland Portugal Singapore
South Korea Switzerland Thailand Turkey Ukraine Vietnam

Oxford is a registered trade mark of Oxford University Press
in the UK and in certain other countries

Published in the United States
by Oxford University Press Inc., New York

British Library Cataloguing in Publication Data

Data available

Library of Congress Control Number: 2010929117

Typeset by SPI Publisher Services, Pondicherry, India
Printed in Great Britain
on acid-free paper by
CPI Antony Rowe, Chippenham, Wiltshire

ISBN 978–0–19–921656–7 (Hbk.)
ISBN 978–0–19–921657–4 (Pbk.)

3 5 7 9 10 8 6 4 2

Preface

Arguably, a piece of research is only as good as the methods used to derive and interpret the data. In this respect, studies of the biology of marine mammals are no different from any other research. However, where they do tend to differ is in the particular challenges presented by the aquatic lifestyle, their large body size and, in many cases, the endangered status of marine mammals. The total amount of research on marine mammals has increased rapidly in recent decades. By any metrics—number of researchers, total funding, relevance to policy or commerce, research outputs, quality of research outputs, or the number of meetings or conferences dedicated to marine mammals world-wide—there has been a rapid increase in the volume and sophistication of research being carried out on marine mammals. This has changed the field from a small, slightly cliquey branch of research that sat at the boundaries between physiology, ecology, mammalogy, behavioural biology, and marine biology to one that is making an important contribution to many of these fields and that is now contributing to the development of ideas in diverse areas of biology. This contribution lies in both the theoretical and empirical spheres of biology. Some 30 years ago, the principal method for studying marine mammals was for researchers to follow in the wake of hunters or become hunters themselves and work with dead tissue samples. Times have changed. Remarkable innovations have led to new ways of investigating the lives of marine mammals. Some of these have relied upon the development of new technology, but many have been driven by genuine ingenuity, invention, and intellectual advances, as represented by examples of novel instruments for obtaining measurements or statistical approaches to analysing and interpreting data.

Marine mammals present particular challenges to ecologists and conservation biologists. They are often difficult to observe, at least during certain critical parts of their life cycle; they occupy a vast environment both in terms of the geographical range of species and their use of the water column. They present difficult conservation challenges both because there may be perceived conflicts between people and marine mammals and because some species are close to extinction. Marine mammals also command a high level of public attention (they can be truly described as 'charismatic species') reflected in specific legislation for their protection and management in many countries.

Although there are numerous books about marine mammals, there has been only one previous attempt to produce a book that summarized techniques in an aspect of marine mammalogy (Laws 1993). Consequently, we approached the task of producing this book with the objective of summarizing the methodological approaches to studying marine mammals in a single, accessible, authoritative volume. Although it is in the nature of scientific writing to provide detailed descriptions of methods, there are several reasons why we considered that it would be useful to produce a higher level description in a single volume. First, we are aware of the new generations of marine mammalogists that will be entering training. Their experience of the methodological pallet available to them in their research careers needs to be broader and more integrative than anything they can obtain from the condensed descriptions of methods found within research papers. Second, we wish to encourage experienced researchers 'to think outside the box'. For many, the methods they use will represent the constraints on what they can study and it is possible that we might be able to encourage innovation by making connections across different approaches. Third, we also see a need to standardize methods once they have become accepted and established, because through standardization one can encourage cross-compatibility of studies (although we also recognize that standardization is tensioned against innovation in many ways). Finally, we want to encourage researchers to think first and then act. This is central to the increasingly important ethical approach to research as discussed in Chapter 1. Research these days is highly regulated, but it is essential for researchers themselves to regulate their activities because, if they fail to do so, other less qualified individuals will do so. The consequences of this will be bad for humanity and for marine mammals. This is why we have opened the book with a chapter on ethics and the reader will find that ethical considerations are an almost continuous thread that is weaved through the fabric of the book. They come to the fore in particular when capturing and handling marine mammals, a subject that is dealt with in Chapter 2.

Methods can change, and in some areas of marine mammal biology they are evolving rapidly—tracking technology (Chapter 10), diet estimation (Chapter 9), modelling (Chapters 3 and 4), and population genetics (Chapter 14) to name but four. Application of new technologies and imaginative new approaches are allowing rapid headway in areas previously well beyond the scope of most science. The rapidity of change is a problem for a book of this type because it can soon become outdated. The intention, however, has not been to produce a manual. Instead, we have attempted to produce an overview that will have a long half-life by pointing out generic methods and illustrating their strengths and weaknesses.

Our intent to make the book authoritative could rightly have included most active marine mammal researchers as contributors, but this was clearly impossible. Nevertheless, a total of 46 people have contributed their knowledge and expertise to writing this book and we are grateful to all for their contributions. We are also grateful to the numerous individuals who provided reviews of chapters after they had been drafted.

The chapter subjects are intended to represent the main areas of marine mammal biology, but doubtless there will be gaps and we acknowledge that it would have been possible to divide the subjects differently. However, we were particularly keen to blur the division that is all too prevalent at times within marine mammalogy between cetacean and pinniped research. While each group does have its special problems, and specific methods have been developed to address these, there is probably much to be learned by mixing across this division because, if innovation is to occur, it could come just as easily from pinniped and cetacean biologists working more closely together than by either finding stimulation from outside the field.

A challenge throughout production was to constrain the book from becoming too large and, therefore, uneconomical. Enabling authors to pitch their contributions at the right level to ensure that they were both comprehensive and also concise was a considerable challenge, and we decided to trade off words for illustrations in many chapters as a way of making sure that the information was provided at an appropriate level of detail and that important areas were not excluded. We hope that we have got this balance about right.

Inevitably, when in the final stages of editing the book we have seen gaps. Perhaps the most serious of these is the way in which we have dealt with the rapidly moving field of genetics. Chapter 14 provides one side of this story at the level of populations, but we are also aware that the methods used to examine gene expression and issues surrounding the apparent importance of heterozygosity for the fitness of individuals has highlighted the need to recognize the dangers of the ecological fallacy, and this is especially true for marine mammals. It could be argued that many misconceptions concerning the management of marine mammals, and that possibly even now reside within legislation used to manage marine mammals, have emerged because individuals are seen in terms of populations and not populations in terms of individuals.

One increasingly important issue not explored in this book is the issue of study design. Given additional space we might have written a specific chapter on this. It is a recurring issue throughout the book and it underpins the ethical features of marine mammal research. The issue of not undertaking studies unless they have clear objectives, and also that the methods used are going to provide the statistical power to come to firm conclusions, is something that cannot be overstated. Chapter 13, which is a description of some long-term studies, contains some important lessons in design. These studies are, by definition, those that have been successful. However, their success has come about for a wide range of reasons. In many cases, serendipity has played an important part, but in others there has been an intentional design from an early stage that has driven the study forward based upon long-term goals. Perhaps the two factors that unite these studies are organization and relevance. In some cases these factors have been promoted by enterprising individuals who have, by dint of the power of their personality, made the study happen and given it staying power, but more often than not the studies have succeeded because higher level organizational structures (such as funding agencies or strategic science institutes) have accepted the importance of the research. This has happened either because of the unique capacity of the research to illustrate important points of general interest to biology or because of its relevance to specific management problems, especially in conservation and bioresource management.

Therefore, the book has not dwelt upon the strategic aspects of research design and management, which will likely become increasingly important parts of the overall methodological repertoire of researchers. Indeed, if we were to move in to the dangerous areas of trying to predict future trends in methods where major gains might be made it might not be in technology or some of the more obvious areas including statistical data analysis, but instead in the higher level organization of research which encourages more efficient use of sparse research funding and more focused advanced planning. This might include the process by which research priorities are set and established and how hypotheses are formulated. The free flow of data and ideas and the establishment of international databases will be needed to effectively capitalize on the growing importance of comparative and meta-analytic approaches to understanding critical habitat, biodiversity hotspots, and the impact of anthropogenetic disturbance on marine mammals. The methodology that needs to be developed and applied in this case concerns the development of incentives to cooperate amongst researchers. Although further discussion of this is well beyond the scope of the current book, the rapid evolution of the approach to issues of ethics by marine mammal researchers gives hope that this is not beyond the scope of practising researchers. Indeed, there have already been some important moves towards addressing this issue and we hope that this book becomes a part of that movement.

Finally, we thank all our authors who have been patient and generous with their time and energy during the long gestation of this book. In addition to their own writing tasks many also reviewed chapters and we are also grateful to them for this assistance. But we also wish to thank other reviewers for their time, including W. Amos, M. Bekoff, A. Cañadas, R. Connor, T. Goldstein, J. Graves, M. Hammill, M. Heithaus, P. Jepson, T. Ragen, R. Reeves, W. Testa, P. Thompson, and P. Wade. Our thanks also go to the staff at Oxford University Press for their guidance and patience throughout the production of the book.

Ian L. Boyd
W. Don Bowen
Sara J. Iverson

Contents

List of abbreviations   xviii
List of contributors   xxiii

1 Ethics in marine mammal science   1

Nick J. Gales, David Johnston, Charles Littnan, and Ian L. Boyd

1.1 Introduction   1
   1.1.1 Ethics: defining the boundaries between right and wrong   3
   1.1.2 Financial and professional costs of compliance   4
   1.1.3 Keeping the research ethics and conservation ethics apart   5
1.2 Guiding principles   6
   1.2.1 Science for the common good   6
   1.2.2 Principle 1   6
   1.2.3 Principle 2   7
   1.2.4 Principle 3   7
   1.2.5 Principle 4   7
   1.2.6 Principle 5   8
1.3 The role of professional societies in developing and maintaining
   ethics in marine mammal science   8
1.4 The 3 Rs (refinement, reduction, and replacement)   9
1.5 Type I and Type II errors   11
1.6 Best practice   12
   1.6.1 Threatened versus abundant   12
   1.6.2 Timing and location   13
   1.6.3 Experimental procedures and equipment   13
   1.6.4 Training   13
   1.6.5 Environmental considerations   13
   1.6.6 Cultural consideration   14
   1.6.7 Decision analysis frameworks and cost–benefit analyses   14
1.7 Summary   15

2 Marking and capturing   16

Tom Loughlin, Louise Cunningham, Nick J. Gales, Randall Wells, and Ian L. Boyd

2.1 Introduction   16
2.2 Applying marks   17
   2.2.1 Flipper tagging pinnipeds   17
   2.2.2 Discovery tags   19

2.2.3 PIT tags   19
2.2.4 Hot-iron branding   20
2.2.5 Freeze-branding   21
2.2.6 Fur clipping   23
2.2.7 Dye marking   23
2.2.8 Dorsal fin tags on cetaceans   24
2.3 Photo-identification   25
2.3.1 Strengths and weaknesses   26
2.3.2 Application   26
2.3.3 Pattern recognition methods   26
2.3.4 Assessing errors   28
2.4 Capture and restraint   28
2.4.1 Techniques for restraint in pinnipeds   29
2.4.2 Chemical restraint and immobilization in pinnipeds   33
2.4.3 Techniques for capture–release of cetaceans   35
2.4.4 Techniques for restraint and handling   38
2.4.5 Risks to cetaceans during capture and handling   39
2.5 Risks to researchers during marine mammal capture and handling   40

3 Estimating the abundance of marine mammals   42

Philip S. Hammond

3.1 Introduction   42
3.2 Extrapolation of counts   44
3.3 Mark–recapture methods   45
3.3.1 Assumptions in theory and practice   47
3.3.2 Data collection   49
3.3.3 Analysis   51
3.4 Line transect sampling   53
3.4.1 Assumptions in theory and practice   55
3.4.2 Extensions and variants of conventional line transect sampling   56
3.4.3 Data collection on visual line transect surveys   59
3.4.4 Analysis   63
3.5 Concluding remarks   65
3.6 Acknowledgements   67

4 The spatial analysis of marine mammal abundance   68

Jason Matthiopoulos and Geert Aarts

4.1 Introduction   68
4.2 The importance of life history   69
4.3 Usage data and sampling design   72
4.4 Basic concepts and challenges   75
4.5 Pre-processing   81

4.6 Analysis techniques   82
4.7 The spatial analyst's software toolbox   91
4.8 Prospects and trends   95

5 Morphometrics, age estimation, and growth   98
W. Don Bowen and Simon Northridge

5.1 Introduction   98
5.2 Standard morphometrics   98
     5.2.1 General issues   98
     5.2.2 Body condition   100
     5.2.3 Standard measurements   101
5.3 Age determination   107
     5.3.1 Dental growth layers   108
     5.3.2 Other age-structured material   114
     5.3.3 Chemical methods   115
5.4 Growth rates   116
5.5 Conclusions   117

6 Vital rates and population dynamics   119
Jason D. Baker, Andrew Westgate, and Tomo Eguchi

6.1 Introduction   119
6.2 Reproductive rate   120
     6.2.1 Description of parameters   120
     6.2.2 Techniques   123
6.3 Survival rate   126
     6.3.1 Cross-sectional age structure analysis   126
     6.3.2 Capture–recapture   127
     6.3.3 Study design considerations   131
     6.3.4 Model selection   132
     6.3.5 Multi-state models   134
     6.3.6 Recoveries from dead animals   134
     6.3.7 Robust design   135
6.4 Population models   135
     6.4.1 Exponential and geometric models   136
     6.4.2 Matrix models   137
     6.4.3 Incorporating uncertainty: deterministic versus stochastic models 138
     6.4.4 Density dependence   139
     6.4.5 Individual-based models (IBM)   140
     6.4.6 Population viability analysis (PVA)   141
     6.4.7 Bayesian approach to modelling   142
     6.4.8 Model fit and selection   142

7 Epidemiology, disease, and health assessment   **144**

Ailsa J. Hall, Frances M.D. Gulland, John A. Hammond, and Lori H. Schwacke

7.1 Introduction   144
     7.1.1 Exposures and responses   146
     7.1.2 Confounding factors   147
7.2 Effects, responses, and diagnostic techniques   148
     7.2.1 Measuring disease occurrence   148
     7.2.2 Responses, health panels, and disease classification   149
     7.2.3 Sample and data collection   152
     7.2.4 Disease diagnosis   155
7.3 Epidemiological study designs   158
7.4 Risk assessment   160

8 Measurement of individual and population energetics of marine mammals   **165**

Sara J. Iverson, Carol E. Sparling, Terrie M. Williams, Shelley L.C. Lang, and W. Don Bowen

8.1 Introduction   165
     8.1.1 Definitions   167
8.2 Measurement of metabolism   169
     8.2.1 Direct calorimetry   170
     8.2.2 Respirometry (measurement of gas exchange)   170
     8.2.3 Doubly labelled water (DLW) and isotope dilution   173
     8.2.4 Proxies for assessing energy expenditure (EE)—heart rate (f_H) and stroking rate (f_S)   176
     8.2.5 Proxies for assessing energy expenditure (EE)—allometry   178
8.3 Estimating body composition   179
     8.3.1 Carcass analysis and the relationship between chemical constituents   179
     8.3.2 Total body water (TBW) measurement   180
     8.3.3 Ultrasound   181
     8.3.4 Bioelectrical impedence analysis (BIA)   182
     8.3.5 Novel approaches to estimating body composition   182
8.4 Energy balance analysis   183
8.5 Energetics of lactation   184
     8.5.1 Milk composition   185
     8.5.2 Milk output and milk energy output   187
8.6 Population energetics   189

9 Diet   **191**

Dominic J. Tollit, Graham J. Pierce, Keith A. Hobson, W. Don Bowen, and Sara J. Iverson

9.1 Introduction   191
9.2 Collection of gastrointestinal tract contents   194

9.2.1 Sampling dead animals 194
9.2.2 Lavage 195
9.2.3 Rectal enema and faecal loops 195
9.3 Collection of faeces and regurgitated/discarded prey remains 196
9.4 Sampling bias 197
9.5 Laboratory processing of prey hard structures 198
9.5.1 Prey extraction 198
9.5.2 Prey identification 199
9.5.3 Prey enumeration using minimum number of individuals (MNI) 200
9.5.4 Measurement of prey structures 201
9.6 Quantification of diet composition using GI tract and faecal analyses 203
9.6.1 Accounting for complete digestion of hard part structures 203
9.6.2 Other factors affecting recovery of hard part structures 204
9.6.3 Accounting for partial digestion of hard part structures 204
9.6.4 Prey length and mass reconstruction 205
9.6.5 Quantification methods 206
9.7 Molecular identification of prey remains 208
9.8 Fatty acid (FA) signatures 210
9.8.1 Tissues for analysis 211
9.8.2 Sample storage and chemical analysis 213
9.8.3 Using predator FAs to qualitatively infer diet 214
9.8.4 Using predator and prey FAs to quantitatively estimate diet 214
9.9 Stable isotopes and other markers 216
9.9.1 Tissues for analysis 217
9.9.2 Trophic modelling 218
9.9.3 Source of feeding and marine isoscapes 219
9.9.4 Other elements and compounds 219
9.9.5 Field and laboratory methods and data analysis 219
9.10 Summary 221

10 Telemetry 222

Bernie McConnell, Mike Fedak, Sascha Hooker, and Toby Patterson

10.1 Introduction 222
10.2 Design considerations 223
10.3 Attachment 224
10.4 Location determination 227
10.5 Sensors 228
10.6 Information relay 230
10.7 Data modelling and compression 234
10.8 Visualization and analysis of individual paths 234
10.8.1 Heuristic approaches to position filtering 236
10.8.2 Statistical approaches to path analysis 237
10.9 Concluding remarks 241

11 Foraging behaviour 243

Mark A. Hindell, Dan Crocker, Yoshihisa Mori, and Peter Tyack

11.1 Introduction 243
11.2 Observation of foraging 243
  11.2.1 Following focal animals 243
  11.2.2 Passive acoustic observations 244
11.3 Cameras 245
11.4 Reconstruction of foraging behaviour 246
  11.4.1 Time–depth recorders 246
  11.4.2 Sampling strategies 249
  11.4.3 Interpretation of 'foraging' dives 249
11.5 Feeding success 252
11.6 Acoustics and echolocation 254
11.7 Foraging tactics 256
  11.7.1 Central place foraging 256
  11.7.2 Search tactics and characterizing the prey field 257
  11.7.3 Functional response 258
11.8 Foraging models 259
  11.8.1 Optimal foraging models 259
  11.8.2 Stochastic models 261

12 Studying marine mammal social systems 263

Hal Whitehead and Sofie Van Parijs

12.1 Introduction 263
  12.1.1 The definition of social structure 263
  12.1.2 How do we study social structure? 263
  12.1.3 Styles of studying social structure 265
12.2 Field research 265
  12.2.1 Identifying individuals 265
  12.2.2 Collecting interaction, association, and group data 266
  12.2.3 Collecting social data without observing animals 268
12.3 Relationship measures 269
  12.3.1 Interaction rates 269
  12.3.2 Association indices 269
  12.3.3 Temporal measures 270
  12.3.4 Matrices of relationship measures 270
12.4 Describing and modelling social structure 271
  12.4.1 Visual displays 271
  12.4.2 Testing for preferred/avoided companions 273
  12.4.3 Network analyses 274
  12.4.4 Lagged association rates 276
  12.4.5 Describing mating systems 277

	12.4.6 Other methods of social analysis	277
	12.4.7 Useful software	278
12.5	Broader issues	278
	12.5.1 Evolutionary forces behind marine mammal social structures	278
	12.5.2 How can we study culture in marine mammals?	279
12.6	Acknowledgements	280

13 Long-term studies — **283**

W. Don Bowen, Jason D. Baker, Don Siniff, Ian L. Boyd, Randall Wells,
John K.B. Ford, Scott D. Kraus, James A. Estes, and Ian Stirling

13.1	Introduction	283
13.2	Grey seal (*Halichoerus grypus*) W. Don Bowen	284
	13.2.1 Motivation	284
	13.2.2 Nesting of short-term objectives within long-term objectives and change in focus through time	285
	13.2.3 Standardization of methods and effects of changing technology and techniques	285
	13.2.4 Challenges	286
13.3	Hawaiian monk seal (*Monachus schauinslandi*) Jason D. Baker	286
	13.3.1 Motivation	286
	13.3.2 Nesting of short-term objectives within long-term objectives and change in focus through time	286
	13.3.3 Standardization of methods and effects of changing technology and techniques	287
	13.3.4 Challenges	287
13.4	Weddell seal (*Leptonychotes weddellii*) Don Siniff	288
	13.4.1 Motivation	288
	13.4.2 Nesting of short-term objectives within long-term objectives and change in focus through time	289
	13.4.3 Standardization of methods and effects of changing technology and techniques	290
	13.4.4 Challenges	290
13.5	Antarctic fur seals (*Arctocephalus gazella*) Ian L. Boyd	290
	13.5.1 Motivation	290
	13.5.2 Nesting of short-term objectives within long-term objectives and change in focus through time	291
	13.5.3 Standardization of methods and effects of changing technology and techniques	292
13.6	Bottlenose dolphin (*Tursiops truncatus*) Randall Wells	292
	13.6.1 Motivation	292
	13.6.2 Nesting of short-term objectives within long-term objectives and change in focus through time	293

13.6.3 Standardization of methods and effects of
changing technology and techniques 294
13.6.4 Challenges 294
13.7 Killer whale (*Orcinus orca*) *John K.B. Ford* 295
13.7.1 Motivation 295
13.7.2 Nesting of short-term objectives within long-term
objectives and change in focus through time 295
13.7.3 Standardization of methods and effects of
changing technology and techniques 296
13.7.4 Challenges 296
13.8 North Atlantic right whale (*Eubalaena glacialis*) *Scott D. Kraus* 297
13.8.1 Motivation 297
13.8.2 Advances in methods and changing objectives through time 298
13.8.3 Challenges 299
13.9 Sea otters (*Enhydra lutris*) and kelp forests *James A. Estes* 299
13.9.1 Motivation 299
13.9.2 Approaches 299
13.9.3 Methods 300
13.9.4 Rewards 301
13.9.5 Challenges 301
13.9.6 Serendipity 301
13.10 Polar bears (*Ursus maritimus*) *Ian Stirling* 302
13.10.1 Motivation 302
13.10.2 Standardization of methods and sampling 303
13.10.3 Two long-term approaches 303
13.10.4 Challenges 305
13.11 Conclusions 305

14 Identifying units to conserve using genetic data 306

Barbara L. Taylor, Karen Martien, and Phillip Morin

14.1 The biology of structure and the role of genetics 306
14.2 Scale—units to conserve 308
14.2.1 Taxonomy 308
14.2.2 Evolutionary significance 309
14.2.3 Demographically independent populations (DIPs) 309
14.3 Genetic markers 312
14.3.1 Mitochondrial DNA sequencing (mtDNA) 313
14.3.2 Microsatellites 314
14.3.3 Single nucleotide polymorphisms (SNP) 315
14.3.4 Amplified fragment length polymorphisms (AFLP) 315
14.3.5 Nuclear locus sequencing 316
14.4 Analytical methods 317
14.4.1 Choosing methods to match questions 317
14.4.2 Assessing the strength of inference 320

15 Approaches to management   **325**

John Harwood

15.1 Introduction   325
15.2 Sustainable use and the importance of economic factors   326
15.3 A brief history of marine mammal exploitation   328
15.4 Lessons from whaling and sealing   330
    15.4.1 The International Whaling Commission   330
    15.4.2 Northern fur seals   332
    15.4.3 Harp seals   332
15.5 Ecotourism   333
15.6 Obtaining indirect benefits from the management of marine mammals   334
15.7 Defining and achieving the objectives of management   334
15.8 The future of management   338

16 Conservation biology   **340**

Andrew J. Read

16.1 Introduction   340
16.2 What is conservation biology?   341
16.3 The road map   342
16.4 Which populations are at risk?   343
16.5 Quantitative assessment of extinction threat   349
16.6 Experimentation   350
16.7 Direct intervention   352
16.8 Fisheries by-catch   353
16.9 Case study: by-catch of harbour porpoises in the Gulf of Maine   355
16.10 Future directions   358
16.11 Conclusions   359

References   360
Index   433

List of Abbreviations

%FO	percentage frequency of occurrence
ADMR	average daily MR
ADP	adenosine diphosphate
AFLP	amplified fragment length polymorphisms
AIC	Akaike's information criterion
APM	age at physical maturity
ARS	area restricted search
AS	aerobic scope for activity
ASM	age at sexual maturity
ATP	adenosine triphosphate
BHT	butylated hydroxytoluene
BIA	bioelectrical impedence analysis
BLAST	basic local alignment search tool
BMR	basal metabolic rate
Bn	binning
BR	reconstructed biomass
CA	correspondence analysis
cal	calorie
CART	classification and regression trees
CATS	comparative anchor tagged sequences
CC	calibration coefficient
CCA	canonical correspondence analysis
CCAC	Canadian Council on Animal Care
CCAMLR	Convention for the Conservation of Antarctic Marine Living Resources
CeTAP	Cetacean and Turtle Assessment Program
CFIRMS	continuous-flow isotope-ratio mass spectrometry
CI	confidence interval/cumulative incidence
CITES	Convention on the International Trade of Endangered Species
CJS	Cormack–Jolly–Seber (survival estimation)
CO-I	cytochrome c oxidase-subunit I
COSEWIC	Committee on the Status of Endangered Wildlife in Canada
cpue	catch per unit effort
CR	capture–recapture
CRW	correlated random walks
DA	discriminant analysis
DCF	digestion coefficients/correction factor
DE	digestible energy
DFA	discriminant function analysis
DFO	Department of Fisheries and Oceans

DGGE	denaturing gradient gel electrophoresis
DIP	demographically independent population
DLW	doubly labelled water
D-tag	digital (acoustic recording) tag
ECS	European Cetacean Society
EE	energy expenditure
ELISA	enzyme-linked immunosorbent assay
EnEn	environmental envelope
ENFA	environmental niche factor analysis
EPA	Environmental Protection Agency
EPBC	Environment Protection and Biodiversity Conservation
EPIC	exon priming, intron crossing
ESA	Endangered Species Act
ESC-IRMS	elemental analysis-isotope ratio mass spectrometry
ESU	evolutionarily significant units
EyeB	eyeballing
F	the rate of flow through a chamber
FA	fatty acid
FABE	fatty acid butyl esters
FAME	fatty acid methyl ester
FBR	fixed (equal weighted) (average amount of) reconstructed biomass
FE	faecal energy
Fe	the fractional concentration in excurrent air
FFM	fat-free mass (aka lean body mass)
fH	heart rate
Fi	the fractional concentration of O_2 or CO_2 in incurrent air
FMR	field metabolic rate
FPT	first-passage time
fS	stroking rate
GAM	generalized additive model
GAMM	generalized additive mixed-effects models
GBR	gross birth rate
GE	gross energy
GER	gross energy requirements
GLG	growth layer group
GLM	generalized linear model
GLMM	generalized linear mixed-effects model
GPR	gross pregnancy rates
GPS	global positioning system
GTS	global telecommunication system
HCR	harvest control rule
HIF	heat increment of feeding
HM	harmonic mean
HPLC	high-performance liquid chromatography
HT	habitat preference

IBM	individual-based models
ICD10-CM	International Statistical Classification of Diseases and Related Health Problems with Clinical Modifications, 10th Revision
IL2	interleukin-2
IM	intramuscular
IP	intraperitoneal
IR	incidence rate
IUCN	International Union for Conservation of Nature
IV	intravenous
IWC	International Whaling Commission
J	joules
Kg	kriging
KS	kernel smoothing
LC	location class
LM	linear model
LMM	linear mixed-effects model
LPS	lipopolysaccharide
LRT	likelihood ratio test
MAMVIS	mammal visualization system
MCMC	Markov chain Monte Carlo
ME	metabolizable energy
MFO	modified frequency of occurrence
MGED	microarray gene expression data
MI	milk intake
MLE	maximum likelihood estimation
MMHSRP	Marine Mammal Health and Stranding Response
MMPA	Marine Mammal Protection Act
MMR	maximal metabolic rate
MNI	minimum number of individuals
MR	metabolic rate
MRT	multivariate regression trees
mS/cm	milliSiemen per centimetre (measure of salinity)
MSY	maximum sustainable yield
mtDNA	mitochondrial DNA
MUFA	mono-unsaturated fatty acids
N^cm2	newtons per square centimetre
NCF	numerical correction factors
nDNA	nuclear DNA
NE	net energy
NIRS	near-infrared spectroscopy
NIST	National Institute of Standards and Technology
NMFS	US National Marine Fisheries Service
NMP	new management procedure
NN	neural networks
NOAA	National Oceanic and Atmospheric Administration

NSF	(US) National Science Foundation
NWHI	North-western Hawaiian Islands
ODBA	overall dynamic body acceleration
PAM	passive acoustic monitoring
PBR	potential biological removal
PCA	principal component analysis
PCB	polychlorinated biphenyl
PCR	polymerase chain reaction
PDER	predischarge event recording
PHIDIAS	Pathogen–Host Interaction Data Integration and Analysis System
PIT	passive integrated transponder
PL	phospholipids
Plg	polygon
psi	pounds per square inch
PUFA	poly-unsaturated fatty acids
PVA	population viability analysis
QFASA	quantitative FA signature analysis
qPCR	quantitative real-time PCR
QUASIMEME	Quality Assurance of Information for Marine Environmental Monitoring
RER	respiratory exchange ratio
RF	radiofrequency
RFID	radiofrequency identification
RHIB	rigid-hulled inflatable boat
RMP	revised management procedure
RMR	resting metabolic rate
RQ	respiratory quotient
RS	remotely sensed
RSF	resource selection function
SARA	Species at Risk Act
SBD	short burst data (service)
SC	subcutaneous
SD	standard deviation
SDA	specific dynamic action
SI	stable isotope
SLTDR	satellite linked time–depth recorders
SLW	singly labelled water
SMM	Society for Marine Mammalogy
SMR	standard metabolic rate
SNP	single nucleotide polymorphisms
SRDL	satellite relay data loggers
SSFO	split-sample frequency of occurrence
SSM	state–space modelling
STAT	satellite tracking and analysis tool
TAG	triacylglycerols
TBE	total body energy

TBF	total body fat
TBP	total body protein
TBW	total body water
TDR	time–depth recorders
TF	trend-fitting
TNZ	thermal neutral zone
TOSSM	testing of spatial structure methods
TRT	take reduction team
TSSC	Threatened Species Scientific Committee
UE	ultimate energy
USFWS	US Fish and Wildlife Service
UTC	units to conserve
VBR	variable (amount of) reconstructed biomass
$\dot{V}CO_2$	CO_2 production
VFDB	Virulence Factor Database
VIF	variance inflation factor
$\dot{V}O_2max$	maximum rate of O_2 consumption of an individual (a measure of aerobic capacity)
WE	wax esters
WHO	World Health Organization
WMO	World Meteorological Organization

List of contributors

Geert Aarts, Sea Mammal Research Unit, Scottish Oceans Institute, University of St Andrews, Fife, KY16 8LB, UK and Institute of Marine Resource and Ecosystem Studies, PO 167, Den Burg, The Netherlands.

Jason D. Baker, Pacific Islands Fisheries Science Center, 2570 Dole Street, Honolulu, HI 96822-2396, USA.

W. Don Bowen, Population Ecology Division, Bedford Institute of Oceanography, 1 Challenger Drive, Dartmouth, Nova Scotia, Canada B2Y 4A2.

Ian L. Boyd, Sea Mammal Research Unit, Scottish Oceans Institute, University of St Andrews, Fife KY16 8LB, UK.

Dan Crocker, Department of Biology, Sonoma State University, 1801 E. Cotati Ave, Rohnert Park, CA 94928, USA.

Louise Cunningham, Sea Mammal Research Unit, Scottish Oceans Institute, University of St Andrews, Fife KY16 8LB, UK.

Tomo Eguchi, Southwest Fisheries Science Center, 8605 La Jolla Shores Blvd., La Jolla, CA 92038, USA.

James A. Estes, USGS Santa Cruz Field Station, Center for Ocean Health, 100 Shaffer Road, Santa Cruz, CA 95060, USA.

Mike Fedak, Sea Mammal Research Unit, Scottish Oceans Institute, University of St Andrews, Fife KY16 8LB, UK.

John K.B. Ford, Fisheries and Oceans Canada, Pacific Biological Station, 3190 Hammond Bay Road, Nanaimo, British Columbia, Canada V9T 6N7.

Nick J. Gales, Australian Government Antarctic Division, Channel Highway, Kingston, Tasmania 7050, Australia.

Frances M.D. Gulland, The Marine Mammal Center, 2000 Bunker Road, Fort Cronkhite, Sausalito, CA 94965, USA.

Ailsa J. Hall, Sea Mammal Research Unit, Scottish Oceans Institute, University of St Andrews, Fife KY16 8LB, UK.

John A. Hammond, Immunology Division, Institute for Animal Health, Compton, Berkshire RG20 7NN, UK.

Philip S. Hammond, Sea Mammal Research Unit, Scottish Oceans Institute, University of St Andrews, Fife KY16 8LB, UK.

John Harwood, Centre for Research into Ecological and Environmental Modelling, The Observatory, Buchanan Gardens, University of St Andrews, St Andrews, Fife KY16 9LZ, UK.

Mark A. Hindell, Antarctic Wildlife Research Unit, School of Zoology, University of Tasmania, Private Bag 05, Hobart, Tasmania, 7001, Australia.

Keith A. Hobson, Environment Canada, Science and Technology, 11 Innovation Blvd., Saskatoon, Saskatchewan, S7N 3H5, Canada.

Sascha Hooker, Sea Mammal Research Unit, Scottish Oceans Institute, University of St Andrews, Fife KY16 8LB, UK.

Sara J. Iverson, Department of Biology, Dalhousie University, 1355 Oxford Street, Halifax, Nova Scotia, Canada B3H 4J1.

David Johnston, Duke Marine Laboratory, 135 Duke Marine Lab Road, Beaufort, NC 28516, USA.

Scott D. Kraus, New England Aquarium, Central Wharf, Boston, MA 02110-3399, USA.

Shelley L.C. Lang, Department of Biology, Dalhousie University, 1355 Oxford Street, Halifax, Nova Scotia, Canada B3H 4J1.

Charles Littnan, Pacific Islands Fisheries Science Center, 2570 Dole Street, Honolulu, HI 96822-2396, USA.

Tom Loughlin, TRL Wildlife Consulting, 17341 NE 34th Street, Redmond, WA 98052, USA.

Bernie McConnell, Sea Mammal Research Unit, Scottish Oceans Institute, University of St Andrews, Fife KY16 8LB, UK.

Karen Martien, Southwest Fisheries Science Center, 8605 La Jolla Shores Blvd., La Jolla, CA 92038, USA.

Jason Matthiopoulos, Sea Mammal Research Unit, Scottish Oceans Institute, University of St Andrews, Fife KY16 8LB, UK.

Yoshihisa Mori, Department of Animal Science, Teikyo University of Science and Technology, 2525 Yatsuzawa, Uenohara, Yamanashi, 409-0193, Japan.

Phillip Morin, Southwest Fisheries Science Center, 8605 La Jolla Shores Blvd, La Jolla, CA 92038, USA.

Simon Northridge, Sea Mammal Research Unit, Scottish Oceans Institute, University of St Andrews, Fife KY16 8LB, UK.

Toby Patterson, CSIRO Marine & Atmospheric Research. GPO Pox 1538, Hobart, Tasmania 7001, Australia.

Graham J. Pierce, Instituto Español de Oceanografía, Centro Oceanográfico de Vigo, P.O. Box 1552, 36200, Vigo, Spain, and Oceanlab, University of Aberdeen, Main Street, Newburgh, Aberdeenshire AB41 6AA, UK.

Andrew J. Read, Duke Marine Laboratory, 135 Duke Marine Lab Road, Beaufort, NC 28516, USA.

Lori H. Schwacke, National Oceanic and Atmospheric Administration, National Ocean Service, Hollings Marine Laboratory, Charleston, South Carolina, USA.

Don Siniff, Department of Ecology, Evolution and Behavior, University of Minnesota, St Paul, MN 55108, USA.

Carol E. Sparling, Sea Mammal Research Unit, Scottish Oceans Institute, University of St Andrews, Fife KY16 8LB, UK.

Ian Stirling, Canadian Wildlife Service, 5320-122 Street, Edmonton, Alberta, Canada T6H 3S5.

Barbara L. Taylor, Southwest Fisheries Science Center, 8605 La Jolla Shores Blvd., La Jolla, CA 92038, USA.

Dominic J. Tollit, SMRU Ltd, New Tecnology Centre, North Haugh, St Andrews, Fife KY16 9SR, UK and Marine Mammal Research Unit, Aquatic Ecosystems Research Laboratory, 2202 Main Mall, University of British Columbia, Vancouver, British Columbia, Canada V6T 1Z4.

Peter Tyack, Woods Hole Oceanographic Institution, Department of Biology MS #50, Woods Hole, MA 02543-1049, USA.

Sofie Van Parijs, NOAA Northeast Fisheries Science Center, 166 Water Street, Woods Hole, MA 02543, USA.

Randall Wells, Chicago Zoological Society, c/o Mote Marine Laboratory, 1600 Ken Thompson Parkway, Sarasota, Florida 34236, USA.

Andrew Westgate, Duke Marine Laboratory, 135 Duke Marine Lab Road, Beaufort, NC 28516, USA.

Hal Whitehead, Department of Biology, Dalhousie University, 1355 Oxford Street, Halifax, Nova Scotia, Canada B3H 4J1.

Terrie M. Williams, University of California Santa Cruz, Center for Ocean Health, 100 Shaffer Road, Santa Cruz, CA 95060, USA.

1

Ethics in marine mammal science

Nick J. Gales, David Johnston, Charles Littnan,
and Ian L. Boyd

1.1 Introduction

This opening chapter in a book on marine mammal research techniques is about ethics, which reflects the increasing importance of ethical considerations to researchers working with marine mammals. Until recently, the subject of research ethics received little attention in the marine mammal literature (Bekoff 2002; N.J. Gales *et al.* 2003) and often was not an explicit component of experimental design.

Consideration of how ethical principles might be incorporated into the manner in which researchers carry out their work is not easy; indeed it is seen by many scientists as a real threat to their ability to conduct good science (e.g. McMahon *et al.* 2007). The debate at the interface between research and ethics has sometimes been polarized, and the perceived threat of unreasonable constraints has slowed the acceptance of ethical considerations into all aspects of research. But engagement from both ends of the spectrum of opinion is essential.

Marine mammal science is now an established international discipline that brings together researchers with a rich diversity of values, attitudes, motivations, and ethnic backgrounds (e.g. Lavigne *et al.* 1999). This richness reflects the cultural diversity of relationships between marine mammals and humans in society as a whole, so it is not surprising that marine mammals occasionally become the subject of ethical debates that sometimes emerge at important political levels. Given that ethics deal with concepts such as responsibilities and the definition of right and wrong, which in turn are defined by the values and customs of each person or group of people, then it is clear that there can be no single defining ethical view with which each marine mammal researcher can be identified.

While the diversity of researchers is challenging to the development of a set of universal ethical guidelines, most marine mammals' researchers would share a number of basic principles:

- Do no net harm;
- Seek to reduce or replace direct experiments on marine mammals with alternative approaches;
- Seek to refine methods in ways that reduce harm;
- Ask important scientific questions;
- Construct experiments that employ best practice throughout;
- Report and interpret results in an objective manner;
- Distribute results and data openly using the peer-reviewed literature;
- Display professionalism while working with colleagues and the public.

These shared goals form a fertile basis from which more specific and broadly agreeable guidelines for the use of marine mammals in research can evolve.

Agreeing on the guiding principles that characterize the way researchers behave to achieve these desires can be difficult. In practice, this may include limiting negative effects on individual marine mammals (e.g. disturbance, stress, or pain), their populations (e.g. involving repeated low-level disturbance in a manner that leads to cumulative effects), and their surroundings (e.g. involving reduction of the habitat quality for marine mammals or other species). An explicit consequence of these principles is that researchers will only use sufficient numbers of animals in their studies to answer the scientific questions being addressed.

In this chapter we explore the practical manner in which ethical principles can guide research on marine mammals. These ethical considerations are relevant to each of the remaining chapters in this book. The selection and application of the diverse techniques described in this book are two of the many important steps that should be influenced by our personal and professional ethical standards.

Before proceeding, some consideration needs to be given to terminology. Throughout this chapter *costs* are considered to reflect the collective, or cumulative, harm caused by carrying out research on marine mammals. This is distinct from *financial costs*, which involve the expense to researchers of conducting their research. Much in this chapter concerns the balance between *costs* and *benefits*. In this case, *benefits* include the benefits to human society, assuming that human society will usually wish to see some form of short- or long-term pay-off from the research it is sponsoring. It is assumed that benefits to animals will usually be aligned with benefits to human society, although clearly this will not always be recognized by all sections of human society, and this can be the source of conflict. Finding ways of resolving these types of conflicts is an important element in the training of researchers, so that they can be seen as individuals within human society who bridge ethical gaps rather than as individuals who tend to exacerbate them. Part of this comes through the development of *best practice*. This is a term that has entered the lexicon of moral philosophy about professional activities, and it reflects less of the actual practicalities of how things are done and more about the awareness displayed by those carrying out activities of their own fallibility, of the need to assimilate and implement new ideas as they arise, and to be considerate of the way in which their activities are perceived by others.

1.1.1 Ethics: defining the boundaries between right and wrong

Ethics is the intellectual process that seeks to address the question of morality and to define the difference between right and wrong, or good and bad. However, since the boundaries between right and wrong or good and bad in areas such as animal research can be unclear, we need a practical mechanism for defining where this boundary lies. This is complicated by the fact that the boundary can shift depending upon circumstances—the ethical boundary may be different for a rare dolphin compared with a common species of seal, for example. It can also change through time and between different sections of human society. These circumstances are defined by societal values and, as we all know, human society can be fickle and diverse. Researchers are then left with the unenviable task of trying to stay on what they perceive as the correct side of the ethical boundary reflective of the social framework they work within. There are three mechanisms that help researchers to achieve this.

First, there are legal instruments that may be established at national scales (however, for the European Union these are now operating at Europe-wide scales). Although these differ greatly among jurisdictions, most are what is termed 'enabling' legislation. This means that the legislation is phrased in a way that defines the underlying principles, but executive authority for making decisions and evolving the standards applied is devolved to federal, national, or state administrations. Such a system has the advantage that the ethical boundaries can evolve without the need to regularly enact new primary legislation.

Second, there is the process of ethical review. Ethical review is often formally embedded within legislation and may exist as standing committees set up by competent bodies, such as universities or research institutions, to ensure that their employees comply with the conditions established under local legislation. Ethical review essentially defines and redefines the boundaries between right and wrong or good and bad. In a sense, these review bodies are expert juries that carry sufficient expertise to understand the rationale for particular proposed research. However, importantly they also usually include 'lay' representation to ask the difficult questions on behalf of society as a whole. Sometimes ethical review is carried out through public consultation by making sure that intentions are made clear in published notices. This is what happens under the US Marine Mammal Protection Act (Baur *et al.* 1999). In the UK the process is kept confidential in order to minimize risk to the health and welfare of the researchers themselves from extreme, minority elements of human society who consider that their views are correct, irrespective of the overall wishes of society as a whole. There is no 'right' way to assimilate the views of society into decisions about whether research should be permitted. All mechanisms have their strengths and weaknesses, and all are flawed to some degree. They are, however, the best available mechanisms and researchers need to work with them and not against them.

Third, there is the process by which researchers themselves develop professional guidelines about how to conduct ethically acceptable research. These guidelines are usually developed by professional societies, such as those of the Society for Marine Mammalogy (N.J. Gales *et al.* 2009). Unlike those associated with national or regional legislation, these guidelines have the advantage that they assimilate the notion of best practice from across the whole research community. A well-founded ethical review process will usually take account of these types of guidelines.

Taken together, this structure provides a robust system that has two advantages for the researchers involved. The first of these is that it provides the researchers with the opportunity to debate the strengths and weakness of their approach. In other words there are no absolutes in the search for right or wrong, and researchers have the opportunity to the make the case for research that may be very costly but may accrue even greater benefits. The second is that this protects researchers both legally and morally. If researchers engage with this process they can proceed with assurance that they are fulfilling their duty as paid-up members of human society and, in return, human society will provide them with legal protection against some who would disagree.

1.1.2 Financial and professional costs of compliance

Consideration of ethics is neither easy nor without financial cost. In competitive academic communities, researchers are under pressure to maximize the economic efficiency of their research and to publish the results. Compliance with ethical guidelines may require change to more accepted techniques involving the purchase of expensive, specialized equipment or the hiring of additional specialized technical assistance. Financial pressures can limit or bias such ethical considerations.

But, can professional guidelines prevent scientists from doing some important science that might, for example, be needed for effective conservation management of threatened species? Some authors have argued that contentious invasive techniques, such as hot-iron branding, have already been banned unreasonably in some cases and that 'it is becoming increasingly difficult to collect good data as ethical scrutiny of research involving wildlife intensifies' (McMahon *et al.* 2007). We can take two important lessons from situations and views such as these; firstly, whether or not the authors are right in their contention, it will not always be science that is given primacy in decisions that attempt to balance science with ethics and, in some cases, politics. It is an important prerequisite for the professional scientist to accept that when engaging in a debate between two disciplines (or value systems) such as science and ethics, neither should have assumed primacy. Secondly, and more importantly, if ethical guidelines for the conduct of research are to be developed, it is in the interest of the scientists and their representative societies to ensure that they are centrally involved, and that they take proper account of the real ethical issues at hand.

Examples like these also show how the ethical review process operates, as well as its potential weaknesses. Review is only as good as those who take part in the debate, and also only as good as the information available to make informed decisions. Typically, high uncertainty in the outcome will lead to high precaution in the ethical review process. New methodologies often take some time and much debate to become acceptable. Recent debates over controlled exposure experiments (a field method in which controlled doses of sound are transmitted to focal animals for the purposes of assessing their behavioural and physiological responses), are typical of this process where, after many years of debate, and disagreement, the first tentative steps have been made to conduct some of these experiments. This exploration of the cost–benefit trade-offs within the ethical review process can be designed to establish what is acceptable to society. Thus, the role and mechanism of updating and evolving the ethical review and guidelines is every bit as important as that of creating them in the first place.

1.1.3 Keeping the research ethics and conservation ethics apart

An acceptance and willingness to comply with research guidelines is an easy step for researchers who do not use the types of techniques that give rise to debate and that might ultimately limit their ability to conduct an experiment. However, much of our knowledge of marine mammals has been acquired using techniques that affect the welfare of experimental subjects. Guidelines that restrict this work unreasonably may leave some populations of marine mammals more vulnerable to poor and ill-informed management, and a consequence could be higher depletion and extinction risks. Therefore, appropriate guidelines should consider how to distinguish science (high and low quality) from ethics (Campbell 2007). To use an example from Campbell (2007): 'It is bad science to claim that reducing environmental protection will not have adverse effects on rare species, but the decision whether we should protect rare species or not is an ethical one'. Thus scientific arguments should be used to support the case for carrying out research, and should not be invoked to argue for a particular ethical position. Similarly, the generally untested assumption by many scientists that the increased knowledge provided by research on their study animal will inform its conservation (Farnsworth and Rosovsky 1993) can unduly weight their perceived importance of a scientific case for research over an ethical one. Clarity in such distinctions goes a long way to optimizing outcomes when balancing scientific and ethical imperatives.

There is considerable potential for the misrepresentation of research results because of understating uncertainty, overstating statistical power, or by exaggerating the need for research in conservation management. These types of problems erode the credibility of research. Misuse of research results is perhaps best exemplified in some of the modern international calls to cull marine mammals to increase fishery yields (e.g. the 'whales eat fish' debate). In this example, there is confusion about how the benefits of research are to be measured and, in particular, for whom the benefits accrue. Some prefer to see the benefits in terms of the short-term productivity of fisheries and increased profits for a focused group of individuals. Others prefer to see the benefits in terms of the long-term ecological health of the oceans and the subsequent general benefits for humanity. Both positions are accompanied by enormous uncertainties and are underlain by the unwritten, unspoken wish for some very small parts of society to resume, maintain, or expand the commercial exploitation of some species of marine mammals. This debate addresses the question about whether minorities can profit at the cost of a larger majority in society—as also seen within the debate about climate change. However, we must be clear that it is not the role of science, or research, to take part in this debate, other than to help to narrow the scientific uncertainty and help those holding different views to reconcile their positions with reliable information.

1.2 Guiding principles

1.2.1 Science for the common good

It is frequently argued that professional scientists are obliged to conduct research for the common good (Shrader-Frechette 1994), and this also applies to professional marine mammal scientists. The investments by society in research connect researchers to society in important ways. Through public funding, special training, and advanced educational opportunities, researchers become the trustees of public knowledge; they are experts with near-monopolies on specific subjects that provide powers and benefits within society. Along with these benefits come significant responsibilities, including the obligation to conduct research that furthers the aims and goals of society in general. There are at least five principles, derived from those put forward by Shrader-Frechette (1994) that should further guide research.

1.2.2 Principle 1

We ought not [to] *conduct research that causes unjustified risks to humans and our non-human study subjects.* Much of the research conducted on marine mammals poses few risks for humans. In situations where it does (e.g. the physical handling of large pinnipeds and cetaceans), the risk should be controlled and mitigated to acceptable levels (e.g. using appropriate assessment procedures and specifically designed nets and/or chemical restraint) or by finding alternative ways of addressing the research question. The welfare of study subjects should also be addressed explicitly in a similar risk-based manner. Determination of where the threshold lies between justified and unjustified risks will vary among researchers, so this is where agreed guidelines, and ethical review, that outline best practice can help. Periodic reviews of research and refinement of research methods minimize risk levels.

1.2.3 Principle 2

We ought not [to] *conduct research that purposefully violates informed consent.* Non-human subjects cannot offer informed consent, so we must first seek the consent of the animal's 'owner'—in a legal sense this will usually be the nation in which we do the experiment—and equally importantly, assume responsibility for animal welfare ourselves. This is explicitly addressed through permitting processes and the consent of legislative bodies responsible for the welfare and protection of animals. Furthermore, this is achieved through periodic review of the research, optimized study designs, and the application of best practices.

1.2.4 Principle 3

We ought not [to] *conduct research that inappropriately converts public resources to private profits.* Most marine mammal research is supported by the public, and the benefits of this research should be passed back to society in a transparent manner, e.g. the publication of results in peer-reviewed journals. Researchers are obligated to publish their work, and promote practices that facilitate this, such as acting as reviewers for scientific journals. Situations should be avoided in which there is asymmetry between the costs borne by society and the benefits gained by individual researchers or institutions. For example, the use of public resources to conduct expensive research that risks the viability of endangered populations of marine mammals should be avoided. In these situations, the public bears the costs of the research (damage to a national or international heritage and the costs of funding the work), while individuals or institutions may gain financially (through grants) and professionally (through publications).

1.2.5 Principle 4

We ought not [to] *conduct research that is biased, and we are obliged to interpret objectively the results of our scientific endeavours.* Researchers should develop, employ, and promote methods designed to maximize objectivity and minimize the effects of personal biases on study results (Wilson Jr 1991). An example of this would be using a blind or double-blind survey design in observational studies of marine mammals (e.g. see Chapter 3 for the use of this in double-team data collection procedures when using distance sampling), where observers are unaware of treatment and control phases of an experiment and therefore unable to consciously or unconsciously direct the outcome of surveys. Uncertainty and variability also have to be taken into account when interpreting results. One of the more seductive temptations arises when the interpretation of the data has management consequences. While it is entirely appropriate for researchers to advocate management options, their arguments should be robust and defendable (see Section 1.1.3).

1.2.6 Principle 5

We ought not [to] *conduct research that jeopardizes environmental welfare.* Marine mammal researchers should strive to limit harmful effects of research on animals and the environments within which they function. This includes effects on non-target species and their habitats, such as those that could occur during controlled sound exposure experiments, limiting disturbance to nesting birds while working at pinniped rookeries, or reducing unwanted effects on the marine environment through the use of quieter and more fuel-efficient engines.

1.3 The role of professional societies in developing and maintaining ethics in marine mammal science

A number of professional societies exist for biologists, varying in focus by discipline, taxonomy, and geographic region. Examples of these societies include the European Cetacean Society (ECS) and the Society for Marine Mammalogy (SMM), but they also include societies focusing on specific disciplines such as the Society for Conservation Biology.

Scientists join professional societies to support the general aims of the society, and because they convey a set of benefits to members not generally available to non-members. For example, joining professional societies provides members with significant opportunities for access to colleagues and to join academic and ethical debates. Joining a society may also provide members with access to educational materials such as newsletters or journals published by the society (e.g. *Marine Mammal Science*), which help to keep them current with respect to techniques and relevant projects. Although applications for membership of a professional society are very rarely vetted, this also conveys a sense of legitimacy on members.

Professional societies are governed by rules, and these rules are expressed as membership criteria, statements of values, ethics guidelines, codes of professional conduct, and guides to best research practices for members. Membership of the society supports individual researchers by allowing them to align their approach to research with those of other members of the society and by encouraging best practice. So long as the society has a legitimate governance structure, this provides a degree of protection to members' interests.

At present, both the ECS and the SMM have established Standing Ethics Committees to provide guidance on ethical issues for their members, although only the SMM currently includes a publicly available set of guidelines on its website for the ethical treatment of animals in field research. The SMM guidelines were drafted by an ad hoc ethics committee and ratified by a vote by SMM members in 2008. These guidelines were subsequently published online on the SMM website (www.marinemammalscience.org) and in the Society's journal *Marine Mammal Science* in 2009 (N.J. Gales *et al.* 2009). Controversy over some research approaches (e.g. Dalton 2005; N.J. Gales *et al.* 2005a) means that these guidelines need to evolve. In addition, the Canadian Council on Animal Care (CCAC) produced internationally peer-reviewed guidelines on the care and use of wildlife in 2003 and will be publishing species-specific guidelines on marine mammals in 2010; all documents are available from the website (www.ccac.ca).

1.4 The 3 Rs (refinement, reduction, and replacement)

A useful tool for researchers to aid in creating ethically sound studies is the concept of the 3 Rs. Originally described by two British scientists, William Russell and Rex Burch, the 3 Rs was initially applied to studying the ethics of using animals in biomedical research (Russell and Burch 1959). The 3 Rs are central to the principles of animal research in the UK.

Refinement refers to keeping any pain, suffering, or other harm that may be caused to a minimum for each animal used in the research. Thus, the question scientists should ask is '*How can I minimize the potential negative impacts of every aspect of this work?*' The invasiveness and severity of any proposed research should be considered during the planning stages of the work, for two reasons. Firstly, it is necessary to anticipate the extent of pain, suffering, distress, or lasting harm that might be caused to the animals by each part of the proposed procedure, in order to work out the best ways to minimize these effects. Secondly, it is necessary to balance the expected levels of costs to animal welfare against the anticipated benefits of the work in a cost–benefit analysis. The aim of this is to make sure that any harm is the lowest that is practically feasible and that the benefits are the greatest that can be reasonably achieved. Refinement is possible in many areas of marine mammal research, and the development of progressively smaller and more capable instruments for attachment to marine mammals (see Chapter 10) is an excellent example of such a process.

At present, there is no convenient and standardized way of objectively assessing animal pain and distress. The assessment is generally based on a subjective interpretation of a suite of clinical and behavioural indicators. The implementation of refinement depends largely on the ability of researchers to observe and understand these indicators, but many experimenters may lack familiarity with the normal behaviour of their study species. A reasonable approach to controlling pain and distress is to assume that a procedure that inflicts pain and distress in humans will inflict at least as much pain and distress in animals unless there is evidence to the contrary. In animals with important social bonds, these may extend to stress caused by alterations in social relationships (e.g. the removal of an animal from a social aggregation). Research communities can be informed about refinement through the regular updating of guidelines produced by professional societies and through peer-reviewed publication.

Reduction involves methods for ensuring that the minimum number of animals is used to satisfy the research objectives. The more animals used, the greater will be the overall costs in terms of the animal welfare. Conversely, if too few animals are used to satisfy the study objectives, then the welfare costs to those experimental animals used in the study will have been wasted. So the second question scientists must ask is *'What is the fewest number of animals needed to successfully achieve the objectives of the study?'*

Knowledge of experimental design and statistical power is a necessity for researchers and they need to understand the interaction between statistical power, sample size, and measurement variance. A general method of increasing power and, hence, the sensitivity of an experiment is to use a large sample of subjects. The more variable the data, the more power is needed to detect an effect and, therefore, the greater the sample size needed. While the degree of variability cannot always be determined prior to an experiment, the amount of variability likely to be encountered can be estimated by conducting small, pilot studies or by examining previous research in the same or related areas.

The assessment of the impact of a technique through the determination of statistical power is an essential part of an ethical evaluation of the cost–benefit balance. In some circumstance the outcome may require researchers to abandon plans for an experiment.

One obvious means of reducing the number of studies being carried out is to avoid unnecessary repetition of animal procedures by improving the sharing of existing data and samples from previous studies. Thus, the ethical responsibility for researchers is not only to look for existing information, but to ensure that the samples and data they collect are archived and made publicly available.

Replacement includes those methods that permit a given purpose to be achieved without conducting experiments or other scientific procedures on animals. Replacement means that animals, generally higher order animals capable of suffering or feeling pain, should not be used at all if the same research objectives can be achieved in other ways. Researchers must ask themselves at the planning stage of a study is *'Do I need to use higher order animals at all?'*

Replacement may not be applicable for many studies of marine mammals, although it may be possible to conduct research on one species of marine mammal as a replacement for another if circumstances mean that there is a greater probability of success for a smaller cost. Mathematical and statistical models are an increasingly powerful, replacement technique that can allow sophisticated examinations of relationships between multiple parameters based on inputs estimated from existing studies (Zurlo *et al.* 1996). Medical imaging of dead marine mammals can lead to anatomical and physiological reconstructions, with important implications for our understanding of the ecology and behaviour of species without any need to undertake research on live subjects.

1.5 Type I and Type II errors

In scientific analyses, we can make two types of errors when testing null hypotheses. If we conclude that there is an effect when none exists, a Type I error results. However, if the null hypothesis is not rejected when an effect does exist, a Type II error results (Zar 1999).

Research scientists are regularly trained to limit the number of false-positives (Type I errors) that arise through their research, by focusing on a level of statistical significance as a threshold for accepting or rejecting a null hypothesis (e.g. Toft and Shea 1983; Anderson *et al*. 2000). Depending on the particular null hypothesis, this may not be the most appropriate approach. In some cases it may be more appropriate to minimize the chances of committing Type II errors when addressing uncertainty associated with the research in question. Such an approach tends to maximize the protection of the public good (the benefits) from any potential harmful effects of the research (Shrader-Frechette 1994). This may be especially applicable in cases dealing with endangered species and where there is often considerable underlying uncertainty whether the research should be conducted at all.

To illustrate this, we present a hypothetical example—assessing the effects of instrument tagging a small number of individuals from a population of an endangered small cetacean. In this situation there is considerable uncertainty about how the animals will react to the capture and surgical procedures required for tagging, and harming or killing the animals is a distinct possibility (e.g. Irvine *et al*. 1982; Wang Ding *et al*. 2000). Here we assume that harming the cetacean is undesirable from the perspective of the research objectives, let alone the conservation objective, because injured animals may behave abnormally.

We can present this as a null hypothesis:

H_0:   *Instrumenting dolphins will not harm them.*

If a Type I error was to occur we would reject the null hypothesis when it is actually true. In other words, we reject the notion that tagging these animals is harmless, even though it actually is harmless and no effects are demonstrated. If a Type II error was to occur we would accept the null hypothesis when it is actually false. In this latter case we accepted the notion that tagging is harmless to dolphins, when in fact it actually does harm the animals. The Type II error is more dangerous to the animals than the Type I error. These issues can be addressed statistically when looking for effects on study subjects, through the use of a power analysis (Burgman *et al*. 2000). Power analyses provide researchers with the ability to assess the likelihood of committing both types of errors, and to set criteria to avoid incorrect and potentially misleading conclusions (Dayton 1998).

When weighing the use of invasive or risky techniques with small populations (as in the example above), researchers can minimize the chance of committing a

Type II error by requiring increased statistical power when testing for effects. The example explicitly acknowledges that attempts to test for harmful effects of tags are likely to have small samples sizes and low statistical power, and the chance of committing a Type II error is therefore elevated. Indeed, an approach that risks Type I errors is clearly preferable, where animals are not tagged—even though it will not cause them harm—as a less invasive technique can then be implemented.

These issues have been addressed previously in the conservation literature and, from an ethical perspective, there appear to be several good reasons for minimizing Type II errors in applied conservation research (Noss 1986; Shrader-Frechette 1994; Dayton 1998). For example, it has been argued that society should strive to prevent acts that cause harm before it promotes acts that enhance welfare (Shrader-Frechette 1994). If so, then for the present example it is more important to protect the public from harmful research than to save it from the harmless impacts of not instrumenting the small cetaceans. Furthermore, these situations may exhibit ethical asymmetry (Shrader-Frechette 1994). Researchers who conduct tagging research receive greater personal benefits (financial compensation and primary literature publications) than the general public regardless of the outcome, yet the general public bears the risks and costs of the research if animals are harmed or killed. Finally, the public has a right to protection from research that harms the public good or environment in ways that cannot be compensated (Thomson 1986; Shrader-Frechette 1994). In other words, since researchers probably could not compensate the public for the research-related loss of animals from the small cetacean population in question, they should seek solutions to minimize the chance of those losses. These balances between the costs and benefits of research are challenging to assess, but will often involve ethical review that includes representation that can reflect public interests.

1.6 Best practice

There are no hard definitions of what constitutes *best* practice. The following are issues that may be evaluated as part of a *best practice* approach. While this list is not exhaustive it provides a starting point for researchers.

1.6.1 Threatened versus abundant

In general, where the level of impact is expected to be significant or completely unknown, researchers should consider selecting species that are not classified as endangered or threatened. This is particularly the case where novel equipment or tools are being tested. Where threatened or endangered species are the subject of study, the work should have the objective of improving the conservation status of this species.

1.6.2 Timing and location

The vulnerability of an animal to disturbance will often vary with its age, sex, reproductive status, social situation, location, and level of exposure to human activities. Researchers should minimize the potential for disturbance by selecting animals that will be least affected by the disturbance, or they should develop protocols to minimize disturbance while still satisfying the requirements of the study design. In particular, researchers should minimize potential disruption of critical social bonds, particularly those of mothers and dependant young.

1.6.3 Experimental procedures and equipment

Marine mammal researchers have an increasingly powerful and broad suite of methods available. These vary in the degree of invasiveness as well as in the types and quality of data they provide. Many factors affect a researcher's choice of field technique, including the aims of the study, economics, logistics, availability of expertise, legislative requirements, and personal experience. Researchers should select techniques that minimize welfare costs to the animals, while still delivering data sufficient to satisfy the aims of the experiment. Much potential pain and distress can be avoided or reduced with the proper use of anaesthetics, analgesics, and tranquilizers. These are critical components of veterinary care that need to be applied, or overseen, by trained veterinary staff.

1.6.4 Training

Most animal handling and sampling of marine mammals requires skill and experience, even for routine procedures such as capture and blood collection. Inexperience generally reduces the quality of data collected, increases the duration of procedures, and reduces animal welfare. Principal investigators should ensure that any activities that might affect animal welfare or experimental outcomes are directly supervised or conducted by personnel with sufficient experience to ensure the welfare cost to the animal is minimized. The concept of recognizing, minimizing, and eliminating pain and distress in experimental animals should be included in training, and all researchers should receive training. Details of refinement and animal welfare considerations, and experimental design (including statistical power) should also be routinely included in scientific papers and publications.

1.6.5 Environmental considerations

Researchers need to take account of their impacts on the environment, including landscape, flora, and fauna, particularly in protected and environmentally sensitive areas. The need to work in environmentally sensitive areas should be balanced with the potential disturbance this could cause. Permitting procedures will often require assessment of potential negative impacts to the environment and implementation of mitigation. Much marine mammal research is conducted in areas that do not formally require this level of scrutiny, but researchers have a duty of care for the environment that is at least as stringent as that placed on other members of the public. This applies equally to those working in the field or the laboratory.

1.6.6 Cultural consideration

Ethical attitudes can vary between nations and also between cultural groupings within national boundaries. Many countries have specific legislation to regulate research activities but this can vary in terms of stringency. Marine mammals often cross national and cultural geographical boundaries, meaning that research can be affected by variations in the rules that are applied. Research sites may have historical or cultural significance and researchers should recognize the need to conduct research using best practice procedures that are aligned with local community values. Normally, researchers would apply the ethical rules or laws associated with the location in which they are operating, unless these are less stringent than either those of one's home nation or the internationally accepted ethical guidance provided by professional societies. Best practice would be to apply the most stringent rules in all circumstances. There will often be inconsistencies between local rules and internationally accepted guidance, but rarely would a more conservative approach be unacceptable to either set of rules. Where there is a clash between these sets of rules researchers would need to enter a period of consultation with local authorities and international referees to resolve differences. The greatest difficulties can arise when different rules are applied by individuals within a research group, as could happen when local researchers are recruited to assist with research activities. Agreements about which rules are applied and who is responsible for implementing these rules within the group need to be in place before any research commences.

1.6.7 Decision analysis frameworks and cost–benefit analyses

Balancing the benefits and costs of invasive techniques and the application of novel technology in conservation research and management is a complex task. In many cases, researchers and managers must deal with scientific uncertainty in quantitative data and account for qualitative and subjective input from stakeholders, including competing value systems. Occasionally there is disagreement among stakeholders about the appropriate methods for conducting a study and this can often involve ethical considerations.

Using a decision analysis framework to direct conservation efforts can be used to resolve disagreements (Maguire 1987). This approach breaks down a complex problem into components that are small enough to be readily understood. These discrete components are methodically recombined to create a conceptual model of the process and highlights critical components (including ethical guidelines, decisions, uncertainties, and endpoints) and the relationships between them. This approach can provide a mechanism for establishing decision rules, such as maximizing expected benefits or minimizing possible costs, which can be understood by all stakeholders (Maguire 1987).

Decision analysis is increasingly being applied across diverse fields including human medicine (e.g. Forrow *et al.* 1988) and conservation biology (Maguire 1986; Peters *et al.* 2001). The decision analysis framework can be mainly qualitative, but increasingly decision analyses can be quantitative, providing detailed statistical advice on how to proceed with management efforts (Peters and Marmorek 2001; Peters *et al.* 2001). Decision analysis can also be applied to the treatment of stranded marine mammals, and in particular to the decision to euthanize animals that are unlikely to survive rehabilitation attempts. To facilitate these decisions, M.G Moore *et al.* (2007) established a qualitative decision-tree approach to help researchers and conservation biologists navigate the complex cultural and legal issues created by stranding events. Wade (2001) provides examples of quantitative decision analyses techniques used to help establish catch limits for exploited endangered species, and addresses issues related to balancing the risks of committing Type I and Type II errors in the conservation of exploited species.

1.7 Summary

Discussions of ethics and the development and implementation of ethical guidelines are not intended to prevent legitimate research. In fact, they should serve to increase the quality of the work being undertaken, increase public trust, and maximize the value of the research. However, the ethical considerations surrounding research are not static. There is a continuous process of refinement of methods as knowledge, technology or techniques improve. This refinement is also present in the organization and implementation of the processes that regulate research activity within ethical constraints. The decisions made by individual researchers effect how marine mammal research is perceived, and it is the responsibility of individuals to align themselves with the expectations of society.

2

Marking and capturing

Tom Loughlin, Louise Cunningham, Nick J. Gales,
Randall Wells, and Ian L. Boyd

2.1 Introduction

One of the greatest constraints on studies of marine mammals involves our ability to capture and follow sufficient numbers of animals in the wild. Until comparatively recently the only feasible way of studying reasonable samples of many marine mammal species was using dead specimens, but there are limits to what this can tell us about critical aspects of the biology of these species. Consequently, advances in the methods used to capture, restrain, and release marine mammals have been fundamental to scientific progress.

The subject of capturing, restraining, and marking marine mammals is the point at which the ethics of research on marine mammals often comes into sharpest focus (Monamy 2007). Researchers have to consider the safety of both the animals and researchers as a first priority. In some cases, capture may be unnecessary because there are remote, non-invasive methods for addressing the research questions being pursued. In other cases, it may be that the research questions will never be satisfactorily addressed by capturing marine mammals because it is impossible to capture enough to provide sufficient statistical power. In any case, the minimum number of animals should be captured to satisfy the objectives of the study. Disturbance at breeding colonies or of social groups also needs to be weighed up for its effects. Researchers should employ risk assessment methods to estimate the likelihood and severity of unforeseen effects. Once the researchers determine that capture and handling is justified, and the numbers of animals required, they take on the responsibility for the health and welfare of the animal concerned and must weigh the benefits of the research on the subjects against the possible harm to it and the population. Ethical review and permitting procedures will often help this process (see Chapter 1).

2.2 Applying marks

Marks should be easily applied, easily recognized in the field, and durable enough to remain functional throughout the study. Marking techniques should cause minimum pain or stress to the animal during application, and should minimize any potential effects on normal activities or the possibility that any equipment could spread infection (Murray and Fuller 2000). Researchers should coordinate with other researchers to minimize any conflicts in terms of the mark characteristics and the system for data management.

2.2.1 Flipper tagging pinnipeds

Millions of tags have probably been applied to pinnipeds around the world during the past 50 years. It is often considered to be an inexpensive, relatively non-invasive research tool and this has led to the indiscriminate use of tags, often of inappropriate design and with no real research objective in mind. In some circumstances tags may not be a benign, non-invasive method of marking. Poorly applied tags can cause lesions or impair animals to an extent that could reduce their survival.

There is no tag specifically manufactured for pinnipeds. Instead, pinnipeds are commonly tagged with cattle and sheep ear tags, e.g. Dalton Animal Identification Systems (Oxford, UK) and Delta Plastic Allflex of New Zealand (Lander *et al.* 2001). Tags are made of plastic or metal (stainless steel or 'Monel' which is a nickel–copper–iron alloy), are embossed with a number or letters that may be a unique combination to allow the future identification of individuals, and they may come in different colours, especially the plastic tags. They are normally applied to the axilla of the trailing edge of the front flipper (otariids) or to the interdigital webbing of the rear flippers (phocids).

Most tags contain a self-piercing 'male' post that penetrates through the entire flipper and attaches to the 'female' element. Placement will affect the longevity of tags, and both post length and diameter need to be matched to the depth of the flipper at the tag application site, with allowances made for growth. The application area should be cleaned and the tagging equipment itself should be disinfected between uses.

Many trade-offs have to be considered when choosing the type of tag. Table 2.1 summarizes the characteristics of different tags. In general, metal tags are more robust but may rip out of the flipper more readily than plastic tags (Bowen and Sergeant 1983; Scheffer *et al.* 1984; Testa and Rothery 1992; Jeffries *et al.* 1993). Although plastic tags can be easier to read from a distance, colours can fade and numbers rub off. Plastic tags can last 10–15 years in ideal circumstances, but bright colours often become brittle and break (Erickson and Bester 1993) and some batches of plastic tags can be brittle in cold conditions depending upon the grade of plastic used in manufacture. The visibility of both plastic and metal tags can be enhanced through the use of streamer markers, such as nylon cloth strips reinforced with vinyl, which may last for several years (Wells 2002).

Table 2.1 *Characteristics of tags used for pinnipeds.*

Tag material	Metal		Plastic		
				Flexible plastic	
Tag type	Monel (nickel–copper–iron alloy)	Stainless steel	Hard plastic	Wide profile	Narrow profile
Durability	Very high	Very high	High	High	Moderate
Longevity	Low	Low	High	Low	Moderate
Number type	Embossed	Embossed	Embossed/printed	Printed	Printed
Number durability	High	High	High (embossed) Low (printed)	Low	Low
Colour range	None (grey)	None (grey)	Very high	High	High
Colour durability	Very high	Very high	Very high	Low	Low
Visibility	Low	Low	Moderate	High	Moderate
Number readability (when new)	Low	Low	Moderate	High	Moderate
Phocid suitability (rear flipper web)	Low	Low	High	Low	High
Otariid suitability (fore flipper trailing edge)	Low	Low	High	High	High
Cost	Low	Low	Low	High	High
Example	Conservation tag, National Band and Tag Company, NY	Conservation tag, National Band and Tag Company, NY	Dalton Roto tag	Allflex	Allflex Jumbo

Double-tagging can be used to assess tag loss (e.g. Testa and Rothery 1992), but tag loss will result in additional damage to the flipper so double-tagging should be minimized (Wells 2002). Re-tagging animals that are observed with only one tag remaining may extend the life of a study (Cameron and Siniff 2004), but requires repeated handling and access to the animal throughout its life and is only possible for small populations (Melin *et al.* 2006). Accompanying tagging with other method of marking, such as clipping the fur (Harcourt and Davis 1997), can increase the probability of re-sighting individuals. The probability of re-sighting can be increased by making the tag more visible. Hall *et al.* (2000) glued the 'hat' tags, made of high-impact styrene and filled with a mixture of buoyant materials, onto the heads of grey seal pups. This made numbers readable at distances of more than 50 m. Multiple marks generally have different longevities and their unequal loss rates will influence re-sight probabilities over time.

2.2.2 Discovery tags

More than 20 000 'Discovery' tags were applied to baleen and sperm whales during 1932–1985. These numbered metal cylinders were shot through the blubber into the muscle of whales and recovered when the whales were captured and rendered. Thus, the tags documented the occurrence of whales at two points within their range—the location of tagging and killing. These tags provided some of the first information on movement patterns for some species of whales, but tag return rates were low (<15%). The effects of these tags on the whales were not assessed, but anecdotal accounts suggest that some serious injuries resulted. Use of smaller versions of these tags with whales less than 4.6 m long has been discouraged because of the risk of serious injury.

2.2.3 PIT tags

Passive integrated transponder (PIT) tags, also called radio-frequency identification tags (RFID tags) or implantable transponders, are low-frequency computer chips programmed with an identification number and are encapsulated within a biocompatible material (Melin *et al.* 2006). A scanner sends a radio signal that excites the computer chip, and the chip then sends back the unique identification code it contains.

The tags are as small as 1 cm and can be injected under the skin, embedded in external tags, or implanted into the body cavity. PIT tags have been used in wild animals for many years and the technology continues to improve. They are easy to apply, inexpensive, and will last the life of the animal. Subcutaneous PIT tags have been used in small populations of New Zealand fur seals and Australian sea lions (McIntosh *et al.* 2006; McIntosh 2007). However, close access to the animals is required to read the tags, and in some applications there have been problems with migration of the tag in the animal, or its loss from the injection site associated with minor infections. Available scanners are limited by only being able to read the tags at a maximum distance of 1.0 m, and the distance is reduced if the reader is in the wrong orientation to the tag. To be effective, PIT tags often have to be accompanied by some form of external mark to signify that a PIT tag is present.

There are restrictions on high-frequency RFID tags because of interference with radio communications. High-frequency (850–950 MHz and 2.4–2.5 GHz) systems allow more distant ranges for reading the tag (>30 m) and high reading speeds. Because these tags are larger than low-frequency tags, they require an internal power source such as a battery. They can only be applied externally or surgically implanted. These tags have been successfully implanted in the body cavities of harbour seals (Lander *et al.* 2001), but this is not a feasible method for tagging large numbers of animals.

2.2.4 Hot-iron branding

The practicality of hot-iron branding as a means of permanently marking pinnipeds in the wild has been demonstrated in several studies (e.g. T.G. Smith *et al.* 1973; Calkins and Pitcher 1982; Pendleton *et al.* 2006). Brands are large enough for animals to be identified unambiguously from a distance (Merrick *et al.* 1996). Hot-iron branding is more practical than freeze-branding in most circumstances because the logistical difficulties required to safely apply freeze brands at remote sites are daunting, and because freeze brands are not immediately readable at the time of application (Harkonen 1987, in NMFS 2007). The correct application of the hot brand to the fur and skin causes a second-degree burn that kills the hair follicles and melanocytes creating a permanent, bald, unpigmented mark. Incorrect application of hot-iron brands can produce a serious injury. The quality of a brand and the likelihood of injury are affected by the temperature of the branding iron at the time of application, the duration and pressure of the application, animal restraint, and the state of moult. Van den Hoff *et al.* (2004) reported that the scars on 98% of branded elephant seal pups were fully healed within one year. However, a review of the effects of hot-iron branding on southern elephant seals reported that 54% of one-year-old seals had one or more unhealed brands and 20% of all brands were not fully healed. Complete healing commonly takes many years and can involve serious scaring that, combined with incomplete healing, rendered 14% of brands of sub-yearling elephant seals unreadable (Gales 2000). Better results of brand healing were reported for New Zealand sea lions (Wilkinson *et al.* 2001) and Steller sea lion pups (Merrick *et al.* 1996). However, unsuccessful attempts to brand other species have generally not been reported. For example, branding fur seals has low success.

Branding equipment includes a forge, propane tank (or other energy source), and burner to heat the branding irons. The irons are made of 1 cm diameter mild steel and each number or letter is attached to a separate shaft. The equipment described by Merrick *et al.* (1996) has been used (with some local variation) to hot-iron brand otariids and phocids, including California sea lions, Steller sea lions, grey seals, ringed seals, and northern and southern elephant seals. The size of the brand may vary, although it is usually about 5 cm wide by 8 cm high. Arabic numerals are often used, but alternative symbolic systems can be used, especially to avoid the need for circular figures (0, 6, 8, 9) that could restrict circulation to a section of fur, and for which healing and readability are often a problem.

Brands are normally applied at a right angle to the animal's body (shoulder or rump). Curved brands (e.g. 0, 3, 6, and 8) are rolled slightly during application. The brand is heated to a ruby-red colour and applied with a light even pressure (so the flesh is not deeply pressed in) at about 5 psi (3.5 N cm^2 for 2–4 seconds, Merrick *et al.* 1996). Contact is maintained until the hair is burned off and the skin is slightly singed. After the brand is removed the character appears golden brown and is distinctly readable. Too much prolonged pressure will burn through the skin into the underlying muscle and blubber and will be slow to heal and difficult to read. Irons should be 'scaled' after 25–30 applications to remove carbon build up. Stainless steel shafts partially alleviate this problem.

Burn injuries are painful, so when possible it is recommended that animals are anaesthetized with gas anaesthesia (Heath *et al.* 1996; N.J. Gales *et al.* 2005a) prior to branding to reduce the pain and brand blurring caused by animal movement during the procedure. Recovery from gas anaesthesia is almost immediate and the animals are released back into the population with no apparent after-effects. Phocid pups (e.g. grey seals) can be successfully sedated with 10 mg of Valium given IV to improve the quality of hot-iron brands (Bowen, pers. comm.).

Mellish *et al.* (2007) studied the physiological effects of branding on Steller sea lion juveniles and found increases in white blood cell count, platelet levels and globulin and haptoglobin concentration up to two weeks after branding. No significant differences were found in serum cortisol levels. These changes were consistent with minor tissue trauma and indistinguishable from baseline levels after 7–8 weeks. Hastings *et al.* (in NMFS 2007) conducted a mark–recapture study in Southeast Alaska that examined survival rates of branded Steller sea lion pups. They found that weekly survival rates of branded pups were nearly identical to estimates from a control group of unbranded pups at an undisturbed part of the rookery. Studies of southern elephant seals also showed that branding did not negatively influence survival in the short or long term (McMahon *et al.* 2000, 2006).

2.2.5 Freeze-branding

Freeze- or cryo-branding also requires the capture and restraint of animals and is another lasting mark that can be observed from a distance. One method for freeze-branding is summarized in NMFS (2007). Branding irons are often made of copper, lead, brass, or stainless steel. The branding iron is chilled in a liquid coolant of alcohol and dry ice (−67 °C to −77 °C) or liquid nitrogen (−190 °C). An area of fur is first removed and the cooled brand then placed against the skin for 15–60 seconds at 10–15 psi (6.9–10.3 N cm^2). Anaesthesia (e.g. with isoflurane gas) is preferable for pain management and to reduce the potential for smudging the brand.

The marks produced by freeze-branding result in either short-term brands which kill the pigment-producing cells (melanocytes), or long-term brands which kill the hair follicles and pigment-producing cells to create a permanent bald brand (Merrick *et al.* 1996; Melin *et al.* 2006). Melanocytes can return on some brand marks in 1–2 years and make the brand difficult to read (Keyes and Farrell 1979; Wells 2002). Freeze-branding has had limited success in pinnipeds, with readable brands obtained in elephant seals, California sea lions, Australian sea lions, and walrus, but the duration of discernable brands rarely exceeds a few years (Wells 2002).

In the past, freeze-marking of pinnipeds was also accomplished by spraying a specially formulated liquid, such as Freon 22 or combinations of chlorodifluoromethane and dimethyl ether, from an aerosol through stencils. However, many of these chemicals are highly damaging to the environment and are not recommended, and may even be illegal to use.

Freeze-brands (see Fig. 2.1) have been used with greater success on small cetaceans, with marks sometimes remaining readable for more than 10 years (Scott *et al.* 1990a; Wells 2002). Cooled metal numerals 5–8 cm high are typically applied to the cetacean's body or dorsal fin for 15–20 sec. Ideally, brands should be held with even pressure for the entire period, but in some cases the curved plane of the dorsal fin will require rocking or multiple placements to bring the entire brand into contact with the skin. Brand application may be followed immediately by the application of water to return the skin temperature to normal. Early in the

Fig. 2.1 Freeze-brand on a bottlenose dolphin. (Photograph courtesy of Randall Wells.)

development of the technique, a slurry of dry ice and alcohol was used to cool numerals (Evans *et al.* 1972; Irvine and Wells 1972; Irvine *et al.* 1982), but better brands were found to result from the use of liquid nitrogen (White *et al.* 1981; Irvine *et al.* 1982; Scott *et al.* 1990a; Wells 2002). Dolphins exhibit little or no behavioural reaction to the application of freeze brands. If brands are applied for too long, then minor skin lesions may occur, but these generally heal quickly. Tissue response from the freezing varies among individual animals, and is also related to the skill and experience of those applying the brands. Brands disappear within seconds of application and then reappear a few days later.

Although most brands fade somewhat over time, even faded brands can remain recognizable for many years in good quality photographs, from partial or re-pigmented numerals. Brands tend to disappear more rapidly and more completely on younger animals. Freeze-branding has been used safely and successfully with bottlenose dolphins, spinner dolphins, short-beaked common dolphins, Pacific white-sided dolphins, short-finned pilot whales, false killer whales, Amazon River dolphins, Risso's dolphins, and rough-toothed dolphins (Evans *et al.* 1972; Irvine and Wells 1972; White *et al.* 1981; Irvine *et al.* 1982; Scott *et al.* 1990a; Wells 2002).

2.2.6 Fur clipping

Shearing of hair to create a temporary mark is useful in situations where short-term identification of a few individuals or a group of individuals is needed (NMFS 2007). It is most useful when the under-fur is a different colour than the guard hair. Shearing a small patch of guard hairs from the head has been used to estimate northern fur seal pup production since 1963 (Chapman and Johnson 1968; York and Kozloff 1987; Melin *et al.* 2006). Shaving a strip of fur on the rump of Australian sea lion pups has also been used (McIntosh *et al.* 2006).

2.2.7 Dye marking

Use of dye, paint, and bleach to mark marine mammals was summarized by Erickson and Bester (1993), Wells (2002), and Melin *et al.* (2006). These are temporary marks lasting only a few weeks to several months, but sometimes up to a year (Gentry and Holt 1982). These marks can be applied remotely or when the animal is under restraint (Fig. 2.2). Pellets containing dye or paint can be used to mark animals using specially manufactured guns or slingshots. They have the advantage of not requiring capture and restraint (NMFS 2007). Optimal performance of the dye, bleach, or paint is achieved when the fur is dry and when the animal remains out of the water for a period after application. Large marks of numbers and letters are used regularly on the black pelage of seal pups of many species using hydrogen peroxide hair dye. Within a few hours a bright beige to dark orange colour appears in the affected area (Gentry and Holt 1982). Light coloured sea lions have been marked with Nyanzol D to obtain a dark mark, and dark coloured fur seals and elephant seals have been marked with Nyanzol D with absolute alcohol added to produce a clearer dye mark (Erickson and Bester 1993).

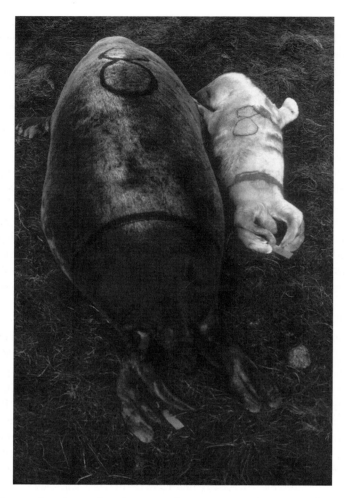

Fig. 2.2 Grey seal mother and newborn pup show temporary dye marking for identification through lactation as well as flipper tags (unseen in this photograph but Roto tags in this case) with orange streamer attached to make the tag more obvious. (Photograph courtesy of I.L. Boyd.)

2.2.8 Dorsal fin tags on cetaceans

The dorsal fin or ridge is the most common site used for small cetacean tagging because it is the site most easily observed. A variety of attachment tags have been tested for the identification of cetaceans (as opposed to electronic tags for telemetry, discussed in Chapter 10), with limited success (Wells 2002). The most successful tags have been the same small plastic cattle ear tags, or roto tags, used in pinnipeds (Table 2.1; Norris and Pryor 1970; Irvine *et al.* 1982; Scott *et al.* 1990a). These have been applied to the trailing edge of the dorsal fin, often through a small hole made under sterile technique with a 6-mm biopsy punch. The tag colour and attachment position on the dorsal fin provide the most useful identification information. Even if a tag is lost, the hole or notch can provide useful identification information. Tag longevities of several years are possible, but most are lost within a few months. Roto tags have been applied to a wide range of delphinids.

Other kinds of dorsal fin tags have proved less useful. Following on from the design of Peterson disk fish tags, button tags have been developed and applied to several species of small cetaceans (Evans *et al.* 1972; Scott *et al.* 1990a). Numbered and coloured fibreglass or plastic discs or rectangular plates on each side of the fin have been attached through the dorsal fin by means of one or more plastic or stainless steel bolts or pins. Because of poor tag retention and the potential for injury to the animals, especially in coastal situations where the tags can be rubbed on the seafloor, the use of button tags has been largely discontinued (Irvine *et al.* 1982).

The application of tags to other parts of cetacean bodies has been discouraged for many years due to problems of tag retention and damage to tissues (Irvine *et al.* 1982; Wells 2002). Coloured vinyl-covered strands of wire cable of variable length with metal dart tips, known as streamer or spaghetti tags, were anchored between blubber and muscle through the use of a jab stick or a crossbow. Abscesses frequently formed at tag attachment sites, with subsequent tag loss. Similarly, tethers or plastic-coated wires or polypropylene or soft rubber tubing have demonstrated poor retention and caused abrasions when attached to the caudal peduncle.

2.3 Photo-identification

The objective of photo-identification is to 'tag' an individual by using natural marks recorded in photographs. Photo-identification is an increasingly important tool for examining individual movement (e.g. Karlsson *et al.* 2005) and for capture–recapture studies (see Chapters 3 and 6) to assess abundance (e.g. Forcada and Aguilar 2000), survival (e.g. Mackey *et al.* 2008), and reproductive rates (e.g. Thompson and Wheeler 2008) in both pinnipeds and cetaceans. Pelage marks in pinnipeds have been used as natural marks for many years, as in leopard seals (Walker *et al.* 1998; Forcada and Robinson 2006) and grey seals (Pomeroy *et al.* 1994), and by using computer-aided pattern recognition of digitized photographs (Hiby and Lovell 1990; Cunningham 2009).

Photo-identification of cetaceans has proved to be one of the most useful approaches to learning about ranging patterns, social structure, life history, and abundance (Hammond *et al.* 1990a; Scott *et al.* 1990a; Wells 2002). Different species offer different opportunities for individual identification. Right whales exhibit distinctive patterns of callosities on their heads (Payne *et al.* 1983; Kraus *et al.* 1986). Humpback whales have distinctive colour patterns on the ventral surface of their flukes (Katona *et al.* 1979) and lateral bodies. Blue whales also have distinctive lateral body markings (Calambokidis *et al.* 1990). Killer whales have distinctive dorsal fin shapes and notch patterns in addition to distinctive light-coloured eye and saddle patches (Bigg 1982). Many dolphins can be identified from patterns of nicks and notches on their dorsal fins, building on the pioneering work of Würsig and Würsig (1977). Details of the application of this technique to specific small cetaceans have been reviewed extensively elsewhere (Scott *et al.* 1990a, Würsig and Jefferson 1990, Wells 2002) and are an important methods for the estimation of abundance (see Chapter 3).

2.3.1 Strengths and weaknesses

Photo-identification has the potential to be used more in the future as imaging technology continues to improve, both in terms of optics (including high-resolution, image-stabilized digital cameras) and the ability to process and match photographs more rapidly because of increasing computer power and electronic storage media capacity. The challenges that have restricted its use in the past include regularly and reliably being able to approach animals to a range that allows imaging of an appropriate part of the body, and which could be repeated consistently at sufficient resolution for dependable identification. Photo-identification is also difficult in species where there is little morphological, pigmentation, or pelage variation (but see McConkey 1999 and Caiafa *et al.* 2005).

In some marine mammal species the area used to identify individuals, known as the pattern cell, needs to be broadly consistent in shape and location, as with the examples of dorsal fins or tail flukes. Pinnipeds do not have a comparable distinctive area and so the pattern cell must be described in relation to morphological features such as the position of the eyes, ears, and nose.

2.3.2 Application

Perhaps the most difficult challenge is to obtain photographs of sufficient quality. Where possible, animals should be photographed from different angles and both sides in order to obtain good quality images for recognition. As a guide, the pattern cell should be no smaller than one-tenth of the frame. Photogrammetic analysis of individuals using two laser spotlights at a known separation distance and trained on the animal at the same time as photographs are being taken is beginning to be used. Photographic identification of cetaceans involves trying to obtain high-quality, high-resolution, full-frame images of identifying features (Würsig and Jefferson 1990). It is important to use a high level of rigor in selecting and matching images (Wells 2002). Photographic quality and fin distinctiveness should be carefully controlled (Read *et al.* 2003).

2.3.3 Pattern recognition methods

Computer-aided matching systems aim to reduce the number of images needing visual matching by describing an area numerically, and then calculating similarity scores between pairs of images and comparing them with all the others of the same side (left or right) and pattern cell. For example, *FlipperMatch*, a computer program developed to assist the photographic identification of individual Steller sea lions, compares the pattern of notches seen in the posterior edge of the flipper (Hillman 2005). Results of a preliminary study testing the utility of the program indicated that identification performance was better than chance but was not highly impressive (Hillman 2005).

The computer program used for grey seals (Hiby and Lovell 1990) and harbour seals (Cunningham 2009) locates the pattern cell, or combinations of pattern cells, within each photograph and extracts a numerical description from the grey-scale intensities (Fig. 2.3). To reduce the effect of gradual shifts in lighting conditions, the 'similarity score' is calculated as a mean from several sub-regions within the pattern cell extract. The program accounts for alignment errors by stretching one identifier array over the other to determine the maximum correlation coefficient. Pairs of photographs are then compared visually to ensure they are of the same individual. This removes the possibility of false-positives (i.e. the matching of different individuals, also known as a Type II error), and so any error only results from two images of the same individual not being matched (false-negative, also known as a Type I error).

Fig. 2.3 Screen shot showing a 3D model of the head of a harbour seal fitted to a photograph. The model allows for correction due to the angle of the head, and is fitted to the head using a fixed point such as the nose, eye, and external meatus. This allows consistent identification of the pattern cell. (Photograph courtesy of L. Cunningham.)

Several computer programs have been developed to facilitate matching individually distinctive cetaceans, especially for species which have the potential for the inclusion of large numbers of individuals in catalogues. Computerized comparisons of colour patterns on the ventral surfaces of flukes expedite matching for humpback whales (Mizroch *et al.* 1990; Mizroch and Harness 2003). Similarly, the trailing edges of sperm whale flukes have been catalogued and can be compared via computer programs (Whitehead 1990a; Burnell and Shanahan 2001; Bas *et al.* 2005). Catalogues of dorsal fins showing distinctive patterns of fin nicks and notches can include thousands of individuals for intensively studied species such as bottlenose dolphins, leading to the development of fin-matching software by several research teams (e.g. Kreho *et al.* 1999; Araabi *et al.* 2000; Hillman *et al.* 2003; Adams *et al.* 2006). While each of these computer-assisted systems can reduce the number of images to be considered, in most cases the final decision must be made through visual comparison of images.

2.3.4 Assessing errors

The effects of image quality need to be assessed and an appropriate system devised to reduce potential biases caused by including poor quality images (and resulting in overestimating population size). Images can be graded depending upon certain criteria, such as focus, resolution, the proportion of the pattern cell visible within the photograph, and the angle of the animal relative to the focal plane; those that do not pass a certain standard may be excluded. However, even poor quality images have the potential to provide useful information, and more work needs to be done to develop statistical methods that take account of quality as a variable within capture–recapture models.

2.4 Capture and restraint

The capture and restraint of marine mammals can be dangerous to the animal and the researcher. Researchers should select methods that limit all risks to the extent possible. In general terms, capture and restraint techniques should (modified from Gales *et al.* 2009):

- limit the duration of pursuit and capture;
- limit the number of capture attempts;
- minimize disturbance to non-target animals;
- ensure airways are unobstructed and breathing is monitored;
- minimize, monitor, and mitigate potential stressors such as thermal stress, physical trauma, and vulnerability from physiological state;
- minimize restraint duration to the minimum level required to conduct the procedure safely and humanely;
- ensure that the system of restraint allows for rapid, safe release of the animal in the event of problems.

2.4.1 Techniques for restraint in pinnipeds

Otariids can be captured on land and in the water, depending on their age, size, and location. Young pups on rookeries can be picked up by hand, while juvenile and sub-adult fur seals and some sea lions may be captured and restrained on land with hoop nets. These nets are typically about 1 m in diameter and are often mounted on handles of various lengths. The hoop of the net is commonly padded to reduce injury to the animal's mouth, gums, or teeth should it bite on the hoop. Net mesh size varies depending on the species and size of the animal being captured, but typically ranges from 1 cm to 8 cm stretched mesh. Woven nets, without abrasive knots, are preferable as they limit potential injury to the animal. Sufficient net is attached to the hoop to allow the full size of the animal to be enclosed, yet be small enough to restrict movement (see illustration in Geraci and Lounsbury 1993). Additional layers of net around the hoop apex, or a breathable, but non-visual barrier, are a useful addition to reduce the ability of the animal to see once netted. This generally quietens the animal and decreases its chances of biting a handler during restraint. If the seals are to be anaesthetized with gas anaesthesia during the restraint (see Sections 2.2.4 and 2.4.2), then a hole at the apex of the net that contains the closed muzzle of the animal facilitates the process.

Noose poles, as described by Gentry and Holt (1982) for the capture of female northern fur seals, are still used to capture individual juvenile and adult female otariids. Both the length and strength of the noose poles are determined by the size of the animal and the distance at which it is to be captured. A 1 cm rope about 95–110 cm long is placed between two holes drilled about 15–23 cm apart to create a loop. The hoop is placed over the animal's head and the pole is then rotated until the loop tightens around the neck. The animal is then pulled from the capture site by one or two people for restraint. Restraint boards are sometimes used, which are wooden platforms fitted with a yoke that closes around the animal's neck. The yoke contains padded 'V' notches. The upper blade is tied in place with a quick-release knot and the pressure adjusted to accommodate the animal's needs while maintaining the safety of the researchers. Additional restraints may be used to inhibit flipper or body movements while the animal is in the yoke. When using noose poles and restraint boards, the researcher needs to ensure that respiratory function is not restricted for a protracted period. The techniques are not designed to subdue the animal through anoxia, rather they are used to access and physically restrain the animal.

Northern fur seals have been captured and restrained in the rookery by using a roving two-person observation blind (approximately $1 \times 1.5 \times 2$ m, and about 4–50 kg). These mobile blinds provide access to almost all pups on the rookery and give researchers protection from attacks by territorial males while pups are processed (Boltnev and York 1998). Pups can be captured by hand, noose pole, or hoop net and brought into the 'blind' for processing. Females are also pulled into

Fig. 2.4 Walkway constructed over a breeding beach of Antarctic fur seals at the island of South Georgia. Individual animals can be marked with paint and dye and can be restrained and lifted on to the walkway to be weighed, measured, and marked. (Photograph courtesy of I.L. Boyd.)

the 'blind' by noose pole and restrained with a modified wooden yolk. A similar method using overhead walkways has been used at South Georgia to gain close access to Antarctic fur seals (Fig. 2.4).

An innovative technique has been developed to capture and restrain otariids while they are in the water. This incorporates a raft with a cage-trap attached to the top and which is then anchored near sites where sea lions regularly occur (Gearin *et al.* 1996). The raft serves as a temporary haul-out site for the sea lions. The 3 m × 3 m raft can be fitted with a 6-foot high steel cage with walls around the perimeter and with a wide drop gate or guillotine trap door on one side (Fig. 2.5). The door is usually left open for some time to allow animals to move freely between the raft and the water and to get accustomed to using the raft. The trap door can then be dropped when desired by the researchers. When ready to process, the animals are then moved one at a time through the trap door into a holding cage on a support barge, then into a squeeze cage for sampling, measuring, and marking. The technique has worked well for California sea lions and to a lesser extent for Steller sea lions.

Another technique used to capture juvenile Steller sea lions involves lassoing these animals underwater, capitalizing on the natural curiosity of these animals. Two or three SCUBA divers, supported by a skiff and a larger support vessel, attract the sea lions with a lure. When within range, a rope loop attached to a floating buoy is placed over its head near the shoulders. The rope is tightened and retrieved by the

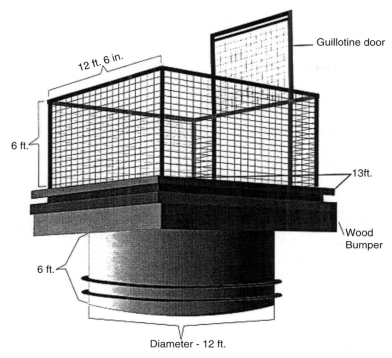

12 ft. 6 in.

Guillotine door

6 ft.

13ft.

Wood
Bumper

6 ft.

Diameter - 12 ft.

Fig. 2.5 Raft trap used to capture California sea lions. (Gearin *et al.* 1996.)

support skiff. The lasso is designed to prevent restriction of air or blood flow. Once on the surface researchers in the support skiff can pull the sea lion next to the skiff where it is wrapped in a restraining 'blanket', pulled into a restraint box in the skiff, then transported to the support vessel where it is immobilized with gas anaesthesia for handling (Heath *et al.* 1996; Curgus *et al.* 2001).

Australian fur seals that interact with fisheries at sea are attracted to catcher/ processor vessels and are captured by researchers on board the fishing vessel (S. Goldsworthy, LaTrobe University, Victoria, Australia, pers. commun.). A pouched net of trawl mesh is strung from a square frame and fish are attached to the bottom of the inside netting. The mesh size is sufficiently small to preclude the insertion of sea lion noses or heads, and made of material strong enough to support the weight of juvenile to adult fur seals. The net assembly can be lowered next to the vessel and suspended just beneath the water surface to a depth encouraging entrance by fur seals, but allowing visibility to the bottom of the net. After a fur seal enters the net it can be lifted out of the water within the net and brought on board.

Many large phocids can be easily approached, but because of their size and strength a restraining technique is needed. The technique described by Stirling (1966) for

Weddell seals is still used today for restraining and handling these and other large phocid seals. This technique consists of a large, stout sack with four ropes attached to each quarter around the sack mouth. The sack is normally tapered to constrain the head of the seal as the sack is pulled down the body. Breathing holes are also normally provided at the end of the sack. Two sizes of sack have been used, one 45 cm in diameter and 61 cm deep, and the other 66 cm by 91 cm (Erickson and Bester 1993). Standing on either side of the seal, two people, each holding two ropes, pull the sack over its head. The smaller sack fits to the shoulders while the larger one can contain the front flippers thereby allowing more control over the animal. The same technique with slight modifications has been used on northern and southern elephant seals. The elephant seal is manually restrained by one or two people (depending on the size of the animal) straddling its shoulders, or on either side, while another person immobilizes the animal by administering a drug into the extradural intervertebral vein (Antonelis *et al.* 1987; McMahon *et al.* 2000). Woven hoop nets which incorporate an apex similar to the smaller of the head bags have been used successfully to capture most Antarctic phocids and this method has also been used on adult male otariids.

Phocids (and otariids) can be captured and managed with stretcher nets, throw nets, crowding boards, and hoop nets (see illustrations in Geraci and Lounsbury 1993). Grey and harbour seal pups have been captured by hand or hoop net and their mothers captured in a net fastened between two 2 m aluminium poles. The net was hinged at the head end and open at the rear and was closed by bringing the poles together under the animal, once the animal was in the net (Bowen *et al.* 1992). This is a very effective method of restraining and transporting large pinnipeds.

Arctic phocid seals have been captured using hoop nets. Three inflatable boats can be used to surround a seal hauled out on ice. Often the seal becomes confused by the approaching boats and hesitates while deciding if it should escape into the water or stay on the ice floe, and this provides an opportunity for capture with a hoop net.

Harbour and grey seals resting on shore can be captured by hoop nets from boats approaching at speed or by tangle nets set in the water immediately adjacent to a haul-out site, or in a narrow coastal passage through which seals are likely to pass (e.g. Jeffries *et al.* 1993; Goulet *et al.* 2001; Small *et al.* 2005). The capture nets can be 120–170 m in length with a 20 or 30 cm stretch mesh. A ground line is weighted at each end and floats keep the net header line towards the surface. The usual method of deployment involves the use of a fast skiff or RHIB (rigid-hulled inflatable boat) with the net folded into a bin at the stern. An approach is made to a haul-out at high speed and the weight at one end of the ground line is dropped overboard. This pulls the net out as the boat moves along the front of the haul-out and deploys the net before most animals have had a chance to escape. The method is only practical in water below a depth of 5–10 m, but floating nets can also be designed for use in deeper water. Nets cannot be left unattended and there should always be sufficient sag in the net for trapped seals to reach the surface to breathe.

Caution should be exercised about capturing too many seals in a single set in case there are insufficient personnel to haul the net and ensure the safety of the animals.

2.4.2 Chemical restraint and immobilization in pinnipeds

Field procedures that are prolonged, painful, require an entirely immobile animal, or are conducted on animals too large for physical restraint are often augmented with chemical agents. The selection of drugs, or combinations of drugs, to be used and the methods and routes of administration will be determined by the desired drug effects (e.g. sedation, analgesia, and/or anaesthesia), the species, sex, and age class of the animal and the field conditions under which they will be applied. These decisions will determine the sophistication of the equipment required and the degree of expertise required to manage the medical procedures.

Many classes of drugs have been used to chemically restrain pinnipeds and it is beyond the scope of this chapter to provide a comprehensive review (see Gales 1989 and Haulena and Heath 2001). The lessons of this history of experimentation is that drug response is highly variable and dependent on species, and that the physiology of diving mammals presents a particular challenge for the safe use of drugs. There is also a clear trend—mediated through the engagement of veterinary specialists, the advent of new and sophisticated drugs, and the modification of medical equipment for field use—which has led to techniques that enjoy wide margins of safety, efficient animal handling, and minimal post-handling, drug-related problems. Consequently, as with other areas of rapid methodological advancement such as telemetry studies (see Chapter 10), the researcher should thoroughly review the most recent literature when determining whether to use chemical restraint and which techniques to select.

There is no simple guide as to when to augment physical restraint with the use of drugs. While chemical use is obvious if pain relief is required for the animal, or when the animal is too large for physical restraint alone, the use of drugs should also be considered when it is appropriate to reduce the level of anxiety of the animal and/or when the procedure can be conducted more safely and rapidly with the use of drugs.

Broadly speaking, drugs used for chemical restraint can be considered in two general groups: injectable compounds, and inhalational compounds. These can be used alone, or in combination, depending upon the experimental needs.

The injectable drugs most commonly used today include the sedatives (usually the benzodiazepines such as diazepam) which assist in handling and the anaesthetics (most commonly the cyclohexamines such as tiletamine) which provide analgesia and full immobilization. The use of sedatives is characterized by relatively wide margins of safety for the animal, few side-effects, and consequently requires less specialized personnel. The modern injectable anaesthetics, although more reliable than earlier drugs, have narrower margins of safety, suffer more unwanted side-effects (generally to the respiratory, cardiovascular and/or thermoregulatory mechanisms), and require more specialized personnel. Injectable drugs offer some very practical advantages. When delivered into the muscle mass they can be deployed remotely via blowpipe, pneumatic gun, or a length of flexible medical tubing. Delivery by injection is also rapid. When given intravenously, which requires prior physical restraint, the drug effect is mediated rapidly. The flipside to these practical advantages is that the weight-dependent dose must be estimated prior to administration and, once delivered, the effect of that drug must be managed until it has been metabolized by the animal, or (in some cases) reversed with the use of other injectable drugs (which might have their own side-effects). Additionally, if the animal is not restrained during drug delivery it is able to escape prior to the onset of the drug effect and may become vulnerable through entering the water (in which case a full anaesthetic dose is like to be lethal) or may be attacked by con-specifics.

The inhalational drugs, most commonly isoflurane, require prior physical restraint of the animal, expensive and specialized field equipment, and experienced veterinary supervision. However, these drugs offer the great advantages of relatively wide margins of safety, manageable side-effects, and, most importantly, they can be delivered 'to effect'. The result is a predictable and safe medical procedure for which the depth of anaesthesia can be controlled and for which recovery is rapid and smooth.

Combinations of injectable and inhalational drugs are being used increasingly, particularly for the more drug-sensitive pinnipeds, which include many of the phocids. Some of the Antarctic ice-seals, such as crabeater, Weddell and leopard seals have suffered highly variable results with injectable anaesthetics, partly due to dive-like physiological responses during anaesthesia, and—particularly in the case of leopard seals—highly collapsible tracheas. Techniques that deliver sedative drugs via remote injection (most commonly midazolam at doses of 0.02–0.05 mg/kg), and then use physical restraint in specially modified hoop nets to facilitate inhalation anaesthesia with isoflurane (at maintenance doses of 1–3%) have proven highly effective in these species (N.J. Gales *et al.* 2005a). For smaller, more tractable pinnipeds, isoflurane can provide rapid and smooth induction and anaesthesia maintenance once the animal has been physically restrained in hoop nets which provide access to the nares at the net apex.

While inhalational drugs are being used increasingly, with great improvements in experimental and welfare outcomes, injectable combinations of a sedative and anaesthetic (most commonly zolazepam and tiletamine) remain widely and successfully applied. For example, over 1000 southern elephant seals were successfully handled with zolazepam and tiletamine administered intravenously with no mortalities or substantial side-effects (McMahon *et al.* 2000).

Overall, the determination of the need, selection, and administration of drugs for the restraint of pinnipeds has become increasingly more specialized. Researchers should ensure they are compliant with the many laws that regulate the use of these drugs, and should be encouraged to consult widely with experienced practitioners to determine if and how to use chemical restraint. A well-trained and dedicated person should be part of any team using chemical restraint, and results, good and bad, should be disseminated to continue the trend of increasingly safe and manageable options for the chemical restraint and immobilization of pinnipeds.

2.4.3 Techniques for capture–release of cetaceans

Many research projects that involve the capture and release of small odontocete cetaceans use variations on the capture techniques described by Asper (1975). Attempts to restrain large, powerful, wild, air-breathing animals in an aquatic environment carries inherent risks, but careful planning, patience, and experience can often make the risks manageable. Selection of an effective and safe technique depends on the size of the animal, the number to be caught at one time, the habitat, and the behaviour of the animals. Dolphins, porpoises, and small whales can be caught one at a time with breakaway hoop nets, tail grabs, and hand and rope. Individuals or small groups of dolphins, small whales, or porpoises may be caught by shallow-water seine, hukilau, fishing weirs, modified gillnets, or purse seines. Larger odontocetes, such as killer whales, have been captured on occasion through very expensive and logistically complex means that are beyond the scope of techniques available to most researchers, and so are not considered further here.

Breakaway hoop nets are useful for catching large individuals, or animals that do not enter shallow waters where other techniques might be applied. As described by Asper (1975), the breakaway hoop net consists of a long pole attached to a large hoop frame, a small-mesh net suspended tightly from the hoop by elastic bands (or equivalent) attached to a long tether (often an elastic cord), with a large float at the end. The hoop is of sufficiently large diameter to allow the entire body of the animal to pass through without contact. This technique requires that the animal rides on the bow wave of a vessel, and surfaces to breathe in front of the boat in a position directly below a pulpit extending out from the bow of the vessel. As the animal passes through the hoop, the net breaks away from the hoop and wraps the anterior half of the animal. The tether and float are released, they are retrieved as the animal tires, and the animal is manoeuvred into a sling and brought aboard the capture vessel. This technique has been used with large animals such as pilot whales and false killer whales, as well as smaller delphinids.

Another technique that takes advantage of bow-riding dolphins is the tail grab (Würsig 1982). This device consists of a padded clamp on a break-away handle that closes around the peduncle of a dolphin surfacing at the bow. When the dorsal surface of the peduncle contacts a trigger on the inside of the open grab, a motorcycle spring powers the heavily padded arms of the grab to close around the peduncle, and the grab, attached to a 15 m long elastic cord, separates from the handle. The animal is retrieved and brought aboard the capture vessel. Because this technique uses the tail instead of the head, no bow pulpit is required, and it can be used from vessels as small as 3.5 m in length (Würsig 1982). Tail grabs tend to be used with smaller species, such as dusky dolphins, Pacific white-sided dolphins, pantropical spotted dolphins, and

Hawaiian spinner dolphins, among others. As the technique will lead to the animal being hauled backwards through the water, and the force of restraint is limited to the peduncle, great care is required to minimize the risk of injury.

The hand and rope technique has been successfully used with belugas for many years (Asper 1975). A small boat can be used to herd individual whales into shallow (<1.5 m deep) water. A catcher then jumps from the bow of the boat when it is alongside a whale, and tries to pass a soft rope over its head. Once the line is around the animal, others enter the water to help secure the whale until it can be placed in a sling. A similar method has been used to capture dugongs (Lanyon *et al.* 2006).

One of the most commonly used techniques for capturing individuals or small groups of dolphins in shallow water is a large-mesh seine net, typically deployed from a fast boat to form a circular corral (Asper 1975; Wells *et al.* 2004). Floats and lead weights cause the net to form a full mesh barrier from the surface of the water to the bottom. The length and depth of the net depend on the species and its habitat. Nets range in length from about 350 to 500 m. Net depth ranges from about 4 m to 8 m, slightly deeper than the water in which it will be set. The nets are designed to be sufficiently strong to stop the cetaceans should they strike the net, sufficiently soft to minimize the possibility of injury from the net, and sufficiently lightweight for the animals to be able to reach the surface even when entangled. For bottlenose dolphins, 48–52 gauge nylon with 20 cm stretched mesh is often used. Floats are spaced to keep the net at the surface, and facilitate detection of entanglements. Enough lead is used in the bottom line to sink the net to the bottom quickly, while being sufficiently light enough for a dolphin to lift it easily. Whenever possible, seine nets are used in waters where handlers can stand with their heads above water, in order to be able to safely provide support for dolphins that entangle in the net.

Once dolphins are selected for capture, and they are in a suitable location (ideally, <2 m deep, minimal current and wave height, no obstructions on the bottom), the seine net is deployed around them from a vessel moving at 30 knots or more. A second vessel often moves at high speed in the opposite direction, closing the circle with engine noise and bubbles until the net circle is complete. The circumference of the circle is checked for overlays of the lead line over the floats, and additional vessels with handlers move into positions around the periphery to monitor the number of animals surfacing within the net corral and to detect float submergences that might indicate dolphins in the net. The net boat may next attempt to create a smaller net circle around the animals within the original corral, or may drag the entire corral into shallower water nearby. Once the net is in final position, handlers may be deployed around the outside of the corral. For bottlenose dolphins, typical teams comprise at least four times as many handlers as the maximum number of dolphins that might be caught at one time. Dolphins may be restrained individually as they swim or float inside the corral, or be removed from the net if they swim into it, and then manoeuvred into a sling for transfer to a boat or held in the water for other procedures. This technique has been used successfully with a variety of dolphins, including bottlenose dolphins, Franciscana dolphins (Bordino *et al.* 2007), and belugas (Ferrero *et al.* 2000).

A variation of the seine net technique has been described for the capture of botos, or Amazon River dolphins (botos: da Silva and Martin 2000). During periods of low water and minimal current, a narrow (<80 m wide, up to 10 m deep) river channel may be closed off with net of 12–15 cm stretched mesh. When a group of botos enters the channel and approaches the net, a second net is set quickly 100 m from the first to block their escape. Additional nets are set to reduce the area of the enclosure, and a smaller-mesh seine net is used to bring one individual at a time to shore for procedures, allowing the remaining animals to swim within the enclosure.

Another relatively shallow water technique takes advantage of existing fishing structures, herring weirs, in the vicinity of the Bay of Fundy, Canada. Harbour porpoises (*Phocoena phocoena*) enter weirs in search of herring, and become trapped inside. The capture and removal of porpoises from the weirs by scientists provides conservation benefits, facilitates the job of the fishermen, and provides opportunities to obtain life history and health information and to tag members of a species that is otherwise difficult to access (Neimanis *et al.* 2004).

Modified gillnets provide a way of capturing small animals, such as harbour porpoises, for tagging when the animals are in deeper water. Hanson *et al.* (1999) describe a technique in which a 182 m long, 9.1 m deep, 30.5 cm stretched mesh monofilament gillnet, with a light lead line (360 g/m) and floats spaced every 1.1 m, is set as a driftnet, extending straight out from a 6.1 m vessel equipped with a hydraulic-powered reel. The float line is closely monitored for changes that might indicate entanglement, such as submergence and bunching. A second vessel investigates suspected entanglements, and captured porpoises are lifted aboard the vessel onto a padded deck.

Open ocean dolphins provide serious challenges for research that involves handling. In some cases it has been possible to work with large purse seine vessels designed for catching tuna, and gain access to pantropical spotted and/or spinner dolphins captured along with the tuna (e.g. Jennings *et al.* 1981, Scott and Chivers 2009). Dolphins encircled by the large purse seine net are manoeuvred into special enclosures or alongside specially designed platforms for measuring, sampling, and/ or tagging before being released over the net, prior to completing the pursing process for the fish.

Norris and Dohl (1980) described an experimental approach for controlling the movements of schools of Hawaiian spinner dolphins. Their Hawaiian net, or hukilau, consisted of a 450 m long float line with 20 m long weighted vertical lines spaced 2–3 m apart. Groups of 40–60 spinner dolphins were encircled in 20–40 m of water and were held by this insubstantial barrier for periods of up to nearly four hours. Variations on the hukilau concept have been used to try to move dolphins, but with mixed success.

2.4.4 Techniques for restraint and handling

Wild cetaceans vary in their responses to capture and handling for the first time, but in many cases the initial restraint leads to a brief period (seconds to minutes) of elevated activity as the animals attempt to free themselves. The animals then become more passive, while remaining alert and watching activities around them. In some cases, such as bottlenose dolphins, individuals become increasingly docile with repeated capture–release events (Wells *et al.* 2004). As the animals are brought out of the water and placed on shore or on the deck of a boat, the fact that they are probably experiencing their own body weight for the first time may become a factor in their reduced activity level. For the brief periods involved in most research captures, chemical sedation is neither necessary nor desirable, as the animals need to be fully functional upon release.

Procedures for caring for the animals following capture vary with the research goals, species, and situation. Most species are sufficiently large that specially designed slings are necessary to carefully lift them from the water, by hand or with mechanical assistance. Often, individuals are placed gently on closed-cell foam pads, shaded from the sun, and cooled with clean, sponged water (taking care to keep water from the blowhole during respirations). Many research operations incorporate the services of a veterinarian with marine mammal experience. Trained staff can monitor respiratory patterns and associated heart rate patterns for normal cardiac arrhythmia, and surface and/or core body temperature. Under some circumstances it is also possible to monitor blood chemistry and haematocrit in the field with a portable clinical analyser, which can provide results within 5–10 min.

Behavioural monitoring is crucial. A dedicated person should monitor inter-breath intervals and alert the veterinarian to intervals that exceed a pre-set threshold for concern, depending on the species and the animal's age. Declines in an animal's response to stimuli, the degree to which it is actively watching activities around it, or its rate of sound production (e.g. whistles) can be indicative of potential problems. Escalation from unusual passivity to rapid, erratic, and shallow breaths, or arching of the back with simultaneous extreme elevation of the head and tail, in conjunction with breath-holding are behaviours that should raise serious concern for the health of the captured individual. If the animal does not respond to splashing water over the closed blowhole, then it should be immediately returned to the water to be supported until its condition stabilizes, and the veterinarian may decide that emergency drugs should be administered. The veterinarian's emergency kit might include injectable doxapram hydrochloride for stimulating respiration, diazepam to relax the animal, and corticosteroids for stabilization, but the efficacy of these drugs to restore homeostasis remains poorly determined.

Chemical restraint and immobilization are rarely applied to cetaceans in the field because of a high risk of mortality. Successful anaesthesia has only been accomplished rarely under highly controlled conditions. While the use of some sedatives and other drug classes such as local anaesthetics might be appropriate for some prolonged or invasive field methods, their use should be limited to experienced veterinarians and they are not discussed further in this chapter.

2.4.5 Risks to cetaceans during capture and handling

Risks are inherent in the capture and handling of any wild cetacean. Responses to the capture process vary by species. The cetacean species composition of exhibits at zoological parks likely reflects, at least in part, how different species respond to capture and handling. For example, bottlenose dolphins, the mainstay of many exhibits worldwide, are much more tolerant of capture, handling, and captivity than are more pelagic species such as spinner dolphins, a species rarely maintained in captivity.

Cetaceans occasionally die during the capture process. Death can result if capture gear prevents an animal from reaching the surface to breathe, or if struggles near the surface lead to aspiration of significant quantities of water into the lungs. The research team must be able to quickly detect and assist animals having difficulty surfacing. In situations where more than one animal may be caught at a time, the research team must be alert to the possibilities that multiple individuals may entangle simultaneously, some entanglements below the surface may not be immediately evident, or individuals may be trapped, unseen, below other animals. Reliable knowledge of the number of animals targeted for capture and careful accounting of all of the individuals through each surfacing sequence prior to restraint can reduce the risk of death. Avoiding capturing very young, very old, and obviously compromised individuals can also reduce risks. Using ultrasound to check adult females for pregnancy as soon as possible following capture can minimize risks to the pregnancy that might have resulted from further handling.

Additional risks exist beyond immediate mortality as a result of the capture process. The possibility of physical injuries from capture gear, such as lacerations or flipper dislocations, can be reduced through quick restraint to minimize struggling. Stress from exertion due to capture and handling can potentially lead to muscle damage in extreme cases (St. Aubin and Dierauf 2001). Norman *et al.* (2004, p. 53) reviewed animal health concerns during capture and handling of odontocetes and summarized the situation as follows:

> A general rule of thumb is that the degree to which an animal is compromised increases with the amount of handling and the length of time that the animal is handled. An individual animal may not display any obvious outward signs of being compromised beyond a threshold from which it cannot recover, as evidenced by the occasional sudden death of an animal that otherwise outwardly appears to be tolerating handling. Thus, handling should be kept to the minimum necessary to complete the research objectives.

Careful monitoring by experienced personnel during handling, as described in Section 2.4.4 above, can reduce risks to the animals. Further protection for the animals may be provided through controlling exposure to human pathogens.

Although transmission of disease from humans to cetaceans has not yet been documented, the transfer of microorganisms such as virulent staphylococci, which are common to both humans and dolphins, is possible (Buck *et al.* 2006).

Few data are available to evaluate the potential long-term effects, if any, of capture and handling on wild cetaceans. The long-term resident bottlenose dolphin population of Sarasota Bay, Florida, has been studied since 1970, providing such an opportunity (Irvine and Wells 1972; Scott *et al.* 1990b; Wells 1991, 2003). Using a seine net in shallow water, capture–release for tagging occurred during 1970–1971, and 1975–1976. Capture–release of selected individuals for life history and health assessment has occurred during 22 of the 24 years from 1984 through 2007, with 209 individual dolphins being handled, many of them multiple times. Applying the technique of DeMaster and Drevenak (1988) to data from 1984 to 2007, the annual survival rate for captured and released dolphins included from the time of their first handling was calculated to be 0.95 ($\pm$0.019 SD). Reilly and Barlow (1986) reported annual survival rates for wild bottlenose dolphins of 0.92–0.95. DeMaster and Drevenak (1988) reported mean annual survival rates of 0.93 (0.92–0.94, 95% CI) for captive bottlenose dolphins. In addition, reproductive rates, site fidelity, and social patterns of the resident bottlenose dolphins of Sarasota Bay have not been affected significantly by capture–release (Wells 2003). Similar capture–release studies for the health assessment of bottlenose dolphins are carried out regularly elsewhere in the United States.

2.5 Risks to researchers during marine mammal capture and handling

Most cetaceans and pinnipeds are large, powerful animals that, when handled, wield the potential to cause death or serious injury to humans. Thrashing flukes or swinging rostra in cetaceans and the tendency for pinnipeds to bite can be dangerous. Accidental entanglement of humans with these species in capture gear can lead to broken bones or drowning. In addition to threats from the animals themselves, researchers should try to prepare for the unexpected, including equipment failure, boating accidents, and injuries from other creatures such as sharks, stingrays, Portuguese men-of-war, oyster shells, etc. When working on seal colonies the greatest danger can come from con-specifics including territorial males. In the Arctic, polar bears and walrus are dangers that require high levels of awareness. Experience, training, mock capture exercises, careful planning, development of detailed safety protocols (including formal risk assessments), teamwork, and incorporation of emergency medical professionals into research teams can help to mitigate the risks to personnel.

Norman *et al.* (2004) reviewed human health concerns during the capture and handling of odontocetes. Some pathogenic bacteria and fungi (e.g. *Brucella* spp., *Mycobacterium marinum*, *Erysipelothrix rhusiopathiae*, *Blastomyces* spp., *Loboa loboi*) have been transmitted from cetaceans to humans (Geraci and Ridgway 1991; Norman *et al.* 2004; Buck *et al.* 2006). Where pinnipeds are concerned, the greatest risk to researchers of infectious disease is from 'seal finger'. This can come about from handling dead seal tissues as well as live seals. It causes cellulitis, debilitating joint inflammation, and oedema of the bone marrow. The organism responsible is most likely to be a species of the *Mycoplasma* genus called *Mycoplasma phocacerebrale* (Baker *et al.* 1998). A likely route for the transmission of infection is through open wounds or abrasions. Precautions against infection may include wearing surgical masks and gloves, avoiding exposure to exhalate and bodily fluids, and careful washing with disinfectant between and following contact with marine mammals (Norman *et al.* 2004). However, some of these precautions may be impractical under some field conditions. Pregnant or immunocompromised personnel are advised to avoid close contact with marine mammals.

3

Estimating the abundance of marine mammals

Philip S. Hammond

3.1 Introduction

The introduction to any paper describing the abundance of marine mammals is likely to begin with some general statements about the importance of this information to advance ecological understanding and/or to inform conservation and management. This is not just to satisfy ever more demanding journal editors that the work can be placed in a wider context. In the first edition of their excellent textbook on ecology, Begon *et al.* (1996) wrote:

> Every species of plant and animal is always absent from almost everywhere. But, a large part of the science of ecology is concerned with trying to understand what determines the abundance of species in the restricted areas where they do occur. Why are some species rare and others common? Why does a species occur at low population densities in some places and at high densities in others? What factors cause fluctuations in a species' abundance?

Knowledge of the abundance of a species is indeed at the heart of ecology. Theoretical studies of population dynamics have greatly advanced our understanding of how species interact with each other and the other elements of the environment in which they live. But ultimately, to address the questions posed above, abundance must be determined in the field.

Equally, perhaps more importantly, knowledge of abundance is needed to assess conservation status and thus prioritize management actions. Many species of marine mammals have been (and continue to be) the target of exploitation in the form of hunting or incidental by-catch in fisheries, or otherwise threatened by human activities causing disturbance or loss of habitat. The recently completed IUCN Global Mammal Assessment has assessed the status of all marine mammal species under revised criteria that are heavily reliant on information on abundance

(Schipper *et al.* 2008). Management procedures for determining safe limits to catches or by-catches similarly rely on estimates of abundance (Wade 1998; Punt and Donovan 2007).

This chapter covers practical aspects of estimating the abundance of marine mammal populations. I use the term population here broadly following the definition of Krebs (1972):

> a group of organisms of the same species occupying a particular space at a particular time ... (its boundaries) in space and in time are vague and in practice are usually fixed by the investigator arbitrarily.

The logistics of marine mammal abundance estimation typically necessitate such pragmatism, although we are sometimes able to stretch that definition to something more biologically meaningful.

Abundance of an animal species may on very rare occasions be enumerated by census (when all individuals are counted), but in almost all situations this is impossible. This is especially true for marine mammals, which pose particular problems of access and scale. They live entirely (in the case of cetaceans and sirenians) or mostly (in the case of pinnipeds) in water and, when at sea, typically spend a substantial majority of their time underwater. Populations of marine mammals can be very large and range over very wide areas. Baleen whales typically inhabit whole ocean basins and some species have populations numbering tens or even hundreds of thousands of individuals (http://www.iwcoffice.org/conservation/estimate.htm). Some seal species number in the millions (Laidre *et al.* 2008). Conversely, some populations are very small (e.g. Jaramillo-Legorreta *et al.* 1999; Branch 2007) and this presents problems of a different kind for estimating abundance.

Marine mammal abundance must thus be estimated, which means collecting samples of data and making inferences from them with the aid of some kind of statistical model. There is a limited choice of how to obtain these data: we can sample dead animals, live animals, or the space in which animals live. Using samples of dead animals to estimate abundance via catch-per-unit-of-effort or change-in-ratio methods is now rare in marine mammal studies, if only for the simple reason that hunting is far less prevalent than it used to be. Methods using such data are not covered in this chapter.

Instead, the chapter focuses on the methods most commonly used to estimate marine mammal abundance today. Methods based on sampling live animals include extrapolating from counts of individuals (e.g. Thompson *et al.* 1997; Rugh *et al.* 2005), and mark–recapture analyses of capture histories of individuals marked with tags or photographically (Hammond 1986). The standard method based on sampling space is line transect sampling (Buckland *et al.* 2001); variations on this theme include cue-counting (Hiby and Hammond 1989), density surface modelling (Hedley *et al.* 1999, 2004), and acoustic sampling (Van Parijs *et al.* 2002; Barlow and

Taylor 2005; Lewis *et al.* 2007). Which method is most appropriate depends on the intended use for the estimates, the species and its characteristics, and the available resources (Hammond 1995).

Whichever method is used, collecting data to estimate marine mammal abundance is typically logistically challenging and often expensive. Pinnipeds give birth and may moult or haul out for other reasons on land or ice, and abundance can be estimated from data collected during these phases of their life cycle. Observations may be made from the ground, but more often an aircraft is required. Some cetacean species migrate conveniently close to coasts from where land-based observations can be made, but typically a sea-going vessel or an aircraft is required for surveys to estimate abundance. The area of interest may be relatively small but, for good coverage of the distribution of wide-ranging species, large-scale surveys are necessary (Hammond *et al.* 2002; Gerrodette and Forcada 2005).

To infer population size from the data, we make a number of assumptions about how representative the sample is of the population and how the estimation method works in practice. The robustness of any estimate depends on how well the assumptions of the method are met. Violation of these assumptions typically results in biased estimates of abundance; such biases may be small and therefore unimportant, but if they are large they may render estimates effectively useless. It is thus extremely important to know the assumptions made by a particular method and to understand the implications of their violation. The level of bias acceptable in an estimate must also be considered in the context of its precision. There is little point in expending a lot of resources to ensure an estimate is unbiased if that estimate has a very wide confidence interval. Conversely, it makes no sense to commit all one's resources to collecting a large sample to reduce variance if the resulting estimate is seriously biased. Balancing bias and precision to make efficient use of resources to achieve the best answer to the question posed is a fundamental part of estimating abundance.

3.2 Extrapolation of counts

Conceptually, the simplest way to estimate abundance is to obtain an accurate count of a readily identifiable segment of a population and to extrapolate this count to the rest of the population. There are a variety of ways in which this can be done.

For some species of seal, e.g. harbour seal (*Phoca vitulina*), one method is to count the number of animals at haul-out sites at a particular time and to correct the count by an estimate of the proportion of the population that is not on land at that time using telemetry data (e.g. Thompson *et al.* 1997; Ries *et al.* 1998; Boveng *et al.* 2003; Sharples *et al.* 2009). Important features of this method are the need for accurate counts from land, sea, or air and the need for accurate telemetry data relevant to the time of the counts. Variation in haul-out behaviour must be appropriately accounted for when determining or modelling counts and when correcting counts for the proportion of the population at sea.

Another method, used for grey seals (*Halichoerus grypus*) around Britain (Thomas and Harwood 2008), is to count the number of seal pups born at breeding colonies and to estimate the number of adult females in the population needed to produce that number of pups by fitting a population model to the annual pup production estimates. Total adult population size is obtained via a simple sex-ratio calculation. At the base of the method, the quality of the aerial photographs is critical to the accuracy of the pup counts. The model used to estimate pup production from the pup counts needs to account for variability in the timing and length of the pupping season. Any uncertainty in structure of the model used to estimate population size from estimates of pup production and in the data used to estimate fixed parameter values in the model must be accounted for appropriately to obtain accurate and precise estimates of population size. Thomas and Harwood (2008) used a state–space model within a Bayesian statistical framework.

For some whale species—particularly grey (*Eschrichtius robustus*), bowhead (*Balaena mysticetus*), and humpback (*Megaptera novaeangliae*) whales—the number of animals migrating past a convenient point on land can be counted and corrected for the proportion missed due to various factors (Brown *et al.* 1995; George *et al.* 2004; Rugh *et al.* 2005). The key elements of this method are: obtaining an accurate count of animals detected; correcting for periods during which searching did not occur; correcting for animals missed while searching; and correcting for the proportion of the population within range. These last two features can be termed perception bias and availability bias, respectively, see Section 3.4 *Line Transect Sampling*). The estimates obtained from this method are typically precise because a relatively high proportion of the estimated population is actually observed. Attention to all the major sources of error in counts means that the estimates should also have low bias.

3.3 Mark–recapture methods

Mark–recapture techniques are widely used in ecology, particularly to estimate survival rates and population size (see Chapter 6 for their application to the estimation of survival rates). For marine mammals, they have proved useful for studying some species that can readily be captured (tagging and branding of pinnipeds on haul-out sites) and some species that cannot but can be 'marked' and 'recaptured' in other ways (photo-identification of cetaceans). In this chapter, we focus on the use of mark–recapture for estimating abundance, but there is a growing literature on using these methods to estimate the survival rates of marine mammal populations, as summarized in Hammond (2009).

The difficulties associated with physical handling and tag attachment makes the application of mark–recapture techniques more challenging for cetaceans than for pinnipeds. General texts on using these methods in studies of marine mammals have therefore tended to focus on cetaceans, especially the use of photo-identification (Hammond 1986; Hammond 1990a; Hammond *et al.* 1990). Although somewhat dated, much of this material is still relevant; a more recent summary is given in Hammond (2009). Inevitably, much of this section also focuses on the practical aspects of estimating cetacean population size using photo-identification data. Studies of cetaceans have focused on a number of species, including: humpback, blue, sperm, and northern bottlenose whales; and bottlenose, humpback, and Hector's dolphins (Whitehead *et al.* 1997a, b; Chaloupka *et al.* 1999; B. Wilson *et al.* 1999a; Bejder and Dawson 2001; Read *et al.* 2003; Stevick *et al.* 2003; Calambokidis and Barlow 2004; Larsen and Hammond 2004; Stensland *et al.* 2006).

For pinnipeds, examples include studies of Weddell, leopard, elephant and grey seals, and Australian fur seals (Shaughnessy *et al.* 2000; Cameron and Siniff 2004; Garcia-Aguilar and Morales-Bojorquez 2005; Forcada and Robinson 2006; Gerondeau *et al.* 2007).

Mark–recapture methods rely on sampling and re-sampling individual animals, rather than sampling the area in which the animals live (see Section 3.4, *Line Transect Sampling*). The estimate of abundance obtained is therefore the number of animals using the study area during the study period—not, as estimated by line transect sampling, the density of animals in the study area. This seemingly subtle but fundamental difference may or may not be important in terms of the objectives of the study.

To estimate abundance using mark–recapture methods, the basic data required are the of capture histories of individually identified animals. A capture history simply describes whether or not an animal was captured in a series of sampling occasions, discrete periods of data collection, usually represented by a series of 1s (captured) and 0s (not captured). Depending on the species and the objectives, duration, and location of the study, a sampling occasion may be anything from a day to a season.

Estimating abundance using mark–recapture is based on the idea of marking a number of animals in a population, and then using the proportion of marked individuals recaptured in a subsequent sample of animals as an estimate of the marked proportion in the population at large. The simplest case of a single marking occasion and a single recapture occasion is commonly known as the Petersen estimator. A sample of individuals (n_1) is captured, marked, and released. On a subsequent occasion, a second sample of individuals (n_2) is captured, of which a number (m_2) are already marked. Conditional on meeting a number of assumptions (see below), the proportion of marked individuals in the second sample should equal the proportion of marked animals in the population at large (N):

$$\frac{m_2}{n_2} = \frac{n_1}{N}$$

so that population size can be estimated as:

$$\hat{N} = \frac{n_1 n_2}{m_2}$$

The proportion of marked animals in the second sample $\frac{m_2}{n_2}$ is an estimate of the probability of capturing an individual, $\hat{p}$, so that:

$$\hat{N} = \frac{n_1}{\hat{p}}$$

This same general idea can be extended to multiple sampling occasions, in which each sampling occasion (except the first) generates recaptures as well as first captures. If multiple samples are taken, the length of the study will determine whether the study population can be considered closed or open, i.e. whether or not births, deaths, and permanent immigration or emigration are occurring.

3.3.1 Assumptions in theory and practice

Mark–recapture methods use the data for animals that have been captured to estimate the number of animals that have never been captured. This inference makes mark–recapture estimates of population size sensitive to violation of their assumptions, which can cause substantial bias.

It is not uncommon for estimation of population size to be attempted using data from photo-identification studies of cetaceans that initially began without this objective. Data from such studies are particularly prone to violation of assumptions. Although analysis can sometimes help to recover such situations, there is no substitute for a well-designed field study to ensure that assumptions are met to the greatest extent possible.

Assumptions about the marked individuals

Three assumptions are made about the accuracy of the data on marked individuals.

MARKS ARE UNIQUE

The uniqueness of a mark should be assured with the application of a numbered tag/brand, but for photo-identification studies this depends on the amount of information contained in the natural markings and, to some extent, the number of animals in the population. If there is more than a remote possibility of there being 'twins' in the population, then bias will result. Examples of markings typically considered to be unique include the black and white pattern and shape of the trailing edge of tail flukes of humpback whales, the pattern of nicks and notches on the trailing edge of the dorsal fin of some small cetacean species, and the pelage patterns of several species of seal. Inevitably, markings of some individuals will contain more information than others. An animal should not be considered 'marked' unless one is convinced that it could later be recognized from a photograph of acceptable quality; we return to this issue later.

A common situation is that not all animals in the study population possess acceptable markings. Where skin/fur patterns are used as marks, all animals in the population should be markable. But nicks, notches, and scars are acquired with age and young delphinids tend to be unmarked. It is important to define what is meant by the markable population; this has implications for data collection and analysis.

MARKS CANNOT BE LOST

A marked animal must be recognized on recapture, so it is important to ascertain whether mark loss could occur. Tags may fall off or fade. A natural marking used as a mark may change, although markings should be chosen to avoid this. A good natural mark must last at least as long as the study period (see B. Wilson *et al.* 1999a). If marks are known or suspected of being lost, the rate of mark loss should be estimated, for example by appropriate double-tagging studies (e.g. Stobo and Horne 1994; Bradshaw *et al.* 2000; Pistorius *et al.* 2000).

ALL MARKS ARE CORRECTLY RECORDED AND REPORTED

Making mistakes in the field, or in transcribing field notes, or in the laboratory processing of data is inevitable to some extent. However, researchers should pay particular attention to avoid making such errors in mark–recapture studies because they can potentially have a big effect, especially in datasets with few recaptures. Correct identification of recaptures is critical. Errors in matching numbered tags should be small, but they can potentially be large in photo-identification studies. False-negative errors (missing matches) will cause abundance to be overestimated; false-positive errors (calling two or more different individuals the same) will lead to underestimates. The effects of false-negative or -positive errors in matching are most severe in the analysis of datasets with few recaptures. Stevick *et al.* (2001) found that decreasing photographic quality led to higher rates of false-negative errors in humpback whale photo-identification.

Assumptions about capture probabilities

In basic mark–recapture models, capture probabilities are assumed to be equal for all animals within each sampling occasion, whether marked or not. This is unlikely to be the case for a number of reasons.

BEHAVIOURAL RESPONSE TO CAPTURE (TRAP DEPENDENCY)

If marking affects future catchability, animals can become 'trap-happy' or 'trap-shy' after first capture, which causes bias unless accounted for in estimation models. This is not uncommon in studies where animals capture themselves in traps, but is likely to be less of a problem in studies of marine mammals, which must be actively captured either physically or, especially, photographically.

UNEQUAL PROBABILITY OF CAPTURE

Sample (study) areas are almost always smaller than the area inhabited by the target population, so it is often stated that all animals in the population must mix completely between sampling occasions so that each will have an equal chance of being captured. However, marine mammals commonly have preferences (sex-specific, age-specific, or individual) for particular areas. Animals that spend more time in the sample area will have a greater chance of being captured. Over time, this leads to more recaptures than expected, an overestimate of capture probability and an underestimate of abundance. Note that sampling in different areas on different occasions can potentially have the opposite effect. Consider a situation where some animals in a population prefer one area and other animals prefer another. Sampling in the first area on one occasion and the second area on the next occasion will lead to fewer recaptures than expected and an overestimate in abundance.

A second factor is that some individuals may be inherently more difficult to catch than others because they may avoid capture; conversely, some animals may be easier to catch. This leads to underestimates of abundance. A third factor in photo-identification studies is that some animals may have more distinctive marks than others making them easier to recognize, while others may be inherently difficult to recognize. The effect on estimates of abundance is the same as described above.

Failure of the assumption that all animals have an equal chance of being captured in each sampling occasion is often referred to as heterogeneity of capture probabilities. If unaccounted for, the bias that this causes in population estimates can be severe. It is good practice to minimize heterogeneity through good sampling design, but to be prepared to look for it and take steps to mitigate its effects in analysis.

3.3.2 Data collection

Capturing and marking individuals

Pinnipeds are available for capture at pupping colonies and other haul-out sites. Conventional ways to mark them are flipper tagging and branding, methods that are commonly used on terrestrial mammals. Hall *et al.* (2001) glued plastic numbered 'hat-tags' on the heads of grey seals in a short-term study. Increasingly, photo-identification is being used for seals (e.g. Forcada and Robinson 2006; Gerondeau *et al.* 2007).

Some species of small cetacean can sometimes be physically captured and branded or tagged, but the associated logistical and welfare issues mean that this is not usually a viable method. Instead, the most commonly used technique for capturing and marking cetaceans is photo-identification, in which photographs are taken of the natural markings of individual animals. Clearly, the individuals of the study species need to be sufficiently well-marked for this to be a viable method, but this is the case for quite a wide range of species. Photo-identification avoids physical capture, handling, and marking, a significant advantage, but it also makes it more challenging to meet the assumptions of mark–recapture methods.

Another means of marking that does not require physical capture or tagging is to take a biopsy sample (using a biopsy dart fired from a rifle or bow) and identify the animal genetically. This is a common technique for studying population structure, but the need to biopsy more than once to obtain recaptures means that this method has rarely been used for estimation of abundance (but see Palsbøll *et al.* 1997; T.D. Smith *et al.* 1999). Mark–recapture data can also be obtained from telemetry devices (e.g. Ries *et al.* 1998; McConnell *et al.* 2004).

Sampling design

The design of a mark–recapture study includes consideration of the number, frequency, and duration of sampling occasions. If the study is sufficiently short that births and deaths (and permanent immigration and emigration) can be ignored, multi-sample closed population models can be used to estimate population size (e.g. B. Wilson *et al.* 1999a). Analytical methods are more flexible and robust for populations that can be considered closed; in particular, there is scope for modelling and accounting for heterogeneity of capture probabilities. Notwithstanding this, the ideal sampling design gives every animal in the population an equal chance of being captured on each sampling occasion. This is unachievable in practice, but the closer it can be attained the better. Generally speaking, the larger the study area, the less likely that heterogeneity caused by area preferences will occur (e.g. Hammond 1990b).

It is worthwhile considering the interaction between the duration, separation, and number of sampling occasions. The prudent researcher will try to maximize capture probability within each occasion to ensure that there are sufficient recaptures among them, whilst also ensuring that there is sufficient time between occasions for the population to mix. The number of sampling occasions may be determined by the study objectives and/or logistics, but if there is flexibility there is scope for maximizing the effectiveness of analysis. A reasonable number of occasions will allow the effective application of multi-sample models. But a series of many sampling occasions with low capture probabilities will generate fewer recaptures and be less effective than a shorter series with higher capture probabilities. Exploration of the data to find the best balance of the above considerations can be profitable in terms of obtaining the least biased, most precise estimates of population size.

Maximizing capture probabilities within a sampling occasion helps to minimize heterogeneity because the larger the proportion of animals in the population that is captured, the less scope there is for variation in capture probabilities among individuals. How well this can be achieved is a matter of the size and extent of the population, data collection logistics, and the resources available.

In photo-identification studies, all animals encountered should be photographed, regardless of how evasive they are or how distinct their markings are because this reduces the heterogeneity of capture probabilities resulting from differences in behaviour and distinctiveness of natural markings. If not all animals in the population have sufficient natural markings, data on the proportion of well-marked animals in each school encountered can be used to estimate the proportion of markable animals in the population (B. Wilson *et al.* 1999a). Note that there may be a logistical conflict between keeping track of and focusing on animals that are difficult to photograph in order to minimize heterogeneity, and photographing animals 'blind' to their markings in order to obtain data for an unbiased estimation of the proportion of markable animals (Read *et al.* 2003).

Data processing

Processing data from tags and other numbered applied marks should be straightforward.

In photo-identification studies, the quality of the photograph can affect the likelihood of recognizing an individual. Not all photographs will be of sufficient quality to be certain that a marked individual will be recognized. Photographs should be graded according to quality and only the best used in analysis. There is an interaction between the natural marking chosen as the mark and the quality of photograph that is required to be sure of recognizing a marked animal; less distinct marks require better quality photographs. In a study of bottlenose dolphins, Read *et al.* (2003) found that well-marked animals were more likely to appear in better quality photographs and recommended that studies be undertaken to investigate the impact of this on estimates of abundance. Friday *et al.* (2008) explored the implications of this in an analysis of humpback whale data.

A critical part of data processing is determining matches between photographs of individuals—the recaptures. In most studies, this is done manually, assisted by structuring the photographic catalogue of existing individuals by features that make the process more efficient. A computer program may also aid this process (e.g. Hiby and Lovell 1990; Hillman *et al.* 2002). The final decision about a match should always be made by more than one researcher. Ideally, any uncertainty should be recorded so that the effects of this on population estimates can be explored in analysis.

3.3.3 Analysis

Closed populations

TWO-SAMPLE ESTIMATORS

The simple Petersen two-sample estimator was introduced above. A more commonly used variant that reduces small sample bias and has a more robust estimate of variance is Chapman's modification:

$$\hat{N} = \frac{(n_1 + 1)\,(n_2 + 1)}{(m_2 + 1)} - 1$$

with estimated variance

$$\hat{var}_N = \frac{(n_1 + 1)\,(n_2 + 1)\,(n_1 - m_2)\,(n_2 - m_2)}{(m_2 + 1)^2\,(m_2 + 2)}$$

These estimators can readily be implemented in a spreadsheet. In calculating confidence limits, it can be assumed that the estimates are log-normally distributed. The lower and upper limits can then be calculated as N/C to N*C, where:

$$C = e^{1.96\,\sqrt{\ln(1 + CV^2\,_N)}}$$

If the number of marked animals is greater than the lower confidence limit, it can be used as the lower limit in place of the estimate from the above equation.

Multi-sample estimators

If multiple samples are taken from a population that is assumed to be closed, the data allow a range of models to be fitted that make different assumptions about capture probabilities. It is good practice to explore how complex a model is needed to best describe the data. Program MARK (Cooch and White 2008) provides a comprehensive and effective framework for this, including information to allow robust model selection and goodness-of-fit testing.

A good first approach is to call up program CAPTURE, first developed over 30 years ago (Otis *et al.* 1978), that fits models in which the assumption of equal capture probability for all animals in all sampling occasions (called model M_0) is relaxed in various ways. These are that capture probabilities can vary amongst different sampling occasions (M_t); for behavioural reasons, i.e. trap dependency (M_b), and due to heterogeneity amongst individual animals (M_h); and combinations thereof (M_{tb}, M_{th}, M_{bh}, M_{tbh}). In CAPTURE terminology, the large majority of datasets will likely be best described by model M_t, or model M_{th}, depending on how strong the heterogeneity is in the data. Variants of model M_b should not be needed; however, trap dependency may act as a proxy for some other feature of the data (e.g. Ramp *et al.* 2006).

For users of the software R (R Development Core Team 2009), the library RMark provides an effective interface (Laake and Rexstad 2008).

Open populations

If the population is not closed the most obvious option is to use an open population model to estimate abundance. Program MARK provides an extensive range of options based around the Jolly–Seber model. These models are very flexible and also provide estimates of survival, recruitment, and population growth rates.

However, they are unable to take into account individual heterogeneity of capture probabilities, which is so often an important feature of marine mammal mark–recapture data, so they have been used less often for estimating population size of these species. Another option is to use a series of two-sample estimates (e.g. Stevick *et al*. 2003).

Temporary emigration

One feature of mark–recapture studies of marine mammals that is difficult to control is temporary emigration, where some animals in the population are not available to be captured on some sampling occasions. This is not really a problem if animals are temporarily unavailable because of random mixing, but non-random temporary emigration can lead to bias (Whitehead 1990b). Pollock's robust design is an effective framework for investigating and accounting for temporary emigration in studies that allow the data to be organized into a number of secondary sampling occasions within each primary sampling occasion. For example, primary occasions could be years and secondary occasions could be months within years. Closed population models use the data from the secondary sampling occasions to estimate population size, so they must satisfy the relevant assumptions. Survival rate is estimated between primary sampling occasions (e.g. Bradford *et al*. 2006). These models are readily implemented in program MARK.

3.4 Line transect sampling

Line transect sampling from ship or aerial surveys is used extensively to estimate the size of cetacean populations, and is also used less commonly for studies of seals (e.g. Southwell *et al*. 2008) and sirenians (e.g. Preen 2004; Pollock *et al*. 2006). Much of the development of line transect methodology in data collection and analysis since the late 1970s has been driven by the need to overcome the practical difficulties of applying these methods to the study of cetaceans. There is a large amount of published material on line transect sampling as applied to cetaceans; a comprehensive review is clearly impossible. This section draws particularly on a number of comprehensive texts including: Hiby and Hammond (1989); Garner *et al*. (1999); Buckland *et al*. (2001, 2004); IWC (2005).

The basic idea behind line transect sampling is to estimate the density of the target species in strips sampled by surveying along a series of transects, and to extrapolate this sample density to the entire survey area. Line transect sampling thus provides an estimate of the number of animals in a defined area at a particular time or over a period. Note that this is different from the quantity estimated from mark–recapture analysis.

Line transect sampling is commonly described as an extension of a strip transect, in which all animals are assumed to be detected within a given distance of the transect line. But, because the probability of detecting anything declines as a function of distance, not all animals will be detected within any strip unless it is very narrow. In practice, the only way to find out if a strip transect is acceptable is to collect data to confirm constant detection probability across the strip—which is effectively line transect sampling. Strip transects will rarely be an acceptable alternative to line transect sampling; surveys of dugongs are an exception (Preen 2004; Pollock *et al.* 2006).

The use of distance data in line transect sampling has led to an increasing tendency for it to be known as distance sampling, a category which also includes point sampling (not generally used in marine mammal studies, but see *Cue counting*, p. 57). Measurements of the perpendicular distance from the transect line to each detected animal are used to estimate the effective half-width (i.e. on one side of the transect line) of the strip that has been searched. On shipboard surveys, perpendicular distance is typically calculated from measurements of the distance and the horizontal angle relative to the transect line from the ship to the animal. Because animals occur in schools in most cetacean species, the target for detection in a line transect survey is typically a group of animals. Data on the number of animals in each group must therefore also be collected.

The basic equation that relates estimated density, D, to the collected data is:

$$\hat{D} = \frac{n\bar{s}}{2\,e\hat{s}w\,L}$$

where n is the number of groups detected, $\bar{s}$ is the mean group size, L is the total length of transect searched, and $e\hat{s}w$ is the effective strip half-width; n/L is often referred to as the encounter rate. The effective strip half-width is essentially the width at which the number of groups detected beyond that distance equals the number of groups missed within that distance, assuming that everything is seen at a perpendicular distance of zero, i.e. on the transect line. Variance of $\hat{D}$ can be estimated from the variances of n/L (from replicate transects), $e\hat{s}w$ and $\bar{s}$ using the delta method, or by a re-sampling method such as bootstrapping or jackknifing (Buckland *et al.* 2001).

Additional data on sighting conditions are required for additions to, or variants of, conventional line transect sampling as applied to cetaceans, and acoustic data can be used for some species.

For analysis of line transect sampling data to estimate abundance, the very large majority of practitioners need look no further than program DISTANCE (Thomas *et al.* 2006). This software is statistically robust, comprehensive, well-supported, continually updated, and freely available (http://www.ruwpa.st-and.ac.uk/distance/). DISTANCE can generate survey designs and has excellent data management capability so that it can be used from beginning to end in a line transect sampling project.

3.4.1 Assumptions in theory and practice

Sample density is representative

The simple extrapolation to estimate abundance in conventional line transect sampling requires that sample density is representative of density in the entire study area. Because animals are rarely, if ever, distributed randomly in space, this necessitates a survey design in which transects are placed to ensure that every point in the survey area has the same theoretical probability of being sampled; known as an equal coverage probability design. Density surface modelling (see below) does not require the assumption of equal coverage probability because abundance is estimated via a model relating density to environmental covariates rather than a simple extrapolation of sample density.

Detection is certain on the transect line

Line transect theory requires the assumption that all animals/groups are detected with certainty on the transect line itself, that is, at zero perpendicular distance. If the probability of detection on the track line is less than one, sample density and therefore abundance is underestimated proportionally. Clearly, not all cetaceans will be detected on the transect line. They may not be detected because they are underwater, so-called availability bias, or because they are on the surface but simply missed, so-called perception bias (Marsh and Sinclair 1989). These biases can be accounted for, but this involves additional data collection, either as part of the survey or independently, and analysis. Many studies of cetacean abundance using line transect sampling have therefore either assumed that this bias is negligible or accepted that the estimates are negatively biased.

Animals do not move

The assumption that animals do not move relative to the transect line stems from the requirement in line transect theory that, within searching range, animals/groups are distributed homogenously with respect to the transect line. This is to avoid confounding changes in density and detection probability as a function of perpendicular distance. Any spatial variation in density will typically be at a much larger scale, so this will generally be true. However, if animals move in response to the survey vessel before they are detected by observers, their distribution may no longer be homogeneous and bias in sample density may result. Even random movement causes a positive bias because this leads to an increase in detection rate, but its magnitude is small unless the animal's speed is more than about half the speed of the survey platform.

Aircraft clearly travel too fast for movement to cause any difficulties; more problematic is movement in response to a survey ship. Cetaceans may hear an approaching survey ship long before they are available to be detected visually. If unaccounted for, movement by individuals away from the survey vessel, i.e. avoidance, leads to negative bias, while movement towards the vessel, i.e. attraction, leads to positive bias in estimated abundance. Analysis of data from the North Atlantic showed that harbour porpoise, white-sided dolphins, and minke whales displayed avoidance but that white-beaked dolphins were attracted to the survey ships (Palka and Hammond 2001). Attraction, if unaccounted for, can cause large bias. Cañadas *et al.* (2004) found that uncorrected estimates of common dolphin abundance were six times greater than corrected estimates. Responsive movement is often overlooked or ignored in line transect surveys, but it is clearly important to discover whether or not it is occurring.

Data are recorded accurately

The obvious assumption that data should be accurate is problematic in line transect sampling for cetaceans because the 'hands off' nature of data collection means that some measurements are difficult to take. Even unbiased errors can cause problems. For example, rounding angles to convenient values for detections close to the transect line can substantially affect the distribution of recorded perpendicular distances, and thus make it difficult to fit a detection function and estimate *esw*. Unbiased errors in distance measurements lead to negative bias in estimates of abundance using cue counting methods. The importance of collecting data as accurately as possible cannot be overemphasized.

Observations are independent

A common assertion is that the same animal/group may not be detected more than once—'double counting'. This is true if multiple detections of the same animal/ group are on the same unit transect, but this should not be possible in a properly conducted cetacean survey. It is not a problem if the same animal/group is detected on different transects because each transect is treated as an independent sample.

Non-independence of detections occurs when sightings are detected non-randomly in clusters; also known as serial autocorrelation. This leads to negative bias in theoretical estimates of variance. It does not affect point estimates of abundance, nor estimates of variance calculated from replicate independent transects, which is how variance should always be estimated (Buckland *et al.* 2001).

3.4.2 Extensions and variants of conventional line transect sampling

Accounting for uncertain detection on the transect line and responsive movement

Failure to detect all groups on the transect line (zero perpendicular distance) causes negative bias. For marine mammals, these biases are rarely negligible and need to be accounted for if abundance is not to be considered as an underestimate. There are various ways to approach this.

Perception bias (and availability bias, depending on the application) can be accounted for by collecting sightings data from two teams of observers on separate platforms, recording the duplicates (sightings made by both teams), and using the data within a mark–recapture analysis framework to estimate the probability of detecting a group on the transect line (Borchers *et al.* 2002; Buckland *et al.* 2004). The detections of groups by each team serve as 'trials' at which the other team may succeed (duplicate sighting) or fail (no sighting). If the two teams of observers search the same area of water, which they must do independently of each other (two-way independence), availability bias cannot be fully accounted for because some animals will be unavailable to both teams. This is a case of heterogeneity of capture probabilities (as described in Section 3.3) some animals remain unavailable for capture/sighting and are excluded from the calculation.

However, if the search areas of the two teams are separated appropriately then heterogeneity can be eliminated or minimized and availability bias can be accounted for. In this configuration, independence is one-way because the trials are only set up by the team that searches farther ahead. One method of implementing this is to use an aircraft (helicopter) searching ahead of a ship. More commonly, two observation teams are deployed on the same ship. The one that searches farther ahead keeps track of each detected group as the ship approaches and records whether or not it is sighted (duplicated) by the other team. An advantage of this configuration is that the data can also be used in analysis to take account of any responsive movement if the 'tracker' team searches sufficiently far ahead of the vessel.

In aerial surveys, double-team data collection allows perception bias to be estimated (e.g. Marsh and Sinclair 1989; Heide-Jørgensen *et al.* 2007), but additional information is needed to estimate availability bias because it is not possible to separate the area searched by two teams on the same aircraft. Data on surfacing rates have been used to estimate percentage time at the surface and therefore availability bias for a number of species (e.g. Barlow *et al.* 1988; Laake *et al.* 1997; Forcada *et al.* 2004; Heide-Jørgensen *et al.* 2007). Laake *et al.* (1997) used data from a separate independent platform on land to estimate both availability and perception bias for harbour porpoise. Surfacing rate data can also be used to estimate availability bias on ship surveys.

A novel way of separating two teams of observers on aerial surveys is to use either two aircraft flying in tandem or a single aircraft circling back over part of a previously surveyed transect (Hiby and Lovell 1998; Hiby 1999). Duplicates are determined probabilistically from a model incorporating data on speed of movement of the target species. This method was used to estimate harbour porpoise abundance in the eastern North Atlantic (Hammond *et al.* 2002; SCANS-II 2008).

Cue counting

Cue counting is a variant of line transect sampling developed to provide unbiased estimates of abundance in species of cetacean for which a distinct cue (e.g. blow, surfacing) can be counted (Hiby and Hammond 1989). It can be applied to shipboard surveys, but has mostly been used in aerial surveys. As described by Buckland *et al.* (2001), it is actually a form of point sampling conducted continuously along a line. The idea is first to estimate the density of cues in the sample area per unit time. Effective search area is estimated by fitting a detection function to the radial distances of detected cues, analogous to the estimation of *esw* from perpendicular distance data in line transect sampling. The equivalent assumption to that of certain detection of animals on the transect line is that all cues are detected immediately adjacent to the observer (zero radial distance). If this is violated, double-team data can be used to estimate this probability; note that there is no availability bias because cues are, by definition, always available.

The density of animals is then estimated by dividing the estimated density of cues by the duration of the survey and an appropriate estimate of cue rate that must be obtained independently. Cue rate can be estimated from visual observations using similar data to those used to estimate surfacing rate for correction of availability bias (see p. 57). The best cue rate data come from telemetry studies of multiple animals over extended periods in the same season and place as the survey.

The method is sensitive to errors in radial distance measurement; even random errors can cause substantial positive bias. If double-team data are collected, this bias can be estimated and corrected for (Hiby and Hammond 1989). Analysis of cue-counting data can be conducted in program DISTANCE.

Using acoustic data

Data from acoustic recordings of cetacean vocalizations from hydrophones towed behind a ship can be used in a line transect sampling framework (Hiby and Hammond 1989). The estimate of abundance is for vocalizing animals, so either all animals in the population must be assumed to be vocalizing continually or a correction for the proportion of animals vocalizing is needed from independent data. The data for estimating the detection function are measurements of the perpendicular distance to sounds made by distinct animals, so therefore it must be possible to allocate tracks of sounds to individuals. To date, the method has been used to estimate the abundance of sperm whales (Barlow and Taylor 2005; Lewis *et al.* 2007), but ongoing technological developments may expand the range of species to which this method can be applied.

Density surface modelling

The use of spatial modelling of line transect sampling data to estimate abundance was introduced by Hedley *et al.* (1999) and expanded in Hedley *et al.* (2004). The method is based on fitting a model that describes density along the transect line as a function of physical (e.g. depth, slope, distance from shore) and environmental (e.g. sea surface temperature, chlorophyll) covariates, and then using the model to predict density over the whole study area based on the value of the covariates in a spatial grid. Latitude and longitude could also be included as covariates. The idea is for the model simply to account for variability in the data to obtain the best estimate of abundance, not to try to explain how animals use their environment (see Chapter 4).

Spatial modelling does not require that transects be placed to achieve equal coverage probability over the survey area, and is thus an appropriate, indeed the only, way to estimate abundance from surveys for which this has not been possible. However, it does require that the covariates used in the model are well sampled.

The method is sometimes referred to as a model-based approach to abundance estimation, as distinct from the design-based approach of extrapolating sample density directly to the whole area on the basis of an equal coverage probability survey design. It is also known as density surface modelling because the output is a surface of density predicted by the model; density can thus be integrated under any defined area. This latter point is both a blessing and a curse. It provides a means of obtaining estimates of abundance for sub-areas other than predefined survey blocks, but the robustness of such estimates depends on the appropriateness of the fitted model. If a sub-area is in a region where the relationship between density and covariates is different to that represented by the model, the estimate will be wrong. The idea of drawing strength by using data from a wider area to help estimate abundance in a sub-area is appealing, but it must be applied with considerable caution, particularly if the sub-area is small relative to the modelled area. Extrapolation of density beyond the surveyed area can only be justified in an exploratory context.

The application of spatial modelling to estimate abundance is still in its infancy, but there is an increasing literature (Cañadas and Hammond 2006, 2008; Williams et al. 2006a; Gómez de Segura et al. 2007; SCANS-II 2008) which is likely to grow at an increasing rate.

3.4.3 Data collection on visual line transect surveys

General considerations

The objectives of a line transect study will determine the area and time of year to survey, the focal species, the target precision of the estimate, etc.

Resources for conducting surveys are inevitably limited and there are typically logistical constraints. Equally relevant, therefore, are questions relating to the availability of suitable survey platforms, permission to survey the desired area, ability to collect the data adequately without violation of important assumptions, and the availability of appropriately experienced and trained personnel. The answers to these and other questions will dictate the way the survey is conducted.

The platform available for a survey may be predetermined, or there may be a choice between a ship/boat and an aircraft. Aircraft are good for surveying areas with complex coastlines and for taking advantage of good survey conditions because their speed allows a large area to be surveyed quickly. Movement of animals causes no problems and accurate data collection should be possible. Not flying in poor weather and a small crew can also make them cost-effective. However, there are important logistical constraints of which safety—sufficient airports, a reliable aircraft, and an experienced pilot—is clearly paramount. Suitable aircraft should have long endurance but the area of operation will still be limited. They should have high wings and bubble windows to allow the transect line to be observed, have the necessary equipment, including GPS and radar altimeter, and be able to fly relatively slowly at low altitude. Disadvantages are relatively low detection rates (the probability of detection on the transect line typically being substantially less than one) and an inability to collect ancillary data. Most cetacean line transect surveys are conducted from ships. Offshore areas can only be surveyed from ships, and suitable survey vessels are often more available. Encounter rates are higher and more observers can be accommodated in greater comfort. Ships also allow ancillary data (environmental, biopsy, photo-identification) to be collected. Dawson *et al.* (2008) discuss the use of smaller boats for surveys in coastal areas and rivers.

Any line transect survey should aim for the best balance between minimizing bias and maximizing precision. Bias is kept to a minimum by avoiding violation of assumptions. Precision is maximized by minimizing the estimated variability in the component elements of the estimation calculation: encounter rate, effective strip half-width estimated from the fitted detection function, and mean group size. Variation in encounter rate is a function of the distribution of animals, the precision of *esw* improves with a detection function that is 'flat' at perpendicular distances close to zero (which can be influenced by searching behaviour), and variation around mean group size is a feature of the animals. Generally speaking, the more sightings made, the better precision will be.

Good data result in good estimates of abundance, so the importance of training observers in data collection cannot be overemphasized. This should include instruction and practice in species identification, estimation of group size, distance and angle measurement, searching behaviour, and recording of data. Dawson *et al.* (2008) review available data recording software.

As line transect sampling surveys benefit from prior knowledge, it can be highly beneficial to conduct a pilot survey. This provides information on species, encounter rates, and the variability expected, but it also provides an excellent opportunity for testing methods and equipment and for training. There is often a reluctance to spend resources collecting data that will not be used in final results but relative to the investment in a main survey or series of surveys, such investment is typically well worth it.

Survey design

The design of a line transect study starts with the definition of the survey area, which should be determined by the objectives of the study but tailored to minimize logistical problems.

On many surveys, the study area is stratified into blocks for logistical reasons, to generate abundance estimates for different sub-areas/habitats, or to increase precision. A good general design is for each block to have the same coverage so that data can be pooled across blocks for a single estimate of abundance. In cases where it is known that density varies in different parts of the survey area, allocation of searching effort in proportion to expected density can increase precision in the overall estimate. However, if such expectations prove false, unequal allocation of effort can be counterproductive.

Generally, survey design should aim for equal coverage probability of the area, which is obtained with a design in which transects are set out randomly. In practice, a systematic design with a random element, e.g. starting point, is equivalent. If the design does not ensure equal coverage probability, estimated sample density cannot be extrapolated to give an unbiased estimate of abundance in the survey area or blocks within it. The placement of transect lines to achieve equal coverage probability is thus essential unless abundance estimation is model-based (see Section 3.4.2; p. 58, *Density surface modelling*). Good designs typically involve sets of parallel or zig-zag lines with random starting points. The former are often appropriate for aerial surveys, while the latter are typical for larger scale shipboard surveys to avoid or minimize time spent transiting between transects.

As far as practicable, transect lines should run perpendicular to any known or suspected density gradients. For example, it is good practice for transects to be approximately perpendicular rather than parallel to the coast. If possible, surveys are best conducted during periods when there is no directed animal movement. If this is unavoidable, the survey should progress across the direction of migration or against it, not in the same direction.

Which exact design is the most appropriate will depend on local topography and prior knowledge, as well as the objectives and logistical considerations mentioned above. The final design should try to balance all the above considerations using a generous application of common sense. Buckland *et al.* (2001) and Dawson *et al.* (2008) provide good practical advice for the placement of transect lines by hand for a variety of situations. However, the combination of GIS and program DISTANCE now makes automatic survey design relatively straightforward.

Searching behaviour

Conducting a survey involves teams of observers searching forward and to the side of the platform and recording the data they collect. The searching pattern should provide good coverage of the transect line directly ahead of the platform, but not to the detriment of good angular coverage to avoid problems when fitting the detection function. On shipboard surveys, observers should search sufficiently far ahead of the ship to minimize the possibility of responsive movement prior to detection (Barlow *et al.* 2001).

Measuring and recording visual data

Accurate measurements require appropriate equipment. Advances in technology have seen the means of collecting line transect data become more sophisticated to maximize accuracy. Some of these developments are relatively costly and may only be within the resources allocated for large-scale surveys. But inaccurate data typically lead to poor abundance estimates, so investment in the best data collection methods available should be a high priority.

Data on searching effort data are relatively easy to collect accurately. All survey platforms should have GPS for navigation, thus providing an easy means to record the position and length of transects. Data affecting sighting conditions, such as sea state, swell, glare, etc., should be collected regularly and whenever conditions change to allow stratification of effort data when estimating *esw*. Data collected when a sighting is made are more prone to errors.

SIGHTING ANGLES AND DISTANCES

On shipboard surveys, it is extremely important to obtain accurate angle measurements to sightings close to the transect line because these have a large effect on small perpendicular distances, which have most influence on obtaining an accurate and precise estimate of *esw*. Angle boards are commonly used so that angles can be recorded to the nearest degree; this should avoid any rounding to convenient values. Compasses within binoculars can also be used. An automatic method using images from a small camera mounted underneath the binoculars and directed at lines drawn on deck has proved highly effective (SCANS-II 2008).

Distance to groups of cetaceans is difficult to measure. On shipboard surveys, methods employed have ranged, in order of increasing accuracy, from naked eye estimates, calibrated measuring sticks, reticules in binoculars, and video range-finding equipment (Gordon 2001; Williams *et al.* 2007; SCANS-II 2008).

Whichever method is used to collect distance data on ships, experiments using appropriate objects at known distance should be conducted during the survey to provide calibration data for each observer, and also to generate data on variability in distance measurement (e.g. Williams *et al.* 2007). The results of these allow any systematic deviation to be estimated for each observer and used to correct distance.

On aerial surveys, perpendicular distance is most often obtained by measuring the declination angle with an inclinometer when an animal or group comes abeam. The calculation requires height to be known so an accurate altimeter is essential. On cue-counting surveys, accurate radial distance data can be collected by back-calculating from perpendicular distance using the difference in time, recorded to the nearest second, from when the cue was first seen to when it came abeam. Accurate distances and times are also important in cue-based shipboard methods (Schweder *et al.* 1996; Schweder 1999).

SPECIES IDENTIFICATION AND GROUP SIZE

The best way to obtain accurate data on the species and number of animals detected is to break track and approach, or circle above, each detected group, especially for sightings not close to the transect line. This is standard procedure on aerial surveys. On shipboard surveys it is known as 'closing mode'. The advantages of more accurate data are clear. But the disadvantages are that searching time is reduced, and that there is a potential for bias depending on survey protocols. If the vessel resumes searching without returning to the original transect, there is a potential for searching to be drawn into areas of higher density. However, breaking track continually in high density areas can lead to a disproportionate number of missed sightings and the encounter rate being underestimated.

The alternative is 'passing mode', in which searching continues along the track and species identification and group size data are collected without approaching the group. This mode is not prone to bias in encounter rate, but it is more likely to generate data inaccuracies; the proportion of groups unidentified to species is typically higher in passing mode surveys and there is a tendency to underestimate group size. As such errors are more likely for groups detected far from the transect line, truncation in analysis is one way to reduce them. Some surveys alternate passing and closing mode, using data from the former to estimate the encounter rate and the latter to estimate the group size (IWC 2005) but sample size is reduced as a result. Closing mode may be better for surveys in areas of low density or for multi-species surveys. Ultimately, the choice of survey mode may depend on balancing a number of practical and logistical factors.

3.4.4 Analysis

Conventional line transect sampling

Data exploration is an important starting point in any analysis. It serves to identify obvious errors, and also allows you to get to know your data and identify any problems that may lie ahead during analysis. An essential first step is to plot histograms of perpendicular distance data. This can highlight potential problems such as the effects of rounding measurements to convenient values, evasive movement, outliers, etc. If rounding is a problem, angle and/or distance data can be 'smeared' (Buckland *et al.* 2001).

The next step is to consider truncation of the dataset by limiting analysis to those data within a fixed perpendicular distance. This removal of outliers almost always serves to improve the fit of the detection function. As a basic 'rule of thumb', Buckland *et al.* (2001) recommend the removal of 5–10% of observations. However, the more the data are truncated the smaller the sample size.

Following truncation, a model of the detection function is fitted to the perpendicular distance data to estimate *esw*. Perpendicular distance distributions come in different shapes, so best practice is to fit a range of appropriate models. Probability of detection may also vary with the characteristics of the survey vessel

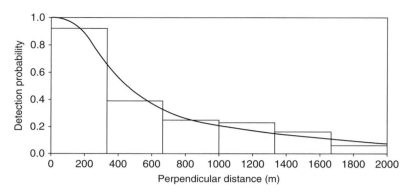

Fig. 3.1 A detection function fitted to perpendicular distance data for bottlenose dolphins in the Alborán Sea (generated by program DISTANCE). (Figure courtesy of Ana Cañadas.)

such as: platform height; sighting conditions, such as sea state, swell, and glare; and with features of the sighting itself, such as group size, cue, etc. Fitting the detection function should thus involve models that include these features as covariates. Fig. 3.1 shows an example detection function fitted to perpendicular distance data.

As mentioned above, truncation of the data to remove observations made at large perpendicular distance can improve model fit. If none of the models fits well, further truncation may improve model fit. An additional 'rule of thumb' suggested by Buckland *et al.* (2001) is to truncate those observations for which the estimated probability of detection is <0.15. Such iteration between truncation of the dataset and model fitting is not uncommon in analysis of line transect data.

Accounting for uncertain detection on the transect line and responsive movement

Analysis of double-team data is a combination of line transect sampling analysis and mark–recapture analysis, and is not something to be undertaken lightly (Laake 1999; Borchers *et al.* 2002; Laake and Borchers 2004). The form of the analysis depends on the configuration of the two teams: (1) both teams operate separately and set up trials for each other; or (2) one team ('tracker') searches further ahead, tracks detections, and sets up trials for the other. There is then a choice of two models: full independence, in which detections by each team are assumed to be independent from each other at all perpendicular distances; and point ('trackline') independence, in which detections by each team are assumed to be independent from each other only on the transect line at zero perpendicular distance. Which of these is the case is typically unknown.

Laake (1999) has shown that the restrictive and strong assumption of full independence results in bias in estimated abundance if there is heterogeneity in the mark–recapture component of the analysis that is unaccounted for (unmodelled) by the covariates used to fit the detection function. Conversely, the weaker assumption of point independence generates estimates that are less biased, so this is generally the best model choice. However, to account for responsive movement one must assume that all detections made by the second team are independent of detections made by the first 'tracker' team, and therefore the full independence model must be used (Laake and Borchers 2004). A more detailed consideration of this issue is beyond the scope of this chapter; the inexperienced practitioner should seek guidance from the experienced.

Another way to correct for responsive movement is to use data on the direction of the travel of animals at the time of first sighting (Palka and Hammond 2001). This method uses differences in the proportion of sightings in different travel direction quadrants ($0°$–$90°$, $90°$–$180°$, etc.), compared to those expected if there were no responsive movement, as a function of radial distance to estimate the distance from the ship at which responsive movement is significant. The data are then stratified into two 'teams', near and far, and two-way independence double-team analysis methods are used to calculate corrected abundance.

Density surface modelling

The analysis of data to estimate abundance through density surface modelling follows a series of steps as described by Cañadas and Hammond (2006, 2008). The first of these involves fitting the detection function as described above for line transect sampling to estimate the density of groups in defined segments along the transect line. Density of groups is then modelled as a function of spatial and environmental covariates, typically using Generalized Additive Models (GAMs). If there is evidence of spatial or environmental variation in group size, this can be modelled in the same way. Finally, values of covariates from a spatial grid covering the survey area are used to parameterize the model(s) to estimate density in each grid cell and, therefore, overall abundance. Variance can be estimated using bootstrap (e. g. Cañadas and Hammond 2008) or jackknife (e.g. Williams *et al*. 2006a) re-sampling. Density surface modelling analysis is available in DISTANCE but only for modelling abundance in one step. Modelling variation in group size separately from abundance of groups requires the use of other software, such as R.

3.5 Concluding remarks

The science of estimating the abundance of marine mammal populations has evolved rapidly in the last 20–30 years, driven primarily by the use of photo-identification as a means of marking cetaceans and data collection and analytical developments in line transect sampling to survey cetaceans. Marine mammal scientists now have at their disposal a variety of effective techniques to generate accurate and precise estimates of population size to inform ecological understanding, conservation policy, and management actions. These methods have developed from, and are therefore very similar to, those used to estimate the abundance of terrestrial mammals and birds. Estimation of the size of fish stocks has also made use of survey and mark–recapture data (Petersen was a fisheries scientist) and also samples of dead animals (catch data). It is tempting whilst reviewing recent developments also to look towards future work that would improve the state of the art.

Buckland *et al.* (2000) did exactly this in a review of wildlife population assessment. In terms of abundance estimation they highlighted GIS, automated survey design, model-based estimation, integration of mark–recapture and line transect in a unified framework, trends in abundance, incorporating biological processes into statistical inference models, and model selection uncertainty as developing areas for the near future. Significant developments have indeed been made in most of these areas in recent years, as outlined above. One aspect that has not moved forward much is the estimation of trends in abundance. However, as monitoring programmes generate longer and longer datasets, there will be an increasing need to estimate trends robustly to provide information on conservation status.

What other developments lie ahead in the next few years? Data collection has become more technically advanced (e.g. digital photography for mark–recapture, video measurement of distance, and angle data on line transect surveys) and will continue to do so. One area that is set to develop rapidly is the use of acoustic data. Work is already in progress to estimate abundance using data collected via towed hydrophones for odontocete species other than sperm whales, and also using data collected via fixed hydrophones for mysticete whales. Detection of cetaceans via video cameras on aerial surveys has been used successfully for beluga and narwhal for some years (e.g. Heide-Jørgensen 2004). Improved technology may give this data collection technique wider applicability; it is increasingly being trialled for other species, e.g. minke whales off West Greenland and in the Antarctic pack ice, and for dugong and humpback whales off eastern Australia via unmanned aircraft.

We should also expect developments in data processing. For example, computer-based systems for comparing images of naturally marked animals will need to become commonplace time savers as datasets grow ever larger. The continued development of software such as MARK and DISTANCE will continue to bring the more complex analytical methods within reach of more and more biologists. For example, density surface modelling of abundance is now available in version 6 of program DISTANCE. Such technical developments will enable the biologist to become increasingly sophisticated in data collection, processing, and analysis.

The increasing availability of physical and remote sensing data on the environment will facilitate the use of density surface modelling as a valuable tool for estimating abundance in situations in which design-based surveys are difficult to achieve. But the implementation of these methods is still relatively novel and requires close collaboration between biologists with existing data or new projects and statisticians who are interested in improving analytical methods. This is also true for other methods, and recent history is encouraging in this respect.

Meanwhile, well-informed biologists can use their ingenuity to balance the effects of violations of assumptions and thus minimize the impact of imperfect data or analytical methods on estimates of abundance and their variance. For example, as discussed above under mark–recapture sampling design, judicious consideration of the interaction between the duration, separation, and number of sampling occasions can help to minimize bias and maximize precision.

More generally, the best practical advice on estimating abundance is the same as for any experiment—consider the method of analysis and its assumptions carefully before embarking on data collection. Good practice in the field minimizes problems in analysis and results in better estimates of abundance.

3.6 Acknowledgements

This chapter is the distillation of many years of discussions and collaborations with many people, too numerous to mention, all of whom I thank. I am especially grateful to Ben Wilson and Ana Cañadas for their valuable comments on earlier drafts, and to Judith Hammond and Ian Boyd for putting up with its overlong gestation.

4

The spatial analysis of marine mammal abundance

Jason Matthiopoulos and Geert Aarts

We must admit with humility that ... space has a reality outside our minds, so that we cannot completely prescribe its properties a priori.

Carl Friedrich Gauss

4.1 Introduction

Space is everywhere in ecology: migration and the contact between individuals has important implications for local population dynamics (Chapter 6) and the spread of disease (Chapter 7). The distribution of marine top-predators drives and is driven by the distribution of their prey (Chapter 11), and quantifying rates of encounter is a prerequisite to understanding the diets of marine mammals (Chapter 9). Detecting and explaining usage hotspots is essential for the design of reserves and the conservation of marine biodiversity (Chapter 15), but even aggregations of animals in space that cannot be explained by environmental drivers may hint at intrinsic factors such as social grouping (Chapter 12).

Heterogeneity in the spatial distribution of animals results from basic biological priorities: the need to mate gives rise to breeding aggregations, risk-avoidance may drive animals to cluster at refuges, competition may lead to geographical niche-partitioning, and the exploitation of heterogeneously distributed prey propagates heterogeneity onto the spatial distribution of the predators. Arguably, complex spatial patterning is the symptom of complex individual behaviour (Taylor *et al.* 1978; Perry *et al.* 2002) and this implies that studying the spatial distribution of marine mammals is particularly challenging and potentially counter-intuitive (Matthiopoulos *et al.* 2005).

Representations of space-use vary in information content. The position of an individual, at any given time, is a set of coordinates: latitude, longitude, and depth (Fig. 4.1a). Multiple observations of the same animal yield temporally discrete descriptions of that animal's trajectory through space (Fig. 4.1b). As observations from the same, or different, individuals accumulate, they reveal their spatial range

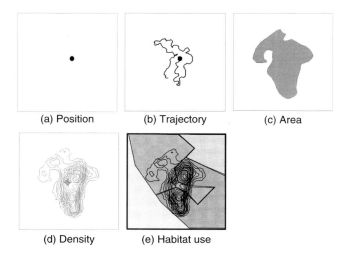

(a) Position   (b) Trajectory   (c) Area

(d) Density   (e) Habitat use

Fig. 4.1 Descriptions of the use of space by animals in order of increasing information content.

(Fig. 4.1c). Significant challenges arise in estimating a population's spatial density (Fig. 4.1d) and in relating usage to habitat preference (Fig. 4.1e). Chapters 3 and 10 describe how such observations can be collected. In this chapter, we critically review the analytical techniques used to infer space-use and its determinants.

We deal with three nested approaches: *range estimation* delineates regions in which animals can be readily found, such as migration lanes or home ranges; *density estimation* quantifies variations in the abundance of animals over space; and *habitat modelling* estimates space-use through its correlations with environmental variables. This can greatly improve the estimates and also hint at why animals go to certain places.

We begin by emphasizing the relevance of life history (Section 4.2) and the limitations of different types of spatial data (Section 4.3). Next, we examine some of the basic concepts and challenges involved in spatial analysis (Section 4.4). We consider the various steps involved in data-preparation (Section 4.5), the main analytical approaches (Section 4.6) and the available software implementations (Section 4.7). Finally, we discuss the more promising developments in this area (Section 4.8).

4.2 The importance of life history

Marine mammals have varied life cycles, and individual priorities depend on their sex and life-history stage. Behaviours such as resting, foraging, migrating, and mating are often best performed in particular habitats so that habitat preferences and the resulting space-use may differ with time and between different types of individuals (see Section 4.5). Also, the features of population space-use emerge from the relative proportions of the different types of individuals that it contains. This has implications for population-level inferences drawn from studies with unbalanced sampling, either across seasons or population components (see Section 4.4). Conversely, the makeup of the environment will favour certain life history traits. Changes in habitat availability can lead to changes in behaviour, differences in growth, survival, and reproduction, and even long-term evolutionary change.

Box 4.1 Glossary of terms useful in spatial analysis

Accessibility: The ease with which an individual can reach a point in space. This is determined by characteristics of the environment (e.g. obstacles), the speed and directedness of movement, and behavioural constraints, such as the need to return to a haul-out/rookery site.

Autocorrelation: Random variables in space and time can be self-similar, in the sense that their value here/now is similar to their value at a nearby location/time. The rate at which similarity declines with distance or the time-lag between two points quantifies the strength of autocorrelation.

Availability (of habitats): The proportion of time that animals are expected to spend in a habitat under a *null model of usage*. For a free-ranging animal, this is equivalent to the proportion of an area occupied by that habitat within the study region.

Empirical/Phenomenological models: A model that tries to emulate the behaviour of a system by using some generic and flexible mathematical function informed by data obtained directly from the system (compare with *mechanistic model*).

Environment: A combination of *environmental conditions*—a single point in *environmental space*.

Environmental condition: A particular value of an *environmental variable*.

Environmental space: Comprises multiple dimensions, each representing a biotic or abiotic *environmental variable*.

Environmental variable: A continuous, discrete, or qualitative random variable representing a spatial attribute (e.g. temperature, sea depth) or resource (e.g. prey, pupping sites). Environmental variables may or may not affect the geographical distribution of the study species. Those that do are called its *covariates*.

Extrapolation: The act of generating model predictions beyond the range of observed values. Spatial extrapolation gives predictions that, geographically, lie outside the regions in which observations have been made. Environmental extrapolation predicts outside the range of environmental conditions occurring in the data to which the model was fitted (compare with *interpolation* and see discussion in Mysterud and Ims 1999).

Geographical space: Comprises the three dimensions of latitude, longitude, and altitude/depth, often projected onto a Cartesian system of coordinates. Studies of space usually focus on the two horizontal dimensions, but marine mammals spend a large proportion of their time diving and for some species, such as pinnipeds and river dolphins, near-one-dimensional environments (coasts and rivers) have a prominent role (compare with *environmental space*).

Habitat preference: The ratio of the use of a *habitat* over its *availability*, conditional on the availability of all habitats to the study animals. Other definitions exist (see Manly *et al.* 2002).

Habitat: A collection of *environments*, a region in *environmental space* (e.g. polar habitat). Alternatively, habitat can be defined in a species-dependent way as the region in geographical space in which an organism lives (e.g. polar bear habitat). The two definitions are not interchangeable (see Hall *et al.* 1997). We opt for the first definition because it allows objective comparisons between species and gradations of preference for single species.

Interpolation: The act of generating predictions within the range of observed values (compare with *extrapolation*) and through the actual observations (compare with *smoothing*). It consequently places a high degree of confidence in the accuracy and precision of the observations and implicitly assumes that the observed process is not stochastic.

Mechanistic models: A model that can describe the function of a system on the basis of the fundamental properties and function of its components. No model in biology is entirely mechanistic in the reductionist sense of Descartes (1997).

Model fitting: The formal statistical process of using data to estimate the parameter values of a model and their associated errors. Models traditionally fitted to data were *empirical* but methodology now exists for fitting *mechanistic models*.

Model selection: The formal statistical process of choosing between different candidate models purporting to explain a given set of data. Selection is done on the basis of how well the model fits the data (goodness of fit) and how simple it is (parsimony).

Multi-colinearity: Correlation between two or more environmental variables that act as candidate covariates in *regression models* (Cramer 1985).

Null model of usage: The expected distribution of *space-use* under the null hypothesis of no habitat preference. For nomadic animals, this is ultimately spatially uniform within the study region. For animals with movement restrictions such as colonial breeders or central-place foragers this model may need to account for *accessibility* (Matthiopoulos 2003b).

Observation effort: Time spent observing a particular group of animals, single individuals, or regions of space.

Over-fitted model: A model that is too complex, i.e. has more parameters than can be estimated accurately and precisely from the data. Over-fitted models usually give a good fit to existing data but predict new data poorly.

Parametric methods: Statistical analyses that assume a known functional form for the distribution of animals in space (e.g. bivariate Gaussian home ranges) or for the dispersion of data around a model expectation (e.g. Poisson distributed count data).

Pseudoequilibrium: The assumption that the spatial distribution of animals is stationary, either permanently or for the duration of the study.

Regression model: An *empirical model* that is used to relate one or more *response variables* to one or more explanatory variables. The functional form of the model is usually specified by the user. The parameters of the model are estimated from data by the process of *model fitting*.

Resource: An *environmental variable* whose abundance can be depleted by an organism.

Response variable: The random variable that is modelled as a function of *environmental variables*. In spatial analyses, response variables are usually equal or proportional to either usage or preference.

Smoothing: Estimation method used with error-prone observational data, possibly obtained from stochastic processes. Smoothing generates a new function from the data that tries to capture the salient features of the underlying process while removing systematic stochasticity and observation error. This means that the resulting surface/curve does not necessarily pass through the observed values at the coordinates of the observations (compare with *interpolation*)

Space-use/Usage: The proportion of time that animals spend in a given region of geographical space. It can be thought of as the rate parameter of a spatial random process that generates the data. It is often modelled as a heterogeneous spatial Poisson process.

4.3 Usage data and sampling design

Although the different types of usage data are treated in Chapters 2, 3, and 10, a quick comparison is useful here (Fig. 4.2). Transect data (Buckland *et al.* 2002) can be collected from various platforms (visual or acoustic, ship, aerial or shore-based) and by various methods (strip-, line-, point-transects, cue-counting). Surveys can cover large areas depending on the type of transect and platform used, and because the regions of observation are subject to survey design they can theoretically achieve representative spatial coverage. However, this is not easy in practice and the challenge for survey design is to reduce bias caused by coastline geometry and imprecision due to variation in detection probability (Strindberg *et al.* 2004). Presence-only transect data (e.g. strandings and incidental sightings) result from surveys with no design and no records of effort. These may be usable for spatial modelling but only if they are augmented with pseudo-absence data (Lütolf *et al.* 2006) or assumptions about the spatial distribution of sampling effort (Gregr and Trites 2001).

Telemetry methods range from visual tracking and radio-telemetry to geolocation and real-time satellite telemetry. By definition, they focus on the particular individuals being tracked. Hence, although there are multiple observations for each study animal, population-level inferences may be based on comparatively few animals. Furthermore, although spatial coverage is potentially global, it ultimately depends on where the animals are tagged and where they decide to go. Thus, most telemetry data-sets are biased representations of the population (e.g. resulting from predominantly tagging animals of a particular age or sex) and the environment. In addition, the observation errors inherent in tracking technologies (e.g. triangulation error) give rise to imprecision in the data.

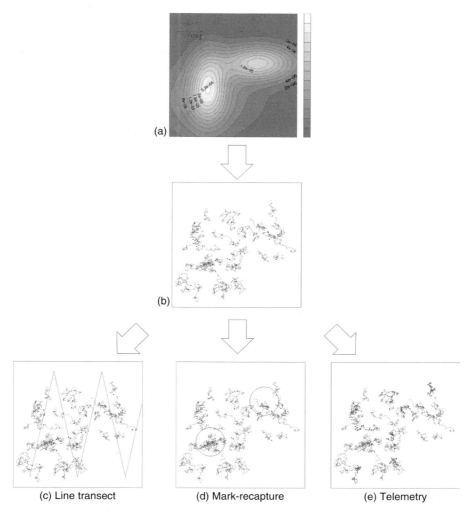

Fig. 4.2 Observations from three different sampling methods obtained from the same population. The expected distribution of animals in space can be thought of as a heterogeneous surface (a). Stochastic realizations from a random process with this expectation, represent possible arrangements of animals in space (the black dots in (b)). These realizations are only partially observed (grey dots in (c), (d), (e)). Line-transect surveys (c) generally observe many animals but make one or a few observations per individual. Mark–recapture (d) has intermediate properties between (c) and (e). In this example, there are two mark–recapture stations (e.g. oil platforms). Wildlife telemetry (e) yields detailed data on the movement of few individuals.

Mark–recapture methods (trapping, photo-ID, transponder tags) have features from both transects and telemetry. Hence, although getting more than one observation for some animals is essential for mark–recapture, the number of repeat observations is never as high as with telemetry. The spatial resolution of the observations is usually error-free because the recapture locations are known, but spatial coverage is usually poor. Representation of the environment at first capture/sighting is within the observer's control but the recapture/re-sighting probability depends on where individuals go. Furthermore, the number of recapture/re-sighting stations are usually fewer than the observation points in a transect survey.

Predictably, all three types of data fall short in some way (Fig. 4.3) and these deficiencies need to be addressed statistically. Some methods are more appropriate for particular ecological questions and species, so it is important to consider the

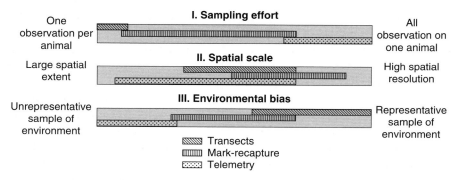

Fig. 4.3 The fundamental trade-offs associated with different types of spatial data. This is a schematic only, obeyed by most, but not all, experimental setups. (I) The distribution of sampling effort across different animals depends on whether the study is primarily interested in population-level estimation or in quantifying individual behaviour. By actively attempting to minimize replication, transect surveys occupy the left-most extreme of this spectrum. In contrast, mark–recapture methods cannot function without some recaptures, and some telemetry studies collect thousands of locations from a few individuals. (II) All three methods are subject to a trade-off between spatial extent and spatial resolution. For example, satellite telemetry methods can achieve large spatial coverage but their location errors also tend to be large. Transect surveys can increase their spatial coverage by increasing boat speed, hence incurring a cost in observer efficacy. Mark–recapture methods tend to be very precise because the locations of capture (e.g. traps or transponder stations) are known with little error, however the extent of spatial coverage is restricted by the number of capture stations. (III) The representation of the environment in the final data is also potentially different. Telemetry data are, by construction, animal-biased. Transect and mark–recapture data are only potentially observer-biased.

relative merits of each type of data for any particular study well in advance of data collection.

4.4 Basic concepts and challenges

Censusing the positions of an entire population is practically impossible except in very specific and rare circumstances. Furthermore, data on individual locations, no matter how extensive, are only stochastic realizations of an underlying spatial point process (see *Usage*, Box 4.1). Hence, even if we could instantaneously dry up all the oceans, locate and identify all the animals (and then re-float them), we would still only have a sample, not a census, of population usage.

We therefore need to ensure that our sample is representative, i.e. that each individual has the same probability of being observed at any given position and every instant of the study period. This is a tall order for spatial data sets, most of which are unbalanced. For example, transect surveys may locate groups of female whales more easily than singular males, with detection probabilities that may vary with weather (Marques and Buckland 2004). Similarly, tagging studies may capture more seals at one haul-out than another and limitations in tag life may result in partial seasonal coverage (Aarts *et al.* 2008). Data analysis must account for such geographical, seasonal or individual biases.

We also need to consider the ability of different types of data to generate measures of absolute abundance: mark–recapture is ideal for population estimation, but transects have difficulty achieving this objective if the intercept of the detection function is not 1. Analyses of telemetry data need to rely on independent population estimates to produce maps of absolute abundance. These estimates are usually accompanied by their own uncertainty requiring the calculation of compounded sampling variability (e.g. see the Appendix in Matthiopoulos *et al.* 2004).

Another important consideration is whether to treat geographical space as continuous or discrete (gridded, otherwise tessellated or replaced by a patch-network). Some discretization is unavoidable—either at the stage of data collection, data analysis, or prediction—but it represents loss of information. Discussions concerning the scale of discretization can be found in Levin (1989) and Dungan *et al.* (2002).

Yet another issue is whether to treat the properties of space as static or dynamic. Local attributes may vary with time and the desirability/accessibility of different locations will change with changing life history priorities. Depending on the biological question, the analyst may choose to include or ignore these features using *pseudoequilibrium assumptions*.

Similar considerations apply to the treatment of *environmental space*. In some cases, it is possible to identify discrete environmental envelopes. This has been common practice in descriptive studies of marine mammal distribution, but it is difficult to do unambiguously. There has been much discussion in the ecological literature on how to discretize environmental space, and some authors consider it an essential requirement for analysing habitat preference (Hill and Binford 2002) but this is not necessarily true (Aarts *et al.* 2008).

Data on where animals were observed can only be interpreted in the knowledge of where they were not. Therefore, sampling effort is crucial for the estimation of spatial abundance and its omission from the analysis can give grossly misleading results. Two approaches exist to incorporate effort. The first involves binning the data by length or time. Line transects can be divided into unit lengths, while telemetry records can be divided into unit time intervals. The observations in each segment can be counted and analysed directly as counts or be converted into presence/absence data. Binned data from sparse and cryptic animals, such as marine mammals, contain many absences. Hence, the variability of counts around the estimated spatial trends often fails to conform to distributions, such as the Poisson, classically assumed by parametric methods. This can be remedied by using more flexible alternatives like the negative binomial (e.g. Nielsen *et al.* 2005) or over-dispersed Poisson (Welsh *et al.* 1996; Fox 1997). Yet another alternative is to use a 'hurdle' approach (Welsh *et al.* 1996) which first models presence/absence and subsequently quantifies density, but only for the regions where the species occurs. Hurdle approaches are particularly useful in modelling both the occurrence and size of social groups.

The main drawback of binning is that it involves an arbitrary decision of scale. The alternative is to supplement the unbinned data with pseudoabsences, i.e. representatively selected locations where the animals could have been observed, but were not. In line transects, pseudoabsences are placed along the track-line, while in telemetry studies they may be placed uniformly in space (but also see Aarts *et al.* 2008).

Counts, relative frequencies, or presence/absence can be used directly to address space-use questions. Questions related to habitat use may be addressed by modelling such measures as a function of environmental variables. This has led to the development of resource selection functions (RSFs—Boyce and McDonald (1999); Manly *et al.* (2002)). RSFs are regression models relating a species' spatial distribution to environmental variables that may or may not be depleted by the animals (so, environmental selection function might have been a more intuitive name). The response variable in RSFs is said to be proportional to the probability of use (Boyce and McDonald 1999) but, in practice, RSFs model the disproportionality between usage and availability, not usage itself. The linkages between space use and habitat preference in regression models is the subject of active research (Keating and Cherry 2004; Aarts *et al.* 2008).

For the analysis of habitat preference, covariate data are required in addition to usage data. These may (ideally) be obtained concurrently with the collection of the distribution data or they may (pragmatically) be borrowed from other surveys. Covariate data may be biotic, abiotic, obtained by visual, acoustic, or satellite surveys (Redfern *et al.* 2006). They may even be the output of other models (oceanographic, atmospheric, or models of the distribution of other species such as prey, competitors, or predators).

How many environmental variables to include in the analysis can be addressed by formal model selection (Buckland *et al.* 1997; Burnham and Anderson 2002). A large number of candidate covariates means that some are likely to be correlated, a property known as *collinearity*. Such dependence between candidate covariates implies that the data may have insufficient information to support the dimensionality of the proposed model. Collinearity yields unstable parameter estimates with large standard errors and a sensitivity to outliers. Simple collinearities between explanatory data are detected as pair-wise correlations, but this ignores multi-collinearities. A better alternative is to use variance inflation factors (VIFs) (Fox 1997) to detect which combinations of explanatory variables will lead to heavily affected models (e.g. Dendrinos *et al.* 2007). Traditional treatments of collinearity may involve either judiciously dropping environmental variables to obtain a lower dimensional model, or transforming them so that they are uncorrelated. Judicious rejection of variables may be based on data availability. For example, certain explanatory data that are both dynamic and difficult to collect (e.g. prey distributions) may be available as a one-off and incorporating them in the model will be of limited value for prediction. Treating collinearity by covariate transformation is achieved with techniques such as principal components analysis (Jolliffe 2002). In practice, this also leads to a lower dimensional model because the last few principal components usually have low predictive abilities. A disadvantage of this technique is that principal components are difficult to interpret biologically.

Several explanatory variables may survive this initial censoring. The question then is whether to proceed with automatic model selection using methods such as the change in deviance, approximate F-tests, or information criteria (McCullagh and Nelder 1989; Hastie and Tibshirani 1990; Augustin *et al.* 1996), or to apply biological intuition to further reduce the model's dimensionality. Inclusion of too many candidate covariates in empirical models is said to lead to 'a subjective and iterative search for data patterns and significance' (Burnham and Anderson 2002). Instead, statistical inference may be conducted among models with a small number of covariates that are believed to be causally related to the response variable (Burnham and Anderson 2002). Although this is good advice in general, the exploratory nature of most studies of marine mammal distribution makes it necessary to examine as many candidate covariates as permitted by sample size and computer power. Admittedly, this increases the potential for overfitting and, therefore, the need for stringent validation of the resulting models.

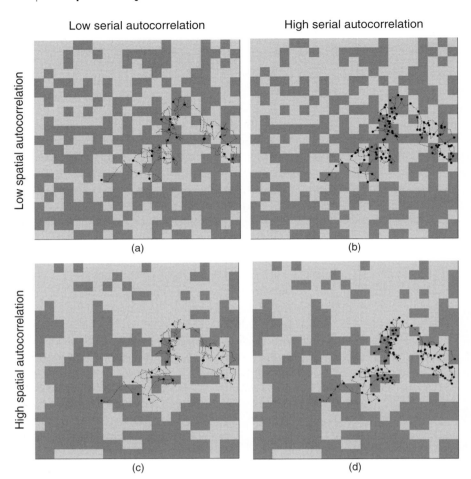

Fig. 4.4 The extent of interdependence in telemetry data is determined by how fast the animals move and how often they are observed (serial autocorrelation), and by the degree of similarity in conditions between neighbouring points in space (spatial autocorrelation). In this example, the same track is sampled more or less frequently (right and left column) and the environment comprises two habitats with high and low randomization (top and bottom row). The assumption of independence is most severely violated when frequently sampled data are regressed against strongly autocorrelated environmental variables (d). (After Aarts *et al.* 2008).

Model selection may be adversely affected by spatial, serial, and temporal *autocorrelation*. Spatial autocorrelation is the similarity in animal density and its environmental covariates between neighbouring locations. It is often the consequence of underlying autocorrelation in the environmental attributes that shape distribution, but may also result from social aggregation.

Serial autocorrelation describes the dependence between the successive positions of individuals. It mostly affects individually referenced data (e.g. telemetry) and is caused either by physical limitations in movement (animals unable to traverse the study region within a single sampling interval) or by repetitious behavioural patterns (e.g. regular returns to a rookery). The interplay between spatial and serial auto-correlation (Fig. 4.4) is a distinctive feature of some spatial data, and researchers ignore it at their peril. For example, Kernohan *et al.* (1998) suggested that using kernel-smoothed density instead of the raw observations avoids problems of serial autocorrelation. Although this removes the need to define the effective sample size of animal observations, it necessitates a decision on the resolution of the spatial grid to be used for smoothing thus transforming a problem of serial into one of spatial autocorrelation.

Temporal autocorrelation refers to the fact that animal distribution only changes gradually in response to seasonal or inter-annual change. This is particularly important for geographically referenced data sets (e.g. transects) where effective sample size may be determined by the frequency of surveys.

The problem with autocorrelation pertains to the scale of sampling: increasing sampling effort within the same study region/period does not necessarily lead to a proportional information gain. When the spatial density and temporal frequency of observations is very high, new observations contribute practically no new information because they replicate information from neighbouring/recent observations. Statistically, the effective sample size of a data set (its degrees of freedom) is discounted by various forms of autocorrelation and the standard errors of the parameters estimated from the data are deceptively small. Hence, during model selection, more variables will be retained than are strictly necessary and the resulting models' predictions will not be as reliable as the standard errors suggest.

Treatments of autocorrelation vary from crude to precise, with several prag-matic solutions in between. Crude solutions involve some type of spatial/tem-poral data-thinning to approximate independence (Rooney *et al.* 1998). At the other extreme, computationally expensive treatments of autocorrelation involve modelling it explicitly by specifying models with an autoregressive structure (Pinheiro and Bates 2000; Wood 2006), or using spatially autoregressive models (Augustin *et al.* 1996; Keitt *et al.* 2002; Lichstein *et al.* 2002). A widespread misconception among ecologists is that using mixed effects models automatically accounts for all within-individual effects. However, accounting for serial auto-correlation with a mixed effects model needs the explicit specification of an autoregressive structure (Pinheiro and Bates 2000, pp. 226–49; Wood 2006, pp. 321–4). More pragmatic solutions may involve post-hoc variance inflation by estimating the effective degrees of freedom in the data (Kruggel *et al.* 2002), or model-selection and estimation techniques relying on cross-validation instead of likelihood-based information criteria (Aarts *et al.* 2008).

Table 4.1 *The fundamental questions of 'where' (space-use) and 'why' (habitat-use) can be addressed with frameworks of increasing complexity that often rely on the biological insights obtained from simpler approaches. For example, the candidate environmental covariates included in a habitat model may be motivated by visual inspection of usage maps obtained by density estimation methods. In turn, the empirical relationships between usage and its covariates may motivate a more mechanistic model of space use.*

Space-use	Qualitative	Select response variable
		Choose representation of space
		Explore response data
		Pre-process response data
	Quantitative	Select analysis quantitative approach
		Parameter estimation
		Spatial interpolation/smoothing
		Validation
Habitat-use	Empirical	Select explanatory variables
		Investigate exploratory data
		Pre-process exploratory data
		Choose empirical model
		Fit empirical model
		Goodness-of-fit
		Model-selection
		Biological inferences
		Spatial extrapolation
		Validation
	Mechanistic	Construct mechanistic model
		Environmental extrapolation
		Validation

Spatial ecology enquires where organisms are and why they are there. These two fundamental questions correspondingly give rise to analyses of space-use and habitat-use. The former may be qualitative (derived from eyeballing data in a GIS) or quantitative (obtained by a range or density estimation method). Habitat-use analyses may be empirical (based exclusively on regression) or more mechanistic (using biological principles to determine the form or parameters of a model). These different analysis paradigms can be arranged into a flowchart of spatial modelling (Table 4.1).

Methods of spatial analysis are 'nested': the range can be obtained from density by thresholding (Heegaard 2002), and density can be obtained from habitat preference by mapping environmental space onto geographical space. So, why use a less general method such as range estimation when a flexible model of habitat preference can also provide range and density? Clearly, in many cases, less powerful methods are sufficient for the questions asked. Space-use methods have been successful in measuring overlap between populations (Fieberg and Kochanny 2005), defining core areas (Wray *et al.* 1992b; Kenward *et al.* 2001), comparing usage to theoretically derived home ranges (Sjöberg and Ball 2000), and comparing the distributions of different species (Garvey *et al.* 1998). Often, habitat models cannot be used due to lack of covariate data. In other cases still, the less general methods are used in pre-processing. For example, a density estimation method (such as kriging or kernel smoothing) can be used on prey data to estimate prey distribution, which is then used as a covariate in a model of predator distribution.

4.5 Pre-processing

Preparation of the data before the main analysis usually involves generating suitably projected GIS layers, importing and organizing them into appropriate data structures, error-correcting or otherwise filtering the usage data, and dealing with inadequacies in the environmental data. Although not essential for the analysis, it is generally more practicable to generate planar representations of the study area for both the usage and environmental data. Since spherical surfaces cannot be flattened without distortion, an appropriate projection must be chosen. Individual animals experience the planet by travel distances so, for studies of animal distribution, equidistance projections are most appropriate. Since many marine mammal species live in remote regions of the planet, remotely sensed (RS) imagery is an invaluable data source. Projecting entire RS data layers may be impractical, so most GIS software programs extract and project environmental conditions only for those regions and periods with some sampling effort.

Regarding data structures, a single spatial observation from an individual represents the highest level of disaggregation; and, therefore, a minimally specified data sheet consists of a column of serial numbers and two columns of spatial coordinates (assuming 2D space). It is a good idea to attach the date and time at which the observation was made. Additional information about each observation may allow more detailed modelling of sampling effort and of the different components of variability in the data. For example, each observation could be referenced with information about an individual's weight, sex, age, behavioural state, etc.

All types of data are affected by observation error, but each in different ways. In telemetry data, location accuracy and precision can be degraded by the process of triangulation (White and Garrott 1990), weather patterns (Musyl *et al.* 2001), and reception coverage and interference (Vincent *et al.* 2002). Early solutions to this problem involved discarding low-quality data. Non-destructive alternatives apply non-parametric smoothing of polygonal animal trajectories based on subjective or objective criteria (Chapter 10). More recent papers (Patterson *et al.* 2008) have employed state–space models which generate error-corrected individual paths as a by-product of the estimation of a movement model. Mark–recapture data may be affected by variation in recaptures (Hearn *et al.* 1991; Smith *et al.* 2000), which may be modelled explicitly as part of the recapture probability. In transect data, error results from biased/imprecise measurements of detection distance (Williams *et al.*

2007). This can be approached either by stratification and estimation of separate detection functions, or by the incorporation of environmental covariates in the detection function (Marques and Buckland 2004).

Some spatial questions can be better addressed with regard to the animal's specific activities. Analysis may then be conducted only for the coordinates at which these behaviours have occurred. At surfacings (particularly in transect surveys) it may be possible to carry out a subjective classification from visual inspection (Karczmarski *et al.* 2000). In telemetry studies, it may be possible to use the geometry of movement to classify locations as 'fast' or 'slow' by methods such as first-passage time (Fauchald and Tveraa 2003; Bailey and Thompson 2006), hazard rate models (Freitas *et al.* 2008), outlier exclusive cores (Kenward *et al.* 2001), or dive profiles (Biuw *et al.* 2007).

The work required to prepare environmental data for habitat modelling is considerable. Environmental data come from different sources, rarely offer complete spatial coverage, and are almost never collected simultaneously with usage data. Consequently, the conditions to which the animals are responding at any point in space and time are not always known, especially if these are highly heterogeneous and dynamic (Isaaks and Srivastava 1990) such as the distributions of other marine species. One solution to this problem is to discard regions of space and periods that come with incomplete environmental information. However, this impoverishes the study's coverage and subjects the model's results to the sampling biases of each environmental layer. Instead, it may be possible to estimate prevailing conditions. Temporal interpolation has been attempted (Stephenson *et al.* 2007), but modelling space-use and preference in changing environments still presents analytical challenges (Arthur *et al.* 1996; Hjermann 2000; and discussions in Boyce *et al.* 2002). Perhaps a more pragmatic approach is to use environmental variables whose spatial distribution remains relatively constant. Incomplete spatial coverage can be treated either by *interpolation* or *smoothing.* Interpolated surfaces are constrained to pass through the observations at the survey locations and are useful for environmental variables that change little and are measured precisely (e.g. geophysical variables— Kafadar and Horn 2002; Ripley 2004). If, the environmental data are imprecise (e.g. data on prey density) or dynamic (e.g. meteorological variables), techniques such as smoothing (Ripley 2004; Silverman 1986) and kriging (Isaaks and Srivastava 1990) are more appropriate.

4.6 Analysis techniques

Here, we attempt a near-comprehensive listing of techniques. For each, we cite: (1) a brief description, (2) variants/extensions/applications on the basic idea, (3) advantages, (4) disadvantages, (5) general technical references, and (6) case study references, preferably on marine mammals. We distinguish between space-use and habitat-use methods. We have commented liberally on the pros and cons of different methods because we want to give as clear a guide as possible. We recognize that this may have contaminated the presentation with our personal preferences so this section should be read in conjunction with other extensive reviews like Andreassen *et al.* (1993); Guisan and Zimmermann (2000); Kenward *et al.* (2001); and Redfern *et al.* (2006).

Space-use methods

These either try to delineate areas of occurrence/high usage (EyeB, Plg, described below) or to estimate the density of use over geographical space (Bn, HM, KS, Kg, TF, described below). There is a separate and extensive class of methods, not covered here, that deals with the detection of spatial patterns in distributions (see Fortin and Dale 2005).

1. EYEBALLING (EYEB)

   Basic idea: View the raw data on a map and identify prominent features of the resulting 'cloud'. Due to its limitations, it is most often used at the initial stages of analysis as a method of acquaintance with the data. Despite its limitations, surprisingly many publications in marine ecology present it as the end-result of spatial analysis.

   Variants: Eyeballing habitat preference by overlaying usage data on environmental layers.

   Advantages: Requires little effort and, in small data sets, may offer no less insight than the results of more elaborate analyses.

   Disadvantages: Ignores sampling effort (by geographical region or population component) and therefore can be grossly misleading. Although the presence of the animals is clearly visible in scatter plots of large data sets, their density is not, giving undue emphasis to outlying observations. Often, the correlational patterns with multiple environmental covariates are difficult to see and features that are completely random can appear significant.

   References: Any GIS manual. Some specific examples of data-basing and visualization are in Fedak *et al.* (2002) and Halpin *et al.* (2006).

   Case studies: Fiedler *et al.* (1998); Ersts and Rosenbaum (2003); Laidre *et al.* (2004a); Martin and da Silva (2004); Croll *et al.* (2005); Johnston *et al.* (2005a); Weir *et al.* (2007).

2. POLYGON (PLG) METHODS

   Basic idea: A single polygon or a tessellation of geographical space is constructed from the outermost positions of the animal(s). These positions are either defined by their distance from some geographical centre (e.g. home range) or by the behavioural interactions between individuals (e.g. breeding territories).

   Variants: Tessellations (Wray *et al.* 1992a), minimum convex polygon, peeled convex polygons, concave polygon, cluster polygons (reviews in White and Garrott, 1990; Andreassen *et al.* 1993).

Advantages: They clearly communicate the spatial extent of usage with only limited assumptions about its shape.

Disadvantages: Ignores the fact that usage within a home range is heterogeneous. The basic method can be sensitive to outliers but correcting this may require arbitrary thresholding.

References: White and Garrott (1990); Andreassen *et al.* (1993).

Case studies: Gregr and Trites (2001).

3. BINNING (BN)

Basic idea: A 2D histogram of usage in latitude and longitude. The relative frequency of observations in a unit area is the estimate of local density.

Variants: Relative frequencies can be standardized by sampling effort. Hence, fishing effort maps have been used to convert by-catch data into usage maps for small cetaceans.

Advantages: Quick, easy, and useful as a first stage of more elaborate analyses.

Disadvantages: Results are sensitive to the scale used for binning.

References: Chapter 6 in Ripley (2004).

Case studies: Lusseau and Higham (2004), C.D. MacLeod *et al.* (2004), Bailey and Thompson (2006).

4. HARMONIC MEAN (HM)

Basic idea: Several home range studies use the concept of a centre of activity. Estimating this as the arithmetic mean of all the locations of an animal has several disadvantages (Dixon and Chapman 1980) that do not afflict the harmonic mean. Usage of a given point is estimated as the HM of the data for that point. The minimum value of the HM over space is defined as the centre of activity.

Variants: Used as a first stage of more elaborate analyses (e.g. Tollit *et al.* 1998).

Advantages: More conservative than most space–use methods.

Disadvantages: Can underestimate the extent of space–use. Harmonic mean is functionally and parametrically specified. The method therefore involves no fitting to the data.

References: Dixon and Chapman (1980); Andreassen *et al.* (1993).

Case studies: Wray *et al.* (1992b); Tollit *et al.* (1998).

5. KERNEL SMOOTHING (KS)

Basic idea: Animal distributions are spatially autocorrelated: observing an animal somewhere implies a heightened probability that it and its conspecifics use neighbouring locations. KS methods place the centre of a kernel (a bivariate probability density function) at the coordinates of every observation. The expected usage of any given unit of area is then estimated as the sum of the contributions from all the kernels. The kernel's spread (also known as the smoothing window or bandwidth) is usually determined by cross-validation (Stone 1984).

Variants: Occasionally, information additional to the distributional data is available (e.g. natural history facts or historical data) that can be used to supervise kernel smoothing (Matthiopoulos 2003a). Simple KS ignores the order of occurrence of individually referenced data. This can be addressed by modelling animal movement in between observations by Brownian bridges (Horne *et al.* 2007). In adaptive KS (Silverman 1986), areas with low densities of observations receive more smoothing. Some studies use KS as a first stage of habitat preference analyses (e.g. Sjöberg and Ball 2000; Hastie *et al.* 2004; Breed *et al.* 2006). A more recent method, non-parametric multiplicative regression (McCune 2006) uses KS to model habitat-use. Finally, anisotropic KS, a method from medical imaging, offers the ability to incorporate an absolute directional bias in the spatial dependence between locations.

Advantages: A flexible, non-parametric method.

Disadvantages: Not all regions of the animals' distribution are necessarily characterized by the same degree of spatial autocorrelation (Osborne and Suárez-Seoane 2002). By using a single smoothing parameter, KS may be smoothing too aggressively in some regions and too leniently in others. This may partly explain why KS methods can perform worse with real rather than simulated data (Hemson *et al.* 2005). Cross-validation is computationally expensive.

References: Silverman (1986); Wand and Jones (1995); Seaman and Powell (1996).

Case studies: Wray *et al.* (1992b); Kernohan *et al.* (1998); Sjöberg and Ball (2000); Matthiopoulos *et al.* (2004); Johnston *et al.* (2005b); Breed *et al.* (2006); Parra (2006).

6. KRIGING (KG)

Basic idea: Like HM and KS, kriging yields local estimates of a random variable as distance-weighted combinations of all observations. The function (called semi-variogram), which quantifies how the similarity between observations should change with the distance between them, is fitted to the raw data prior to spatial estimation. Although kriging is often applied to geophysical variables, it can also be employed for usage estimation, as long as the usage data do not comprise only presences (e.g. telemetry). In this, it differs from HM and KS.

Variants: There are many possible shapes for the variogram and these have barely started to be considered for animal density estimation. Anisotropic kriging enables the modulation of directional dependence between spatial locations, so that points lying along a given directional axis are similar over greater distances. An extension of Kg, called co-kriging also makes use of covariate information to determine similarity between two points, giving it potential as a habitat-use method.

Advantages: Thanks to the method's pedigree in geostatistics the background theory is well-developed. The method is designed to give unbiased estimates with minimum associated uncertainty. It can operate both as a *smoother* and a spatial *interpolator*.

Disadvantages: Assumes normally distributed response variables. In practice, observations of usage are presences/absences, counts, or frequencies that do not adhere to the normality assumption. Similar geostatistical methods for non-normal data do exist (Diggle *et al.* 1998), but are computationally demanding. The methodology is still influenced by engineering terminology and mineralogical motivation, making it inaccessible to most ecologists.

References: Isaaks and Srivastava (1990).

Case studies: Monestiez *et al* (2006).

7. TREND-FITTING (TF)

Basic idea: Use longitude and latitude in a regression model as the covariates of usage. This does not give insights about habitat preferences. Can be thought of as a parametric form of spatial smoothing.

Variants: TF can be implemented using spatially global functions such as polynomials (Buckland 1992) or trigonometric functions (Anderson 1982). Alternatively, local smoothers such as splines can be used (see GAMs, below).

Advantages: With simple, linear models, the method can pick out prominent spatial gradients. More complex models (e.g. GAMs) can be used to describe complex features of the distribution, while at the same time removing sampling stochasticity. It is straightforward to test for the significance of spatiotemporal trends.

Disadvantages: Like all usage-only methods, trend fitting has no predictive power beyond the spatial range of the data. Patterning in geographical space is usually more complex than patterning in environmental space. Hence, more flexible models are needed to describe geographical trends.

References: Ripley (2004); Fortin and Dale (2005).

Case studies: Hedley *et al.* (2004).

Habitat-use methods

Overall, habitat-use studies rely on one of three types of statistical inference, hypothesis testing (HT), multivariate statistical modelling (EnEn, LM, GLM, GAM, described below) and multivariate ordination and classification (PCA, CCA, DA, CART, MRT, NN, described below).

8. TESTING FOR HABITAT PREFERENCE (HT)

Basic idea: Environmental space is divided a priori into a few (usually <10) habitats. The usage of each habitat is standardized by its availability within the study region, and hypothesis tests are conducted to decide whether some habitats are preferred (or avoided), i.e. whether they are used disproportionately to their availability.

Variants: Simple hypotheses can look at the homogeneity of the distribution (e.g. Parra *et al.* 2006) or compare usage between spatial regions of interest (e.g. Hastie *et al.* 2003). Differences in use between many predefined habitats can be tested parametrically by MANOVA (Arthur *et al.* 1996), or non-parametrically by permutation as in Mantel tests (Fortin and Dale 2005), or distance-based redundancy analysis (Legendre and Anderson 1999). Habitat availability is often calculated over an arbitrarily defined study area, but it can also be based on the animal's home range or the species' observed range (but see discussion in Aarts *et al.* 2008).

Advantages: Familiar to most biologists.

Disadvantages: Hypothesis testing has severe drawbacks as a method of inference (Hobbs and Hilborn 2006), often making unrealistic parametric assumptions about species occurrence, usage, and preference. Habitat preference tests often rely on arbitrary habitat classification and are essentially statistical comparisons, so quantitative spatial predictions for usage or occurrence are not possible.

References: Manly *et al.* (2002); Chapter 8 in White and Garrott (1990).

Case studies: Arthur *et al.* (1996); Tollit *et al.* (1998); Schick and Urban (2000); Sjöberg and Ball (2000); Ersts and Rosenbaum (2003); Baumgartner and Mate (2005); Bearzi (2005a); Azevedo *et al.* (2007); Doniol-Valcroze *et al.* (2007).

9. ENVIRONMENTAL ENVELOPES (ENEN)

Basic idea: A region is defined in environmental space in which an organism is expected to occur. For each environmental variable this usually consists of a bracket (min–max) value. The aim is to make use of anecdotal and large-scale information on worldwide distributions (e.g. Harwood and Wilson 2001).

Variants: When using rectilinear envelopes (hypercubes in environmental space), the regions close to the apexes are less likely to be used. Also, animals may have multimodal preferences in environmental space. These problems may be addressed by using convex polytope envelopes (Walker and Cocks 1991). Burgman *et al.* (2001) attempted to associate a 'possibility level' with different values of the environmental variable to provide ad-hoc quantification of prediction uncertainty.

Advantages: Straightforwardly informed with expert knowledge. More suitable for deriving coarse, large-scale, multi-species descriptions.

Disadvantages: May require prior definition of habitat types and involves no formal fitting to data. Hence, results are mostly user-driven rather than data-driven. *EnEn* methods that are data-driven, implicitly assume spatially uniform sampling effort, and consequently perform worse than methods incorporating effort or pseudoabsence data (Engler *et al.* 2004). Functional associations of suitability indexes with spatial usage or the probability of occurrence are obscure.

References: Guisan and Zimmermann (2000).

Case studies: Kaschner *et al.* (2006).

10. LINEAR MODELS (LM)

Basic idea: A multivariable extension of simple linear regression. Used if animals are believed to respond linearly to environmental covariates (a preference gradient). Like univariate linear regression, assumes that variability around the fitted trend is normal.

Variants: Transformations of the response variable such as the log-transform of count data (e.g. Yen *et al.* 2004). Transformations of the explanatory variables (e.g. use of polynomial terms) or use of interaction terms. LMs can be extended to mixed-effects linear models (LMM—Pinheiro and Bates 2000) to account for variation within and between sampling units (e.g. region or individual animal).

Advantages: The assumption of normality offers considerable computational gains and the theory behind these models is well-established. Any parameter of the resulting model is easy to interpret directly as the slope of the response to the covariate.

Disadvantages: The assumption of normality is usually violated in usage data, which are either counts or presence/absence data.

References: Fox (1997), Krzanowski (1998).

Case studies: Rare in the spatial literature. Littaye *et al.* (2004) present an example in which they incorrectly use pair-wise correlation coefficients as model parameters.

11. GENERALIZED LINEAR MODELS (GLM)

Basic idea: GLMs generalize on the concept of LMs by relaxing the requirement for normally distributed residuals. The response variable can be a count of observations within a unit of area, a binary variable of presence/absence, or the relative frequency of presence.

Variants: Transformations of the explanatory variables (e.g. polynomial terms) can be used to capture non-linearity in their relationship with the response. GLMs can be extended to Generalized linear mixed-effects models (GLMM; Pinheiro and Bates 2000) with an autocovariance structure to take account of *serial autocorrelation* (Pinheiro and Bates 2000).

Advantages: More flexible than LMs with equally well-developed theory and software.

Disadvantages: Simple GLM assumes monotonic relationships between preference/usage and environmental variables. Use of polynomial terms to overcome this introduces non-linearity globally in environmental space (contrast with GAM, below).

References: McCullagh and Nelder (1989), Krzanowski (1998), Boyce and McDonald (1999), Guisan *et al.* (2002), MacKenzie *et al.* (2006).

Case studies: Khaemba and Stein (2000), Gregr and Trites (2001), Cañadas *et al.* (2002), K. MacLeod *et al.* (2004), Cañadas *et al.* (2002, 2005), Jiménez (2005), Panigada *et al.* (2005), Dendrinos *et al.* (2007), Stephenson *et al.* (2007).

1 2. GENERALIZED ADDITIVE MODELS (GAM)

Basic idea: An extension of the GLM which allows locally non-linear relationships between response and explanatory variables to be fully specified from the data.

Variants: GAMs can be extended to generalized additive mixed-effects models (GAMM; Wood 2006) and can be implemented with an autocovariance structure to take account of *serial autocorrelation* (Wood 2006).

Advantages: Flexibility in the fitted relationships is directed at the data-rich regions of environmental space. The shape of the relationship comes from the data, not a-priori perceptions.

Disadvantages: GAMs are prone to overfitting. This can be addressed by a more stringent review of the assumptions of the model and the procedure used for model selection.

References: Hastie and Tibshirani (1990), Guisan *et al.* (2002), Wood (2006).

Case studies: Forney (2000), Hedley *et al.* (2004), Hastie *et al.* (2005), Aarts *et al.* (2008).

1 3. DISCRIMINANT ANALYSIS (DA)

Basic idea: A classification technique that uses observations of known group membership for which covariate data are available. It first attempts to discriminate between these observations on the basis of their characteristics, and then use the resulting discrimination rules to allocate new observations to the same groups.

Variants: DA has had particular application in presence/absence modelling. It produces linear combinations of environmental variables that optimally classify spatial locations as likely/unlikely to contain the study species.

Advantages: One of the simplest classification methods.

Disadvantages: Cannot readily be used for abundance modelling. Spatial predictions can be generated but are not probabilistic.

References: Quinn and Keough (2002).

Case studies: Manel *et al.* (1999), Cañadas *et al.* (2002).

1 4. CLASSIFICATION AND REGRESSION TREES (CART)

Basic idea: The tree is constructed by repeatedly splitting the abundance data, each time on the basis of the response to a single environmental variable. Presence/absence leads to a classification tree, and usage to a regression tree. The method aims to minimize the size of the tree (the final number of groups) while simultaneously maximizing homogeneity within groups. For example, in the case of presence/absence data, the objective is to end up with as few groups of spatial locations as possible each containing mostly presences or mostly absences.

Variants: Multivariate regression trees (MRT—De'Ath, 2002) enable data from many species simultaneously to be related to environmental covariates.

Advantages: Non-parametric method: it makes no assumptions about the functional form of the relationship between the response and explanatory variables.

Disadvantages: This is a classification method. Hence, when used with abundance (rather than presence/absence) data, its predictions are discrete.
References: De'Ath and Fabricius (2000), Vayssiéres *et al.* (2000).
Case studies: Iverson and Prasad (1998), Goetz *et al.* (2007).

15. PRINCIPAL COMPONENT ANALYSIS (PCA)
Basic idea: PCA transforms a set of continuous variables into new variables (the principal components) that are linear combinations of the originals. Although the method yields as many new variables as in the original set, the new variables are linearly independent and ranked in order of the variability that they explain (i.e. the topmost variables in the ranking have the most explanatory power).
Variants: PCA is as a dimensionality-reducing technique: keeping only the first few PCs captures those characteristics of the original data that contribute most to variability (Khaemba and Stein 2000). Hence, PCA has been used as a treatment for multi-collinearity (see above). However, ad-hoc implementations exist (Robertson *et al.* 2001, Calenge, *et al.* 2005) that generate maps of habitat suitability. Related to PCA, environmental-niche factor analysis (ENFA—Hirzel *et al.* 2002) assumes a unimodal response to environmental covariates, representative sampling, and equal accessibility of all space to animals. It works by comparing the statistical properties (position and spread) of the distribution of habitat usage with the distribution of habitat availability.
Advantages: PCA routines exist in all statistical packages.
Disadvantages: The main disadvantage of PCA is that, although the original variables have some physical interpretation (e.g. sea depth or surface temperature), the principal components can be hard to interpret. There are several additional problems with using PCA for mapping species distributions: the variable used for prediction ('a probability of bioclimatic suitability for each locality') is meaningless both biologically and as a probability and the Normal errors associated with it are only weakly justified. With the exception of ENFA, PCA approaches assume that organisms have monotonic (increasing or decreasing only) distributions along environmental gradients. For marine mammals, the assumption of habitat suitability methods (e.g. ENFA), that observations are a representative sample of usage, and that all points in space are equally accessible to the animals, are rarely met.
References: Quinn and Keough (2002).
Case studies: Robertson *et al.* (2003).

16. CORRESPONDENCE ANALYSIS (CA)
Basic idea: Similar to PCA in that new variables are obtained as combinations of the originals. However, CA can also deal with non-linear (unimodal) relationships.
Variants: The main variant of CA in ecological use is canonical correspondence analysis (CCA—Ter Braak 1986). This enables the distribution of one or more species to be directly related to a set of environmental variables.

Advantages: Can be used to compare the habitat preferences of many species simultaneously. In contrast to PCA, it can model environmental optima in habitat preference.

Disadvantages: Assumes that organisms have only unimodal distributions along environmental gradients. As with PCA, the interpretation of the new explanatory variables can be difficult. Spatial predictions of distribution are not straightforward (Hill 1991).

References: Ter Braak (1986), Palmer (1993), Quinn and Keough (2002).

Case studies: Reilly and Fiedler (1994), Ferrero *et al.* (2002), Griffin and Griffin (2003).

17. NEURAL NETWORKS (NN)

Basic idea: An interconnected set of simple processing elements (neurons), usually in three layers. In the context of animal distributions, the neurons in the first (input) layer represent different environmental variables. The third (output) layer represents values of the response variable. So, a NN model of presence/absence would only have one neuron at the output layer, which, once activated, would indicate a high probability of presence at the geographic location having the environmental properties specified at input. The neurons belonging to the middle (hidden) layer generate linear combinations of the inputs that are then combined, again to decide on the excitation state of the output neurons. Procedures such as back-propagation, are used to adjust the network's activation values, the equivalent of model fitting in regression.

Variants: Several variants of the NN concept exist in the computer literature. Only the most basic have been implemented in ecology.

Advantages: Updating the network can be done 'on the run', as new data are obtained. NN approach is adaptive. Complex neuron connectivity potentially allows it to emulate highly non-linear processes.

Disadvantages: Operates as a black box and therefore offers few mechanistic insights about the factors shaping animal distribution. Outputs are only approximately continuous. Relatively little work has been done on quantifying prediction uncertainty.

References: Guisan and Zimmermann (2000).

Case studies: Manel *et al.* (1999), Lusk *et al.* (2002).

4.7 The spatial analyst's software toolbox

The choice of software depends on the fundamental trade-off between flexibility and user-friendliness. The broad-spectrum packages in categories 1 and 2 below can perform any visualization/analysis imaginable but may require some programming. The software in categories 3, 4, and 5 will perform specific analyses easily but with limited user control.

Table 4.2 A subjective assessment of methods. The ideal method is: insensitive to error and outliers (Robustness); unbiased (Accuracy); numerically efficient (Computation) and user-friendly (Software); able to represent highly heterogeneous spatial surfaces (Flexibility); capable of estimating range, distribution, and habitat preference (Generality); and useful in obtaining and incorporating biological insights (Interpretability). Our overall score weights these traits equally but, clearly, they must be ranked according to the needs of the application at hand.

Method	Robustness	Accuracy	Comput.	Software	Flexibility	Generality	Interpret.	Overall
EyeB	★☆☆☆	☆☆☆☆	★★★★	★★★★	☆☆☆☆	☆☆☆☆	☆☆☆☆	★☆☆☆
Plg	★★☆☆	★★★☆	★★★☆	★★★☆	★★☆☆	★★★☆	★★★☆	★★★☆
Bn	★☆☆☆	★★☆☆	★★★★	★★★★	★★★☆	★★☆☆	★★☆☆	★★★☆
HM	★★★☆	★★★☆	★★★☆	★★★☆	★★☆☆	★★★☆	★★★☆	★★★★
KS	★★★★	★★★★	★★★☆	★★★☆	★★★☆	★★☆☆	★★☆☆	★★★★
MSKS	★★★☆	★★★★	★★☆☆	★★☆☆	★★★★	★★★☆	★★★★	★★★☆
Kg	★★★★	★★★☆	★★☆☆	★★★★	★★★★	★★☆☆	★★★☆	★★★★
TF	★★★☆	★★★★	★★★★	★★★★	★★★☆	★★☆☆	★★★☆	★★★★
EnEn	★★★☆	★☆☆☆	★★★★	★★★☆	★★☆☆	★★★☆	★★☆☆	★★★☆

ENFA								
HT								
CoKg								
LM								
GLM								
GAM								
PCA								
CCA								
DA								
CART								
MRT								
NN								

Abbreviations: EyeB, eyeballing; Plg, polygon; Bn, binning; HM, harmonic mean; KS, kernel smoothing; Kg, kriging; TF, trend-fitting; EnEn, environmental envelope; HT, habitat preference testing; CoKg, co-kriging; LM, linear models; GLM, generalized linear models; GAM, generalized additive models; PCA, principal components analysis; CA, correspondence analysis; DA, discriminant analysis; CART, classification and regression trees; MRT, multivariate regression trees; NN, neural networks.

1. Low-level programming languages

C++
http://www.dmoz.org/Computers/Programming/Languages/C++/

Fortran (Various implementations)

Pascal (Various implementations)

2. High level, multi-purpose analysis and programming languages

R (Freeware, statistical and graphical software) http://www.r-project.org/

SAS (Strong in stats) http://www.sas.com/technologies/analytics/statistics/stat/index.html

S-Plus (Strong in stats) http://www.insightful.com/

Systat (Strong in stats) http://www.systat.com/

Statistica http://www.statsoft.com/

Mathematica (Strong in maths) http://www.wolfram.com/

MatLab (Strong in maths) http://www.mathworks.com/

Maple (Strong in maths) http://www.maplesoft.com/

MathCad (Strong in maths)
http://www.ptc.com/appserver/mkt/products/home.jsp?k=3901

3. Software for data preparation and visualization

ESRI (ArcView, ArcGIS, Atlas) http://www.esri.com/

Manifold http://www.manifold.net/index.shtml

IDRISI http://www.clarklabs.org/products/index.cfm

GRASS http://grass.itc.it/

MapInfo http://www.mapinfo.com/location/integration

MAMVIS http://biology.st-andrews.ac.uk/seaos/technology.htm

4. Specialized analysis software for particular applications

Distance (Distance sampling data) http://www.ruwpa.st-and.ac.uk/distance/

Mark (Mark–recapture data) http://welcome.warnercnr.colostate.edu/

~gwhite/mark/mark.htm

Geobugs (Bayesian geostatistics) http://www.mrc-bsu.cam.ac.uk/bugs/winbugs/geobugs.shtml

GRASP (Regression software for spatial prediction) http://www.unine.ch/cscf/grasp/

Transect (Line transect data) http://nhsbig.inhs.uiuc.edu/wes/density_estimation.html

Ranges (Home range estimation software) http://www.anatrack.com/index.html

Random Forests (Classification and regression trees) http://www.stat.berkeley.edu/~breiman/RandomForests

CART (Classification and regression trees) http://www.salford-systems.com/cart.php

CANOCO (Canonical Correspondence Analysis) http://www.microcomputerpower.com/catalog/canoco.html

MjM (Multivariate analysis of ecological data)

http://home.centurytel.net/~mjm/

Many more purpose-specific freeware can be found at the excellent, Clearinghouse for Ecology Software http://nhsbig.inhs.uiuc.edu/

5. General search and optimization algorithms

AD Model builder (Automatic differentiation) http://www.otter-rsch.com/admodel.htm

WinBugs (Monte Carlo Markov Chain sampling)

http://www.mrc-bsu.cam.ac.uk/bugs/

SIS (Sequential Importance Sampling) http://www.statslab.cam.ac.uk/~ben/library_web/sis.html

4.8 Prospects and trends

We collect here what appear to be the most fruitful future directions in spatial ecology, particularly in relation to marine mammal science. Progress should occur in parallel, through improvements in the biological content of models and the estimation/computational features necessary for dealing with this added complexity.

With the exception of work on critically endangered species (vaquita, baiji, monk-seals), most studies pose questions related to larger, but only partially sampled populations. Statistical models are essential for making such population-level inferences, but they are complicated by sampling biases, autocorrelation, and observation error. Methodologically, we need to improve and encode the methods dealing with all three of these features, particularly spatiotemporal autocorrelation. Currently, the lack of a coherent and practical solution to autocorrelation, combined with the careless use of hypothesis tests and information criteria for model selection, makes analyses prone to spurious results. Such methods need not only be statistically correct but also computationally efficient to deal with the typically large spatial data sets.

Another desirable development would be the disaggregation of spatial data into biologically defensible subsets (locational observation, foraging trip, individual, type of individual, social group, colony, sub-population, metapopulation), within hierarchical models. This will enable us to correctly represent the sources of variability in the data by taking into account regional, temporal, and individual effects.

Making these technical improvements will reduce the risk of overfitting/bias and, therefore, benefit the predictive abilities of empirical models. However, the best way to improve predictive ability is to increase a model's biological detail (Austin 2002; Robertson et al. 2003). This involves the replacement of various flexible phenomenological features of the model by more specific, mathematical relationships derived from first principles or controlled experiments (Perry et al. 2002). Empirical modelling can instigate the development of these relationships by hinting at causal interactions. Although highly desirable, this transition from empirical to mechanistic is an arduous and slow process, because any potential gains in predictive ability must be constantly weighted against the risk of model mis-specification. In this light, it is hard to see the applied value of publications that develop detailed, individual-based models in a biological knowledge-vacuum; or the scientific merits of black-box approaches, such as neural networks, that aim to replicate patterns but are, by construction, resistant to the incorporation of mechanism (Manel *et al.* 1999).

More mechanistic models may require information on how marine mammals use the water column and how their distribution changes in time. Surprisingly little has so far been done to link dive profile data with synchronous location data and to relax the *pseudo-equilibrium assumption* (Guisan and Zimmermann 2000). Preliminary attempts to develop dynamic models (Rushton *et al.* 1997; Whitehead 2000) have relied on sensitivity analyses rather than model fitting. On the other hand, due to computational restrictions, state–space models for population dynamics (Buckland *et al.* 2007) have been limited to formulations that are, at best, spatially implicit.

Taking yet another step up in complexity, community modelling presents a worthwhile objective. Spatial hotspots of biodiversity and niche partitioning are becoming increasingly important in the literature (Bearzi 2005a,b, Balance *et al.* 2006, Parra 2006); particularly because exclusion of human competition (Matthiopoulos *et al.* 2008) through the introduction of marine special areas of conservation (Hastie *et al.* 2003, Cañadas *et al.* 2005) is becoming politically palatable. We have encountered three ideas for achieving this: the number of species in a unit area, or some other measure of biodiversity, can be modelled as a spatial function of environmental covariates (e.g. Guisan and Theurillat 2000). Alternatively, the distributions of different species can be compared using non-parametric tests (Syrjala 1996; Garvey *et al.* 1998) or correspondence analysis. More adventurous approaches (Guisan and Zimmermann 2000) propose the development of simultaneous (coupled) regression models for different components of the ecosystem.

Extending the scope of the models will require a more inclusive approach to data. Previously unused types of data, such as acoustic detection data (Lewis *et al.* 2007), can be combined to map space-use. In addition, multiple types of spatial data may be combined. Examples include the simultaneous use of acoustic and visual data (Matthiopoulos and Frantzis, unpublished), coastal surveys with offshore telemetry (Matthiopoulos *et al.* 2004), and marine transects with telemetry (Johnston *et al.* 2005b).

Most of the above forecasts are subjective and uncertain, but it is probably safe to venture that spatial statistics will continue to become more complex. This will be accompanied by increases in the reliability and user-friendliness of the various data-analysis software platforms, making it easier to perform analyses that are ever-harder to understand. Faced with the risk of committing serious methodological flaws, the basic dilemma for every practitioner in this area will be whether to re-train or delegate. Re-training as a spatial-modeller places unreasonable demands on the time of marine mammal scientists who are, after all, already having to deal with their own discipline. Delegating to a spatial analyst is only practicable if the biologist and the modeller have common ground in both research interests and terminology. Finding and establishing a rapport with an interested and suitably skilled statistician is not always easy, but taking just such a 'middle-road' approach to the study of space is better than taking the 'third way' of ignoring it altogether.

5

Morphometrics, age estimation, and growth

W. Don Bowen and Simon Northridge

5.1 Introduction

Body size, age, and growth are fundamental animal characteristics that are needed to study diverse areas of biology. Given their large size and adaptations to aquatic living, taxon-specific suites of measurements are used to characterize body size in marine mammals. Morphometrics have special importance in the study of marine mammals because, in some cases, the animal's large size means that body mass, for example, is often estimated from its length or some combination of length and girth, which are more easily taken. Studies of life-history variation and behavioural modelling typically involve age- or size-related metrics, including those referred to as condition indices. Marine mammalogists pioneered the use of dental layering to estimate the age of individuals and have continued to develop novel approaches for species without dentition. By contrast, statistical methods for studying the growth of marine mammals follows closely the methods used in other taxa and therefore we devote relatively little text here, but point to the pertinent literature.

5.2 Standard morphometrics

5.2.1 General issues

Morphometrics are intended to provide quantitative information on the size and shape of an animal's body (Oxnard 1978). These measurements are taken to study biomechanics, growth of specific body parts or, most frequently, of the whole animal, or to compare variation in body form within or between species. They may also be used to derive an index of body condition. In the past, the most common reason for taking morphological measurements was to inform the taxonomy of species, sub-species, or populations of a marine mammal. Although there is a propensity for taxonomists and museum curators to take many measurements, from the taxonomic perspective, the development of techniques to investigate the genetic relatedness of individuals and populations has reduced some of the need to take detailed morphological measurements. Many researchers studying taxonomy now take only a small number of external measurements but use a sample of tissue for genetic analyses.

Nevertheless, morphometric studies can still provide useful insights into the taxonomy or population structure of marine mammals, and can also enable field identification where genetic or osteological differences cannot be assessed (Perryman and Lynn 1993; Wang, J.Y. *et al.* 2000). External measurements are often more variable and less likely to reveal new population level differences among groups of marine mammals, though some, including pectoral fin length in porpoises (Jepson 2003), tooth counts, and rostral length measurements in common dolphins (Heyning and Perrin 1994), have been important in highlighting population structure.

Often it is impossible to measure the body mass of a marine mammal carcass in the field, and in these cases length, or some other combination of measurements, may provide a means of estimating body mass. So called length–weight relationships have been estimated for many species (see for example Burns, J.J. 1981; Lockyer and Waters 1986; Markussen *et al.* 1989; Laws, R.M. *et al.* 2002; Perrin *et al.* 2005), though better estimates of body mass are obtained by including girth measurements or some other measures of shape as well (Castellini and Calkins 1993; Laws, R.M. *et al.* 2002). Photogrammetric methods have also been used to estimate the body mass of marine mammals (Haley *et al.* 1991; Bell *et al.* 1997; Ireland *et al.* 2006; Proffitt *et al.* 2007).

A recurring issue with morphometrics concerns standardization of the measurement across studies, individual observers, and even through time within long-term studies (see Chapter 13). For example, the apparent body length of a pinniped can vary depending upon whether it has been measured when alive or dead, or whether it has been chemically immobilized or physically restrained (see Chapter 2). There is also a need to ensure appropriate and regular calibration of measurement equipment, especially scales used for weighing.

All measurements are subject to error. It is important to have an appreciation of the magnitude and sources of this error to minimize random error and to increase the accuracy and precision of measurement. The magnitude of random error is represented by the coefficient of variation (standard deviation/mean). As a rule of thumb, a coefficient of variation >15% should lead to an investigation of ways of improving the precision of the measurement. The existence of bias within measurements is much more difficult to deal with than random error. Bias occurs when a measurement is consistently smaller or larger than it should be, and occurs in addition to random error. Careful calibration of measurement methods against a known standard is about the only way to eliminate or calibrate bias. However, bias can also be non-linear. Examples of non-linear bias may occur when the spring of a scale changes its calibration depending upon the test weight because the spring becomes stretched, or when a tape measure becomes stretched over one particular part of its length.

5.2.2 Body condition

Length, girth, and body mass provide information about the size and condition of individuals, but the age of the animal also usually needs to be estimated or known if these measurements are to be used to make inferences about growth rates or the influence of ecological covariates on body condition. Several indices of condition are used for marine mammals (Table 5.1), the oldest of which is simply a measure of the degree of fatness (Smirnov 1924). Given the importance of stored lipids for most marine mammals, the objective of such indices is to provide a measure of the fat content of individuals relative to total body mass. Often there is a trade-off between the simplicity and complexity of a set of morphometric measurements and the extent to which they reflect condition. Ideally, condition indices would take account of the season (or the stage of the annual reproductive cycle) when measurements were made, and the age and sex of individuals if ecological inferences are to be made from the measurements (McLaren and Smith 1985). Although blubber depth is commonly used, the Smirnov, and Boyd and McCann indices (Table 5.1) of condition are often poor predictors of fat content. To attempt to overcome this problem, other indices such as the LMD index (Ryg et al. 1990), which is a more precise measure of blubber content, have been developed. R. Gales and Renouf (1994) compared condition indices and found that both blubber depth and the Smirnov index were poor predictors of percentage fat in harp seals, but the LMD index performed better and explained 59% of the variation in blubber mass. Nevertheless, the application of some of these more complex methods also depends upon the ease with which measurements can be made. Some situations make it impossible to obtain these measurements or, even if they can be made, the errors and biases associated with the measurements may be so large as to make them almost worthless, and measurements that are severely biased are potentially misleading.

Table 5.1 *Commonly used indices of body condition in marine mammals.*

Condition index (CI)	Source
$G*100/L$	Smirnov 1924
B	Anon. 1967
M_t/L	Boyd and McCann 1989
$\sqrt{\frac{L}{M_t}}*B$	Ryg et al. 1990
M_t/M_t where $M_t = aL^b$	Trites and Bigg 1992
$V = L^{2}*G*10^{-5}$	Nilssen et al. 1997
$V = (L*G^{2})/6000\pi$	Chabot et al. 1996

Abbreviations: G, axillary girth (cm); L, standard body length (cm); M_t, body mass (kg); D, xiphosternal blubber thickness (cm); V, body volume (cm^3).

Trites and Bigg (1992) suggested another approach to measuring body condition, whereby condition is estimated as the ratio of measured body mass of an individual over that predicted from a population regression of body mass on length (Table 5.1). This index can be calculated by age, year, or other covariates of interest and has the advantage of clearly indicating whether condition is good or poor relative to the longer term.

Measurement of blubber mass from carcasses, combined with estimates of the percentage of lipid in the blubber, provides a direct estimate of the quantity of the stored lipid (Lockyer *et al.* 1985; Bowen *et al.* 1992; Koopman *et al.* 2002). However, there are situations where it will not be feasible to make these measurements. Gales R. and Renouf (1994) evaluated four non-invasive geometric methods of estimating blubber mass using combinations of blubber depth and girth measurement at multiple sites. Of these, the truncated cone method (Gales N. J. and Burton 1987) performed best but required 31 measurements, whereas a method requiring only 3 measurements performed almost as well and thus would be preferred in most field situations.

Because body mass can change for reasons other than a change in lipid content, total body lipid or energy of an individual, relative to body mass, provides a more accurate measurement of condition. This can be estimated from carcass dissection or from isotope dilution methods on live individuals, whereby body mass and estimates of total body water can be used to estimate body fat and protein content (See Chapter 8). For example, Beck *et al.* (2003) used hydrogen isotope dilution to estimate total body energy of adult male and female grey seals to investigate seasonal dynamics of energy deposition and expenditure.

5.2.3 Standard measurements

The American Society of Mammalogists recommended taking 35 standard external measurements from small cetacean carcasses (Norris 1961) and 14 standard measurements from pinnipeds carcasses (Anon. 1967), in addition to some other biometric and descriptive data. The main recommended external measurements are listed in Figures 5.1 and 5.2 and Boxes 5.1 and 5.2, respectively in cetaceans and pinnipeds. A priority list (items highlighted in **bold text**) was proposed when time is limited. Included in these were standard length, maximum girth, anterior and posterior flipper lengths, and body mass. One or more measurements of blubber thickness were also recommended, as well as seven additional measurements of the prepared pinniped skull. Life history data were also recommended, including counts of ovarian corpora and dimensions of testes, but these issues are dealt with in Chapter 6. Standard body measurements of sirenians were described by Heinsohn (1981), and these are illustrated in Figure 5.3 and Box 5.3.

Although these measurements were considered fundamental, they were not intended to discourage other measurements needed for particular studies. For example, several other body-length measurements are often taken for pinnipeds,

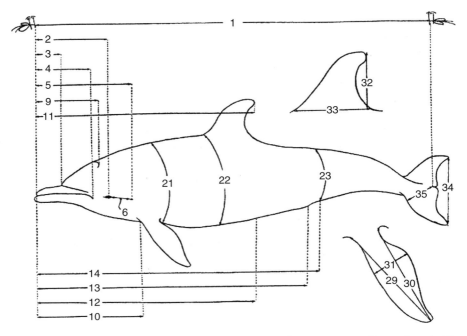

Fig. 5.1 Standard external measurements to be taken from cetaceans. (Reproduced with permission from The Committee on Marine Mammals, American Society of Mammalogists (1966–67) and Kenneth S. Norris, *Journal of Mammalogy*, **42** (No. 4, Nov. 1961), pp. 471–476; published by The American Society of Mammalogists.)

Box 5.1 Description of measurement in Figure 5.1

1. Weight
2. Number of teeth (or baleen plates): right upper; right lower; left upper; left lower
3. Diameter of largest tooth
4. Number of throat grooves
5. Colour (and pattern diagram)

Body measurements:

1. **Length, total (tip of upper jaw to deepest part of notch between flukes)**
2. **Length, tip of upper jaw to centre of eye**
3. Length, tip of upper jaw to apex of melon
4. **Length of gape (tip of upper jaw to angle of gape)**
5. Length, tip of upper jaw to external auditory meatus
6. Centre of eye to external auditory meatus (direct)
7. Centre of eye to angle of gape
8. Centre of eye to centre of blowhole(s) (direct)

9. **Length, tip of upper jaw to blowhole along mid-line, or to mid-length of two blowholes**
10. **Length, tip of upper jaw to anterior insertion of flipper**
11. **Length, tip of upper jaw to tip of dorsal fin**
12. Length, tip of upper jaw to midpoint of umbilicus
13. Length, tip of upper jaw to midpoint of genital aperture
14. **Length, tip of upper jaw to centre of anus**
15. Projection of lower jaw beyond upper (if reverse, so state)
16. Length, tip of upper jaw to posterior extremity of throat creases
17. Thickness of blubber, mid-dorsal at anterior insertion of dorsal fin
18. Thickness of blubber, mid-ventral at mid-length
19. Thickness of blubber, mid-ventral at mid-length
20. Length, throat creases: maximum; minimum
21. Girth, on a transverse plane intersecting axilla
22. **Girth, maximum (describe location as distance from tip of upper jaw)**
23. Girth, on a transverse plane intersecting anus

Apertures

24. Dimensions of eye: height; length;
25. Length, mammary slits: right; left.
26. Length, genital slit, anal opening
27. Dimensions of blowhole(s): width; length(s)
28. Diameter of external auditory meatus: right; left; absent

Appendages

29. **Length, flipper (anterior insertion to tip)**
30. **Length, flipper (axilla to tip)**
31. **Width, flipper (maximum)**
32. **Height, dorsal fin (fin tip to base)**
33. Length, dorsal fin base
34. **Width, flukes (tip to top)**
35. Distance from nearest point on anterior border of flukes to notch
36. **Depth of notch between flukes (if none, so state)**

All measurements, except girths, should be taken in straight lines and not over the curvature of the body. However, measurements of large baleen and sperm whales are most conveniently taken as direct lines (along the curvature of the animal) and not as measurements parallel to the major body axis. This practice was established during the early years of whale science conducted on board whaling vessels, and is often continued to this day as a matter of convenience.

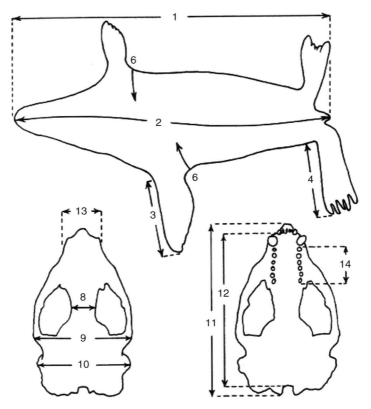

Fig. 5.2 Standard external measurements for pinnipeds. (Reproduced with permission from The Committee on Marine Mammals (1966–67). *Journal of Mammalogy*, **48** (No. 3, Aug., 1967), pp. 459–462; published by The American Society of Mammalogists.)

Box 5.2 Standard measurements of seals illustrated in Figure 5.2

1. Standard length
2. Curvilinear length
3. Anterior length of front flipper
4. Anterior length of hind flipper
5. Thickness of blubber
6. Axillary girth
7. Weight

8. Interorbital width
9. Zygomatic width
10. Cranial width
11. Condylobasal length
12. Basilar length of Hensel
13. Rostral width
14. Length of upper postcanine series

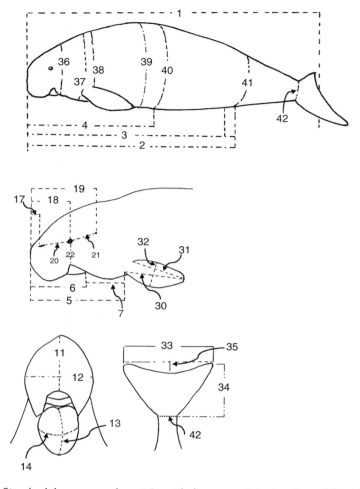

Fig. 5.3 Standard dugong morphometrics with diagrammatic instructions. (After Heinsohn, 1981.)

Box 5.3 Standard measurements of sirenians illustrated in Figure 5.3

1. Total body length
2. Snout to centre of anal opening
3. Snout to centre of genital opening
4. Snout to centre of umbilicus
5. Head plus neck length
6. Head length
7. Neck length
8. Genital opening length

9. Anal opening length
10. Teat length: from thorax to teat
11. Snout length
12. Snout width
13. Chin length
14. Chin width
15. Distance separating centres of tusks at tips

16. Tusks length (from gum to tip)	31. Posterior flipper length
17. Snout to nostrils	32. Flipper width
18. Snout to eyes	33. Fluke width
19. Snout to external auditory meatus	34. Fluke length
20. Eye to nostril	35. Fluke notch depth
21. Eye to external auditory meatus	36. Anterior neck girth
22. Eye to gape	37. Posterior neck girth
23. Gape length	38. Axillary girth
24. Eye height	39. Maximum girth
25. Eye length	40. Umbilicus girth
26. Nostril height	41. Anal girth
27. Nostril width	42. Tail stock girth
28. Total nostril width	43. Dorsal skin thickness
29. External auditory meatus diameter	44. Lateral skin thickness
	45. Ventral skin thickness
30. Anterior flipper length	46. Total body weight

including ventral curvilinear length, dorsal curvilinear length, and back-up standard length (McLaren 1993). Leatherwood *et al.* (1988) listed 51 recommended measurements from cetaceans based on a synthesis of measurements being recorded at that time by six major US biological institutions.

Overall body length is the single most useful measurement to represent the size of an animal, and therefore is needed for a wide variety of physiological, behavioural, and ecological studies. This is usually measured as the straight-line distance from nose to tail for pinnipeds, or tip of the snout to the notch of the tail fluke in cetaceans. Among pinnipeds this measurement is taken with the animal lying on its back, while among cetaceans the carcass is usually dorsal side uppermost (back-up). The easiest way to determine this for smaller animals is to align them with their longitudinal axis at right angles to a vertical surface—a wall or a bulkhead for example—with the tip of their jaw touching that surface, then marking the location of the notch or tip of the tail fluke or tail and, after moving the animal, measuring the distance between the vertical surface and the marked location of the notch of the fluke. For larger beached animals that cannot easily be moved, overall length can be measured by taking the distance between perpendicular markers from the tip of the snout to the notch or tip of the tail. When this is not possible due to the orientation of the animal, the curvilinear length provides an alternative measure. Indeed, this is the way that whale length was usually measured in the early days of whale science.

With increasing attention to live-capture research among pinnipeds, standard length measured with the animal on it's belly (i.e. back upwards) often replaces standard length. Back-up standard length is also often taken from large pinniped species that are difficult to role over onto their back. Curvilinear length is not usually recommended as repeatability of measurements is lower than standard length. Nevertheless, all length measurements of marine mammals (alive or dead) suffer from some degree of imprecision because of differences in the degree of fatness, relaxation of the animal, and rigor mortis.

Typically, body length is not measured from the same set of individuals throughout their lives (i.e. longitudinally), but instead a cross-section of the population is measured within a relatively short period. Exceptions do exist where individuals from a population have been measured repeatedly, for example among bottlenose dolphins in Sarasota (Wells R.S. 1991; Read *et al.* 1993). Thus a sample of lengths, with corresponding ages, enables an average growth rate for the population to be determined, and an age–length key to be constructed if needed. Repeated cross-sectional samples spanning several years or decades may also be used to determine trends in individual average growth rates. Comparisons of the strengths and weaknesses of longitudinal and cross-sectional sampling are given in Table 5.2.

Girth is another metric often recorded, and by convention this is measured just in front of the dorsal fin in cetaceans, where the animal's girth is at a maximum (see Figure 5.1). Other girth measurements may include those at the anus and at the axilla of the pectoral fins or flippers in pinnipeds. Girth measurements are useful in estimating body mass, if this cannot be measured, and in describing condition (see Section 5.2.2 above).

Blubber thickness is now usually measured by convention at the dorsal, lateral, and ventral extremes of the line described by the maximum girth measurement in front of the dorsal fin in cetaceans and at the posterior end of the sternum for pinnipeds. Ultrasonic measurement of blubber thickness can made at multiple sites on live animals (Gales and Burton 1987). Ventral blubber depth over the sternum is another standard measurement. This measurement was also intended to be taken from carcasses. However, it is now possible to routinely take ultrasound measurements from live-captured and released pinnipeds (e.g. Hall A.J. and McConnell 2007) and some cetaceans (Moore M.J. *et al.* 2001). Given differences in the distribution of blubber among species, ventral blubber depth may not provide the best single-measurement index of condition within and between species.

5.3 Age determination

While a great deal can be learned about organisms without knowing the age of individuals, the ability to determine the age of individuals enables a deeper understanding of physiology, behaviour, and population biology than would otherwise be possible. Although, age of marine mammals is typically derived from persistent structures (e.g. teeth and bone) that record seasonal variation in the annual growth cycle, several other methods of determining the age of marine mammals have been developed based on other morphological structures or changes in chemical concentration with age. Increasingly, long-term studies of permanently marked individuals with known birth dates are also providing animals of known age (see Chapter 2).

Table 5.2 *Summary of the strengths and weakness of longitudinal (measurements taken from the same individuals through their lives) versus cross-sectional (measurements taken from different individuals at a single point in time) methods for obtaining information about the morphometrics of populations from individuals.*

Type of study	Strengths	Weaknesses
Longitudinal	• Provides information about underlying processes; • Powerful methods using mark–recapture (see Chapters 2 and 3) for estimating growth rates and condition dynamics; • Useful for building individual-based models; • Recognizes differences between individuals; • Growth rates are estimated directly from observation.	• High long-term effort required to collect the data; • May involve relatively small samples of populations; • Sample bias may be present and difficult to control because of (1) the need to focus on a specific geographical region or (2) the need to focus upon one segment of the population, such as adult females during the breeding season or particular social groups, in order to make the method tractable.
Cross-sectional	• May include large, comparatively unbiased samples of the population; • Ancillary data may be available, including estimates of vital rates; • Simple growth models are applicable that also reflect experience with fisheries-type models.	• Treats all animals as being the same; • Often requires invasive sampling, including culling/harvesting; • Simplified growth/condition models used with these data may provide a poor representation of the growth/ condition dynamics; • Since growth is cumulative, instantaneous measurement of growth using this method can reflect historical conditions rather than current condition, although estimates of body condition are less constrained in this way.

5.3.1 Dental growth layers

Counting annual layers in teeth resulting from the incremental growth of both cementum and dentine is the standard method used to determine the age of most marine mammals (Perrin and Myrick 1980; Klevezal 1996). It was not until about 1950 that the value of dental layers for determining the age of marine mammals was recognized (Scheffer 1950; Laws R. M. 1952; Nishiwaki and Yagi 1953). In addition to endogenous processes that seem to be responsible for the deposition of annual layers, other layers are also deposited within annual layers in dentine and cementum

(Hohn 2002). With respect to age determination, a growth layer group (GLG) is defined as layers (usually two) that occur with a cyclic and predictable pattern. The use of growth layers' groups (GLGs) in teeth as a means of age determination has been validated, with reference to individuals of known age, in northern fur seals (Scheffer 1950; Anas 1970), cape fur seals (Oosthuizen 1997), harp seals (Bowen *et al.* 1983), grey seals (Mansfield 1991), bottlenose dolphins (Hohn *et al.* 1989), harbour porpoises (Grue Nielsen 1972), and Hawaiian spinner dolphins (*Stenella longirostris*; Myrick *et al.* 1984). Bjorge and Donovan (1995) edited a volume on age determination in odontocetes, with particular reference to phocoenids. The use of GLGs has been extended to the use of pinniped incisors which can be easily removed from living animals, enabling age to be determined for animals used in long-term studies (e.g. Arnbom *et al.* 1992; Bernt *et al.* 1996). Similar long-term studies have been conducted on the bottlenose dolphins in Sarasota Bay (Hohn *et al.* 1989).

Teeth grow incrementally and are not remodelled retrospectively, and thus retain any growth layer groups that have been deposited. In the case of dentine, these GLGs are laid down from the outside of the pulp cavity toward the inside (Figure 5.4). Cementum on the other hand is laid down on the external surface of the tooth, but in some species the cementum layer is thin making GLGs difficult to resolve. Thus, dentine is generally used in age estimation of odontocetes, excepting the franciscana, *Pontoporia blainvillei*, and the beaked whales where the cementum is well developed (Hohn 2002). In pinnipeds, sea otters, and polar bears, age is most often derived from cementum, but in some pinniped species dentine GLGs can also be used (Bowen *et al.* 1983; Stewart *et al.* 1996). In species with homodont dentition (i.e. odontocetes), all teeth contain the same layering except for anterior and posterior underdeveloped teeth. Therefore any normally developed tooth will provide good estimates of age. In species with heterodont dentition (e.g. pinnipeds), the canine is usually the tooth of choice for counting dentine GLGs, but cementum GLGs tend to be better developed in postcanines (Hohn 2002). In sirenians, teeth (excepting the tusks of the dugong) are replaced throughout life (so-called 'marching molars') and are therefore not a reliable method of age estimation after the first few years of life (Marsh 1980).

Accurate age estimation from counts of dentine GLGs requires the identification of the neonatal line (see Figure 5.4), which is deposited at birth and therefore represents age zero. In species with limited tooth growth, the neonatal line remains visible throughout life, but in species with continuously growing teeth, such as the walrus, bearded seal, narwhal, sperm whale, and the tusks of dugongs, the neonatal line will at some point erode, which means that only a minimum age can be determined (Hohn 2002).

Preparation and sectioning

Teeth need first to be removed from the jaw. When sampling from live individuals, veterinary support will be required for any species that has not been sampled before. Some use of local anaesthesia and postextraction antibiotic treatment would normally be used. Specialist dental elevators designed to a size that will fit into the tooth socket can be obtained. So far as it is possible, these should be kept sterile and washed in a sterilizing agent between uses. The normal method used to extract a small incisor or a postcanine tooth is to use the dental elevator to free the tooth from the socket with a light twisting motion. The mouth of the animal can be kept open by placing an appropriately sized rod or pole covered in rubber between the jaws. Using this approach tooth extraction can take as little as 30 seconds.

Extracting teeth from the jaws of dead animals can be done in the field with a knife or, more effectively, again with a dental tooth extractor or elevator. Sometimes in smaller cetaceans and pinnipeds, and especially when animals are decomposing and only limited sampling is possible, it may be simplest to remove all or part of the jaw if time does not permit the removal of individual teeth. In cetaceans, usually three to five teeth are removed from each animal, whereas in pinnipeds a canine or postcanine is usually taken, although incisors can be used in some species (e.g. Bernt *et al.* 1996; Blundell and Pendleton 2008; Chambellant and Ferguson 2009). The largest and least worn teeth should be chosen, and by convention this usually involves several teeth from the middle of one or the other side of the lower jaw for odontocetes. Teeth should be cleaned of any remaining soft tissue before sectioning or decalcifying. This can be done mechanically, though some workers use proteolytic enzymes or ultrasonic cleaners.

Teeth can be stored in one of a number of ways. Although storage in 70% ethanol is preferred, teeth can be stored in a hypersaline solution for weeks or months with no evident problems. Teeth can be frozen or stored dry, but this is not a recommended procedure for long-term storage due to problems associated with the teeth cracking as they dry out. Storage in neutral buffered formalin is not recommended as counts of GLG may be affected (Perrin and Myrick 1980), and there may be a subsequent inhibition of staining in those teeth that are subsequently decalcified and thin-sectioned.

Large teeth, such as those from sperm whales and seals, can be prepared either by grinding or by cutting them with a low-speed, diamond-edged saw. Longitudinal sections 50–200 μm (depending on species) thick through the centre of the tooth should bisect the pulp cavity. Grinding to produce a half tooth should continue until the pulp cavity is fully exposed. Etching the half tooth or tooth section with acid (HCl or formic acid) for several hours, and then washing it thoroughly in running water, can help to increase relief in the section and visibility of GLGs. Large teeth can then be read (i.e. analysed) by eye, smaller ones will need magnification under a binocular microscope. The use of polarized (Figure 5.5), transmitted, and reflected light provides several different options for the person counting the growth layers to adapt the observation conditions in order to obtain the best viewing condition for an individual tooth.

Small teeth, such as those from most dolphins and some pinnipeds, are usually decalcified before thin-sectioning with a microtome followed by staining to optimize the resolution of the growth layer groups. Stained, thin sections provide better resolution of GLGs and often reveal detail that would otherwise have been missed (Perrin and Myrick 1980; Stewart *et al.* 1996; Hohn and Fernandez 1999).

Several proprietary rapid decalcifiers as well as slower decalcifiers are available. Individual workers have their own preferences. Rapid decalcifiers include products such as RDO Rapid Decalcifier (by APEX Engineering, IL, USA), RDC (by Cellpath plc, UK), and Shandon TBD-1 decalcifier (Thermo Electron Corporation, Basingstoke, Hampshire, UK). Slower decalcifiers include RDF (also by Cellpath plc, UK) and Formical-4, both of which are based on formic acid rather than hydrochloric acid, and EDTA (ethylene-diamine-tetra-acetic acid). The use of nitric acid, a rapidly decalcifying strong acid, can lead to cell damage. Proprietary decalcifiers include various stabilizers to protect proteins and membranes, which maintain staining characteristics. Decalcifying may take several hours to several weeks depending on the decalcifier used and the size of the tooth.

Some trial and error will be required with the particular teeth and decalcifying agent, to see how well it performs and to judge how long teeth need to be left to decalcify. Initial trials should therefore use teeth from one or more animals for which there are numerous samples available. The endpoint of the decalcification process can be determined chemically using an endpoint test. This requires the use of ammonium oxylate and ammonium hydroxide. A 5 mL sample of the decalcifying reagent is removed from the specimen container and then 5 mL of 5% ammonium hydroxide added. The solution is left for ten minutes, and if a precipitate forms then decalcification is incomplete and the tooth should remain in the decalcifying agent. Constant agitation can speed up the decalcification process. Once the endpoint is reached the tooth must be washed thoroughly in running water for up to 48 h (some authors recommend shorter washing times, but the aim is to ensure thorough washing). A weak phenol solution after washing can also be used.

There is some dispute as to which is the best plane for sectioning teeth. Longitudinal sections may be oriented so that the cut is either across or with the line of the jaw, known as the porpoise or the dolphin plane or cut, respectively. This is something that can be tried and tested in both orientations for any species being studied. Teeth should be sectioned soon after decalcifying, but can be stored in distilled water for a few days. Sectioning can be done using a cryostat or freezing-stage microtome, but more precise handling and thinner sections are possible if paraffin wax-embedding is used to section the tooth. Sections of teeth can be 5–30 μm. Luque *et al.* (2009) compared age estimates obtained from paraffin-embedded sections with cryostat sections and found no significant difference. As there are various techniques for sectioning and then staining sections, it is often simplest to get the help of an experienced histologist and use their preferred method. It is important that sections are cut through the pulp cavity of the tooth, and here the importance of choosing the largest and straightest teeth become clear. Twisted, misshapen, or worn teeth do not yield clear sections.

Staining is another area where there is no overall agreement on the best method, and to some extent this is a matter of personal choice. Frequently used stains include Toluidine Blue, Eosin, Methylene Blue, Harris' haematoxylin, Ehrlich's haematoxylin and Mayer's haematoxylin. Staining times must be found by trial and error, and depend on the thickness of the section and the temperature and the age of the stain. After staining, sections may be washed and blued using ammonia vapour or Scott's tap water (a weak ammonia solution). After further washing the sections are mounted. Slides are dried and can then be read using a binocular microscope.

Reading tooth sections

The basic principle of reading tooth sections is simple. As with trees, alternate dark and light bands within the dentine denote a GLG (see Fig 5.4). But in many cases these bands are indistinct and may also contain other finer scale growth layers ('ancillary layers'). Reading teeth is therefore something that needs to be learned from other people with experience.

Even though age readers attempt to follow consistent and objective rules for identifying GLGs, there is a subjective component to age determination with the

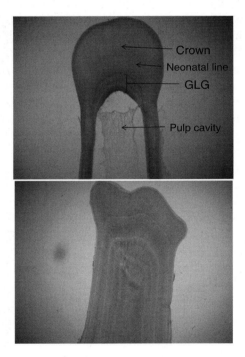

Fig. 5.4 Decalcified and stained thin sections of the teeth of a harbour porpoise. As animals become older, the pulp cavity becomes occluded as dentine is laid down from the inside of the cavity. GLGs become smaller towards the centre of the tooth and more difficult to count. (Photo courtesy of Simon Northridge.)

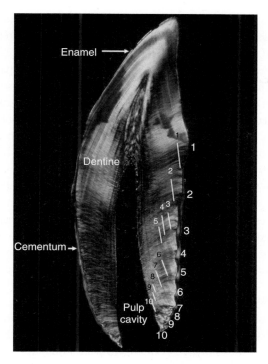

Fig. 5.5 Longitudinal, un-decalcified, thin section of the upper canine of a male Antarctic fur seal (*Arctocephalus gazelle*) mounted in DPX medium on a glass slide and viewed using transmitted polarized light. White lines have been used to show each of the GLG transitions and each of these are numbered by year. The numbers down the right indicate the external ridges in the root of the tooth of these species which can be used to obtain a reasonably accurate age without having to section the tooth. Close inspection of the substructure in the GLGs of this species shows numerous sub-lines which when examine with scanning electron microscopy average 12 per GLG, suggesting that the substructure may be related to the lunar cycle. See Boyd and Roberts (1993) and Hanson *et al.* (2009) for examples of the application of GLG measurements. (Photo courtesy of Ian Boyd.)

result that errors do occur. These errors can lead to bias in the estimation of age-specific birth and death rates or add variance in estimates (Doubleday and Bowen 1980). The example illustrated in Figure 5.5 is about the best that can be achieved for any species. Considerable effort has been directed to ensure consistency in age determination and to correct for bias resulting from inconsistent readings (Doubleday and Bowen 1980; Perrin and Myrick 1980; Lawson *et al.* 1992; Campana 2001). Nevertheless, it is important to remember that consistency (precision) is not the same thing as accuracy. Accuracy can only be established against teeth of known age. Campana *et al.* (1995) examined some graphical and statistical methods for ascertaining the consistency of age determination and found that the coefficient of variation was a robust measure of precision, but the commonly used percent agreement statistic was not. Age-bias plots, which show the age determined by one reader plotted against the results from another independent reader, is a useful graphical method of detecting bias among readers. W.G. Clark (2004) provides a non-parametric method to estimate age misclassifications from paired readings.

Compiling a reference collection of known-age teeth or consensus-derived ages of teeth should be an ongoing element of age determination in any research project, both to ensure consistency over time and readers and to serve as a tool for training new readers (Campana 2001). To the extent that the reference collection accurately reflects the absolute age of individuals, it serves as a quality control for both accuracy and precision of age estimation. Campana (2001) recommended for fish that a reference collection of 200–500 samples was needed. The size of the reference collection in marine mammals will be more limited by the availability of known-age material. The sample sizes suggested by Campana (2001) are largely required to minimize the effects of non-independence of samples caused by observers remembering particular samples and the past results. Careful use of a reference collection with this problem in mind will allow the calibration of observers using smaller samples.

5.3.2 Other age-structured material

Teeth cannot be used to estimate the age of baleen whales, nor in manatees whose teeth are worn down and replaced during life, so other tissues have been investigated instead. Incremental layers are known in bone, baleen, and mysticete earplugs.

Earplugs are a secretion in the external auditory meatus of the ear in baleen whales, composed of alternating layers of light and dark keratinized epithelial cells in a crystalline cholesterol matrix. Earplugs were established as a means of estimating baleen whale ages by Laws and Purves (1956), and are a good recording structure as layers appear to be continuously deposited and do not appear to be subject to resorption or remodelling (Klevezal 1996). Earplugs are best collected as soon as possible after death, and the entire plug must be collected right down to the proximal end of the auditory canal. This is a difficult task (Anon 1974). After fixing in 10% neutral formalin, Lockyer (1984) suggested storage in fresh neutral 5–10% formalin. The entire core of the earplug needs to be exposed, by cutting or grinding along its longitudinal axis down to the mid-line, from the fetal zone to the most recently formed zone at the germinative epithelium. Some workers have cut and stained earplug sections to count growth layers of the plug core, while others have used their naked eye or a low-powered binocular microscope. Earplugs have been used to determine the age of individuals in several populations (Anon. 1974; Hohn 2002) with varying degrees of success (reviewed by Lockyer 1984). The use of earplugs in age determination is difficult, in part because the earplugs themselves are not easy to collect and they are fragile, but also because they are not always present. Furthermore, the pattern of annual deposition of layers has been disputed (Klevezal 1996), although the technique was widely used to determine the ages of Antarctic baleen whales (Chittleborough 1959; Lockyer 1972) and, where earplugs can be collected, is still the most widely used method of estimating baleen whale age.

Baleen plates have also been used to estimate mysticete whale ages, and were often used in the past (Ruud 1940; Tomilin 1945; van Utrecht 1965). However, baleen plates tend to abrade with age, which limits the accumulation of layers and will bias any estimation of age. For this reason baleen plates are no longer widely used in age estimation, although they can be useful for ageing younger animals.

Since there is active turnover of bone tissue, bone does not generally provide layers of incremental deposition that are useful for ageing. Nevertheless, Marmontel *et al.* (1996) used GLGs observed in histological preparations of the periotic bone (the bone surrounding the ear) to estimate the age of manatees. These estimates were validated against individuals of known-age, minimum age, and those whose teeth had been marked previously with oxytetracycline. Oxytetracycline is incorporated quickly into actively growing mineralized tissue and fluoresces under ultraviolet light, thus acting as a time marker. Several investigators have also attempted to use changes in bone mineral density to estimate the age of small odontocetes (e.g. Butti *et al.* 2007), but this method appears to provide little information beyond that contained in body length and the predicted ages are rather imprecise. Tympanic bullae, which are among the densest of all bones, have been used to estimate the age of some baleen whales. However, there have been inconsistent results among different readings of the same bullae sections, poor agreement with other age estimation methods for the same animals, and it is thought that the poor precision and potential biases associated with the use of tympanic bullae make them a poor resource for indicating ages in baleen whales (Olsen, 2002).

5.3.3 Chemical methods

The age of some marine mammal species has been estimated based on changes in specific amino acids, fatty acids, and elemental isotope ratios. Bada *et al.* (1980) pioneered the use of an amino acid, aspartic acid, found in teeth and eye lenses to age cetaceans. The method takes advantage of the fact that aspartic acid can exist as two forms, referred to as D- and L-optical isomers. Living organisms produce only the L-isomer, but in metabolically inert tissues such as teeth and eye lens, racemization occurs whereby the L-isomer is converted into the D-isomer until an equilibrium ratio of 1.0 is achieved. Given that the rate of racemization is known, the D/L ratio of aspartic acid can be used to estimate age. D/L ratios are determined using ion-exchange chromatography or more commonly by high-performance liquid chromatography (HPLC). This method has been used to estimate the ages of bowhead whales (George *et al.* 1999) and minke whales (Olsen and Sunde 2002). Both these studies provide a useful discussion of potential sources of bias and analytical challenges.

Regular variation in elemental isotope enrichment has been used to estimate age. Schell *et al.* (1989) and Lubetkin *et al.* (2008) used the seasonal variation in $\delta^{13}C$ from baleen, muscle, and fat samples to infer the ages of bowhead whales up to about 20 years old. After this, baleen wear limits the reliability of baleen-based ageing techniques. The seasonal variation in $\delta^{13}C$ reflects tissue responses to geographic variation in the $\delta^{13}C$ of prey consumed during the annual migration. $\delta^{13}C$ is measured using elemental analysis-isotope ratio mass spectrometry (ESC-IRMS). These analyses can be performed in appropriately equipped university and commercial laboratories (see Chapter 9). Heavier isotopes have also been investigated as a means of age determination. Kastelle *et al.* (2003) used the disequilibrium between ^{210}Pb and ^{226}Ra in tympanic bullae of grey whales and bowhead whales to estimate age. The results of this preliminary study suggested that further study of grey whales could be useful, but that the method would seem inappropriate for longer lived species such as bowhead whales.

Bomb radiocarbon dating has recently been used to resolve the question of the number of GLGs associated with the annual growth of beluga whale teeth. Stewart *et al.* (2006) used ^{14}C from the fallout from atmospheric nuclear tests in the 1950s and 1960s as a dated chemical marker to determine the age of a small sample of archived teeth. ^{14}C from individual, or several GLGs, was assayed with accelerator mass spectrometry. This study provided strong evidence that each year of growth is represented by only one GLG, but more generally it points to the value of this approach in validating the ages of other marine mammal species.

Finally, in several species it appears that the concentration of certain non-dietary fatty acids in blubber is positively correlated with age, suggesting another approach to age estimation. Koopman *et al.* (1996) found that the concentration of isovaleric acid in the outer layer of the blubber of male harbour porpoises was linearly related to age over the age range 0–16 years. Several groups of fatty acids were also linearly related to age, but did not account for as much of the observed variation as did isovaleric acid. Recently, Herman *et al.* (2008) used endogenous fatty acids and fatty acid ratios in the outer blubber layer to estimate the ages of free-ranging killer whales. The authors developed a multi-linear bivariate predictive model in which age accounted for 89% of the variation in fatty acid content of known-aged whales. This model appears to be independent of sex and ecotype, thus providing a way to estimate the ages of individuals from less well-studied populations and also potentially from blubber samples taken from live individuals. When combined with methods of photo-identification (see Chapter 2), this could be used to examine the long-term, age-related dynamics of social groups and populations.

5.4 Growth rates

To determine growth rates it is necessary to describe the relationships between the measured dimensions of the animals (usually body length, sometimes actual or estimated body mass) and the estimated age. This can be done graphically as a scatterplot of lengths against estimated ages, but it is generally more useful to describe the growth rate—that is the growth rate of an average individual within a population—by means of an expression with relatively few estimated parameters.

Several standard models of growth can be applied to marine mammals that achieve this objective (McLaren 1993; Winship *et al.* 2001). These are generally fitted using a non-linear least-squares method. They usually represent growth as an exponential process that is ultimately limited by some form of upper size. Four commonly used models are the logistic, von Bertalanffy, Gompertz, and Richards' models. All are standard models that are available in most up-to-date software programs or that can be programmed into the software quite easily. We do not intend to describe them in detail here. However, in general, the more parameters within the model then the better fit will be obtained. There are different versions of these models that contain different numbers of parameters (e.g. depending upon whether the model is forced through the origin), so it is wise to be aware that additional parameters will provide better fits, even though they may not provide more biological insight.

The Gompertz and Richards' models often provide a slightly better fit to marine mammal growth data than other options, but, in contrast to the von Bertalanffy model, they provide a purely empirical description of the data. However, the better fit for marine mammals may be because the Gompertz and Richards' models can better accommodate two phases of growth, a feature that is often most obvious in male pinnipeds that have high sexual dimorphism in body size (McLaren 1993). In this case, rapid initial growth is followed by a slowing of growth until the animals become sexually mature when there can be a further acceleration of growth. In contrast, the parameters of the von Bertalanffy model are derived from the exponential scaling of the balance between catabolic and anabolic processes, and so have some foundation in the physiological processes underlying growth. It is very much up to individual researchers to choose the most appropriate type of growth model.

It may also be useful to examine the deviations of individuals from the predicted size-at-age as a measure of the growth performance of that individual relative to the rest of the population (Trites and Bigg 1992). Fitting growth models also allows a statistically valid approach to comparing the growth performance of different populations or of the same population through time. For example, Calkins *et al.* (1998) compared the growth of Steller sea lions estimated from two population samples, one from the 1970s and the other from the 1980s and, based on the statistical fit of growth models, concluded that there had been a change in the overall growth of individuals within the population.

5.5 Conclusions

Measurement of size- and age-related features of marine mammals provides fundamental information that, if appropriately analysed with due attention to potential random errors and biases, can be applied to a broad range of problems. Prediction of one morphological character from another is a common problem but many morphometric characters can also be used to describe the state of an individual, and the combination of these metrics provides an overall measure of its state. How these metrics change, both relative to each other and in relation to environmental factors, can underpin predictions of how individuals and, ultimately, populations will respond to environmental change. This has some clear applications within the context of conservation and the assessment of climate change on marine mammals. Morphometrics can also be used within the context of the comparative approach to ecology and physiology. By comparing the morphometric characteristics of populations, useful information can sometimes be derived about the effects of the contrasting ecology of populations, and can also sometimes provide useful insights into population or stock structure.

6
Vital rates and population dynamics

Jason D. Baker, Andrew Westgate, and Tomo Eguchi

6.1 Introduction

Population dynamics is the study of changes in the number of individuals in populations over time. Chapter 3 describes methods for estimating the abundance of marine mammal populations. Superficial characterization of population dynamics can be obtained by making a series of population estimates over time. Marine mammal ecologists and conservation biologists, however, are typically interested in the underlying issues regarding the mechanics and drivers of trends in population size. The change in abundance during a given period is, quite simply, the sum of additions to the population (births and immigrants) minus losses from the population (deaths and emigrants). Setting aside immigration and emigration for now, we will focus on the *vital rates*—reproductive rate and survival.

Countless environmental factors (including, but by no means limited to, climate–ocean variability, changes in prey, levels of predation, fisheries by-catch, disease, competition, and habitat alteration) affect the chances that individual animals will survival or give birth in a given year. Vital rates are the aggregate manifestation of these individual-level responses to the environment summed over all animals in a population. Marine mammals are large, long-lived vertebrates and their vital rates vary with age and sex. Therefore, we usually attempt to ascertain sex- and age-specific, or at least stage-specific (i.e. calf/pup, immature, adult) vital rates. Moreover, information on temporal and spatial variability in vital rates is critical for understanding both the intrinsic (age-structure, density dependence) and extrinsic environmental drivers of variable vital rates. The latter is a rapidly advancing area of interdisciplinary research where links are being established between marine mammal demography, climate–ocean variability, foraging behaviour, and health parameters. Characterizing these connections is critical at this time when anthropogenic alteration of marine mammal habitats, through, for example, global climate change, are predicted to escalate (e.g. Ragen *et al.* 2008).

Characterization of vital rates is most informative when integrated with other information (such as population age/sex structure) in an appropriately structured demographic model. Even when little data exist, a model can be used heuristically to identify critical data gaps. Given at least some demographic data, modelling may allow a more holistic approach to diagnosing a decline. For example, Holmes *et al.* (2007) inferred temporal changes in both reproductive rates and survival based upon trends in the counts and age structure of Steller sea lions (*Eumetopias jubatus*). A model populated with more comprehensive vital rates estimates affords the opportunity to predict future trends and evaluate management options. Taking this approach to analyse the critically endangered North Atlantic right whale population, Fujiwara and Caswell (2001) concluded that if the deaths of just two adult females could be avoided each year, the population would cease to decline. In this chapter we present methods for estimating survival and reproductive rates in marine mammals, and provide an overview of approaches to marine mammal population modelling.

6.2 Reproductive rate

Reproduction can be studied using either longitudinal or cross-sectional approaches, both of which require large sample sizes. Longitudinal studies involve collecting information on individual reproductive histories, ideally throughout their entire lives. Compared to cetaceans, pinniped populations have been more extensively studied using longitudinal approaches because individuals can more readily be live-captured for marking (Lunn and Boyd 1991; Laws 1993; Le Boeuf and Laws 1994; Bowen *et al.* 2006). Longitudinal studies of most cetacean populations have relied on re-sighting histories of uniquely marked individuals (Hammond *et al.* 1990; Barlow and Clapham 1997; Mann 2000, Hamilton *et al.* 2007), although temporary capture to identify uniquely marked individuals has also been used (Wells *et al.* 1987; Wells 1991). Cross-sectional studies rely on observations collected once from many individuals in the population. Although cross-sectional data can be used to calculate a variety of reproductive parameters, they cannot be used to address individual annual variation or examine lifetime reproductive output. Reproductive parameters from many cetacean populations have been determined using cross-sectional data obtained from directed takes (Marsh and Kasuya 1984) and fisheries by-catch (Read *et al.* 2006). Pinniped populations have also been studied using this approach (Bowen *et al.* 1981; Stirling 2005).

6.2.1 Description of parameters

Fecundity

Fecundity is usually defined as the average number of female offspring born per female per unit time in either the population (gross) or in a given female age class (age-specific) (Gotelli 2001). However, the term fecundity is sometimes used differently depending upon the context. For example, in population models dealing with males and females, fecundity may be used to refer to total offspring per female.[1] Fecundity should not be confused with reproductive rate, which can be defined using natality (the proportion of females giving birth in a defined period), pregnancy rate (the proportion of pregnant females in a sample), or ovulation rate (the proportion of a sample females showing evidence of ovulation) (Perrin and Reilly 1984; Huber *et al.* 1991). Reproductive rates can also be reported for different age groups (Lunn *et al.* 1994). Next we consider the components of fecundity, i.e. pregnancy and birth rate.

Pregnancy and birth rate

The reproductive cycle of females can be divided into seven distinct events: ovulation, conception, implantation, gestation, birth, lactation, and weaning (Boyd *et al.* 1999a). Rates can be calculated for any of these stages but, because of stage-specific failures, these rates will differ. Although all are important from a life history perspective, the most important to the present discussion are pregnancy and birth rates.

Gross pregnancy rates (GPR) combine ovulation and implantation stages, and measure the proportion of pregnant females in a sample of the population (Read 1990; Laws, R. M. and Sinha 1993):

GPR = number of pregnant females/number of sexually mature females

Gross birth rate (GBR), or natality, relates specifically to parturition (process of giving birth) and specifies the proportion of sexually mature females that actually give birth (Croxall and Hiby 1983):

GBR = number of parturient females/number of sexually mature females

Unlike fecundity, birth rate quantifies neonates of both sexes. Although the birth rate will ultimately be used in most demographic analyses, determining the pregnancy rate is also important because significant differences in these parameters can reveal that there may be environmental constraints on reproduction. Age-specific pregnancy and birth rates can also be estimated by calculating the individual rates for each age class i (Croxall and Hiby 1983; Huber *et al.* 1991):

BR_i = number of parturient females at age i/number of females at age i

Age-specific analysis describes the lifetime pattern of reproduction, which is required for matrix population models (see Section 6.4.2), whereas a gross analysis only provides the average birth rate for the population. Age-specific data are also

---

[1] The fecundity parameter in Leslie matrix population models may be a combination of birth rate and survival of females or neonates (Caswell 1991).

especially useful for comparing reproductive rates among populations where age structure differs (Harting *et al.* 2007).

Although we have referred mainly to age-specific analyses in this chapter, we only use age as an example of any state variable that could be used to subdivide a population into logical units. Caswell (2001) demonstrates stage-based approaches to population modelling. Age-based approaches are one special case of stage-based approaches. An alternative to using age is to use an indicator of body size. If energy-based approaches to understanding population dynamics are to be used then it may be more productive to structure population models by size rather than age, or at least to use both as state-based vectors.

Age at sexual maturity

This varies among species, as well as among and within populations. Variation in age at sexual maturity (ASM) influences reproductive output both for the individual and the population (Stearns 1992). Age at sexual maturity may also be density-dependent (Bowen *et al.* 1981; Read and Gaskin 1990). Therefore, ASM should not be viewed as a static parameter, especially in exploited populations or those living in unstable environments (Roff 1992).

For females, a sexually mature individual is defined as one which has ovulated at least once and that may have a *corpus luteum* or *corpus albicans* (see 'Ovarian and uterine analysis' in Section 6.2.2) in either ovary [see Harrison (1969) for a description of cetacean reproductive anatomy and ovulation cycle and Boyd *et al.* (1999a) for pinnipeds and sirenians]. For males, sexual maturity has been less precisely defined, but generally is assumed to be an individual having evidence of spermatogenesis in the central portion of the testis and that may have sperm present in the epididymis (Perrin and Reilly 1984; Laws, R.M. and Sinha 1993). These definitions may not always hold in species, such as many pinnipeds, that have highly seasonal reproductive cycles. Male reproduction is usually not considered in most demographic analyses and population models. ASM and age at first reproduction may differ, with the latter referring to the age at which primiparity (first birth event for a female) occurs.

Age at sexual maturity can be estimated from data obtained by longitudinal or cross-sectional sampling (Bowen *et al.* 2006; Hadley *et al.* 2006; Westgate and Read 2006; Danil and Chivers 2007). There are several analytical approaches to determine ASM in populations. The first, called the sum of fraction immature algorithm, is a non-parametric technique that was developed by DeMaster (1978) and subsequently modified by Hohn (1989). ASM is estimated as the sum of the fraction of immature individuals in each indeterminate age class (age classes in which both immature and mature individuals occur) added to the age of the first indeterminate age class. Following Hohn (1989):

$$ASM = j + \sum_{i=j}^{k} p_i x_i$$

where j is the age of the youngest mature animal, k is the age of the oldest immature animal, p_i is the proportion of immature animals in age class i, and x_i is the number of age classes combined in age class i to achieve a sample size of at least 2. The variance can be estimated using:

$$Variance\ (ASM) = \frac{\sum p_i q_i x_i}{N_i - 1}$$

where q_i is the proportion of mature individuals in age class i and N_i is the sample size for age class i.

Another common technique is to estimate when 50% of the population has matured. This is typically done using logistic regression analysis (Caswell et al. 1998; Danil and Chivers 2007). Confidence intervals can be constructed using bootstrap techniques (Caswell et al. 1998; Danil and Chivers 2007).

An important distinction between ASM and age at physical maturity (APM) is that the latter is defined as the age at which an individual reaches its asymptotic size (either standard length or body mass). There can be a considerable lag between ASM and APM in some marine mammal species, especially amongst males (e.g. Olesiuk et al. 1990a; Read and Hohn 1995).

Senescence

Reproductive senescence describes a reduction in fecundity at older ages (Promislow 1991) and is poorly understood in most marine mammal species. This is partly because these long-lived vertebrates have not been studied sufficiently long to collect the necessary data, or when they have, the sample sizes in the oldest age classes are too small to be informative. Also compounding the difficulties in documenting senescence is the possibility that females with lower lifetime productivity may live longer, thereby falsely indicating senescence. Still, evidence of senescence has been reported in grey seals (Bowen et al. 2006), Hawaiian monk seals (Harting et al. 2007), subantarctic fur seals (Lunn et al. 1994), and northern elephant seals (Sydeman et al. 1991). It has also been inferred in cross-sectional studies on odontocetes by documenting decreases in the pregnancy rate in increasingly older age classes (see Perrin et al. 1976, 1977). It is difficult to detect senescence because relatively few females live to the advanced ages when senescence is expressed. This also means it will usually have little impact on population dynamics so, from this narrow perspective, it may not be a particularly important demographic feature.

6.2.2 Techniques

Sighting histories of marked individuals

Long-term studies based on re-sighting marked individuals have provided the most comprehensive estimates of reproductive parameters (Olesiuk et al. 1990a;

Sydeman *et al.* 1991; Barlow and Clapham 1997; Pomeroy *et al.* 1999; Mann 2000; Wells 2003; Kendall *et al.* 2004; Bowen *et al.* 2006; Hadley *et al.* 2006; Harting *et al.* 2007). Ideally, marked animals will be of known age, either because they were first tagged (see Chapter 2) as neonates or by ageing individuals using extracted teeth (Wells 1991; Arnbom *et al.* 1992).

Documenting reproduction is often straightforward; observing a female giving birth or nursing a neonate is sufficient to confirm parturition. However, lack of such observations does not necessarily indicate a female did not reproduce. To obtain reliable estimates of natality, for example, field protocols should be designed to maximize the probability of detecting births. This works well in colonial pinnipeds with highly seasonal parturition. Alternatively, a way must be devised to estimate the probability of re-sighting reproductive and non-reproductive females (e.g. capture–recapture methods which are discussed in Section 6.3 on survival estimation).

Census-based studies

In any situation where the abundance of sexually mature females and young of the year can be estimated, it is possible to obtain an estimate of gross (as opposed to age-specific) reproductive rates (e.g. Johanos *et al.* 1994). This approach has been applied at seal colonies where marked mothers are repeatedly documented with marked offspring. It has also been used in free-swimming cetacean populations where calves are identified based on their swimming position (Perryman *et al.* 2002). The latter application can give biased results if mothers and calves are not associating as predicted, or if there is social segregation of the population where the presence of calves co-varies with the underlying social process.

Determination of pregnancy

Marine mammals have streamlined bodies which can make it difficult or impossible to determine if a female is pregnant by visual inspection, therefore both direct and indirect methods are used to determine pregnancy status.

Ovarian and uterine analysis

The most reliable way to determine pregnancy is to conduct a direct macroscopic examination of the ovaries and uterus after they are removed during a routine necropsy (but see the Ultrasound section, p. 126) (Perrin and Donovan 1984; Laws, R. M. and Sinha 1993). Each uterine horn is dissected longitudinally along its mid-line, starting from the cervix. This should reveal the presence of an embryo or fetus. In both pinnipeds and odontocetes, the early stages of pregnancy are difficult to observe with the naked eye, so the absence of a fetus does not rule out that the individual was pregnant. The dimensions of the uterine horns increase during pregnancy and can therefore be used to support the assignment of a reproductive status to an individual. It is also important to examine the ovaries. These are examined grossly for the presence of a corpus luteum that indicates pregnancy (Perrin and Donovan 1984). As pregnancy progresses the corpus luteum enlarges, approaching the mass of the rest of the ovary. When a corpus luteum regresses there is extensive fibrosis and hyalinization (Harrison *et al.* 1972) and these scars, called corpora albicantia, are believed to remain for life in some species (some cetaceans). Counts of corpora albicantia from individuals may be used in some circumstances to calculate reproductive histories, but it is important to document the life-history (e.g. Boyd 1984) of corpora lutea and corpora albicantia before using this ovarian anatomy to derive information about reproductive rates.

Hormone analysis

The gonadal steroid hormones, mainly progesterone and oestradiol, may be used for determining pregnancy. Both blood and blubber samples (St. Aubin 2001) have been used, but much care is required with the assays for these hormones because of the potential for cross-reaction with irrelevant compounds in the tissue samples. Considerable effort is required to validate the assays being carried out. Remote blubber biopsies can be obtained from wild individuals for hormone analysis (Kellar *et al.* 2006), which means that, assuming assays can be appropriately validated, this is a promising technique for sampling large numbers of individuals with minimal intervention (Krützen *et al.* 2002). Pregnant animals tend to have elevated levels of hormones (Mansour *et al.* (2002), relative to non-pregnant or immature animals, but covariates are sometimes required to fully discriminate pregnant from non-pregnant individuals (Gardiner *et al.* 1996). This approach has several limitations. For example, hormone levels may not be significantly elevated during the early stages of pregnancy, resulting in a false-negative diagnosis; and there may be overlap between the normal range of hormone concentrations in pregnant and non-pregnant individuals. Consequently, individual variation can make establishing a threshold level to indicated pregnancy difficult, and therefore species-specific validation studies are necessary before wild populations can be assessed. Reproductive status may also be determined by quantifying hormone metabolites in faecal samples (Rolland *et al.* 2005). Hormones levels are typically quantified using standard radio- and enzyme-immuno-assay techniques.

In general, hormone determinations of pregnancy based upon gonadal steroids do present significant challenges, and are probably most useful when sampling from individuals sequentially to examine the relative level of these hormones. Interestingly, in common with humans but with very few other mammals, pinnipeds (and possibly also cetaceans) produce a chorionic gonadotrophin (Hobson and Boyd 1984). This means that it may be possible to test for pregnancy using a specific test like the human pregnancy test. However, the protein structure of the gonadotrophins in marine mammals is sufficiently different from the human to make human pregnancy test kits ineffective for marine mammals.

Ultrasound

Ultrasonography (diagnostic B-mode) has been used on both seals (Adam *et al.* 2007) and dolphins (Stone 1990; Brook *et al.* 2001) to visualize the presence of a developing fetus as well as follicular and corpus luteum development. Ultrasound can be conducted on cetaceans directly, but seals require shaving (fur compromises image quality) or transrectal and transvaginal approaches (Adams *et al.* 2007). While this technique has not been widely used in field situations, it is commonly used in captive and husbandry facilities (Brook *et al.* 2001). Ultrasound imaging is becoming more accessible to researchers with the development of portable diagnostic units (Adam *et al.* 2007). Its main limitations are that it requires trained expertise to interpret the sonograms and the animal must be captured.

6.3 Survival rate

As with reproductive parameters, survival can be estimated from cross-sectional and longitudinal observations. The former involves inferring survival from a sample of the population age structure (i.e. number or proportion of individuals in a population by age), whereas the latter involves observing individually marked animals over time. Because marine mammals are long-lived, we are usually interested in *annual* survival rates, rather than shorter intervals. Likewise, population models that incorporate survival rates typically operate on an annual time increment so that survival from one year to the next is the parameter of interest. However, in principle, methods using marked individuals can be used to evaluate survival over any time interval desired.

6.3.1 Cross-sectional age structure analysis

A traditional method for estimating the survival of wildlife populations involves acquiring a sample of animals and evaluating their age composition at a given point in time. The sample might come from a directed research effort, but it could also derive from commercial, subsistence, or recreational hunting.

Age structure analysis infers survival rates by examining the number of individuals in adjacent age classes. For example, a sample with 50 × 5-year-olds and 40 × 6-year-olds would yield an estimated 80% survival rate from age 5 to 6 (that is 40/50 = 0.8). Calculating this ratio for every age available results in a set of age-specific rates. While intuitive and simple, age structure approaches are fraught with problems related to strong and rarely valid assumptions. The first issue is that the age structure of the sample must be representative of the age composition of the population of interest, which is equivalent to assuming that all individuals in the population have the same probability of being selected. In cases of commercial or recreational hunting, it is unlikely that animals taken will be representative of the total population, either due to hunter selectivity or differences in the accessibility of different age groups (Roff and Bowen 1986). However, even if the sample were representative, another issue remains. For an observed pattern of dwindling numbers with increasing age to truly reflect survival rates, additional strong assumptions are required. First, the population must have a stable age distribution, meaning that the proportion of animals in each age group does not change from year to year. Also, total population abundance and survival rates must be constant over time. Why are these conditions required? Consider what happens, in the example above, if survival were to vary. Our estimate of survival from age 5 to 6 years assumed that the proportion of 5-year-olds in the sample we obtained remained constant from the previous year (when our current 6-year-olds were 5-year-olds). That is, had we taken an equal number of animals the previous year, we would have also counted $50 \times$ 5-year-olds. However, what if survival was particularly poor in the past year and there had actually been $80 \times$ 5-year-olds? Then the real survival rate would have been 50% $(40/80 = 0.5)$ rather than the 80% we estimated. Likewise, let's consider what happens if reproduction (and consequently abundance) fluctuated. If there is a bumper crop of offspring one year followed by dismal production the next, one might find a case where there are more yearlings than neonates. That would lead to an impossible estimate of a survival greater than 100%.

It is rarely the case that any wildlife population has a stable age distribution and stationary abundance. Rather, environmental conditions vary, and predator and prey populations fluctuate, which buffets vital rates and keeps age structures dynamic. In fact, it is often precisely because we suspect a change in vital rates has led to a population decline that we are interested in estimating these parameters. If that is the case then using age structure to estimate survival is perilous indeed.

While less than ideal, cross-sectional age structure methods may yield the best available estimates of survival. Realities of funding and logistics often dictate that optimal approaches to estimating vital rates cannot be achieved. If teeth from hunted seals or from dolphins incidentally caught in a fishery are available, these should be used to their fullest potential (e.g. Ferguson *et al.* 2005). A thorough treatment of age structure approaches to estimating survival is presented in Skalski *et al.* (2005).

6.3.2 Capture–recapture

Approaches to estimating survival using marked animals include 'known fate' models (where all individuals marked can be found again and consequently their fate is known), so-called 'band recovery' models where marked animals are re-identified only when they are dead (usually associated with hunting/harvesting), and finally 'live recapture' models whereby, as the name implies, live animals are re-sighted or recaptured. Models which combine both live and dead encounters also exist. Here, we will focus primarily on live recapture, or Cormack-Jolly-Seber (CJS) survival estimation (Cormack 1964; Jolly 1965; and Seber 1965) and its broader application (Lebreton *et al.* 1992), with some discussion of other model types.

Studies that involve observing individually identifiable animals over time are variously referred to as 'mark–recapture', 'capture–recapture', or 'capture–mark–recapture' and additional information about these models applied to estimating abundance is given in Chapter 3. Because many marine mammal species have natural identifying features (dorsal fin shape, fluke colour patterns, body scars, see Fig. 6.1), it is not always necessary to apply a 'mark' in order to recognize unique animals. For that reason, we use the general term 'capture–recapture' (CR) here. Further, the term 'capture' (or 'recapture') connotes any occasion when an individual is identified (or re-identified), which may involve actual physical capture, visually reading an applied tag from a distance, or simply photographing an animal and subsequently matching the photograph to a known identity.

The basics

To demonstrate the concept of CR methods, consider the simplest possible case. We capture, tag, and release 100 seals. A year later we come back, diligently scan the beach and see 35 of our tagged animals. Thus, we conclude the 1-year survival rate was 35% ($35/100 = 0.35$). The problem with this simplistic approach is that, try as we might, we are usually unable to find all the seals that have survived. That is, the *probability of capture* is less than 1, and we have missed some seals that were alive. The key is to estimate how many seals were alive but *not* seen. The objective of CJS analysis, then, is fundamentally to estimate both the probability of survival and probability of capture, given that an animal is alive.

Now let's expand our simple example. We tag our 100 seals and subsequently conduct re-sighting effort at annual intervals. For each of our animals we will compile an 'encounter' or 'capture' history, which is a row of 1s and 0s, indicating whether or not the individual was seen each year. For example:

1101

represents an animal marked in the first year, seen in the second and fourth year, but not seen in the third year. What probabilities underlie this particular encounter history? First, we introduce some notation. It is conventional to denote survival from time i to $i+1$ as Φ_i, and the probability of capture at time i as p_i. In our example, we released our newly tagged seal in the first year of our study, which is indicated by the first '1' above. In the second year, we re-sighted our seal, an event with a probability equal to the product of the chances of surviving and being seen given one is alive: $\Phi_1 \cdot p_2$. On the third occasion, our seal was not seen, which might mean that it died (probability $1 - \Phi_2$, the probability of *not* surviving from year 2 to 3). However, because we saw this wily character in the fourth year, we know he was alive and missed in the third year (probability $\Phi_2 \cdot (1 - p_3)$). Seeing the animal in the fourth year had a probability of $\Phi_3 \cdot p_4$. Finally, the probability of this entire encounter history is simply the product of all of its components:

Fig. 6.1 Examples of marine mammals that have been the subject of studies in to vital rates using natural markings and photo-identification. (a) An approximately 35-year-old southern resident killer whale female and her adult daughter in Washington State (photo courtesy of Kim Parsons, Center for Whale Research, Washington, USA). (b) A sub-adult male West Indian manatee approximately 5 years old in Crystal River, Florida (photo courtesy of US Geological Survey, Sirenia Project). (c) A 4-year-old female Hawaiian monk seal at Kure Atoll with scars from a large shark attack (photo courtesy of Jessica Lopez, US National Marine Fisheries Service). (d) A 10-year-old female bottlenose dolphin in the Moray Firth, Scotland, who has been followed over her lifetime by using natural marking for identification. She has also been shown to give birth to two offspring. (Photo courtesy of Barbara Cheney, University of Aberdeen.)

$$\Phi_1 \, p_2 \, \Phi_2 \cdot (1 - p_3)\Phi_3 \cdot p_4$$

In our four-year study, there are eight distinct possible encounter histories. They are:

1111
1110
1101

1100
1011
1010
1001
1000

CJS survival estimation employs a statistical method called 'maximum likelihood estimation' (MLE). In our example, we would compile the encounter histories for all 100 study subjects and then use MLE to identify which combination of parameter values for the various Φ's and p's is most likely to produce the observed encounter histories. MLE parameter estimates as well as their variances and covariances can be readily computed by a number of computer programs. Program MARK[2] currently implements the broadest range of CR models within a single software package.

Assumptions

The CJS model has a number of associated assumptions. First, it is important to recognize that to estimate true survival, we must assume that our animals do not emigrate from the study area. If they do, then emigrants will appear to have died. Thus, mortality and emigration are confounded and indistinguishable. If emigration cannot be ruled out (or estimated, see p. 135) then in fact CJS models estimate what is termed *apparent survival*, which is the product of the probability of surviving and remaining in the study area.

In addition, the CJS model involves the following assumptions:

- each marked animal has the same probability of capture (p_i) at time i
- each marked animal has the same probability of surviving (Φ_i) from time i to $i+1$,
- marks are permanent and correctly recorded upon capture, and
- all samples are obtained instantaneously, and releases occur immediately.

As with most statistical models, adhering to underlying assumptions is a challenge in practice. There are, however, ways to minimize departure from these assumptions through model structure. The standard CJS model incorporates time-dependent survival and capture probabilities, but can be readily expanded to accommodate a universe of possibilities. The first two assumptions above will be violated if, for example, survival varies with age (as it does in long-lived species), by sex, location, or any number of other factors. Likewise, some animals may be more readily 'catchable' than others because of their age, reproductive status, etc. To account for these differences, age-dependent models can be constructed and

---

[2] http://www.warnercnr.colostate.edu/~gwhite/mark/mark.htm (see http://www.phidot.org/software/mark/rmark for a handy interface to Program MARK through the free software package R).

distinct parameters fit for defined groups (i.e. sex, location). In this manner, the first two assumptions are relaxed such that capture and survival probabilities need only be homogenous *within* groups. Within the CJS framework, individual covariates (e.g. body mass) that might influence the probabilities of survival and capture can also be incorporated into model structure. Not only do these modifications of model structure help comply with assumptions, when coupled with model selection (see Section 6.3.4) they provide a mechanism for testing alternative hypotheses about basic biological and ecological relationships.

The third assumption, that marks are permanent and correctly recorded can be another challenge. Mark loss leads to negatively biased survival estimates as animals that lose their identifying marks, like emigrants, appear to have died. In marine mammal photographic identification studies, mark impermanence may involve such things as scars that fade or dorsal fin marks that are obscured by the acquisition of additional marks. Properly applied hot-iron brands are permanent marks that have been successfully used on pinnipeds, but some brands can also become difficult to read in inclement weather conditions and as animals age (see Chapter 2). These issues can be minimized through study design and accounted for quantitatively. Redundancy of marks (e.g. double-tagging, use of implanted passive integrative transponder (PIT) tags), helps ensure that identities are not lost. Redundant marking also facilitates estimating tag loss rates which can be used to adjust survival rates accordingly (Cameron and Siniff 2004). Periodic re-tagging or regularly updating photographic identification databases will help prevent loss of identities.

The final assumption, that samples are collected instantaneously, is of course not possible. In practice, this means that the duration of capture or re-sight effort should be small relative to the interval for which survival is estimated.

6.3.3 Study design considerations

Studies of marine mammal survival require a minimum of two years just to obtain simple return rates uncorrected for sighting probability. In general, it must be anticipated that it will take several years to obtain reliable estimates. At the outset, then, it is especially important to think carefully about study design to maximize the likelihood of success. Indeed in some cases, consideration of the potential errors and biases, together with determination of the accuracy, or precision, required to address the objectives of the study, might show that a study is not feasible. The precision of estimates increases with both the number of animals marked and the probability of recapture. This makes intuitive sense. Let us imagine that 50 animals are marked and the subsequent recapture probability is only 1%. That would mean that on average only one animal would be encountered every other year and the survival estimates would be very uncertain. With most marine mammals, the investigator will likely have some idea about how many animals can be 'marked' and some sense of recapture probability. In Program MARK, one can simulate datasets with specified sample sizes, survival rates, and capture probabilities. Then models can be fitted to those data and evaluated as to whether the resulting precision is acceptable. We strongly recommend this.

Adherence to CR assumptions can also be improved through proper study design. Heterogeneity of capture probabilities results from both the intrinsic properties of individual animals (e.g. behaviour or distinctiveness of marks) and from field protocols. If animals do not randomly mix within the study area and, due to convenience or proximity, some parts of the area are more thoroughly searched than others, the result will be that some animals will be more or less likely to be encountered. Distributing re-sighting effort equally throughout the study area will ameliorate this source of heterogeneity. A related concept has to do with variable recognizability of marks, something that is common with photographic identification of cetaceans. Imagine a cetacean study that relies upon dorsal fin marks to identify individuals. An animal with the top half of its fin missing is far more conspicuous than one with small subtle nicks. If both appear in a large group near the study vessel, the lop fin is much more likely to be detected. However, it is possible to equalize these animals sighting probabilities by systematically attempting to photograph each animal in the group rather than allowing undue attention to be focused on the conspicuous marks. This concept can be applied broadly to any situation where some animals are easier to 'recapture' than others.

Finally, individual capture probabilities can be made more homogenous by pooling all sightings over an extended field season. To establish that an animal has survived since the previous year, it need only be seen once in the current year. Animals that have differing probabilities of detection on any given survey day will have more similar probabilities of being seen *at least once* during multiple survey days. To illustrate, consider two tagged seals that spend 30% and 15% of their time on the beach and assume that researchers can detect all seals on land during a survey. Clearly, if only one survey is done, then the first seal is twice as likely to be seen. However, after 10 survey days, the first animal has a 97% chance and the second an 80% chance of being detected at least once—far more comparable capture probabilities. A potential problem with extending survey periods too long is that we begin to run up against the assumption that all samples are obtained simultaneously. The result is that when the field season expands, the actual interval over which we are observing survival begins to vary more among animals. That is, an animal seen late in the survey period one year and early the next has been observed to survive a shorter interval than one seen early one year and late the next. A trade-off, then, is to make the survey period long enough to achieve acceptably high (and as homogenous as possible) capture probabilities, but still relatively short compared to the time between survey periods (usually one year).

6.3.4 Model selection

When analysing CR data, it is customary to fit a variety of models with different parameterization and then select the 'best' model. Model selection can help to obtain the most reliable survival estimates and is especially useful for testing hypotheses about ecological factors related to survival. For example, if we want to know whether survival (or capture probability) differs according to sex, colony, or any other factor, we need only build alternative models which fit separate survival (or capture) parameters for each level of the factors of interest and compare them to a 'reduced' model where distinct groups all have the same parameters.

In CR survival analysis, nested models can be compared using the likelihood ratio test (LRT, Lebreton *et al.* 1992). More recently, Akaike's information criterion (AIC) has become a preferred tool for model selection. The theoretical basis and various properties of AIC are beyond the scope of this chapter; however there are several excellent references available (Anderson and Burnham 1999; Anderson *et al.* 2000; Burnham and Anderson 2002). The basic concept of AIC is that it indicates the relative support for different models given the data available. Generally, model fit improves with the number of parameters included, but adding parameters tends to degrade precision. AIC balances the trade-off between model fit and the number of parameters to indicate the most parsimonious model. Note that AIC indicates *relative* support. That is, it does not tell you whether any of your models are good or true; rather it tells you the degree of support *among* the models you have fitted. If you have neglected to include a highly significant parameter in all your models, you will not have anything close to the best possible model, but AIC will still tell you which of your (lousy) models is best.

One advantage of AIC is that, unlike the LRT, non-nested models can be compared as long as the underlying likelihood function is the same. Another feature of AIC is that it frees the analyst from the constraints of arbitrary *p-values*. With traditional methods, we compare two models and declare a winner only if we are 95% certain ($p < 0.05$) that one of the models is superior. Sometimes the critical value is set at 0.01 or 0.10, but in every case this criterion for significance is quite arbitrary. In contrast, AIC provides a measure of relative support with no strict critical value. *Akaike weights* normalize the relative support for each of a set of models based upon their AIC values. The sum of all Akaike weights is 1, and they are denoted as w_i, with i indexing the various models. For example, if you have fitted six different models and there is a clear frontrunner, then the w_i for the best model might be something like 0.92, the next best model might only have a w_i of 0.04, and so on. In cases where several models have similar AIC values, choosing between these is unnecessary. Instead, model averaging can be used, which essentially means obtaining a parameter estimate (and variance) by taking the average value from all the fitted models, weighted by their Akaike weights. By doing so, the uncertainty inherent in model selection is reflected in the model averaged parameter estimates.

The preceding has provided a basic grounding in CJS models for estimating survival in marine mammals. While CJS approaches are most common and can be applied in a wide variety of scenarios, additional methods exist which can provide estimates of parameters other than survival or are better suited to certain types of data.

6.3.5 Multi-state models

Consider you are studying a species which occasionally moves among distinct habitats. Perhaps survival and capture probabilities vary depending upon where the animals are, either because of varying risks (e.g. predators), differences in behaviour, or differences in accessibility of the sites to you, the researcher. You might consider analysing survival rates using a CJS model and stratifying the individuals by location. The problem with that approach is that a single individual may visit more than one location over time, so that each animal cannot be assigned to a stratum. Multi-state, or multi-strata models, as they are also known, are designed to deal with situations where individuals can move between states which have distinct associated capture and survival probabilities (Brownie *et al.* 1993; Schwarz *et al.* 1993; Nichols and Kendall 1995; Lebreton and Pradel 2002). With this approach, we estimate not only those parameters, but also probabilities of transition between strata at each time step. These transition probabilities are also referred to as movement parameters as they may entail physical movement as in the example above. Another application of multi-state models is for species that do not breed every year, which is often the case among marine mammals. In this case, the strata refer to conditions (i.e. breeder versus non-breeder) rather than a location. Beauplet *et al.* (2006) used multi-state models to evaluate survival rates of subantarctic fur seals (*Arctocephalus tropicalis*), as well as the survival costs of breeding.

6.3.6 Recoveries from dead animals

Until now, we have considered only capture–recapture studies where the observations involve live encounters with animals that had previously been marked. There may be situations whereby marked animals are recovered dead. In these cases, rather than estimating survival and capture probabilities, the parameters of interest are apparent survival and probability of 'recovery'. The latter conflates three probabilities—that of dying, being recovered, and being reported. The original approaches to dealing with these type of data are Brownie models (Brownie *et al.* 1985), and have typically been applied to situations such as ringed game birds recovered during hunts. However, it is not required that the animals die as a result of a hunt (Seber 1970). Analogous situations may occur when marked marine mammals are harvested for food, caught incidentally in fisheries, or found stranded.

In some cases, both live re-sightings and dead recoveries of marked animals may be available to estimate survival. Burnham (1993) and Barker (1997) developed methods for jointly analysing both types of data in a single modelling approach. Under some circumstances, it may be possible to separate true from apparent mortality by using both live encounters and dead recoveries. Dead recoveries and joint live encounter/dead recovery studies are rarely conducted in marine mammals. However, Hall *et al.* (2001, 2002) demonstrate the utility of combining a variety of data sources, including directed research surveys, re-sightings by the public, and opportunistically reported deaths of marked grey seals.

6.3.7 Robust design

The most complex and, potentially, the most informative CR model is that of Pollock's robust design (Pollock 1982; Kendall *et al.* 1995, 1997). It involves clustered recapture occasions separated by a longer interval. For example, let's say each year we conduct a re-sighting effort five times within a two-week period. Thus, we will have clusters of five capture occasions separated by a year. The key to the classic robust design is that during the two-week sampling period we assume population closure, meaning there are no births, deaths, or emigration or immigration (except under certain restrictive criteria). In contrast, during the year between periods of clustered re-sight effort, the population is open. The power of this approach is that it allows the estimation of a suite of important population parameters, including apparent survival, abundance, and temporary emigration and immigration. If additional data (dead recoveries) are available, true survival and permanent emigration may also be disentangled in some cases.

Finally, the open robust design (Schwarz and Stobo 1997; Kendall and Bjorkland 2001) can be applied when immigration/emigration occur throughout the periods of clustered sampling. This is the case, for example, in pinniped colonies where individuals arrive and depart from the colony during a breeding season. Schwarz and Stobo (1997) used grey seal data to demonstrate the method of open robust design.

6.4 Population models

Here, we discuss population models as mathematical expressions of population size as a function of population parameters and time. In other words, a population model is a mathematical formula (or a set of formulae) that describes the relationship between the abundance (or density), vital rates, and other factors that affect abundance. Although there are infinite possibilities for mathematically modelling change in population size over time, models can be categorized as either deterministic or stochastic. In deterministic models, all parameters (e.g. survival and fecundity) are constants, whereas stochastic models treat at least some parameters as random variables. Often, stochastic models draw parameter values from probability distributions, which in turn have been derived from other studies (for example, of fecundity and survival as described above).

Models can be used for parameter estimation or projection. In parameter estimation, a model (or series of models) is fitted to data and the parameter of interest, e.g. population growth rate, is estimated. When a model is used for population projection, model parameters must be estimated or conjectured. Ideally, uncertainty in parameter estimates should be incorporated into the projection process so that results reflect the real uncertainty of possible future outcomes.

Although brand new models may be created, many are already available in textbooks and the primary literature (e.g. see Caswell 2001; Williams *et al.* 2001; Clark, J.S. 2007). Rather than reviewing a wide range of models for marine mammal populations, we will discuss key concepts, modelling frameworks, and relevant examples. Models are only as good as the data that go into them. Therefore, we also encourage early consideration of the data needs to support a planned model. If the required data are unlikely to be obtained, consider employing a simpler model structure with lesser data demands.

6.4.1 Exponential and geometric models

One of the first decisions when creating a model is whether or not to treat time as a continuous variable. If time is continuous, all processes are modelled with *differential equations*, whereas *difference equations* are used when time is treated as a non-continuous (or discrete) variable. Although differential equations appear more complex than difference equations, the former involve fewer assumptions about birth and death processes because all vital rates are defined as instantaneous. In contrast, with difference equations in discrete models, we must assume that all births and deaths occur at one point between censuses (Caswell 2001).

The most basic concept of population modelling starts with the following equations:

$$\frac{1}{N}\frac{dN}{dt} = r \qquad (6.1)$$

and:

$$\frac{N_{t+1}}{N_t} = R + 1 = \lambda \qquad (6.2)$$

where r is the intrinsic per capita growth rate, R is the net discrete per capita growth rate, λ is the discrete per capita rate of growth, N is the population size, and t is time. Equation 6.1 treats time as continuous (exponential growth), while equation 6.2 represents discrete time steps (geometric growth). Applying a little calculus and rearranging equations 6.1 and 6.2 yields the simple population models:

$$N_t = N_0 e^{rt} \qquad (6.3)$$

and:

$$N_t = N_0 \lambda^t \qquad (6.4)$$

The per capita intrinsic growth rate (r) is the net change in the number of individuals in the population during a time interval. When the time increment (dt) is small relative to the life span of the species, r is the difference in per capita birth (b)

and death (*d*) rates: $r = b - d$. In the discrete time model, we substituted $\lambda = R + 1$. From equations 6.3 and 6.4, we can see that $\lambda = e^r$.

Population growth rate, even without decomposition into birth and death rates, is a valuable indicator of population status. If *r* is positive (or $\lambda > 1$), the population is increasing, while decreasing populations have a negative *r* ($\lambda < 1$). Although very simple, these models can be used to determine the rate of population change from simple count data. For example, Brown *et al.* (2005) fitted a variety of population growth models to a 27-year time series of aerial counts of harbour seals at haul-out sites in Oregon. Building on work by Lande and Orzack (1988), various analytical methods have been developed for computing the population growth rates from simple count data or a time series of abundance (Dennis *et al.* 1991; Holmes 2004; Staples *et al.* 2004). Sometimes such time series are the only information available for certain marine mammal populations. We caution that these methods require that time series data are collected in a consistent manner to avoid introducing unknown biases during the estimation process.

6.4.2 Matrix models

In the previous section, we considered population models based on abundance or count data. Although such models are extremely useful for estimating population growth rates, in cases where age-specific survival and fecundity estimates are available, these can be incorporated into a more complex and informative model.

When we account for differences among groups (ages or life stages) there are as many equations as the number of modelled groups, and linear algebra is used to concisely express them. Consequently, these are often called *matrix models* (Leslie 1945; Lefkovitch 1965; Caswell 2001). The scalar components of equation 6.2 are replaced by vectors and matrices:

$$\underline{\mathbf{n}}_{t+1} = \mathbf{L}\underline{\mathbf{n}}_t \tag{6.5}$$

where $\boldsymbol{n}_t$ is a column vector of the numbers of individuals in each age or stage class at time *t*, and $\boldsymbol{L}$ is a matrix that specifies birth and survival rates of each age or stage.

As the time subscripts in equation 6.5 suggest, time is treated as a discrete variable in matrix models. The simplest form of the matrix in equation 6.5 is called the *Leslie matrix*, in which only entries for survival rates and per capita fertility are included for each age class. In marine mammals, it is often difficult to estimate these parameters for every age. Consequently, consecutive ages are often combined to form stage classes (*Lefkovitch matrix*, Lefkovitch 1965; Caswell 2001). In these cases, the matrix will contain entries representing the probability of remaining in each stage at each time step. These matrices (Leslie or Lefkovitch) are collectively called projection matrices. Brault and Caswell (1993) used published data (Bigg *et al.* 1990; Olesiuk *et al.* 1990a) to parameterize a stage-based matrix for killer whales (*Orcinus orca*) in the Pacific Northwest. The model stages included yearlings, juveniles, reproductive adults, and post-reproductive adults.

Matrices have associated scalars and vectors called eigenvalue–eigenvector pairs. Their mathematical definitions are beyond the scope of this chapter, but for now it is sufficient to know that eigenvalues and eigenvectors embody key properties of populations represented by the projection matrix. The dominant eigenvalue, or the largest eigenvalue, is the asymptotic population growth rate (λ) and the population age structure will stabilize in the same proportion as the *dominant right eigenvector* (stable age distribution).

The matrix approach can be extended to model metapopulations and multi-state systems (Caswell 2001). In metapopulation models, multiple locations and movements between locations are modelled along with fecundity and survival. Just as with multi-state capture–recapture models discussed earlier, multi-state population models can address situations where individuals may exist in a variety of states, between which they transition over time. Hunter and Caswell (2005) demonstrate a metapopulation model in which the spatial distribution of a population is explicitly modelled within a matrix approach. They demonstrated that the population growth rate can be computed from a multi-state, metapopulation matrix using linear algebra.

Perturbation analysis is a key aspect of matrix modelling. According to Caswell (2001) 'The results of perturbation analysis are often more interesting, more robust, and more useful than the parameter estimates themselves.' Two common types of perturbation analysis are sensitivity and elasticity analysis. Elasticity provides the proportional response of λ to a proportional perturbation to vital rates (Caswell 1978; de Kroon *et al.* 1986), whereas sensitivity provides information about the magnitude of change in λ with respect to each element of the matrix (i.e. additive perturbation). McMahon *et al.* (2005) used elasticity analysis to show that, in general, juvenile survival rates contributed the most to the change in population growth rates of several southern elephant seal (*Mirounga leonina*) populations.

Despite the apparent complexity of linear algebra notation used in matrix models, they really only differ from the geometric model (equation 6.4) in terms of the type of data they incorporate. Mathematically, equations 6.4 and 6.5 are equivalent. Specifically, equation 6.4 is a special case of equation 6.5, where N_t is a scalar rather than a vector.

6.4.3 Incorporating uncertainty: deterministic versus stochastic models

Matrix models are based upon multiple life-history parameter estimates, the precision of which plays a critical role in the precision of the estimated population growth rate derived from the model. When reporting the estimated population growth rate, for example, it is important to convey the uncertainty around the estimate. Some methods for computing the uncertainty for estimated growth rates include series approximations, bootstrapping (e.g. Brault and Caswell 1993), Monte Carlo analysis (Caswell 2001), and Bayesian approaches (e.g. Goodman 2004; Thomas *et al.* 2005).

Recall that when birth and death rates in a population model are treated as constants, the model is considered deterministic. In reality, we know full well that vital rates are not constant over time. A stochastic model accounts for variability in vital rates over time; i.e. process error or process variability. Information about the variability of vital rates can be incorporated directly into a stochastic model by representing parameters with probability distributions rather than constants. Ideally, these probability distributions will be based on vital rates and their variability estimated using methods described earlier in this chapter. For example, Runge *et al.* (2004) used a stochastic matrix approach to model a population of Florida manatees (*Trichechus manatus latirostris*). By incorporating uncertainty, the projection of the population based on the stochastic matrix will include a range of possible outcomes, rather than a single outcome based on a deterministic matrix. Stochastic models are more informative than deterministic models for managers and conservationists because a distribution of possible outcomes will provide a more realistic view of future uncertainty.

When developing a stochastic model, demographic stochasticity and environmental variability may be incorporated. The former refers to the random chance variability in demographic outcomes, whereas the latter involves variability in vital rates due to environmental factors. In general, environmental variability tends to have the greatest influence on the population growth rate, especially for moderate to large size populations. Incorporating these uncertainties (especially environmental variability) into population models becomes critical for estimating the probability of persistence or extinction (see Section 6.4.6 on population viability analysis).

6.4.4 Density dependence

The models discussed so far have had no upper limits on population size. In reality, of course, a population cannot grow forever. Several biological and physical factors, such as food and space limitations, aggression, predation, emigration, and disease, can reduce the population growth rate as the population size increases. Effects of density on population growth rates (or vital rates) are called density dependence.

The geometric and exponential models (equations 6.1 and 6.2) can be extended to include carrying capacity (K), which can be thought of as the equilibrium population level. For example, the following two equations are particular forms of continuous and discrete time models, respectively, with carrying capacity:

$$\frac{1}{N}\frac{dN}{dt} = r\left(1 - \frac{N}{K}\right) \tag{6.6}$$

and:

$$N_{t+1} = N_t + RN_t\left[1 - \left(\frac{N_t}{K}\right)^z\right] \tag{6.7}$$

where χ defines the rate at which the population reaches the carrying capacity. In density-dependent models, the population growth rate becomes a function of abundance (or density).

Note that for the equations above, the rate of population growth decreases as N increases, finally reaching stability (no growth) when $N_t = K$. Although density dependence is an important factor in the population dynamics of large mammals, including marine mammals (Fowler 1981), it is not always included in models. One reason is that the growth rate of a small recovering population may be affected little by its density. In general, if only short-term projections of a small population are performed using models of long-lived species like marine mammals, effects of density dependence may be negligible. Another reason for omitting density dependence in population models is lack of information on how density actually affects vital rates in the species and population of interest. While general patterns may be anticipated (Fowler 1981), obtaining information about the precise functional form of density dependence requires observing vital rates or population growth rates of a population over a wide range of abundances from zero to K, or perhaps by comparing observed rates among multiple populations at different population levels. Alternatively, information from similar species can be used to define some reasonable functional form for density dependence. We caution that model results can be highly sensitive to how density dependence is implemented, especially for long-term projections.

Density dependence usually refers to reduced population growth as abundance increases. There also exists inverse density dependence at low population sizes, often referred to as an Allee effect (Odum and Allee 1954), which describes diminished per capita population growth rates in populations at very low abundance. Allee effects are often associated with difficulty finding mates at low population levels or a breakdown of breeding, foraging, or other key functions that depend upon some minimum density in social species (such as some marine mammals). Dennis (2002) has shown that Allee effects are important to consider for decreasing populations. When modelling a small and declining population, including Allee effects may avoid overly optimistic projections of population growth that could result from including only compensatory density dependence.

6.4.5 Individual-based models (IBM)

We began with modelling an entire population as a group, then introduced matrix models that account for groups (ages, stages) of individuals with distinct associated vital rates. However, the fundamental unit of a population is an individual. The IBM approach explicitly deals with the variability at the individual level. Rather than using a population growth model (e.g. equations 6.1 and 6.2), each individual is modelled with birth/death processes and other factors that affect its survival and reproduction. In essence, the fate of each individual in a population is tracked through time. Such a modelling approach allows incorporation of more detailed information about the population than when a population is treated as a whole. For example, a genetic component may be added to a model so that the survival rate of an individual is based on its fitness, which may be a function of its parents' genotypes. IBMs involve intense simulations and demand much computer processing capacity, as each individual has its own probability functions for demographic rates. Further, and more importantly, a considerable amount of data is required about survival, reproduction, and all other parameters to support an individual-based model (cf. Grimm and Railsback 2005).

Hall *et al.* (2006a) used the IBM approach to examine the effects of PCBs (polychlorinated biphenyls) on the potential population growth rate of bottlenose dolphins (*Tursiops truncatus*) in Sarasota Bay, Florida, perhaps the best studied cetacean population in the world (Scott and Chivers 1999; Scott *et al.* 1990a, b; Wells and Scott 1990). They simulated the process of accumulation and depuration of PCB in the dolphins and modelled first-year calf survival as a function of maternal PCB blubber concentration. They found that they had insufficiently precise data to make definitive conclusions about how PCB affects the population, even though they tentatively concluded that the current environmental levels of PCB were depressing the population's growth rate. The observed population growth rate, however, was found to be outside the 95% confidence interval estimated from the model simulations. This example shows that even for a well-studied population, data may be insufficient to satisfactorily parameterize an IBM. Currently, we do not believe that any marine mammal population has sufficient data to support a meaningful IBM.

6.4.6 Population viability analysis (PVA)

The aim of PVA as originally developed was to evaluate population viability measured in terms of extinction risk within some time horizon (e.g. 100 years). Recently, it has been argued that such estimates of extinction risk may be not accurate or credible (Ludwig 1999; Ellner *et al.* 2002). The reliability of a PVA depends on having information on the population itself (e.g. abundance, vital rates, age–sex structure) as well as such factors as environmental variability, demographic variability, density dependence (including Allee effects), catastrophic events, and anything else that may affect the population. If such factors are not correctly characterized in the model, or uncertainty is inadequately expressed, then PVA results will likely be misleading.

Even with the difficulty in obtaining all this information, there is still a role for PVA in marine mammal conservation and management. While we may never achieve a precise and accurate estimate of extinction risk, a PVA can provide a mechanism for organizing and integrating existing relevant data that bears on extinction risk while characterizing associated uncertainty.

Several books have been written on this subject (e.g. Soulé 1987; Beissinger and McCullough 1992; Ballou *et al.* 1995; Fiedler and Kareiva 1998). Morris and Doak

(2002) discuss a series of considerations for using PVA. These include focusing on uncertainty (e.g. confidence intervals rather than point estimates), considering extinction risk metrics as relative and qualitative rather than absolute, limiting projections into the future to a short period, considering potential effects of overlooked factors, comparing multiple models if possible, and updating a PVA as more data become available. Running sensitivity analyses on a PVA to determine which parameters most influence the results can aid in directing research to fill key data gaps. Finally, we caution that although 'off the shelf' PVA software is available (RAMAS and VORTEX), the output of such programs should be viewed critically and presentation of results should explicitly document model structure, assumptions, and uncertainty about input parameters.

6.4.7 Bayesian approach to modelling

Although discussed separately, we emphasize that the Bayesian approach is not independent of the rest of this chapter. Any model can be analysed with Bayesian statistics, the use of which has been increasing in recent years. A detailed treatment of Bayesian statistics is beyond the scope of this chapter. A brief non-mathematical introduction can be found in Eguchi (2008).

One application of the Bayesian approach to population modelling is state–space modelling. In a state–space modelling approach, the state of the quantity of interest (e.g. abundance over time) is modelled with a mathematical function such as a population growth model. Observed data are treated as samples from the state, where another mathematical function is used to depict the sampling process. For example, Craig and Reynolds (2004) used the first-order Markov process to model the population size of manatees along the Atlantic coast of Florida. They included movement, sighting probabilities, and abundance in the model. Goodman (2004) showed that multiple independent datasets (mark–recapture and stranding surveys) can be used in a model to improve estimates of population parameters, such as abundance and population growth rate. Thomas *et al.* (2005) used a matrix population model and state–space modelling to unify the data collection process and the population growth model. Parameter inference was obtained via the Bayesian approach. They demonstrated the method by modelling the British grey seal metapopulation. Another strength of the Bayesian approach is the ease in which mixed-effects and random-effects models are analysed. Clark (2007) provides an excellent textbook on applied Bayesian modelling for ecological data.

6.4.8 Model fit and selection

Several candidate model formulations may be generated to estimate parameters of interest such as population growth rate. Just as we described with regard to capture–recapture estimation, AIC and its modifications can be used to select the best amongst a suite of models, or to obtain model-averaged parameter estimates where the weight of each model is proportional to its model selection score

(Burnham and Anderson 2002). R.F. Brown *et al.* (2005) provide an example of this process applied to a harbour seal population. They fitted multiple candidate models (density dependent, density independent, site-specific growth rate, and constant growth rate) to seal counts. This process of model selection and parameter estimation is common regardless of one's affiliation to statistical philosophy. In addition to AIC, there are analogous Bayesian model selection criteria (Draper 1995; O'Hagan 1995; Spiegelhalter *et al.* 2002).

7

Epidemiology, disease, and health assessment

Ailsa J. Hall, Frances M.D. Gulland, John A. Hammond,
and Lori H. Schwacke

7.1 Introduction

Understanding marine mammal health and disease and the related impacts on populations is crucial to support effective conservation and management decisions. However, ethical issues involved in conducting experimental studies can limit the scope of marine mammal health research. This forces a focus and reliance on epidemiological studies, similar to those that have been applied to studying factors affecting human health. Marine mammal epidemiology is additionally challenging because most marine mammals are not easily observed for most of their lives, disease states are generally difficult to detect, and reporting mechanisms for disease used in human and veterinary epidemiology (i.e. birth, death, and disease records) are virtually non-existent for marine mammals. Nonetheless, despite these drawbacks, there are many ways in which robust and reliable epidemiological studies can be applied in the field of marine mammal science.

Marine mammal health and disease issues are gaining global attention and a coordinated approach to their study, such as using standard protocols, will enable meta-analyses (combining information from multiple studies) to be carried out in the future. It will then be much more feasible to identify the critical hazards for marine mammals, particularly if the well-defined and accepted epidemiological approaches outlined here are utilized. Advances in capture and sampling methods (see Chapter 2), the expansion of stranding surveillance networks for marine mammals, the centralization of stranding records (Gulland *et al.* 2001a), and follow-up monitoring of mortality and morbidity using direct observation (Wells and Scott 1990; Gulland 1999), all now give us an unprecedented opportunity to adapt methods from human and veterinary medicine. Necropsies of stranded and by-caught animals, as well as visual assessment, remote biopsy, and capture–release

---

Box 7.1 Definitions of useful terminology

Exposure: Disease-causing factors, including infectious, toxic, nutritional, traumatic, genetic, degenerative, physiological, social, and behavioural.

Confounding factor: A factor or variable that correlates with both the exposure and response (i.e. independent and dependent variables in statistical terminology) so that it masks an actual association or falsely indicates an apparent association.

Incidence rate (IR): The number of new cases (disease onsets) divided by the sum of the time over which the individual animals were observed.

Cumulative incidence (CI risk): The proportion of individuals free from disease which *develop* a specific disease over a specified period.

Incidence odds ratio: The ratio of the number of individuals that experience the disease to the number who do not.

Prevalence: The proportion of a population which *has* a disease at a particular point in time.

Cohort: Populations or groups within a single population that are followed over time.

---

health assessment of free-ranging individuals, provide information on causes of death, endemic diseases, emerging diseases, and toxin exposure. This information can then be related to trends in the physical, chemical, and biological environment.

Epidemiology[1] is defined as the study of the distribution and determinants of health-related states in populations (Last *et al.* 2000). As such, it focuses on examining the *occurrence* of disease with the premise that disease does not occur randomly. It is concerned with impacts on populations not individuals. Over the last 50 years or so epidemiological science has evolved into two distinct disciplines that are relevant to wildlife: modern or causational epidemiology (Rothman and Greenland 2005) and infectious disease epidemiology (Hudson *et al.* 2002). Causational epidemiology is the scientific method for investigating potential causal links between exposures and responses. Exposures do not become causes until there is sufficient evidence for a causal link between the agent and the specific response or health state of interest. Additionally, a set of clearly defined study designs to test for causal links have been developed and refined since at least the 1960s.

By contrast, infectious disease epidemiology is concerned with determining the impact that an outbreak of infection (i.e. viral, bacterial, parasitic, or fungal) will have on the dynamics of the host population. Although there is much overlap between these two branches of epidemiology, particularly when investigating the role of confounding variables on disease occurrence, infectious disease

---

[1] Strictly speaking the term epizootiology should be used to refer to diseases in animal populations. However, the word epidemiology is now widely applied to studies of human, veterinary, and wildlife diseases.

epidemiology has largely evolved as a branch of mathematical and statistical modelling (Grenfell and Dobson 1995). Since the causal agent is known, the focus is then to predict the likely spread of infection during an epidemic outbreak, to estimate the cycle of infection and its inter-epidemic interval, and to investigate the potential impact of intervention measures such as vaccination.

In this chapter we will be largely concerned with the science of causational epidemiology. It is our intention to outline the principles and practices of causational epidemiology as it applies to marine mammal science. Additional information can be found in the extensive medical and veterinary epidemiological literature.

7.1.1 Exposures and responses

Epidemiology is premised on the observation that disease has different causal factors that can be determined by comparing disease rates in different populations, or groups of individuals, that vary with respect to their exposures. It is these exposures that are the putative causal factors for a given disease. The term exposure covers a wide range of causal factors: infectious, toxic, nutritional, traumatic, genetic, degenerative, physiological, as well as social and behavioural causes. In some cases the agents or exposures responsible for the specific disease are fairly obvious, as is often the case where infectious agents or toxins are involved. However, for marine mammals the factors involved are not always as apparent as they might be among terrestrial or domestic animals, largely due to the relative inaccessibility of many species.

Although the presence of a particular infectious agent or toxin is 'necessary' for a specific disease to occur, even these diseases are not usually caused by a single factor and additional causal or risk factors will be involved (Rothman and Greenland 2005). The most widespread causes of disease identified to date in marine mammals are certainly infectious, resulting from their exposure to arthropods, helminths, protozoa, fungi, bacteria, and viruses (Dierauf and Gulland 2001; Gulland and Hall 2007). Typically, epidemics have been associated with viral and bacterial infections, whereas the larger infectious agents (such as the helminths) more commonly cause endemic disease and lower mortality. Detection therefore relies upon observation of the pathogen: detecting the presence of pathogen-specific antibodies in the blood, direct culture of the pathogen, or detecting the presence of its genetic material (DNA or RNA). However, the level of antibodies in the blood (termed serology) cannot usually distinguish between current or previous exposure to the pathogen. Non-infectious agents include both the chemical contaminants and the biotoxins produced by harmful algal blooms (reviewed in Landsberg 2002; Vos et al. 2003). Detecting exposure to these relies upon measuring the level of the compound (or its breakdown products, i.e. the metabolites) in the animals' tissues, and new techniques are evolving constantly to improve sensitivity, detect new toxins, and reduce the amount of tissue required for tests. Trauma is also a disease in its broadest sense, and in marine mammals, has long been identified as an important cause of death following interaction with fisheries (Read *et al.* 2006). Although the diagnosis of some traumatic causes can be obvious, such as entanglement in a net (as the exposure), others can be harder to diagnose and rely upon detailed gross and microscopic examination of fresh tissues, e.g. gas embolism (Fernandez *et al.* 2005). Finally, other physiological, degenerative, and genetic exposures as well as causes of disease are currently not well-understood in marine mammals, and their identification requires careful and detailed sampling of individuals.

7.1.2 Confounding factors

Confounding (a term with a specific epidemiological definition, see definitions in Box 7.1 above) occurs when an independent factor is correlated with both an exposure and outcome, making it difficult to tease apart the contribution made by each to the occurrence of the disease (Rothman and Greenland 2005). A confounding factor may thus wholly or partially account for the apparent association between an exposure and a response. To cause confounding in the results however, the factors *must* be associated with both exposure *and* response. For example, primiparous females may have very high contaminant concentrations in their blubber compared to females that have already off-loaded some of their contaminants to their offspring in the milk. A study may find that the females with high contaminant concentrations have high offspring mortality, and conclude that the mortality is related to the maternal contaminant exposure. However, the true cause of increased mortality may be related to maternal inexperience rather than contaminant exposure per se. Thus reproductive status (primipary vs. multipary) is a confounding factor in this example. Potential confounding factors should be considered prior to data collection so that appropriate study design and analysis can be employed.

It is important that exposure measurements and disease diagnoses are carried out in a standardized way both within and between studies. This has certainly been addressed for some well-studied exposures such as the persistent organic pollutants, but unfortunately, for few others. Indeed, many laboratories analysing marine mammal tissues for contaminant levels are participants of the US National Institute of Standards and Technology (NIST) (Kucklick *et al.* 2002) inter-laboratory comparison and calibration scheme, and this includes some marine mammal laboratories outside the US. For example, marine mammal tissue samples collected as part of the NOAA (National Oceanic and Atmospheric Administration) Marine Mammal Health and Stranding Response (MMHSRP) biomonitoring programme for chemical analyses use the NIST protocol. Similar inter-laboratory comparison exercises were conducted in Europe for environmental samples under the QUASIMEME project (D. E. Wells and De Boer 1994). However, schemes such as these are being expanded to include infectious agents and biotoxins, but efforts to implement standardization within the marine mammal community needs to be sustained, especially in the areas of clinical blood chemistry and haematology (Hall *et al.* 2007; Schwacke *et al.* 2009).

7.2 Effects, responses, and diagnostic techniques

7.2.1 Measuring disease occurrence

The goal of all causational epidemiological studies is to evaluate hypotheses about the causation of disease, and to relate disease occurrence to the characteristics of animals and their environment (i.e. their exposure). This requires consistent and standardized classification of disease and pathological findings as well as exposures.

Epidemiologists also use the term 'effect' in two ways. First, in the general sense, where an instance of disease may be the effect of a certain cause (for example, the effect of domoic acid causing hippocampal atrophy in California sea lions, *Zalophus californianus*, Goldstein *et al.* 2008) and second, in a very particular quantitative sense, an effect is the difference in disease occurrence between two or more groups that differ with respect to their exposure (usually termed 'exposed' and unexposed').

Various standard epidemiological measures are used to describe disease occurrence:

1 The **incidence** is the most robust measure of disease occurrence and can also be visualized as the 'flow' of disease. The incidence rate is the number of new cases (disease onsets) divided by the sum of the time over which the individual animals were observed, usually measured as 'animal time'. If 10 animals were observed for 1 year each the denominator would be 10 animal years^{-1}. If 5 were observed for 6 months and 5 for 1 year the denominator would be 7.5 animal years^{-1}. If in each case 5 developed the disease during the study period then the incidence rates would be $5/10 = 0.5$ animal years^{-1} and $5/7.5 = 0.66$ animal years^{-1}, respectively. However, measuring new disease onsets over time in marine mammals is not generally possible except perhaps in isolated cases involving long-term studies of known individuals.

2 The **risk** of a new disease occurring is quantified using the cumulative incidence (also called the incidence risk or incidence proportion). It is the proportion of individuals free from disease that *develop* a specific disease over a specified period, provided they do not die from any other disease during that period. For example, if a group of 20 animals that were initially disease-free were examined again 12 months later and 4 were found to have developed an infection, an individual's chance or risk of becoming infected over the 12-month period would be 20% (4/20).

3 The **prevalence** of disease is the proportion of a population affected at a particular point in time, and is interpreted as the probability of an individual from the same population *having* the disease at that particular point in time. Prevalence is often estimated for infectious diseases from serological data, but this is actually a

measure of an individual's encounter with infection rather than a true prevalence of disease, as the presence of antibody cannot usually distinguish between recovered and carrier animals.

The differences (i.e. the occurrence of disease in the exposed population minus that in the unexposed population) in the incidence rate, the risk (cumulative incidence), or the prevalence of disease between exposed and reference or unexposed control populations is then used to describe 'effects'. Thus the **incidence rate (IR) difference** is calculated as the incidence rate in the exposed population minus that in the unexposed and the cumulative incidence or **risk difference** is the cumulative incidence (CI) in the exposed minus the cumulative incidence in the unexposed.

Relative effect measures (ratios in the exposed compared to the unexposed) are also commonly used where the **relative excess incidence rate** is defined as:

$$\frac{\text{IR exposed} - \text{IR unexposed}}{\text{IR unexposed}} = \frac{\text{IR unexposed} - 1}{\text{IR exposed}}$$

and **the relative excess risk** (also called the risk ratio) is:

$$\frac{\text{CI exposed} - \text{CI unexposed}}{\text{CI unexposed}} = \frac{\text{CI unexposed} - 1}{\text{CI exposed}}$$

Another common relative measure is the **odds ratio**. If the CI is the probability of developing a disease over a specified period then the incidence odds is the probability of not developing the disease (i.e. $1 - \text{CI}$). The odds ratio is then:

$$\frac{\text{CI exposed}/1 - \text{CI exposed}}{\text{CI unexposed}/1 - \text{CI unexposed}}$$

These various relative measures can estimate the magnitude of the association of exposures and responses, whereas absolute measures indicate the potential impact ('effect') on the population. More detail about epidemiological effect measures and their interpretation can be found in many standard epidemiological text books (e.g. Rothman and Greenland, 2005).

7.2.2 Responses, health panels, and disease classification

Diagnostic indicators

Basic haematology and blood chemistry parameters are used commonly in medicine to define health 'panels' to indicate the state of specific organ systems (Table 7.1). This approach assumes that the tests can be validated for the species of interest and that ranges of normal values for wild populations can be established. This has rarely been the case for many marine mammal species. Many studies on wild-caught or live stranded marine mammals include measuring a variety of haematological and clinical chemistry parameters (Lander *et al.* 2003; Boily *et al.* 2006) that might be

Table 7.1 Health panels using haematology and blood chemistry parameters as indicators of the status of specific organ systems.

Renal function	Hepatic function	Haematological status[1]	Nutritional status	Infection/inflammation	Immune status	Skin disease	Endocrine status	Reproductive status	Cardio-pulmonary status	Neurological status
Creatinine	Alanine aminotransferase	Erythrocytes	Body mass index	Leukocytes	Leukocytes	Lesion description	Thyroid hormones / Thyroid-stimulating hormone response	Testosterone	Respiratory questionnaire	Seizures
Phosphorus	Sorbitol dehydrogenase	Nucleated erythrocytes	Glucose	Differential white cells	Differential white cells	Histopathologic results	Aldosterone	Oestradiol	Cytology	Behaviour (attitude, aggression)
Potassium	Aspartate aminotransferase	Haematocrit	Cholesterol	Globulins	Neutrophil phagocytosis		Cortisol	Progesterone	Oscultation	Cerebral spinal fluid aspiration, cytology, culture, serology, biochemistry
Blood urea nitrogen	Gamma-glutamyl transferase	Haemoglobin	Alkaline phosphatase	Erythrocyte sedimentation rate[2]	B- and T-cell proliferation		Adenocorticotrophic hormone response	Ultrsonography[3] of reproductive tract (uterine size, fetus detection, follicles, corpora lutea, testis size)	Culture	Computed tomography, Magnetic resonance imaging[3]

Calcium	Bilirubin (total, conjugated)	Mean platelet volume	Triglycerides	Fibrinogen	Globulins	Semen examination	Radiography, ultrasonography[3]	Electroencephalogram[3]
Cholesterol		Red cell distribution width	Blood urea nitrogen	C-reactive protein	Interleukins		Creatinine kinase	
Triglycerides			Albumin				Blood gases	
Alkaline phosphatase			Electrolytes				Radiography, ultrasonography (Doppler)[3]	
Lactate dehydrogenase			% Lipid in blubber				Lactate dehydrogenase	
Bile acids								

[1] Excludes leukocytes; [2] ESR is not appropriate in lipaemic samples; [3] Other investigative procedures.

used for assessing health status. Some marine mammal species, such as bottlenose dolphins (*Tursiops truncatus*) and harbour seals (*Phoca vitulina*), are being extensively studied at widely geographically dispersed locations and this has led to a number of publications on reference ranges (e.g. Goldstein *et al.* 2006; Hall *et al.* 2007; Schwacke *et al.* 2009). However, once again, it is important that there is appropriate inter-laboratory calibration to eliminate the possibility of artefacts arising because of variations in analytical methods and standards. Many of the currently published 'reference ranges' should also be viewed with caution; these are often from small samples and only report the mean and standard deviations for each parameter, rather than the 95% double-sided reference intervals with 90% confidence limits on the lower and upper bounds as recommended by the International Federation of Clinical Chemistry, Expert Panel on the Theory of Reference Values (Solberg, 1983).

Other diagnostic indicators include: functional immune assays for examining both innate and adaptive immunity, physical examinations, and visual assessment of the skin, looking specifically for infectious agents (arthropods, protozoa, fungi, bacteria, or viruses) as well as neoplastic and traumatic lesions.

Disease classification

Clinical diagnoses can be categorized using disease classification schemes. One such scheme currently being implemented for marine mammals is an adapted version of the WHO International Statistical Classification of Diseases and Related Health Problems with Clinical Modifications, 10th Revision (ICD10-CM, WHO 1992). The ICD system was designed for the classification of mortality and morbidity information for statistical purposes in humans. Each major disease entity is classified in a hierarchical manner. For example, Chapter 1 of the scheme includes 'certain infectious and parasitic diseases' (codes A00–B99): A00–A09 being the intestinal infectious diseases; A07, protozoal; A07.1, giardiasis; and A07.2, cryptosporidiosis. Therefore retrieving data by code can easily identify, for example, all intestinal infections or just protozoal infections, and cases coded in this way can be retrieved for further statistical analysis. In addition, Chapter 18 of the scheme 'Symptoms, signs and abnormal clinical and laboratory findings, not elsewhere classified' (R00–R99) gives the system sufficient flexibility for the purposes of classifying the results from health assessment studies in marine mammals, using both live capture–release biochemical and health assessment data as well as post-mortem strandings' diagnoses. While it is often impossible to give a definitive cause of morbidity or death, general categories at the three-digit ICD code level can often be assigned (for example, R74 is abnormal serum enzyme levels and R74.0 is non-specific elevation of levels of transaminase and lactic acid dehydrogenase).

7.2.3 Sample and data collection

The application of epidemiological study designs requires consistency in sample collection, analysis, and reporting across populations or cohorts. The type of epidemiological framework proposed here for marine mammals (Fig. 7.1) includes standard sample collection and analysis protocols, as well as standard assessment methods. There are three primary methods of sample collection that can be applied to marine mammal health studies, namely the recovery of stranded carcasses, field surveys, and live capture–release.

Stranding recovery

Marine mammal carcasses that wash ashore have long been a principal source of information about marine mammal pathology (Gulland 1999). In many regions formal stranding response schemes are well established, and these involve the examination of carcasses using standard procedures (Kuiken and Garcia Hartmann 1991; Dierauf and Gulland 2001) that allow comparison among different events to be made. Information can also be obtained about exposures to pathogens or toxins by sampling specific tissues that cannot be easily acquired from live animals. Although the interpretation of such data needs to take account of the cause of death, it may then be possible to, for example, estimate the total body burden of contaminants or toxins. Polybrominated diphenyl ethers can concentrate in the adrenal glands as well as the adipose tissue (Klasson Wehler *et al.* 2001), and inorganic compounds are found at higher concentrations in the liver and kidneys when compared to the blood or skin (Marcovecchio *et al.* 1990). This type of information is clearly important when estimating exposure using just a single tissue sample for monitoring.

In addition to the important disease pathogenesis, pathology, histopathology, microbiology, and other disease process information that will be gleaned during a necropsy, it is very valuable to find out why the animal died and, if possible, to determine both the primary and secondary causes of death. In reality however, this is often difficult due to decomposition but it should be the ultimate goal when fresh carcasses are examined.

Live capture–release

Very few marine mammal studies are able to live-capture large numbers of individuals at any one time. However, with recent advances in techniques and an expansion of the skills base, many pinniped and some small odontocete cetaceans are now routinely and safely captured, assessed for their health status, and then released (Harwood *et al.* 1989; Wells *et al.* 2004). The impetus for these studies spans all scientific fields and, although some are carried out specifically to determine the health or disease status of the population, a health assessment component could be carried out for many at a low additional cost. Particularly when combined with strandings and photographic or remote-sensing follow-up of the same population, these types of studies will allow inferences to be made about the health status of the population and its potential role in its population dynamics.

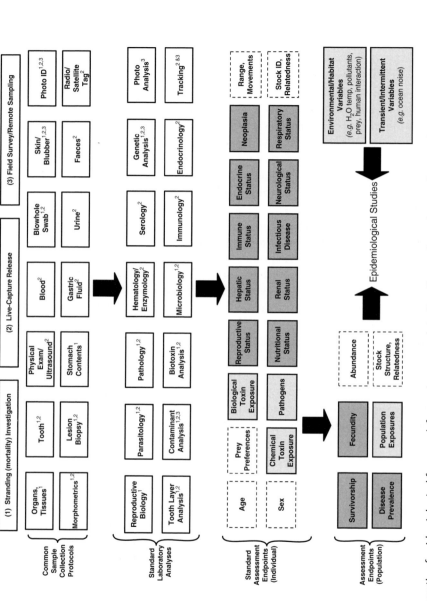

Fig. 7.1 Schematic of epidemiological framework for marine mammal studies. Types of samples are shown with superscripts indicating if they are generally collected through: (1) stranding (mortality) investigation; (2) live capture–release; and (3) field surveys and remote dart sampling. Light-grey assessment endpoint boxes indicate an exposure endpoint, darker grey indicate an effect endpoint. Dashed boxes represent important covariates for analyses.

Field survey

Visual observation, including photo-identification (see Chapter 2), has been used to a limited extent as a tool to investigate changes in endemic disease in cetaceans (B. Wilson *et al.* 2000; Pettis *et al.* 2004; Hamilton and Marx 2005). Many such photo-identification monitoring studies are conducted in conjunction with biopsy darting to collect skin and blubber samples (R. S. Wells and Scott 1990). DNA from skin samples can provide other fundamental covariate data such as sex or stock identification. In addition, the relationships between sampled animals and gene expression for particular proteins could be used in future functional genomics applications (see Section 7.2.4). Certain contaminants (e.g. persistent organochlorines, mercury) and other exposure biomarkers (such as the enzymes important in the breakdown of the contaminants, the cytochrome P450s) can also be measured from biopsy samples (Fossi *et al.* 1997). However, care must be taken when interpreting contaminant measurements in blubber samples if only the superficial blubber layers are sampled.

For some species, additional types of sample collection are also available to complement monitoring studies. For example, a visual health assessment model has been proposed for use in the endangered North Atlantic right whale (*Eubalaena glacialis*) (Pettis *et al.* 2004), with additional efforts being directed towards acquiring baseline health information from faecal samples. Faeces can provide supplementary exposure information, such as the prevalence of specific parasites (Hughes-Hanks *et al.* 2005) as well as reproductive status from hormone metabolites (Rolland *et al.* 2005).

7.2.4 Disease diagnosis

Diagnosing disease in marine mammals and necropsy sampling can be highly sophisticated. It is beyond the scope of this chapter to give details, but these subjects are meticulously covered in the books *Marine Mammal Medicine* (Dierauf and Gulland 2001) and *Marine Mammals Ashore* (Geraci and Lounsbury 2005).

Genomics

Rapid advances in genomics have created new opportunities for understanding the role of gene function in the health of marine mammals and in the diagnosis of disease. As emerging and resurging diseases become more of an issue for marine mammal conservation and disease outbreaks of unknown aetiology more common (Gulland and Hall 2007), molecular methods are likely to prove to be particularly important.

The increasing number of genome assemblies currently available may make it possible to interpolate gene function in species that are, as yet, relatively poorly studied. Functionally homologous genes in other species that are activated in response to infection or toxin exposure may be useful for studying marine mammal health. Although no complete marine mammal genome has been described, a low coverage (×2) assembly of the bottlenose dolphin genome is available as part of the Broad Institute's Mammalian Genome Project (http://www.broad.mit.edu/mammals/). While many of the genomic tools that exist for model organisms are also not yet available for use in marine mammal species, some of the powerful, now-routine methods for screening tissues for differential transcript expression during disease states have enormous potential to identify differences in gene regulation. Some of these may already be applicable to marine mammals. However, knowing about gene expression does not necessarily lead to a functional understanding. Microarrays are enormously powerful tools that enable the activity of a vast number of genes to be determined in a target sample. DNA microarrays containing thousands of gene probes can be simultaneously exposed to a target sample. The probes, selected from cDNA fragments, are spotted onto a solid support and the expression of the corresponding messenger RNA molecules in the target sample determined. However, interpretation of this vast amount of information requires careful management and the availability of reference databases to allow proper analysis. Bioinformatic techniques have been developing simultaneously, but careful study design and statistical modelling is needed to ensure meaningful results (Kerr and Churchill 2001; Churchill 2002). The one array from a marine mammal, again the bottlenose dolphin, has been produced from lipopolysaccharide (LPS) and human interleukin-2 (IL2) stimulated peripheral blood leucocytes (Mancia *et al.* 2007). Such an array is especially useful for detecting leucocyte stimulation and specific gene upregulation. However, genes that have been downregulated in the experimental sample or genes that have been upregulated through other factors may not be represented on such an array. Dogs are the closest relatives to the pinnipeds for which an array is available. If used for marine mammals, controlling for cross-species hybridization would be critical but manageable with thorough analysis and follow-up experimentation to validate the results. However, as more microarrays from different marine mammal species become accessible this will be an extraordinarily powerful tool.

Whatever the final samples and procedures used it is essential that the collection, methods, and data reporting should be standardized. For microarrays in particular, guidelines have been established by the International Microarray Gene Expression Data (MGED) group (http://www.mged.org/Workgroups/MIAME/).

One related field that will considerably advance our understanding and identification of marine mammal disease processes is known broadly as pathogenomics (Pompe *et al.* 2005). This is the examination of both the pathogen *and* the host and how they interact with each other. The study of pathogens in humans and domestic animals is undergoing major changes, largely due to the availability of whole genome sequences, new screening technologies, proteomics, comparative genomics, and bioinformatic methods for both the host and pathogen (Lederberg 2000). Molecular fingerprinting, single-nucleotide polymorphism analyses, and molecular epidemiology all allow the study of the molecular processes during infection in humans, and these methods could also be applied to the study of disease processes in marine mammals, particularly during rehabilitation or *in vitro*. Genetic approaches can also be used to identify novel diseases in combination with pathology and histology. Isolation and characterization of a novel papillomavirus from a bottlenose dolphin has used a pathogenomic approach (Rehtanz *et al.* 2006). This illustrates the power of isolating genetic material from a particular pathogen and using this to achieve a thorough understanding of a new pathogen.

These technological advances and their application in marine mammal microbiology will improve our understanding of host–microbe interactions and immune responses. Web-based resources such as the Virulence Factor Database (VFDB, Yang *et al.* 2008) and the Pathogen–Host Interaction Data Integration and Analysis System (PHIDIAS; Zuoshuang Xiang *et al.* 2007), coupled with the fact that complete genomic sequences of all the major pathogens from humans, plants, and animals are now available (Pallen and Wren 2007), will increase the potential for comparative genomics in marine mammals—including the determinants of virulence and other hidden aspects of disease pathogenesis.

Toxicogenomics

Toxicogenomics involves investigating the influence of a chemical compound on genes and their expression. Genomics has revealed that very few single inherited human alleles are directly associated with a specific disease risk. Over 95% of diseases are caused by a repertoire of genes that modify each other and are influenced by some kind of environmental exposure. This is a particularly pertinent area of investigation for marine mammals due to their well-documented exposure to environmental contaminants and the detrimental effects of such contaminants on their health (O'Shea 1999). There is considerable crossover in methodology and data from pathogenomics, but disease causation cannot be fully understood until the contribution of both genetics and the environment are considered.

These studies rely on a detailed understanding of the environment that the study species lives in and how and when it has been exposed to particular toxins. To some extent this can be accomplished by measuring exposure in the individuals themselves (e.g. lipid-soluble contaminants from blubber biopsy samples), but for a risk assessment approach this will often need to be combined with additional exposure information such as prey type, quality, and quantity. The ultimate aim of such research is to discover susceptibility genes and alleles to particular toxins. This will allow the testing of individuals or population screening for additional susceptibility from chemical exposure.

Web-based public databases are also available for data mining and submission such as the Comparative Toxicogenomics Database (http://ctd.mdibl.org/). This is a searchable multispecies database collating toxicogenomic data to discover relationship between chemicals, diseases, and genes.

7.3 Epidemiological study designs

The design of epidemiological studies from which robust effect measures (see Section 7.2) can be estimated is essential for determining causal links between exposures and responses. Examples of the types of study and the issues involved in each are given in Table 7.2. These approaches have been developed in the fields of human and veterinary epidemiology to test causal hypotheses (including the effect of environmental and nutritional exposures which are highly comparable to the exposure routes for marine species). Despite the extensive application of these methods in other fields and knowledge of the specific definitions of the various epidemiological terms, they have not received the kind of attention they deserve in marine mammal disease research. Many of the limitations often argued for establishing causality in marine mammal health are also problematic in human epidemiology. For example, as with humans, marine mammal health studies usually have to use exposure-response data from surrogate species.

The primary objectives of an epidemiological study are to: (i) investigate the temporal and spatial distribution of disease within the different groups, and (ii) demonstrate a causal link between one or more specific factors and the frequency of occurrence of disease. The first objective is met through the application of descriptive studies such as those involving correlation and cross-sectional studies (see Table 7.2). Studies involving correlation are aimed at identifying groups that do, or do not, develop a disease, which in turn provide clues that can lead to the formulation of causal hypotheses through the identification of differences in exposures.

In a few cases, correlation study designs have already been used to investigate associations between environmental variables and marine mammal disease. For example, B. Wilson et al. (1999b) conducted a correlation analysis to identify factors associated with the occurrence of epidermal disease in bottlenose dolphins. They compared the prevalence of skin disease among dolphin populations in diverse geographical areas and then attempted to correlate the disease prevalence with anthropogenic factors such as organochlorine and trace-metal contaminant exposure, as well as environmental factors such as water temperature, salinity, and UV radiation. Their analysis indicated that lesion prevalence and severity were most strongly correlated with water temperature and salinity, and were not significantly correlated with the contaminants included in the analysis.

More intensive capture–release efforts can also support such correlational analyses, but these can also be used to conduct cross-sectional studies in which exposure and health effects are assessed simultaneously in the same individuals. While these studies still suffer from an inability to distinguish whether the exposure precedes or results from the observed disease, they enable correlation between exposure and effect on an individual level, and as 'snap-shots' they provide information on the prevalence of disease and the overall health of the population for further hypothesis generation.

Table 7.2 *Epidemiological study designs.*

	Description	Issues/disadvantages
Descriptive design	Describes general characteristics of disease distribution; useful for generating hypotheses	Cannot prove causality
Case series	Detailed report on condition of single individual; can suggest the emergence of new diseases	No controls, so cannot assess differences in exposure between diseased and non-diseased individuals
Correlational	Compares exposure and effect (e.g. disease frequency) on a population basis; measures of exposure and effect are not necessarily on the same individuals within a population	Does not link exposure to effect within the same individual; 'ecologic-fallacy', i.e. correlations at the population level may not hold at the individual level
Cross-sectional	Exposure and effect (e.g. presence of disease) is assessed within the same individuals at the same point of time	Generally cannot determine if exposure preceded or resulted from the observed effect
Analytical design	Explicit comparison of exposure and disease status; can be used to test epidemiologic hypotheses	Generally requires more effort and often long follow-up periods
Case-control	A group of individuals with disease and a group of individuals without disease are chosen and their exposures are compared retrospectively	
Longitudinal	A cohort of individuals is chosen based on the absence/presence of exposure and then followed for a period of time to assess the outcome (e.g. development of disease) of interest; may also be conducted retrospectively	Often requires long periods of follow-up

Classical cohort or longitudinal studies remain among the basic analytical study designs for human and veterinary epidemiologists. Here, the investigator defines two or more cohorts (populations or groups within a single population) that are free of disease but that differ according to their exposure to the potential cause of a disease (Rothman and Greenland 2005). Usually one cohort is defined as the reference or unexposed, while one or more additional cohorts are exposed (see Section 7.2). These may represent a gradation of exposure as might be possible in, for example, the study of the impact of contaminants on animals inhabiting a pollution gradient (Reijnders *et al.* 1999) or the extremes of exposure. Following individuals is often difficult for free-ranging marine mammals and it is often impossible to diagnose disease at the time it occurred. However, for some species it is feasible to estimate the likelihood that death or overt disease has occurred within the follow-up period. For example, weekly or monthly photo-identification follow-up of live-capture released individual bottle-nose dolphins in a well-characterized population, such as the Sarasota Bay, Florida population (Wells and Scott, 1990), would enable the timing of mortality or emigration or skin lesion development to be estimated. Thus, by following the fate of known, marked individuals, it is quite feasible to estimate the relationship between disease occurrence or vital rates and exposure status. Such mark–recapture studies can be used to estimate the effect of individual or group covariates on survival probabilities (Hall *et al.* 2001), and simple modifications to these designs can be envisaged that would allow the assessment of disease occurrence instead of mortality using the same framework.

A recent study by Hall *et al.* (2006b) also demonstrated the applicability of case-control studies (in which animals are defined as cases and controls and are then stratified by their exposure status as exposed or unexposed, see Table 7.2) of marine mammal strandings data, allowing estimates of relative risks using odds ratios (see Section 7.2.1) to be determined. As with case-control studies in human medicine, careful control selection and accurate retrospective exposure assessment are required to avoid bias. However, with a sufficiently large sample size it is quite feasible that case-control studies would be equally applicable to other marine mammal strandings datasets.

7.4 Risk assessment

Once a relationship between an exposure and response has been established, the next stage is to answer the 'so what' question—in other words, what is the biological significance? How important is the relationship for the population? The key questions are how to set acceptable risks and understand how risks change under different management approaches. What is 'acceptable' is clearly outside the scope of this chapter and remains an issue that must be directly gauged for each situation and species within its context. For example, what is acceptable for a large, exponentially increasing population may be unacceptable for a small, fragmented population or an endangered species.

Risk assessment, as defined by the US Environmental Protection Agency (EPA) and widely adopted across many different organizations and in various forms, is 'the process in which information is analyzed to determine if an environmental hazard might cause harm to exposed persons and ecosystems'. And since the early 1990s this process has been formalized into a risk assessment paradigm (Fig. 7.2), largely developed within the EPA and the National Research

Council but then taken up by many different groups across a very wide range of applications.

The first step in any risk assessment is the hazard identification (or problem formulation) stage. This includes the development of a conceptual model that sets out the impetus for the risk assessment and the description of the problem, including the interactions of a particular pathogen or toxin (exposure), within a defined population and a defined exposure scenario. The conceptual model thus helps to focus the risk assessment, outline its goals, breadth, and often the policy context in which it is being conducted. The hazard identification process may be triggered by differing types of observations or events, such as: the occurrence of one (or many) unusual mortality events (Gulland and Hall 2007); information from dedicated pathogen surveys, such as those carried out by Gaydos *et al.* (2004) for southern resident killer whales (*Orcinus orca*); information on a chemical spill or pollutant concentrations measured in sediments or biota (Pulster *et al.* 2005); or the observation of a harmful algal bloom from remote sensors (Flewelling *et al.* 2005).

The characterization of exposures and responses make up the analysis phase of the disease risk assessment process, for which various factors need to be considered. These include information on the temporal and spatial distribution of exposure as well as data for disease risk assessment (see Fig. 7.2) about the survival, persistence, and amplification of the agent and its concentration in the environmental media or prey. In addition, the routes of exposure and the size, demographics, and behaviour of the exposed population must be considered. These components are brought together in an exposure profile, which ideally provides a *quantitative* evaluation of the magnitude, frequency, and patterns of exposure to a toxin or disease agent for the scenario developed during the problem formulation stage. It should also include an indication of the underlying assumptions that have been made (based on scientific judgement) and a quantification of the uncertainties associated with each element (i.e. the errors around each factor). Exposure profiles could also be estimated from modelling the movements of animals in relation to point sources of exposure. For example, Littnan *et al.* (2006) evaluated the distribution and movements of Hawaiian monk seals (*Monachus schauinslandi*) in relation to coastal waters and sources of land-based, water run-off and sewage dispersal as sources of pathogens.

The second arm of the exposure–response characterization stage (see Fig. 7.2) requires information about toxicity and about the host to enable a disease risk assessment, as this will determine the impact of the agent or organism on the individual and ultimately the population (for example the age structure, immune and nutritional status of the population, its social or behavioural traits, and its foraging behaviour).

While correlational and cross-sectional studies being carried out on marine mammals are helping to identify hazards and provide information on mechanisms of effect, analytical study designs are helping to define quantitative relationships

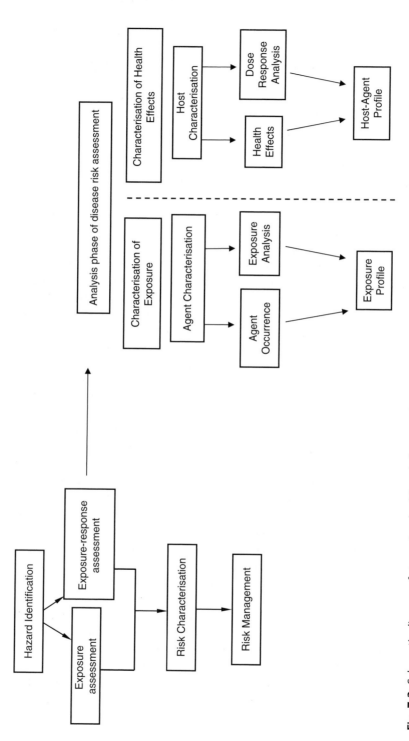

Fig. 7.2 Schematic diagram of stages in a health and disease risk-assessment procedure.

between exposure and various 'endpoints' (i.e. fecundity and survival) relevant to management, and are therefore supporting exposure–response characterization.

Dose–response relationships, as determined from laboratory studies of experimental species and widely applied for human health risk assessment, can also contribute. For example, Schwacke *et al.* (2002) took a novel approach to integrate measured tissue concentrations of polychlorinated biphenyls (PCBs) from bottlenose dolphins with a surrogate dose–response relationship (from laboratory studies in mink) to predict the health risks and associated uncertainties for dolphin populations. The results indicated a high likelihood that reproduction, primarily in primiparous females, was being severely impaired by chronic exposure to PCBs. Excess risk of reproductive failure, measured in terms of stillbirth or neonatal mortality, for primiparous females was estimated to be between 60 and 80%, whereas females of higher parity, which had previously off-loaded a large proportion of their PCB burden, had a much lower risk. These types of risk measures then become very tangible parameters that can be readily used by conservationists and managers. In another study, longitudinal data on maternal PCB tissue burdens and reproductive outcomes from the long-term Sarasota Bay dolphin study (Wells *et al.* 2005) were combined with similar data from a captive dolphin cohort (Reddy *et al.* 2001) to estimate a dose–response function linking maternal PCB burden with calf survival (Hall *et al.* 2006a). The number of individuals included for this analysis was limited and much more data are needed to increase the confidence in the dose–response model predictions in this study. But, with continued follow-up of individual dolphins as well as longitudinal study of additional populations with potentially higher PCB exposures, the necessary data are gradually becoming available.

These examples highlight one of the most difficult aspects of exposure–response assessment in marine mammal disease risk assessment. Statistical models are needed to quantify exposure–response relationships (ideally in the species of interest; but if this is not ethically or logistically possible, in surrogate laboratory model species, a situation that is the same for human risk assessments). These data are often unavailable, as many dosing studies in laboratory species do not report the final tissue concentrations that are needed for comparison with tissue concentration information collected from marine mammals. If the agent of interest is a pathogen or algal toxin, then data also required are the route of exposure, the source, and preparation of any challenge material used, as well as the organism, duration, and multiplicity of exposure.

The final phase of the risk assessment procedure then ties all the previous steps together to define population level impacts that are important for management (Fig. 7.2). The assessment then starts with a risk description of event or events, followed by a risk characterization that must include the risk magnitude and probability of potential impacts. Most importantly, the uncertainty in the risk is characterized and the confidence limits around the various estimates are reported. Sensitivity analyses can also be carried out to evaluate the most important variables and determine the information needs and control measures. Their effect on risk magnitude and profile can then be determined. Finally, decision analyses evaluate alternative risk management strategies.

In all likelihood, the health of marine mammal populations is currently being affected by an aggregation of stressors. Identification of the most critical factors and their interactions will require relatively complex analytical approaches. A recent paper by Plowright *et al.* (2008) also highlights this issue. In a comprehensive discussion of the relationship between ecology and epidemiology they explore how causal inference may be approached by bringing together the tools and techniques that have evolved in each of these separate scientific fields. The efforts to acquire the necessary data to implement such approaches are worthwhile because they promise an opportunity for us to really understand the impacts of anthropogenic actions, both positive and negative, on marine mammal populations. Much as epidemiological studies currently guide the identification of hazards, recommendations for exposure limits, implementation of preventive measures, and allocation of resources for disease responses in human populations, they can ultimately do the same for marine mammals.

8

Measurement of individual and population energetics of marine mammals

Sara J. Iverson, Carol E. Sparling, Terrie M. Williams,
Shelley L.C. Lang, and W. Don Bowen

8.1 Introduction

The overriding currency of all animal life is energy. Animals have evolved strategies of energy acquisition and use, but these strategies also experience tradeoffs between energy allocated to maintenance, activities, growth, and reproduction and are central to our understanding of life histories and fitness. Thus 'energetics'—the study of the metabolic requirements, energy use, and output of animals—underpins many areas of physiology, ecology, evolutionary, and population biology, and even ecosystem dynamics.

Marine mammals pose many challenges for the study, interpretation, and comparison of individual and population energetics. For instance, the ability to study captive animals is often restricted to a few of the smaller marine mammal species. Although opportunities in the wild may be greater, there remain serious limits to our abilities to study species in remote locations (e.g. polar bears, *Ursus maritimus*, in the high arctic), in unstable habitats (e.g. ice-associated pinnipeds in the Bering and Chukchi Seas), or due to endangered status (e.g. monk seals, *Monachus* spp., southern sea otters, *Enhydra lutris nereis*, and vaquita, *Phocoena sinus*). As a group, cetaceans pose further difficulties because of their limited accessibility and large size. Adaptive insulation (blubber) is important in temperature management (Iverson 2009a) and the presence of this comparatively inert tissue can add complexity to the issue of defining the metabolically relevant body mass in marine mammals. Conversely, the sea otter (which does not have blubber) and some fur seals are small species with little internal insulation that rely mainly on thick pelage and the strategy of using heat generated from continual activity, and specific dynamic action (SDA) from frequent feeding, to offset thermoregulatory costs (Costa and Kooyman 1984; Mostman-Liwanag *et al.* 2009).

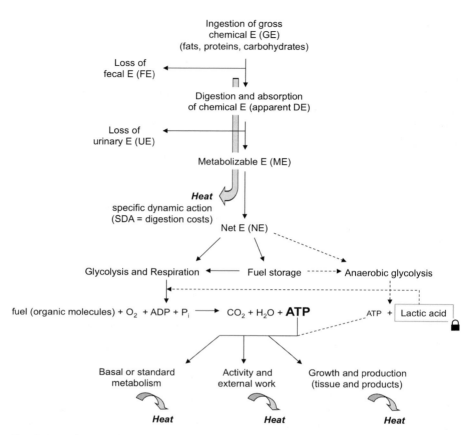

Fig. 8.1 Pathways of energy (E) acquisition and use in animals, illustrating the components that are generally measurable and form the basis for existing methods used for determining aspects of animal energetics. SDA occurs as a summed consequence of digestion and absorption processes, including synthesis of nitrogenous wastes to be exported in urine. Thus, although SDA occurs along the whole process, because its term is subtracted from ME to obtain NE, it is depicted as arising after ME. Digestible E (DE) is referred to as apparent DE because faeces contain both metabolic and undigested nitrogen. Ultimate E use can be divided into three major types of physiological work— biosynthesis, membrane transport, and external mechanical work—but are here divided into the three general categories of summed costs which are most pertinent to evaluating whole animal energetic budgets. 'Respiration' refers to the entire mitochondrial oxidative catabolic pathway, which includes the Krebs' citric acid cycle, electron transport chain, and oxidative phosphorylation, and results in the complete oxidation of fuels. The amount of adenosine triphosphate (ATP) produced from oxidative (aerobic) metabolism is about 19-fold greater than that produced during anaerobic glycolysis alone, and is thus represented by the larger font size under respiration. During anaerobic glycolysis,

Measurement and interpretation of 'maintenance metabolism' (i.e. BMR, SMR, see Box 8.1) also remains problematic for marine mammals. The rigorous conditions required for determining BMR are quite difficult to achieve for living marine mammals (including how 'resting' is even defined, as marine mammals can rest while completely submerged at depth and breath-holding, while floating at the surface intermittently breathing, and while lying on solid substrate in air). Overall, interpretation of BMR is complicated by how we define body mass, by the reliance of many species on extended periods of fasting, by characteristic apnoea and bradycardia when submerged, and by the fact that metabolic rate (MR) while actively diving may actually be lower than MR measured under criteria specified for BMR (e.g. Sparling and Fedak 2004). Boyd (2002a) rejected the idea that BMR, as classically defined for terrestrial mammals, can be realistically quantified in marine mammals. Additionally, the extent to which marine mammals use oxidative versus anaerobic metabolism (Fig. 8.1), and manage lactate production, especially under conditions of pushing physiological limits while diving (e.g. Kooyman *et al.* 1980), also warrant further study.

8.1.1 Definitions

All animals must exchange energy with their environment. The process begins in animals with energy input through the ingestion of food (fats, proteins, and carbohydrates), followed by digestion (chemical and enzymatic degradation of foodstuffs) and cellular absorption of the smaller energy-containing organic molecules produced, such as fatty acids and glycerol, amino acids, nucleotides, and monosaccharides. At several early points following ingestion, energy is lost from the animal—some directly to the faeces (undigested material) and some, after absorption, to the urine (e.g. waste metabolites) (see Fig. 8.1). The remainder of the absorbed energy-containing molecules comprises metabolizable energy (ME). However, the processing costs of both digestion and absorption result in some additional energy being expended and lost as heat (SDA, see Fig. 8.1). The remaining net energy (NE) is what the animal has to use for all physiological processes. Some of this NE is stored directly into tissues as fuel reserves, primarily as fats and proteins, with limited storage of carbohydrate as glycogen. The remainder undergoes catabolism, which comprises the cellular processes that harvest chemical free energy. During catabolism some energy is lost as heat to the surroundings, while the rest is used primarily to generate and store ATP, the

---

Fig. 8.1 (continued)
ATP is regenerated in the absence of O_2 through the reduction of pyruvate to lactic acid (lactate), which not only contains a large potential E source 'locked' in the incompletely oxidized compound, but requires sufficient O_2 to get rid of it (that is, metabolize it via the respiration pathway) and, until such time, the animal incurs an 'O_2 debt' during which it may be debilitated from further activity.

main energy-transferring molecule in the cell. The newly generated ATP then powers cellular work (mechanical, transport, chemical, and synthetic) by coupling exergonic reactions to endergonic reactions. Stored fuels can later be mobilized for catabolism during periods when nutrients are not available from the digestive tract. Once again, this cellular work produces heat, which is lost to the surroundings. In general, the most prevalent and efficient pathway used in all these ATP-driven catabolic processes is oxidative cellular respiration, in which cells consume oxygen (O_2), and produce carbon dioxide (CO_2) as a direct consequence of metabolism (see Fig. 8.1). The critical shuttle mechanism is:

$$ADP + P_i \quad \rightleftarrows \quad ATP$$

whereby energy from foodstuff bonds is used to drive the reaction to the right. When the reaction goes to the left, ADP is produced and energy is released for physiological work (P_i, inorganic phosphate).

The characteristics of this exchange of energy between an animal and its environment, outlined in Fig. 8.1, form the basis for the different ways in which animal energetics are measured, both directly and indirectly. In principle, measurement of any of the components illustrated can provide insight into some aspect of an animal's metabolism and energy budget. The MR of an animal (Box 8.1) is defined as the rate at which it converts chemical energy to heat and external work. Thus, quantification of an animal's heat production is a direct measurement of MR. Since most external work is usually also degraded to heat, heat remains the most direct measurement of MR, accomplished using calorimetry with an animal enclosed in an insulated chamber. Although feasible for smaller animals (as discussed below), such measurements are progressively more difficult for large species including most pinnipeds and cetaceans. Therefore, it is rarely used. Rather, indirect methods of

---

Box 8.1 Acronyms and definitions of terms used in energetics studies

(See also Fig. 8.1 for further terms)

Allometry: the systematic change in MR (or body proportions) with increasing body size, usually expressed as $Y = Y_o M^b$, where Y is a process rate (e.g. BMR, MMR), Y_o is a normalization constant, M is body mass, and b is a scaling exponent.

AS (aerobic scope for activity): measure of an individual's upper metabolic limits, defined as the relative increase in an animal's MR between rest and maximal MR under the same conditions: VO_{2max}/VO_{2rest} (i.e. VO_{2max} being MMR, VO_{2rest} being SMR or RMR).

BMR (basal metabolic rate): the rate at which energy is used by an organism at complete rest (i.e. 'maintenance', assumed to be the lowest stable level MR of an individual). In order to appropriately compare MR across animals, Kleiber (1975) put forth the comprehensive criteria that the animal must be at rest but awake, within its TNZ, stress-free, post-absorptive, mature (not growing), and not reproducing.

EE (energy expenditure): the rate of energy consumption per unit time.

FMR (field metabolic rate): the rate of metabolism that includes the total cost of all activities of a free-living animal.

MMR (maximal metabolic rate): the rate of metabolism at maximal activity and muscle work. May involve both aerobic and anaerobic biochemical pathways.

MR (metabolic rate): the rate of energy consumption by an organism, which is the rate of its conversion of chemical energy to heat and external work. This is measured in calories (cal) as units of heat (the energy required to raise 1 g of water from 14.5 °C to 15.5 °C) or in joules (J) as units of work (the work done to move 1 kg through 1 m), and expressed per unit time as a rate (1 J $= 1$ kg $\cdot$ m$^2 \cdot$ s^{-2}). Joules are now the standard international (SI) unit of energy measure, but can be readily interconverted with calories: 1 cal $= 4.184$ J.

RMR (resting metabolic rate): the MR when at rest, but otherwise not strictly meeting all criteria for Kleiber's BMR (e.g. when an animal is still digesting food).

RQ (respiratory quotient): the ratio of the volume of CO_2 produced to the volume of O_2 consumed during aerobic catabolism. The range of respiratory quotients for animals in energy balance varies with the substrate being metabolized. Fat and carbohydrate catabolism have an RQ of 0.71 and 1.00, respectively. Protein has an RQ of 0.8–0.9, depending on the amino acids being catabolized. RQ (measured at cells) is sometimes called the respiratory exchange ratio (RER, measured at respiratory organs).

SDA (specific dynamic action): the increase in metabolic rate as a result of digestion and absorption processes. The magnitude and duration of SDA varies with the size and composition of the meal. (Also called the heat increment of feeding, HIF.)

SMR (standard metabolic rate): synonymous with BMR, but SMR refers to measurement at a standard temperature.

TNZ (thermal neutral zone): the range of external temperatures at which an animal's requirements for metabolic heat production are at minimum. Above this zone energy must be expended to keep cool, and below more metabolic energy must be used for heat production.

VO_{2max}: maximum rate of O_2 consumption of an individual; a measure of aerobic capacity.

measuring metabolism, although in principle accompanied by greater uncertainties or errors, in reality provide better information on how an animal uses energy under natural or varying conditions. Measurements of O_2 consumption, CO_2 production, water (H_2O) turnover, food energy intake, nutrient and energy balance, fuel storage and subsequent mobilization, tissue growth, synthetic products (e.g. fetus, milk), as well as lactate production (see Fig. 8.1), can all provide valuable insights into animal energetics (see also definitions in Box 8.1).

8.2 Measurement of metabolism

This section details some of the methods available for the estimation of MR of marine mammals.

8.2.1 Direct calorimetry

Quantification of energy released as heat over a given period is the most direct method of measuring MR. Direct calorimetry is performed by placing an animal in a sealed insulated chamber or calorimeter and measuring the heat given up to the circulating air and to water flowing through the chamber walls (e.g. McLean and Tobin 1987; Walsberg and Hoffman 2005). In practice, direct calorimetry is cumbersome and calorimeters are expensive and difficult to maintain. Hence, direct calorimetry is rarely used, especially for large animals, and has not been used in marine mammals. Furthermore, it is impossible to measure the energetic cost of specific activities of wild animals in their natural environment using direct calorimetry. An alternative to direct calorimetry is the minimum heat loss method, which uses physical measurements of heat transfer across the blubber and body surface to infer total heat loss, but it is based on assumptions that are likely inappropriate (reviewed in Boyd 2002a).

8.2.2 Respirometry (measurement of gas exchange)

MRs of animals can be estimated indirectly from the measurement of other variables related to energy utilization, such as O_2 consumption and energy balance (e.g. see Fig. 8.1). 'Indirect calorimetry', although in principle representing any indirect method, most often refers to the measurement of rates of gas exchange (where $VO_2 = O_2$ consumption, $VCO_2 = CO_2$ production) and the translation of these quantities into a heat or work equivalent. Because steady-state metabolism consumes O_2 and produces CO_2 in known amounts (see Fig. 8.1) depending on the type(s) of food molecules being catabolized (i.e. the RQ, Box 8.1), their measurement can be used as an indirect assessment of MR. Respirometry refers to the measurement of these gases. Most relevant for use in marine mammals is open-flow (or open-circuit) respirometry, where air is pumped through a metabolic chamber at a rate that constantly replenishes the O_2 depleted by the animal while removing the CO_2 and water vapour produced by the animal (e.g. Renouf and Gales 1994; Boily and Lavigne 1995; Sparling *et al.* 2006) (Fig. 8.2 and Box 8.2).

There have been several innovative approaches to adopting this technique for use on marine mammals, which take advantage of the fact that they perform many activities underwater yet must return to the surface to breathe. This constraint has enabled researchers to quantify the metabolic costs of swimming and diving in both captive and free-ranging animals. Thus, open-flow respirometry has been applied to sea otters diving in a simulated environment (Yeates *et al.* 2007), otters and pinnipeds swimming in flumes (Davis *et al.* 1985; Fedak *et al.* 1988; T.M. Williams 1989), and exercising cetaceans (Worthy *et al.* 1987; T.M. Williams *et al.* 1993). Examples for free-ranging marine mammals include an isolated ice-hole model developed in the Antarctic for the study of Weddell seals (*Leptonychotes weddellii*, Kooyman *et al.* 1973; T.M. Williams *et al.* 2004), an open-ocean Steller sea lion

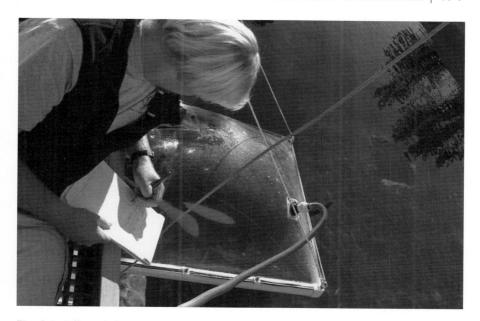

Fig. 8.2 Killer whale (*Orcinus orca*) resting in a metabolic hood for open-flow respirometry. Note that the water surface provides an airtight seal with the hood, thus creating a respirometer for the animal. Hoses connect to a vacuum pump that provides air into the hood at 500 L per minute and samples to the gas analysers. For this and other species studied, the air flow rate must be high enough that the animal is never in danger of being hypoxic, yet low enough so that the amount of O_2 consumed (VO_2) by the animal can be detected with precision, ideally between 20 and 21% O_2. Expired air from this open-circuit respirometer is continuously sampled by gas analysers for determination of the fractional concentration of O_2 and CO_2, which is recorded with a laptop computer. Calculation of the rates of VO_2 or VCO_2 is reliant on knowledge of flow rates into and out of the chamber or hood, plus the fractional concentrations of the gas mixtures in and out of the system. The exact equation used also depends on where the flow meter is relative to the chamber and on whether or not water vapour and CO_2 are present or removed from the air stream prior to analysis (see Withers *et al.* (1977) for a general discussion, and Fedak *et al.* (1981) for details regarding marine mammal respirometry). Energy expenditure (EE) can be calculated from measured VO_2 using the energy equivalent of O_2. The simplest method is to assign a mean energy equivalent for O_2 of 4.83 kcal per litre of O_2. However, this value will vary with the substrate being oxidized (from 4.73 at an RQ of 0.71 to 5.04 at an RQ of 1). In the absence of a measured RQ, one can assume an RQ based on diet. The equation of Weir (1949) allows the calculation of EE if both O_2 and CO_2 are measured: EE (kcal/day) $= VO_2(3.9) + VCO_2(1.1)$. (Photo courtesy of Terrie M. Williams.)

(*Eumetopias jubatus*) laboratory (Hastie *et al.* 2007), and a combined respirometry/ simulated foraging approach (Sparling *et al.* 2007).

The size of some marine mammals, especially the larger odontocetes and mysticete whales, prevents the use of many of the open-flow respirometry methods described in the preceding section. Instead, expired gases may be collected as the whale breathes into a non-permeable bag or balloon. This has been done success-fully for trained killer whales (Kriete 1995), as well as grey whales (*Eschrichtius robustus*) in captivity (Sumich 2001) and in open water (Sumich 1983). A cautionary note when attempting this method concerns the assurance of normal breathing patterns by the subject. Forced exhalations due to obstruction of the blowhole or individual animal responses can result in elevated end-tidal gas levels, which lead to an overestimation of MR (T.M. Williams, personal observation).

Despite the difficulties, there are several advantages for using respirometry methods for energy assessments. Most importantly, this technique allows the detailed quantification of energy costs related to specific activities. Thus, the effects of temperature regulation, exercise, digestion, age, body size, at-sea behaviours, moult-ing, pregnancy, and lactation among others can be evaluated. In addition, if both O_2 consumption and CO_2 production are measured, then the RQ (see Box 8.1) can be calculated, which provides information about the fuel being metabolized. If only O_2 consumption is measured (common in marine mammal studies) then information on the fuel substrate used is needed to translate VO_2 into actual EE (i.e. RQ in Box 8.1 and Fig. 8.2 legend). Furthermore, some calculation techniques are relatively insensi-tive to variations in RQ (Fedak 1986) in cases where only VO_2 is known.

The disadvantages of respirometry are that the measurement of gas exchange can be used in few situations, and results must be carefully interpreted in light of the experimental design. In general, the behavioural repertoire of captive marine mam-mals relative to their wild counterparts cannot be reproduced; this is especially apparent for costs related to interactions between animals. Perhaps, most important for marine mammal energetics is the difficulty of measuring a true diving MR. Open-flow respirometry can only measure gas exchange while the animal is at the water surface. Partitioning the submergence and surface MR post dive is complicated at best. During the surface interval, MRs will include the O_2 consumed for surface resting as well as for recovery from the previous dive and for replenishing O_2 stores for subsequent dives. Different analytical approaches have been used to solve this problem. The simplest is where the VO_2 taken up during a surface period is assumed to have been consumed over the entire previous dive plus the surface recovery period; this is expressed as diving MR, but it does not attempt to partition between O_2 consumption while submerged versus at the surface (Castellini *et al.* 1992; Reed *et al.* 1994; Sparling and Fedak 2004). Another approach is to first measure the RMR (generally determined at the onset of the trial) and then calculate the diving O_2 consumption from the post-dive O_2 uptake minus the volume of O_2 equivalent to the RMR (Hurley and Costa 2001; T.M. Williams *et al.* 2004a; Hastie *et al.* 2007). If the latter approach is used, it is important that the resting VO_2 during the post-dive recovery period is equivalent to the pre-dive resting VO_2. Also, there is the problem of delayed metabolic responses and recovery in diving marine mammals. Sparling *et al.* (2007) provided evidence supporting the hypothesis that diving seals can defer the metabolic costs of digestion until after periods of active foraging, and thereby maximize the O_2 available for extending the duration of dives. This highlights the importance of the timescale over which measurements are made. If diving animals are deferring the metabolic costs of certain processes then it is crucial that any study attempting to quantify energetic costs does so over an appropriate period of time .

8.2.3 Doubly labelled water (DLW) and isotope dilution

Doubly labelled water (DLW)

Gas exchange can also be measured indirectly using radio- or stable-isotopes of hydrogen and oxygen. Thorough accounts of the theory, history, techniques and advantages and disadvantages of the DLW technique have been published (Speakman 1997; Butler *et al.* 2004). This technique involves quantitatively dosing weighed animals with either deuterium- or tritium-labelled water (^2H_2O or ^3H_2O, respectively) and isotopic O_2 (^{18}O), and then sampling the body water (usually blood, but also urine or saliva) after complete isotope distribution (equilibration) throughout the body water pool. A blood sample may need to be collected before isotope administration to determine background levels, and is absolutely required if the individual has previously been dosed. In principle, the isotope mixture can be administered intramuscularly (IM), intravenously (IV), intraperitoneally (IP), subcutaneously (SC), or orally (gastric intubation). During equilibration the animal must not consume food or water. Equilibration can take about 1–6 h, depending on the species and body size, and this should be confirmed with either serial sampling to confirm a plateau in isotope concentration or waiting for the appropriate established maximum period. The animal is then released and later recaptured once, or in a series of recaptures, to collect a blood (or other) sample for the measurement of subsequent isotope elimination over a specified period (days).

The decision to use ^2H_2O or ^3H_2O will depend largely on the size of the study animal and analytical methods available, and on regulations concerning the use of radioisotopes (^3H_2O). Usually, ^2H_2O is most practical for individuals weighing less than about 130 kg, especially if infrared spectrometry is used for analysis (Oftedal *et al.* 1987b; Oftedal and Iverson 1987; Iverson *et al.* 1993); however, if analysed using isotope-ratio mass spectrometry (e.g. Speakman 1997, 2001; Sparling *et al.* 2006), ^2H_2O has been used in walrus (*Odobenus rosmarus*) weighing up to 1597 kg (Acquarone and Born 2007). ^3H_2O is used in relatively small quantities and analysed using liquid scintillation spectrometry (e.g. Ortiz *et al.* 1978; Reilly and Fedak 1990). Samples are analysed for ^{18}O using isotope-ratio mass spectrometry (e.g. Speakman 1997, 2001; Sparling *et al.* 2006).

Isotope concentrations measured at equilibration allow calculation of the body water pool size (see Section 8.3.2). EE is inferred from estimated CO_2 production, which is calculated from the differential elimination of the hydrogen and oxygen isotopes (i.e. labels) from the body water pool (see Box 8.2). The oxygen isotope is eliminated from the body by continuous flux through the body of both water and expired CO_2, but the hydrogen isotope is only eliminated by water flux. The difference between the two elimination rates is correlated with CO_2 production (Lifson *et al.* 1955). Multiplying the difference in the gradients of the exponential declines in isotope enrichments over time by the size of the body water pool gives a quantitative estimate of CO_2 production. There are many complexities involved in correcting for differential distribution spaces of the labels and fractionation during elimination, involving a number of alternative calculation methods (Speakman 1997). Sparling *et al.* (2008) found a good correspondence between DLW-derived estimates of daily EE and those measured using continuous open-flow respirometry, although unusually high MRs have been observed in some studies (e.g. Boyd *et al.* 1995a). Several equations are available for calculating the rates of CO_2 production in DLW studies, varying between one- and two-pool models (separate calculations for hydrogen body pool and oxygen body pool) and differing in the treatment of fractionation effects. Sparling *et al.* (2007) described these different approaches and concluded that two-pool models, which use the measured dilution space ratio and include a correction for fractionation effects, are most appropriate for marine mammals. Speakman's model (1997; eqn 7.43 therein) was judged to be the most appropriate for future marine mammal studies (see Box 8.2).

The principal advantage of DLW is that it provides a measure of FMR, integrating the costs of all activities over the measurement period. This is the most ecologically relevant measure of metabolism. Although expensive, the costs of isotopes, in particular ^{18}O, have declined and developments in mass spectrometer technology have enabled labelling at lower dosages. Thus, DLW studies of marine mammal MR are now feasible for more species, but larger doses of isotope are still required for the same level of precision and accuracy in larger animals. Methods are not well-suited to free-ranging cetaceans or sirenians due to the general capture and restraint requirements. Additionally, field studies can be logistically difficult, with the need to recapture the same animal within a relatively short, specified time window (e.g. about 7–10 days). Although this period is long compared to that needed for techniques such as respirometry, there remain limits to the ecological relevance of individual measures of EE made using DLW. However, useful estimates of FMR can be gained for many species if field studies are carefully designed and executed at several times of the year. An important issue that is often overlooked in DLW studies is that of generating measures of uncertainty in estimates of FMR. Large random errors can occur when using DLW to measure metabolism in individuals, so it is usually important to use grouped data when describing MR using DLW. A Monte Carlo simulation and bootstrapping approach incorporating uncertainty in input parameters (e.g. analytical variability, measurement error in isotope injection volume)

Box 8.2 Calculations of gas exchange in studies of metabolism

Respirometry

There are numerous configurations of open-flow respirometry systems and the basic equations for calculating VO_2 and CO_2 vary depending on which is used. Generally in marine mammal studies, flow rate is measured downstream of the chamber and a subsample of the flow is dried (using a chemical drying agent) before going to the gas analysers—the equations below assume this is the case.

- If CO_2 is not measured and is absorbed after flow measurement but prior to O_2 measurement, RQ has to be assumed and VO_2 can be calculated as follows:

$$VO_2 = (FiO_2 - FeO_2) \cdot F/(1 - FiO_2 + RQ \cdot (FiO_2 - FeO_2))$$

- If CO_2 is measured:

$$VO_2 = F \cdot ((FiO_2 - FeO_2) + FiO_2 \cdot (FiCO_2 - FeCO_2))/(1 - FiO_2)$$

where Fi = the fractional concentration of O_2 or CO_2 in incurrent air, Fe = the fractional concentration in excurrent air, and F = the rate of flow through chamber.

The above equations require the regular calibration of flow meters and gas analysers. However, the following is a simple formula for use where extreme accuracy is not required, RQ is assumed, gas is dried, and CO_2 is absorbed before the gas analyser.

$$VO_2 = (0.2094 VN_2/0.8)(\Delta C/\Delta C^*)$$

This involves using the nitrogen (N_2) calibration technique of Fedak *et al.* (1981) where ΔC and ΔC^* refer to the deflection of the O_2 analyser during measurement and calibration, respectively, and VN_2 is the volume (or flow rate) of N_2 used in the calibration. This has the advantage of eliminating the need to calibrate the O_2 analyser or measure the flow past the animal, thus producing increased accuracy. This is also particularly useful for situations of high ambient humidity as it is relatively insensitive to errors due to water vapour.

Doubly labelled water (DLW)

Sparling *et al.* (2008) suggested that the most appropriate calculation model for marine mammals was the double-pool model from Speakman (1997):

$$rCO_2 = (N/2.078)(k_o - R_{dilspace} \cdot k_d) - 0.0062k_d \cdot N \cdot R_{dilspace}$$

where: rCO_2 = rate of CO_2 production; $k_{o,d}$ = rate of turnover of O_2 and deuterium

$(^{2}H_{2}O)$ isotopes; $R_{dilspace}$ = ratio of the two dilution spaces N_{d}/N_{o}; N = body water pool calculated as $N_{o} + (N_{d}/R_{dilspace})/2$.

There are also equivalent equations when tritium $(^{3}H_{2}O)$ is used as a label (see Speakman 1997). Tritium has the advantage of being easy to measure with great accuracy and precision by detecting the beta particles it emits. However, this radioactivity also means that there are safety implications, and hence require very strict safety procedures for its use and disposal. A further disadvantage is the absence of background levels of tritium—covariation in drifts in the background levels of ^{18}O and deuterium tend to cancel each other out—this covariation cannot be used to minimize error when using tritium.

has been used in previous studies using DLW to generate confidence limits for estimates of EE (Boyd *et al.* 1995a; Speakman 1995; Sparling *et al.* 2008).

Singly labelled water (SLW)

Hydrogen isotopes of water $(^{2}H_{2}O$ or $^{3}H_{2}O)$ alone, can also be used to estimate MR, EE, and food intake and at a substantially lower cost for both label and analyses. Quantitatively dosing an animal and sampling (again, usually blood) after equilibration allows measurement of the dilution space and total body water pool, and from this total body energy can be estimated (see Sections 8.3.1 and 8.3.2). A second dose and equilibration after a specified period allows estimation of change in total body energy over the interval between doses. If the animal is fasting throughout this interval, which occurs for routine periods in many marine mammals, this change equates directly to total EE. In addition, serial sampling of the body pool after equilibration allows measurement of water flux over time, which can be used to estimate EE and energy/food intake, given knowledge about the food type consumed and calculation of metabolic water production (MWP, see Section 8.5.2, Box 8.3 for an estimation of milk intake; for food intake see the adaptation of this equation in Bowen *et al.* (2001) and Muelbert *et al.* (2003) which accounts for the digestible energy (DE) of prey).

8.2.4 Proxies for assessing energy expenditure (EE)—heart rate (f_{H}) and stroking rate (f_{S})

The cryptic behaviours of most wild marine mammals make energetic assessments by respirometry and DLW methods challenging. New tagging technologies coupled with knowledge concerning the correlation between many physiological parameters and MR enable the use of proxies for free-ranging marine mammals. Depending on the species and question to be evaluated, f_{H} provides such a useful proxy. Under steady-state conditions, there is a good linear relationship between f_{H} and MR in most vertebrates studied to date (Butler *et al.* 2004). This is based on the relationship (Fick 1870) between VO_{2} and f_{H}:

$$VO_{2} = (C_{a}O_{2} - C_{v}O_{2}) \cdot V_{S} \cdot f_{H}$$

where V_S is the stroke volume of the heart and $C_aO_2 - C_VO_2$ is the difference in O_2 concentration between arterial and mixed venous blood. Although there are large variations in f_H due to diving activity, Fedak (1986) demonstrated a linear relationship between VO_2 and f_H in a grey seal (*Halichoerus grypus*) if values were averaged over complete dive cycles. Similar results have been reported for Californian sea lions (*Zalophus californianus*), harbour seals (*Phoca vitulina*), and bottlenose dolphins (*Tursiops truncatus*) (T.M. Williams *et al.* 1991, 1993; Butler *et al.* 1992; Boyd *et al.* 1995a).

The general configuration of an f_H recording system consists of electrodes that detect the heart rate signal connected to a logger which stores the signal for later processing. The logger may be directly attached to the electrodes or the signal may be relayed via radiotelemetry to an externally mounted logging device. The f_H signal has been measured in marine mammals using the R-wave detector (most commonly used), the Holter monitor, and the digital electrocardiogram recorder (see Ponganis 2007 for a detailed review). Off-the-shelf tags (tested on free-ranging marine mammals) are currently manufactured by Wildlife Computers (Redmond WA) and UFI Technologies (Morro Bay, CA), using surface-mounted or implantable electrodes. Implantable electrodes require surgery and have been known to induce substantial inflammatory responses in some species (Green *et al.* 2009). Artefacts due to muscle-generated electrical impulse must be avoided; the ability to assure this will dictate the selection of an R-wave detector versus a digital electrocardiogram recorder. In the latter case, skeletal muscle artefacts can be screened and corrected. Surface attachments will only last a finite amount of time and it is also important to recognize that surface electrode attachments have a finite life, with the subsequent decay of f_H signals leading to a falsely assumed decrease in f_H.

Butler *et al.* (2004) gives a detailed account of the advantages and disadvantages of the use of f_H to estimate MR. The greatest advantage of measuring f_H is that it can potentially provide estimates of EE of free-living animals at varying temporal scales. MR can be estimated for specific types of behaviour (Butler *et al.* 1992), with the potential to examine longer term variation in the FMR of free-living individuals, as demonstrated in penguins (e.g. Green *et al.* 2005).

A difficulty of this method is that the relationship between f_H and MR must be established for each species. Realistically, this can be achieved in the laboratory for only a few marine mammals. Additionally, the f_H/VO_2 relationship may differ under conditions of digestion (McPhee *et al.* 2003) and extreme bradycardia (Boyd *et al.* 1995a). Studies on birds have shown that the f_H/VO_2 relationship can vary under differing conditions (e.g. exercise, feeding, thermoregulation) and therefore calibration studies need to be tailored to the likely range of activities undertaken and conditions experienced by free-ranging animals (Green *et al.* 2006). As with many other field methods like DLW, high levels of individual variation have been demonstrated and need to be recognized when interpreting the data (Butler *et al.* 1992; Boyd *et al.* 1995a; McPhee *et al.* 2003).

A comparatively new method for determining the energetic demands of free-ranging marine mammals involves the use of instrumentation to determine f_S. Like heart beats, marine mammals expend energy for each individual swimming stroke taken. With knowledge about the RMR of an animal, the cost of individual swimming strokes, and the number of strokes taken during specific activities, it is possible to calculate the EE for discrete activities. Animal-borne cameras (Davis *et al.* 1999, 2001; Bowen *et al.* 2002) and 3-axis accelerometer microprocessor tags (T.M. Williams *et al.* 2004a; R.P. Wilson, *et al.* 2006) have been used for recording individual strokes. Although a powerful method for detailing the energetic cost of discrete activities including the energetic impact of living in different habitats, the f_S method requires knowledge about the RMR and swimming metabolism of each species. Fortunately, with increased use of this method, stroke-cost libraries for marine mammals are being developed which will provide a database for converting f_S data into an energetic demand (e.g. see T.M. Williams *et al.* 2008 for a comparison of locomotor costs in marine and terrestrial mammals).

8.2.5 Proxies for assessing energy expenditure (EE)—allometry

The logistics of measuring the metabolism of marine mammals, especially under free-ranging conditions, have instigated the use of allometry to predict both short-term and long-term EE. A wide variety of predictive equations are available, but they are only as good as the data comprising them (see Table 9.1 in Boyd 2002a for a compilation of marine mammal metabolic data). Of particular concern is the extrapolation of allometric regressions to include exceptionally large marine mammals for which few data are available. Regardless, the use of allometry can provide a starting point for predicting EE. The main advantage of using allometry is that it is relatively simple and relies only on a measure (or estimate) of body mass. Estimates of metabolism using allometry may be the closest one can get to measuring the MR of the largest cetaceans.

The allometric relationship between body mass and MR is one of the most frequently used energetic relationships. Among marine mammals:

$$BMR = 1.93 \text{ mass}^{0.87}$$

in phocid seals resting submerged (Lavigne *et al.* 1986) and

$$FMR = 30.43 \text{ mass}^{0.524}$$

in pinnipeds and cetaceans (Boyd 2002a), where MR is in watts and mass is in kg. Although some debate exists as to the exact value of the exponent for mammals

MR is generally thought to be proportional to body mass raised to ¾ power. The link between allometric predictions of BMR and estimates of FMR is generally a single multiplier. A general rule of thumb is that the RMR of marine mammals averages 2 times the predictions for terrestrial mammals based on Kleiber (1975), and that FMR is 3–4 times BMR (*see* Costa and Williams, 1999; Boyd 2002a). Exceptions do occur and include comparatively low resting MR for manatees, and comparatively low FMR for deep-diving phocid seals.

The simplicity of using allometry to predict MR can be misleading, and may be complicated in the case of marine mammals. The effects of varying amounts of metabolically inert blubber, especially for species that undergo large seasonal changes in body composition, and uncertainties over the true scaling exponent (which may differ between marine and terrestrial species) will affect the applicability of generalized allometric regressions for individual species. Consequently, allometry may be more applicable to the estimation of MR at the group or population level than amongst individuals.

8.3 Estimating body composition

Body composition analysis divides body mass into components on the basis of differing physical properties. The energy content and nature of the major constituents of an animal's body (lipid, protein, and carbohydrate) can be used to estimate the quantity of stored energy and the nutritional status of an individual, but it can also be used to understand energy gain and expenditure, as well as nutrient utilization during fasting. Body mass is divided into fat and fat-free mass (FFM, sometimes referred to as lean body mass). However, FFM is heterogeneous, including protein, water, and bone mineral. A number of approaches are available to estimate the body composition of marine mammals, and the approach used depends largely on the size and type of animal and on the nature of the question being investigated.

8.3.1 Carcass analysis and the relationship between chemical constituents

The most direct method of estimating the energy content of an individual is by means of bomb calorimetry of the whole carcass or, more frequently, of subsamples of the homogenized carcass (e.g. Blaxter 1989; Estes *et al.* 1998). Given that stored carbohydrate is negligible in mammals, direct determination of the fat and protein content of such subsamples (e.g. Oftedal *et al.* 1987b) can also be used to convert to body energy equivalents using standard values for the energy density of fat (39.5 MJ/kg) and protein (23.5 MJ/kg) (Schmidt-Nielsen 1980; Worthy and Lavigne 1983). However, the need to homogenize the entire carcass in a whole-body grinder generally limits this approach to fairly small marine mammal species or juveniles.

Given the large size of marine mammals, dissection is more generally used to prepare the various body components (i.e. carcass, viscera, blubber, and skin) for analysing separately by either bomb calorimetry or chemical analysis. Care must be taken to reduce evaporation and fluid loss during dissection and to correct for these losses in the final calculations (e.g. Oftedal *et al.* 1989; Reilly and Fedak 1990; Arnould *et al.* 1996). Measurement of the total mass fraction of these body components combined with an estimate of their energy content, has been used to study energy storage and depletion in pinnipeds, cetaceans, and polar bears (e.g. Bryden 1972; Lockyer *et al.* 1985; Bowen *et al.* 1992; Pond *et al.* 1992).

Chemical carcass analysis also provides the standard from which other methods of measuring body composition (see Section 8.3.2) can be validated (e.g. Reilly and Fedak 1990; R. Gales *et al.* 1994b; Arnould *et al.* 1996). That is, from the early empirical studies of Pace and Rathbun (1945), the composition of lean body mass or FFM in mature mammals is assumed to contain a relatively constant proportion of water ($\sim$73%) and protein ($\sim$20%). Thus, in principle in the two-compartment model, if the total body water (TBW) of an animal is known, from this FFM and thus total body protein (TBP) content can be calculated, with the remainder of the body mass comprising total body fat (TBF). From these values, total body energy (TBE) can be estimated as above. Although this principle remains generally valid, the precise values for the relationship between TBW and other body constituents have been re-examined many times and there is some evidence for species-specific relationships. For instance, Reilly and Fedak (1990) empirically established the relationships between body components and TBW in grey seals and derived the following predictive equations:

$$\%TBF = 105.1 - 1.47(\%TBW)$$

$$\%TBP = 0.42(\%TBW) - 4.75$$

$$TBE\ (MJ) = 40.8(\text{body mass, kg}) - 48.5(TBW,\ kg) - 0.4$$

These regression equations accounted for 95–99% of the observed variation in gross body components and have been widely used for pinnipeds, but may be best applied to other phocids. Species-specific equations have also been determined for harp seals (*Phoca groenlandica*, R. Gales *et al.* 1994a) and Antarctic fur seals (*Arctocephalus gazella*, Arnould *et al.* 1996). In the absence of species-specific data, application of the most appropriate equation for a given family or age-group is recommended.

8.3.2 Total body water (TBW) measurement

The most accurate method for non-destructively estimating body composition in marine mammals is based on hydrogen isotope (^2H_2O or ^3H_2O) dilution to determine TBW from the resulting isotope dilution space (Speakman 2001). Other body components and TBE can then be estimated from their empirically derived relationships with TBW (see Section 8.3.1). As with DLW and SLW (see Section 8.2.3), knowing the amount of isotope administered, the isotope concentration at equilibration, and the body mass of the animal allows the dilution space to be calculated, which can then be converted to TBW. That is, while ^2H_2O or

^3H_2O do mix almost entirely with TBW, a small fraction of isotope is lost to rapidly exchangeable hydrogen atoms in organic constituents of the body and thus dilution space will slightly overestimate TBW (reviewed in Bowen and Iverson 1998). A comparison of a number of studies on phocid and otariid pinnipeds, in which both dilution space (using either ^2H_2O or ^3H_2O) and carcass desiccation were measured in the same individuals, revealed a single predictive regression equation which fits all species and can be used to calculate TBW (Bowen and Iverson 1998):

$$TBW = 0.003 + 0.968 \cdot (\text{dilution space})$$

Using TBW and empirically derived equations (see Section 8.3.1), TBF, TBP, and TBE can then be calculated. A number of detailed descriptions of the field and analytical methods, as well as calculation procedures used in hydrogen isotope dilution studies of body composition, are available in the literature and serve as useful examples of specific procedures (e.g. Reilly and Fedak 1990; Iverson *et al.* 1993; Arnould *et al.* 1996; Speakman 1997, 2001; Mellish *et al.* 1999; Sparling *et al.* 2006; Acquarone and Born 2007; Hall and McConnell 2007).

8.3.3 Ultrasound

Imaging ultrasound is also used to measure blubber depth as an index of body condition. B-mode ultrasound comprises a linear array of transducers that simultaneously scan a plane through the animal; this is viewed as a two-dimensional image on a screen from which measurements of internal structures can be obtained. Using a portable ultrasound scanner, blubber depth is measured at a series of dorsal, lateral, and ventral points along the body of a sedated or restrained animal (e.g. Mellish *et al.* 2004). The blubber depth measurements are combined with measurements of length and girth taken at the same points along the body, and the animal is then modelled as a series of truncated cones from which volume (Vs) of each blubber section can be calculated:

$$V_s = 1/3\pi h \left[(2r_x d_x + r_x d_y + r_y d_x + 2r_y d_y) + \left(d_x^2 + d_x d_y + d_y^2 \right) \right]$$

where h is the length of each body cone, $d_{x,y}$ is blubber depth and $r_{x,y}$ is radius of the animals at site x and y (*see* N.J. Gales and Burton 1987; McDonald *et al.* 2008). The total volume of the blubber can be converted to total mass and fat using estimates of blubber density and lipid content; FFM can then be estimated by subtraction (see Section 8.3.1). The method has been validated against carcass analysis and isotopic methods in several species of phocid seals (Slip *et al.* 1992; Worthy *et al.* 1992; N.J. Gales and Burton 1994; Webb *et al* 1998; McDonald *et al* 2008) and has more recently been used in free-ranging bottlenose dolphins (Noren and Wells 2009). To assess reduction in measurements taken, Hall and McConnell (2007) found that a single blubber depth measured at the dorsal midpoint between the foreflippers was the best predictor (explaining 72.5% of variability) of total body fat in juvenile grey seals.

8.3.4 Bioelectrical impedance analysis (BIA)

Bioelectrical impedance analysis (BIA) was initially developed and validated as a rapid and non-invasive means of estimating TBW and, thus, body composition, in humans (*see* Lukaski 1987). It has since been adapted for use in domestic species and a wide range of free-ranging mammals. Predictive relationships between TBW (estimated by isotope dilution) and impedance measurements have been developed in grey seals (Bowen *et al.* 1999), but results in other pinniped species have been less promising. Tierney *et al.* (2001) found significant, positive relationships between TBW in southern elephant seals and BIA variables, but the level of accuracy was inadequate for BIA to be more useful than the other methods. Similar results were reported in harbour seals (Bowen *et al.* 1998) and in female Antarctic fur seals (Arnould 1995).

BIA is based on the conduction of a known, low-level, alternating electric current through an organism. Conductivity is related to water and electrolyte distribution. Because FFM contains most of the body water and electrolytes, conductivity is greater in fat-free tissues than in fat and, therefore, the impedance of the electrical current is dependent on the body composition of the organism (see Lukaski 1987 for a review of the principles and equations). For each individual species a predictive relationship between TBW (estimated either by carcass analysis or hydrogen isotope dilution) and impedance must be developed. Because electrode configuration and placement on the body can have significant effects on both the magnitude and repeatability of measurements (R. Gales *et al.* 1994b; Farley and Robbins 1994; Arnould 1995) they should be tested to ensure that (1) total body impedance is being measured and (2) that the measurements are reproducible. Complete protocols, including electrode configurations and placements have been described in detail for phocids (Gales *et al.* 1994b), otariids (Arnould 1995), polar bears (Farley and Robbins 1994) and small carnivores (Pitt *et al.* 2006). Any movement of the animal during measurement will significantly affect the repeatability of measurements and, thus, the best results are obtained from anaesthetized individuals (Farley and Robbins 1994; Bowen *et al* 1999; Pitt *et al.* 2006). Although BIA is less accurate and precise than carcass analysis or hydrogen isotope dilution, it is relatively fast and non-invasive and, therefore, may provide a valuable alternative for estimating mean differences in TBW among groups (Bowen *et al.* 1999). Thus, investigators will have to weigh the merit of speed vs. cost and precision for a particular study.

8.3.5 Novel approaches to estimating body composition

Several novel approaches have been used to estimate body composition in marine mammals. One involves examining changes in the diving behaviour of seals instrumented with electronic data loggers. Elephant seals regularly perform dives during which they spend a large proportion of time drifting passively through the water column (Crocker *et al.* 1996). The rate of drift depends on the buoyancy of the seal, which, in turn, depends on their body composition, with fatter seals drifting more slowly (Irvine *et al.* 2000). Biuw *et al.* (2003) used this observation to examine the theoretical relationships between drift rate and body composition, and carried out a sensitivity analysis to quantify uncertainty caused by varying model parameters. Using data from Argos satellite tags in the model, they were able to estimate the relative lipid content of individual seals to within about $\pm 2\%$ of that estimated by hydrogen isotope dilution, and to estimate changes in body composition by recording changes in the rate of drift during diving in foraging seals. This approach has recently been extended to changes in drift rate estimated from changes in the swimming speed of diving elephant seals recorded by archival electronic tags (Thums *et al.* 2008a). Although drift dives have been recorded in both northern and southern elephant seals and have recently been described in adult male New Zealand fur seals (Page *et al.* 2005b), it is not known how widespread this type of behaviour is among marine mammals and, therefore, how general this approach to estimating body composition might be.

8.4 Energy balance analysis

Measurement of the chemical energy of food entering and leaving an animal's body will provide an indirect estimate of MR, referred to as analysis of energy or material balance. Since not all GE consumed by an animal is available for metabolism, estimates of DE, ME, and NE provide increasingly better representations of what an animal actually has available for physiological work. If an animal is in a stable state (i.e. maintaining its biomass and not shedding products, such as feathers, fur, fetus or milk), then subtracting FE and UE from GE (i.e. ME, Fig. 8.1) will be a measure of its MR in an absorptive state (e.g. if resting, it will be a measure of RMR, Box 8.1). More complicated scenarios are introduced when the animal is not in a steady state, in which case all significant outputs of organic material (including tissue growth) must be measured.

The principle of energy balance analysis is straightforward, but in practice requires controlled feeding studies in captive situations and thus is restricted to the smaller species of marine mammals. In addition, extended acclimation periods of days are required to allow clearance of previous dietary regimes, and feeding study measurements must be taken over periods of at least 2–3 days or more to ensure that average steady-state energy input and output can be measured (e.g. Keiver *et al.* 1984; Lawson *et al.* 1997; Trumble and Castellini 2005). Finally, given that the collection of urine is required for measuring ME (e.g. see Keiver *et al.* 1984; Ronald *et al.* 1984), methods are not well-suited to the study of marine mammals in water and preclude measurement in cetaceans. Hence, even in pinnipeds, most studies have assessed apparent DE only (reviewed in Costa and Williams 1999).

Experimental approaches to energy balance studies and the estimation of DE require the quantitative collection of faeces (uncontaminated by urine). This can be done using either whole faecal collections (requiring collection of all faeces associated with the food consumed) or, more frequently, by feeding an indigestible inert dietary marker and comparing the changes in energy (or a given nutrient) and concentration of the marker in the faeces relative to that in the food consumed (Kleiber 1975). Markers may be naturally occurring in the food or added manually, but should be non-absorbable, non-toxic, have no appreciable bulk, mix thoroughly with digesta, and be accurately analysable. Examples of markers used in marine mammal studies include chromium sesquioxide (Cr_2O_3), naturally occurring manganese (Mn^{2+}), cobalt-ethylenediaminetetraacetic acid (Co-EDTA), silicon tubing pieces, dried kernal corn, unpopped popcorn, and dyed corncob grit (e. g. Keiver *et al.* 1984; Fadely *et al.* 1990; Mårtensson *et al.* 1994; Lawson *et al.* 1997; Rosen and Trites 2000; Trumble and Castellini 2005; Larkin *et al.* 2007). In current practice, the marker is usually fed just at the beginning of the tested diet trial and faecal collection is begun at first appearance of the marker and continued for 24–72 hours (e.g. Keiver *et al.* 1984; Trumble and Castellini 2005). Diet and faeces are collected, homogenized, and analysed for total water, nitrogen, lipid, and energy content by standard procedures and for marker concentration, according to the marker used (see above references for procedural details).

Energy and nutrient balance studies provide important insight into understanding animal MR, digestive efficiency, and food and energy requirements. These methods also allow the assessment of differing nutrient and energy availability from different food or prey types. This is because diet quality, quantity, and digestive tract morphology together determine the effectiveness of nutrient and energy extraction from food (e.g. Stevens and Hume 1995). For example, in seals fed different species of fish, or diets of several fish species, digestive efficiency and nutrient/energy extraction differs with diet, meal size, and feeding regime (e.g. Rosen and Trites 2000; Trumble and Castellini 2005). Thus, inferences for evaluating MR from such studies should consider these effects.

8.5 Energetics of lactation

The ability of females to efficiently transfer milk energy to their neonates can have significant consequences for both maternal and offspring fitness. Understanding variation in the patterns of energy transfer both within and among species requires detailed knowledge of both the proximate composition of the milk and the rate of milk production. Substantial individual variation among females in both milk composition and milk output has been noted for a variety of free-ranging mammals (see Lang *et al.* 2009). Thus, when attempting to characterize these components for a species, care should be taken to ensure that a sufficient number of individuals are sampled to provide representative data. In addition, to produce accurate estimates of milk output and milk energy output of an individual, values for the milk components of that individual should be used rather than the species averages (see Section 8.5.2).

8.5.1 Milk composition

Sample collection

Milk composition can change substantially over the course of lactation (e.g. Oftedal *et al.* 1987a) and attendance bouts (e.g. Costa and Gentry 1986; Georges *et al.* 2001). Consequently, the timing of sampling is an important consideration. Exogenous oxytocin in IM doses of approx. 0.10–0.15 IU kg^{-1} in pinnipeds helps to initiate the milk ejection reflex and facilitates milk collection. Because circulating levels of oxytocin can remain elevated for several hours, a single dose is sufficient to achieve complete evacuation (Mačuhová *et al.* 2004). Repeated doses of oxytocin can alter mammary secretory processes and should be avoided (Oftedal 1984). Milk composition does not appear to change with evacuation of the gland in pinnipeds (Oftedal *et al.* 1987a; Iverson *et al.* 1993), but the same may not be true for cetaceans, polar bears, and otters and thus requires confirmation.

For large and/or generally inaccessible species of marine mammals, obtaining milk samples represents a significant challenge. Although post-mortem sampling may be the only source for some species, it can be very difficult to obtain samples that are representative and uncontaminated (e.g. from blood; see Oftedal 1984, 1997). Manual expression of milk from intact glands immediately post-mortem has been successful in some species (Peaker and Goode 1978; Ponce de Leon 1984; Oftedal *et al.* 1988). However, due to the loss of the milk ejection reflex, complete evacuation of the gland is probably not possible. Milk samples obtained from the stomachs of neonates should never be used to estimate proximate composition, as they will overestimate the water and carbohydrate content and underestimate the fat and protein content of the milk (Oftedal and Iverson 1995).

Samples should be kept frozen until analysed. To avoid exposing collected samples to repeated cycles of freezing and thawing for individual analyses, it is advisable to aliquot fresh samples prior to freezing wherever possible. Thawed samples should be homogenized before analysis.

Methods of analysis

The analysis of marine mammal milk samples follows standard methods for the analysis of dry matter (water), protein, lipid and carbohydrate (reviewed in Oftedal and Iverson 1995). The gross energy (GE) content can be determined directly using bomb calorimetry, or accurately estimated using standard values for the energy density of milk lipid and protein (see Box 8.3). Analyses should be performed independently and in duplicate. Values for individual components should never be calculated by subtraction (e.g. protein = dry matter − [lipid +

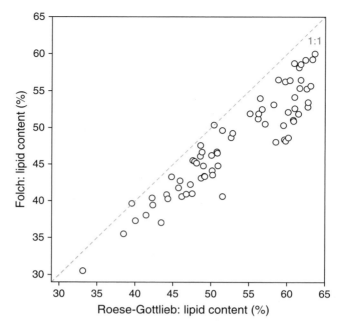

Fig. 8.3 Comparison of milk lipid content values obtained for 71 samples of grey seal milk analysed by the Roese–Gottlieb method (AOAC 2000) and by a modified Folch method (Folch *et al.* 1957; solvent to sample ratio increased from 20:1 to 50:1). Values are averages of duplicate analyses. Differences among duplicates (data not shown) were significantly less for the Roese–Gottlieb (0.14%) compared to the Folch (0.76%; $p < 0.001$, paired *t*-test) method. Estimates of milk lipid content determined by Roese–Gottlieb were significantly higher than those obtained by Folch ($p < 0.001$, paired *t*-test), with the Folch method underestimating milk lipid content by an average of 5.1% and a maximum of 11.6% (S.L.C. Lang and S.J. Iverson, unpublished data).

carbohydrate + ash]), as this will compound errors in measurements. Carbohydrate and non-protein nitrogen are normally found in only trace amounts in the milks of marine mammals. However, because marine mammal milk can be very high in lipids, which contribute most of the GE, lipid analysis warrants special consideration.

The standard and most accurate method for the quantitative determination of milk lipid content is the Roese–Gottlieb method (AOAC 2000). This is a gravimetric method which uses sequential diethyl ether and petroleum ether extractions after pre-treatment of the sample with ammonium hydroxide to disrupt the milk fat globules and break down the hydrophobic casein micelles. Any new method for determining milk fat content should be verified against the Roese–Gottlieb method prior to use. The chloroform- and methanol-based Folch (Folch *et al.* 1957) and Bligh and Dyer (1959) methods have also been used, but these methods are unsuitable for the quantitative determination of total milk lipid content. Direct comparison of the Folch and Roese–Gottlieb methods demonstrates that the values obtained by Folch are more variable and are underestimated compared with the Roese–Gottlieb method, even when the solvent to sample ratio has been substantially increased (Fig. 8.3). There are similar problems with the Bligh and Dyer method (Iverson *et al.* 2001).

8.5.2 Milk output and milk energy output

Measurement of the rate of output by females (i.e. milk intake, MI, by the neonate) using either DLW or SLW techniques (see Section 8.2.3) has been previously described in detail by Oftedal and Iverson (1987). The daily total water intake (TWI) of the neonate(s) over the period of interest is estimated from the elimination of labelled hydrogen from the body water pool of the neonate (see Section 8.2.3, Box 8.3). MI is then calculated from either a measure of MWP and the water content of the milk (DLW method) or from data on the proximate composition of the milk and the rates of fat and protein deposition of the neonate(s),

---

Box 8.3 Calculations for estimation of milk energy output

Formula for determining the gross energy (GE) content of milks from proximate composition

L = lipid, CP = crude protein, S = sugar (i.e. carbohydrate):

$$\mathbf{GE}(\text{MJ kg}^{-1}) = \frac{(39.3^{\dagger} \cdot \%L) + (23.6^{\dagger} \cdot \%CP) + (16.5^{\ddagger} \cdot \%S)}{100}$$

Alternatively, if non-protein nitrogen (NPN) has been determined (where TP = true protein):

$$\mathbf{GE}(\text{MJ kg}^{-1}) = \frac{(39.3^{\dagger} \cdot \%L) + (23.6^{\dagger} \cdot \%TP) + (31.0^{\ddagger} \cdot \%NPN) + (16.5^{\ddagger} \cdot \%S)}{100}$$

Total water intake (TWI)

In a young, growing animal, changes in hydrogen isotope concentration over time reflect not only water loss and the intake of unlabelled water but also dilution as a result of an increase in the size of the body water pool (Nagy and Costa 1980; Oftedal and Iverson 1987).

To account for the changing body water pool size, hydrogen isotope concentrations at time t (H_t) are first corrected (H_t^*) for changing pool size according to the formula:

$$H_t^* = H_t \cdot (N_t/N_i)$$

where N_t is the pool size at time t and N_i is the initial pool size. Note that the correction for changing N requires that N be estimated for each time period, either by a repeated isotope administration or by interpolation from initial and final values for N (e.g. Mellish *et al.* 1999).

The isotope fractional turnover rate (k) is then estimated as the slope of the linear regression of the natural logarithm of H_t^* on time elapsed since isotope administration (see Oftedal *et al.* 1987b).

Total daily water intake (TWI), the sum of water loss (L) and water storage or gain (G), is then calculated as:

$$TWI = L + G = (k \cdot N_{1/2}) + \Delta N$$

where $N_{1/2}$ is the pool size at the mid-point of the study period (assuming a linear change in N with time) and ΔN is the daily change in N. Although changes in N may not be exactly linear, an assumption of linearity will result in relatively small errors unless N changes by more than 40% over the period of study (Nagy and Costa 1980).

Daily milk intake (MI)

The TWI of the suckling neonate(s) includes both the intake of preformed water in milk (MWI) and metabolic water production (MWP) from the catabolism of milk constituents and/or body nutrient stores:

$$TWI = MWI + MWP$$

If neonates have been given DLW (Section 8.2.3), MWP can be estimated from MR and subtracted from TWI. Using the percent water content of the milk, MWI can then be converted to MI.

For SLW (Section 8.2.3), MI is calculated from TWI using the daily fat (F_D) and protein (P_D) deposition rates of the neonate over the sampling period, and data on the percent water (W_M), fat (F_M), protein (P_M), and sugar (S_M) contents of the milk over the corresponding period, and assuming that the oxidation of 1 g of fat, protein, and carbohydrate yields approximately 1.07 g, 0.42 g, and 0.58 g of water, respectively (from Oftedal *et al.* 1987b):

$$MI = 100 \times \frac{TWI + (1.07 \cdot F_D) + (0.42 \cdot P_D)}{W_M + (1.07 \cdot F_M) + (0.42 \cdot P_M) + (0.58 \cdot S_M)}$$

If the values for W_M, F_M, P_M, or S_M change over the period of study, average values should be calculated for these components.

† from Blaxter (1989), ‡ from Perrin (1958).

which allow estimation of MWP (SLW method, see Box 8.3). Daily milk energy intake can then be calculated using the GE content of the milk. Because solid food intake or the drinking of water by the neonate(s) will cause an overestimation of MI, this method is limited to the period that neonates are solely dependent on milk. The optimal frequency and number of recaptures of the neonate(s) for blood sampling over the period of study will depend on factors including suckling pattern and lactation duration, expected rate of water turnover (i.e. MI), patterns of change in the proximate composition of the milk, ease of recapture, and the effects of disturbance on female–offspring interactions (e.g. Oftedal and Iverson 1987; Mellish *et al.* 1999; Crocker *et al.* 2001). Milk output/intake cannot be inferred from suckling behaviour, as neither time spent suckling nor the duration of individual suckling bouts accurately reflects milk intake (reviewed in Cameron 1998).

8.6 Population energetics

Up to this point, the focus of this chapter has been on estimating the energy requirements, expenditures, and body composition of individuals. In this section, we extend this to estimating the energy requirements of populations. Population energetics is important in the study of prey consumption, which tends to be an important input into models of ecosystem structure and functioning (e.g. Bundy *et al.* 2009). Population energetics is also important in the assessment of the vulnerability of populations to changes in prey availability and other threats.

The simplest approach to estimating the gross food or energy requirements of a population is to multiply the estimated ration of individuals by the total population size. However, the requirements of individuals will depend on the size, diet, and perhaps reproductive status of individuals and sex. Therefore, in practice, more useful estimates of population consumption are derived from representing both the population and ration as vectors of numbers at each age and the corresponding age- or size-specific rations (Boyd 2002b).

Bioenergetic models typically estimate the gross energy requirements (GER) of a population as the sum of the energy requirements of individuals in the population, according to the following:

$$\text{GER} = \sum_{a=0}^{a=n} n_a (P_a + (A \cdot \text{BMR}_a)/\text{ME})$$

where a = age, n = oldest age, P is the energy used in production (growth), A is a metabolic multiplier to account for activity, BMR (see Box 8.1), and ME (see Fig. 8.1). This recognizes faecal and urinary losses (E_{f+u}) and loss associated with the energy cost of digestion (E_{SDA}). GER can be calculated on any timescale depending on the resolution of the data and the purpose of the model. P is estimated from growth models (usually Gompertz or Richards models fitted to body mass at age data, see Chapter 5). ME is estimated from empirical laboratory studies (e.g. Ronald *et al.* 1984; Rosen and Trites 2000), BMR (in watts) for mammals is estimated using the allometric equation ($3.4 \times$ body mass$^{0.75}$, Kleiber 1975) and the A multiplier is estimated from empirical studies of MR in relation to various types of activity (e.g. Costa *et al.* 1989; Castellini *et al.* 1992; Sparling and Fedak 2004). Again depending on the data and purpose of the model, A can account for the proportion of time in and costs of different activities. For example, Winship and Trites (2003) used the formulation: $A = water \cdot A_{\text{water}} + (1 - water) \cdot A_{\text{land}}$ where *water* is the proportion of time spent in the water, A_{water} is a multiplier of BMR for water, and A_{land} is the multiplier of BMR for land.

Using a multiplier in this way assumes that active MR is a constant multiplier of BMR (i.e. that FMR scales in the same way as BMR).

Where the data will support a more complex representation, the $(A \cdot BMR)$ term may be expressed with a season-specific estimate of FMR (or average daily MR, ADMR) (e.g. Trzcinski *et al.* 2006). GER may also be indexed for differences among sexes and for different periods (e.g. daily compared to seasonal requirements). There are a number of useful examples of this general approach to estimating population GER (Olesiuk 1993; Hammond and Fedak 1994; Mohn and Bowen 1996; Nilssen *et al.* 2000; Winship *et al.* 2001; Boyd 2002b; Trzcinski *et al.* 2006). In each case, the specifics of the models are tailored to the available data. For example, tracking data for grey seals indicate seasonal changes in population distribution (Breed *et al.* 2006). These seasonal differences in distribution can be easily incorporated into estimates of population consumption to better reflect the predation rate on localized prey species (Trzcinski *et al.* 2006).

These models contain three types of errors. The first is measurement error, which is how well we measure input data. The second, process error, is how well parameters of the model fit the data, and the third is model error, the relationships among the processes in the model (Mohn 2009). To date, most models attempt to account for measurement and process error, but little attention has been given to how different models affect estimates and inferences. Uncertainty tends to increase with model complexity. For example, Trzcinski *et al.* (2006) estimated a coefficient of variation of 4–8% associated with an age-structured model of grey seal populations dynamics, but when the energetics were added to the model the coefficient of variation increased to 30%. Modelling with fewer parameters reduces this inaccuracy. However, a more complex model with increased demographic, spatial, and temporal resolution may provide greater insight as to how prey consumption may change as a result of changes in population size and structure and how it may respond to environmental variability.

Measurement error (variability in input parameters) is often incorporated in model estimates of state variables using Monte Carlo techniques to calculate confidence intervals around the predicted values (Mohn and Bowen 1996; Hammill and Stenson 2000; Boyd 2002b; Winship and Trites 2003; Trzcinski *et al.* 2006). All model inputs have an error distribution associated with them based on their empirical standard errors or an estimate of the upper and lower boundaries of plausible values for each variable. Running the model many times by resampling from these distributions provides a distribution of estimates, which can be used to calculate confidence intervals and the coefficient of variation. Parameters with the most amount of uncertainty have wider ranges of possible values and thus contribute more to overall model uncertainty. The implications of uncertainty and bias in the input data on model estimates can be assessed by examining the sensitivity of model predictions to changes in each parameter or group of parameters (Mohn and Bowen 1996; Boyd 2002b).

9
Diet

Dominic J. Tollit, Graham J. Pierce, Keith A. Hobson, W. Don Bowen, and Sara J. Iverson

9.1 Introduction

Marine mammals are major consumers within marine food webs and probably have a key role in determining food web structure (Bowen 1997). Understanding their diets is important for quantifying trophic interactions and for supporting a broad range of ecological research. Diet estimation in marine mammals has relied on indirect observation because there are limited opportunities to observe directly what marine mammals eat. Indirect methods of observation estimate the diet from samples using a variety of analytical methods that are subject to bias and measurement error. Thus, the quantitative estimation of diet uses statistical inference and is not merely a description of what kinds of prey are eaten.

Historically, diet has been studied by identifying prey remains recovered from stomachs, intestines, and faeces (also termed scats). This can produce a bewildering array of material from almost every type of marine organism above the size of micro-zooplankton, and these are often represented as prey fragments of varying size and stage of digestion. Methods developed more recently include the comparison of stable isotope (SI) and fatty acid (FA) signatures in the tissues of predator and prey (i.e. biochemical-based methods), and molecular identification of prey using DNA. Visual observation and remote sensing (e.g. animal-borne video and sensors) have also been used (Chapter 11).

Each method has its own strengths and weaknesses and there have been considerable efforts to document these, and especially to account for systematic biases (see Table 9.1; Pierce and Boyle 1991; Santos *et al.* 2001; Pierce *et al.* 2004; Stenson and Hammill 2004; Budge *et al.* 2006; Tollit *et al.* 2006). These efforts have included captive feeding studies and computer simulations (e.g. Harvey 1989; Hammond and Rothery 1996; Hobson *et al.* 1996; Tollit *et al.* 1997; Iverson *et al.* 2004; Deagle *et al.* 2005). None of the current methods can be universally recommended, so the use of combinations of complementary methods is advisable

Table 9.1 Main methods used to describe diet in marine mammals, with key information on each method's requirements and products with associated strengths and limitations. 'Regression' refers to the relationship between hard part measurement and length or mass of individual prey.

Method	Animal impact[1]	Diet history	Cost	Identification of prey species	Prey size estimate	Mass percentage (needed for consumption models)	Requirements	Other limitations
Faeces, hard remains	None	Few meals[2]	Low (mod. if all HP used)	Yes (reference collection needed)	Yes (length regressions needed)	Yes (mass regressions needed)	• Reference collection • Regressions • Correction factors for loss and size reduction of HP	• Sex/age of individual generally unknown • Differential prey digestion and retention (prey without HP not represented, prey with fragile HP underestimated, large beaks underestimated) • Special identification skills often required
Regurgitates, hard remains	None	Few meals[2]	Low	Yes (reference collection needed)	Yes (length regressions needed)	Yes (mass regressions needed)	• Reference collection • Regressions	• Sex/age of individual generally unknown • Differential prey digestion and retention (overestimation of large HP)
Prey DNA faeces (stomachs)	None (high[3])	Few meals[2]	Mod.	Yes (genetic data on prey needed)	No	Possible, but expensive and little validation	• Prey primers	• Prey DNA not extractable from all samples • Optimization of primer sets necessary for all prey species • Sensitivity levels poorly known
Stomachs, hard remains	High[3]	Few meals[2]	Low (mod. if all HP used)	Yes (reference collection needed)	Yes (length regressions needed)	Yes (mass regressions needed)	• Reference collection • Regressions • Correction factors for loss and size reduction of HP	• Differential prey digestion and retention (prey with fragile HP underestimated, opposite with large robust HP or beaks) • Information from dead animals, so representation of whole population uncertain • Empty stomachs often reduce sample sizes

Method								Limitations
Lavage (enema), hard remains	Moderate (low)	Few meals[2]	Mod.	Yes (reference collection needed)	Yes (length regressions needed)	Yes (mass regressions needed)	Reference collection - regressions	• Large HP difficult to recover • Empty stomachs (colons) • Capture and handling issues
Stable isotopes	Moderate (capture or dart projector)	Days to years depending on the tissue	Low	No (only estimation of trophic level)	No	No	Isotopic signatures from lower trophic level	• Limited in estimation of prey species • Interpretation of comparison between different environments and time-scales difficult without isotopic signatures from lower trophic level
FA signatures	Moderate (capture or dart projector)	Days–months, depending on the tissue	Mod.	Possible with QFASA (prey FA library and accurate calibrations needed)	No	Possible with QFASA (prey FA library and accurate calibrations needed)	Full prey FA library and appropriate calibrations for QFASA	• Time frame for blubber only approximate • Variability in FA metabolism not fully understood, potentially resulting in imprecise calibrations and diet estimates • Species not in prey FA library are not identified
Animal-borne camera	Moderate (capture)	Days–weeks	High	Possible	Possible	No	Suitable-sized camera	• High cost and unit recovery required • Limited feeding events captured
Direct observation	No	Immediate	Low–mod.	Possible	Possible	No	Suitable viewing platform	• Typically limited to prey brought to the surface • Typically limited to near-shore interactions

[1]Excluding impacts of disturbance; [2]number of meals depends largely on level of retention; [3]stomach analysis requires carcasses.

Abbreviations: HP, hard parts; IS, isotopic signatures; FA, fatty acid; and QFASA, quantitative fatty acid signature analysis; mod., moderate.

(e.g. Hooker *et al.* 2001a; Hammill *et al.* 2005; Herman *et al.* 2005; Tucker *et al.* 2008; Tollit *et al.* 2009). Approaches that provide quantified descriptions of diet (e.g. by mass) are most valuable.

Many marine mammals range widely and exhibit seasonal movements. Thus both the sampling design and the method used should be selected to reflect the temporal and geographical variability in diets and the fact that diet is dynamic, responding in non-linear ways to both intrinsic (e.g. age, sex, condition) and extrinsic (e.g. prey abundance, distribution, energy content) factors.

9.2 Collection of gastrointestinal tract contents

9.2.1 Sampling dead animals

Whenever possible, collection of stomach and intestinal tract contents (Bigg and Fawcett 1985; Pierce and Boyle 1991; Croxall 1993; Ridoux 1994; Santos *et al.* 2001) should be accompanied by morphometric data (see Chapter 5); Geraci and Lounsbury 1993). Appropriate care about infections and other potential health risks needs to be taken when handling gastro-intestinal tracts.

Gastrointestinal tracts often are empty or contain little identifiable food remains, so reducing their value. This can be partly overcome by sampling specimens at sea near foraging grounds. Samples from stranded and by-caught animals have been used for diet analysis, but may not be representative (see Table 9.1; Santos *et al.* 2001). Although these days samples rarely come from hunting, they may be obtained in association with harvesting or from animals being killed as part of a management activity.

Stomachs or the complete digestive tracts should be collected. Individual sections can be ligatured and stored frozen (normally at $-20\,°C$, but $-70\,°C$ is preferred if DNA is to be extracted at a later stage). Food remains are commonly recovered separately from the oesophagus, stomach, and colon, but the remainder of the intestinal tract often yields little additional material. The stomach contents of large cetaceans are normally removed *in situ*.

Ideally, digestive tracts should be examined while the contents are fresh. However, resistant skeletal structures of prey (termed 'hard parts') can be extracted from stomachs even when carcasses are moderately decomposed, although soft tissues of prey will be degraded. Frozen storage in sealed containers will not harm hard structures, but soft tissues (and fatty acids) will continue to degrade, especially at the temperature used in normal domestic freezers. Storage in alcohol or formalin is not recommended since fixed prey remains are harder to separate, fish otoliths (ear bones) may dissolve or become more friable, and such treatment generally prevents biochemical and molecular analyses.

Stomachs and intestinal tracts should be weighed before and after all material has been removed to determine the mass of contents. To collect the prey remains, a complete median longitudinal incision is made from anterior to posterior. The stomach should be thoroughly rinsed. Longitudinal folds should be individually reflected and rinsed, the lining should be examined for cephalopod beaks (or sharp bone fragments). For the remaining sections of the digestive tract, remnants should be squeezed out, cut longitudinally, inverted, and washed thoroughly. Prey should be separated into major types (e.g. fish, cephalopods, crustaceans). The digestion condition of each organism or type may be scored (Meynier 2004; Pusineri *et al.* 2007) to allow interpretation of differential susceptibility of prey to digestion (see Section 9.5). Samples should be processed (identified, weighed, and measured) as soon after collection as possible, with a representative sample of macerated flesh stored frozen at $-70\,°C$ for future DNA analysis. Parasites should be stored in 95% non-denatured ethanol.

9.2.2 Lavage

Stomach-flushing (lavage) has been used in several pinniped species (Boness *et al.* 1994; Harvey and Antonelis 1994). The percentage of samples with food remains can range between 0 and 92% (Rodhouse *et al.* 1992; Boness *et al.* 1994; van den Hoff *et al.* 2003). To perform lavage, animals are restrained and immobilized. A block of wood or stiff rubber (40 cm × 4 cm × 7 cm) with a 3.5 cm hole is placed into the animal's mouth and the lubricated (surgical lubricant), rounded end of a gastric feeding tube (foal size) or clear semi-flexible PVC tube (internal diameter 2.5 cm, wall thickness 0.5 cm) is carefully inserted in the animal's mouth and gently pushed past the oesophagus into the animal's stomach. A canvas strap can be wrapped around the upper canines to ensure safety and to maintain alignment. If the tube is properly in the stomach it should reach the length measured from the mouth to about the end of the sternum; the end of the tube should be placed deeper in the stomach rather than shallower to avoid reflux into the oesophagus. Approximately 2–3 L of seawater can be passed by gravity (Rodhouse *et al.* 1992) and the free end of the tube subsequently lowered below the seal's head and the sample collected in a fine-mesh sieve. An active suction pump can also be used (Boness *et al.* 1994). Multiple lavages may be needed in some species as <50% of recently fed cephalopod beaks were collected in captive tests after two lavages (Harvey and Antonelis 1994). Cephalopod beaks may be difficult to dislodge. As the size of the remains recovered is limited by the diameter of the tube, the results will generally provide a biased view of the diet. Overall, lavaging is subject to unquantified biases and works best for small or heavily digested prey. The use of emetics to provide regurgitated stomach samples is an approach that is not recommended, both because of the animal welfare issues and due to the biases that can result from incomplete recovery of contents.

9.2.3 Rectal enema and faecal loops

Rectal enemas or faecal loops both aim to obtain a sample of faeces from live-captured animals. Staniland *et al.* (2003) found prey remains in 93% of 149 samples from recently returned female Antarctic fur seals, with no significant differences in the mean krill sizes collected by enema compared to scat sampling. Animals are typically physically restrained during the procedure (Gentry and Holt 1982). A soft polyethylene hose (12 mm diameter) is connected to a plastic bottle, filled with approximately one litre of warm water, and inserted into the animal's colon. The water is then introduced *via* the hose and one-way valve by gently squeezing the bottle. Once the bottle is empty or the resistance becomes too great, the hose is removed and the material is naturally expelled by the animal into a large plastic tray.

9.3 Collection of faeces and regurgitated/discarded prey remains

The collection of prey remains from scats is the most widely used method to estimate pinniped diets, but it does not appear to be useful for sampling all species of otariids (because many hard remains are regurgitated) (N.J. Gales and Cheal 1992). Scats can often be collected easily and in large numbers. Disturbance can be minimal or mainly short term (Kucey and Trites 2006).

Although easy to collect, without additional DNA analysis (Reed *et al.* 1997), the sex, age, and potentially even the species of the source animal will usually be unknown. Scats represent relatively recent feeding (last few days) and thus presumably feeding in relatively near-shore areas. This may not pose a problem for coastal species, such as the harbour seal (*Phoca vitulina*), but will bias diet estimates of more wide-ranging offshore species, such as elephant seals (*Mirounga angustirostris*), where only a small portion of foraging effort might be near haul-out sites. An indication of the effective sampling area near a collection site may be estimated from data on food passage rates and swimming speeds (Prime and Hammond 1987), or by tracking animals (see Chapter 11). Interspecific differences in prey passage times (Fea and Harcourt 1997; Tollit *et al.* 2003) may also affect the probability of some prey being recovered in scats at haul-out sites. Integrating passage time information together with foraging trip durations and foraging location data can provide an assessment of the level of potential collection biases (Smout 2006).

The collection of cetacean faeces is more challenging and requires boat-based focal follows (see Chapter 12). Dogs have been used to detect samples from up to a nautical mile from the source, and could locate three to five times the number of samples per unit effort than were collected by human observers (Rolland *et al.* 2006). Faeces usually produce a cloud plume of material in the water (pink coloured in the case of krill consumers), and material can be picked up using a fine (500–1000 μm) mesh net (N.J. Gales and Jarman 2002) and extendable pole or, in the case of dolphins, snorkellers can collect sinking faeces in plastic vials (Parsons *et al.* 1999). Collection of fish scales and tissue from killer whale (*Orcinus orca*) predation events has also been successful (Ford and Ellis 2006). Deecke *et al.* (2005) used underwater sounds of kills made by transient killer whales as the cue to collect their prey remains by net.

Scats range in consistency from semi-liquid to solid and can be broken into fragments and spread over a wide area by currents or animal movements. Ultimately, judgement about which fragments belong together must be based on their location, size, consistency, and colour. Ideally, only entire recently deposited faecal samples should be collected. If dried up, older or partial samples are collected, this should be recorded. Regurgitated samples or prey discards may be collected using the same methods as faecal samples, but should be analysed separately. Scat samples are normally stored frozen at $-20\,°\mathrm{C}$, as soon as possible.

Sub-sampling scats for predator or prey DNA (or hormones) is ideally undertaken within 48 h of collection to minimize degradation. If only information on the defecator is needed then scraping 2–3 ml from the outer surface of the scat with a sterile spoon should suffice, otherwise scats should first be homogenized, ideally by mixing the sample with distilled water ($\sim$50 ml) in a jar, leaving overnight, and gently shaking the resulting slurry. If only soft-tissue prey remnants are being used for identification, approximately 10 g of scat homogenate/slurry is then removed and gently pressed through a <0.5 mm plastic mesh. Approximately 2–3 ml of soft scat material scraped from the underside of the mesh can then be stored at either $-70\,°\mathrm{C}$ or refrigerated with 4–5 times the sample volume of 95% non-denaturing ethanol.

9.4 Sampling bias

The number of samples collected will depend on the questions asked, the method used, and the spatial and temporal scales of interest (Hayes and Steidl 1997; Reed and Blaustein 1997). The number of samples required will also vary in relation to the diet breadth and variability (Arim and Naya 2003; Trites and Joy 2005). Sampling design and the methods used should be selected to reflect the demographic, temporal, and geographical variability in diets. The optimal sample size can be determined if some prior information is available on diet breadth and variability (Lance *et al.* 2001), but conducting a pilot study is recommended. Monte Carlo simulation on the results from pilot samples can be used to determine the sample size required to achieve a predetermined level of precision (Lance *et al.* 2001). Assuming constant proportions of species pass into the scats, Trites and Joy (2005) estimated that 59 scats should be collected to be 95% confident of collecting at least 1 scat containing a species with a 5% probability of occurrence. To statistically distinguish between populations, they found that collecting 59 scats would suffice for diets containing 12 or more exponentially distributed (in terms of frequency of occurrence) species of prey, 94 scats for diets containing 6 or more species, and 179 scats for diets with 3 species. Hammond and Rothery (1996) also used re-sampling techniques to estimate the confidence limits of grey seal (*Halichoerus grypus*) diet estimates and the relative magnitude of different sources of error. Their analysis indicated that a minimum of approximately 100 scats should be collected in each area/season combination, and additionally they highlighted systematic errors due to measurement errors in estimating fish weight from partially digested otoliths.

9.5 Laboratory processing of prey hard structures

The contribution of each prey species to the diet is ideally based on the mass and number of individuals consumed. The reconstructed biomass of prey can then be converted to energy consumed using estimates of prey energy density. Identification of prey remains is time-consuming and requires experience, access to good reference material, and, in some cases (e.g. identifying cephalopod beaks; Clarke 1986), specialist training.

Occurrence indices only require structure identification. A major variant in protocol is whether to use otoliths and beaks alone or to include alternative diagnostic skeletal structures (i.e. all hard parts). The latter approach increases the detection rates for many prey species (Olesiuk *et al.* 1990b; and see Section 9.5.2 below) but it requires excellent taxonomic identification skills and extensive reference material. Few field studies have attempted biomass reconstruction using all hard part structures (Laake *et al.* 2002; Sigler *et al.* 2009), largely due to the scarcity of appropriate allometric regressions and correction factors that aim to take account of the loss in size and number of items due to digestion (Harvey 1989). Laake *et al.* (2002) compared the results of occurrence and biomass reconstruction estimators and found ten-fold differences in species consumption estimates between the two indices for the smallest and largest prey. Sigler *et al.* (2009) highlighted two- to threefold differences. Captive, mixed-diet feeding studies show that it is possible to obtain good biomass estimators using all hard part structures (Tollit *et al.* 2007; Philips and Harvey 2009). Together, these studies highlight the weakness of occurrence indices to quantify diet when diet/scat diversity is high (i.e. generalist predators) and when prey sizes consumed vary considerably (see also Section 9.6).

The processing of GI tract contents, scats, and regurgitates is complicated by the fact that the digestive state of prey is highly variable. Prey remains may come from an unknown number of meals over an unknown period, and different types of food are digested at different rates. For GI tracts, the best approach is one that assesses diet based, initially, on the fresh food fraction alone, as well as a comparison with all prey remains whatever their digestive state. Data from both *in vivo* and recent *in vitro* digestion rate studies can be integrated into analyses to account for the states of digestion (see Sections 9.3 and 9.6; Murie 1987).

9.5.1 Prey extraction

Extraction of prey structures from GI contents, regurgitates, and scats follows the same general procedures. The contents of each section of the GI tract should be poured through up to three or four nested sieves with a minimum mesh size of 0.2–0.355 mm, and washed with water and a soft brush to remove as much soft residue as possible.

Thawed scats should be weighed or a volume noted by measuring the settled level of samples that have been suspended in jars with water, after removing or accounting for any substrate in the case of pinnipeds. Soaking samples in a mild 1% detergent solution can reduce odour and clean up hard parts. Strong detergents may erode or dissolve more friable hard parts and aliquots for any future DNA or hormone analysis should be removed first.

The extraction of prey hard part structures from scats can be done manually, using a water bath, sieving, elutriating, or (with the faecal material inside a mesh bag) a washing machine. Material can be suspended (with 1% detergent) in a sorting tray (35 × 45 cm) and examined. Buoyant parts such as crustacean remains and fish scales can be skimmed, while the dense residue can be examined under a binocular microscope and prey remains picked out manually (Reid 1995). Spray-washing scats through stacked sieves of decreasing mesh size (10, 4.75, 1.0, and 0.5–0.25 mm mesh) and using brushes or soft spatulas to break up hardened material can damage fragile structures. Consequently, Murie and Lavigne (1985) suggested using nested sieves in conjunction with flowing water baths, while Bigg and Olesiuk (1990) developed an elutriator, which is a semi-closed system that separates prey remains from soluble waste material using differences in their densities. The elutriator is efficient, but uses large volumes of water and secondary processing in nested sieves is frequently also necessary. A top-loading washing machine, set on a gentle wash, is recommended (Orr *et al.* 2003) for the bulk processing of approx. 25 scats. Loss rates (∼5%) and size reductions of otoliths (∼1%) using this method were found to be similar to nested sieves, but the processing time was reduced by more than half. Individual scats are placed into one or two labelled, tightly closed, 3.8 L × 124 μm mesh bags.

Some fish have extremely small otoliths—e.g. pipefish (Syngnathinae) otoliths may be lost even using a 0.2 mm sieve, and many other small fish, including gobies (Gobiidae) and sand eels (Ammodytidae), may have otoliths less than 1 mm in breadth. Hard structures for species identification (and enumeration) should be cleaned (immersion in 70% alcohol for some hours will effectively sterilize the material), air-dried on absorbent paper, and then transferred to glass storage vials. Cephalopod beaks, statoliths, lenses and pens, crustacean carapaces and telsons, and other invertebrate remains (which may include polychaete jaws) should be stored in 70–95% non-denatured ethanol or isopropyl alcohol to prevent distortion.

9.5.2 Prey identification

Prey should be identified under a binocular microscope to the lowest possible taxon by comparing with reference material. There are a number of good identification guides of fish structures (Newsome 1977; Härkönen 1986; Rosello Izquierdo 1986; Cannon 1987; Hansel *et al.* 1988; Smale *et al.* 1995; Prenda *et al.* 1997; Watt *et al.* 1997; Harvey *et al.* 2000; Leopold *et al.* 2001; Campana 2004; Tuset *et al.* 2008), as well as extensive archaeozoological (see Casteel 1976) and osteological (e.g. Norden 1961; Mujib 1967; Boschi *et al.* 1992) literature. Less information is available to assist the identification of crustaceans (e.g. Mori *et al.* 1992) or cephalopods (e.g. Clarke 1986). Croxall (1993) provides good identification sources for Antarctic prey.

Teleost fish are typically identified using sagittal otoliths. These are also are used to determine the size (length) of fish, allowing estimates to be made of prey mass using length–mass regressions (www.fishbase.org). Identification of fish species with fragile otoliths (e.g. salmonids, clupeids) or cartilaginous structures (e.g. elasmobranchs) can be improved/achieved using alternative diagnostic skeletal structures (e.g. vertebrae, jawbones, angulars, radials, otics, gill rakers, branchials, operculums, scutes, quadrates, teeth). The approach of using all hard parts is particularly useful when prey heads (i.e. otoliths) are not consumed, as sometimes observed for seals eating salmon. Lance *et al.* (2001) provided a list of key structures commonly used to identify and enumerate prey consumed in the North Pacific.

9.5.3 Prey enumeration using minimum number of individuals (MNI)

The minimum number of individuals (MNI; White 1953) is typically calculated to estimate the number of prey eaten from prey structures recovered. MNI attempts to avoid counting the same prey item more than once, by estimating the smallest number of individuals needed to account for the recovered structures of that taxon. Theoretically, MNI yields a better estimate of the actual number of individuals consumed when the meal consists of relatively few individuals (Nichol and Wild 1984; Joy *et al.* 2006).

Different methods of calculating MNI are needed because skeletal structures can be individual, paired, or multiple (see Lance *et al.* 2001). Fragments from broken structures are matched where possible. For paired unique structures like sagittal otoliths, it is simplest to assume that the number of fish is half the number of otoliths, except when sagittae look like other otolith pairs as in gobies. Refinements include determining separate left-, right-, and unknown sides, and calculating MNI as the greatest number of left or right elements (upper or lower beaks in the case of cephalopods). For species that vary greatly in size, the size range or type of the structure may be used to refine the MNI estimate. When using multiple structures, prey numbers are determined from the structure that yields the highest MNI. Direct counts can be used for individually occurring structures such as the atlas (first vertebrae) or vomer bones. Recovered vertebrae counts can be divided by the actual number of vertebrae recorded for that species (e.g. Ford 1937; Hart 1973; www.fishbase.com). Typically, for non-unique structures such as gill rakers, teeth, and fragments of cephalopod gladius ('pen'), an MNI of one is applied. Pens may be used to indicate the presence of squid, but can only be used for enumeration if they are intact (which is rare, given their fragility). The eye lenses of cephalopods are slightly flattened unlike fish eye lenses which are spherical, and are useful elements to determine recent cephalopod ingestion (Staniland 2002). The number of individual crustaceans is based on eye, telson, or carapace counts.

9.5.4 Measurement of prey structures

Otoliths are the most widely used structure to determine fish size. Otolith length or width is typically recorded. Some authors have used otolith thickness, although this is generally more difficult to measure. In general, otoliths used for diet estimation refer to the sagittae, as they are usually bigger and their shapes show consistent interspecific variation. In the Gobiidae, lapillae otoliths are very similar to sagittae and so the number of gobies should be determined as a quarter of the number of otoliths. Unbroken otolith length should be measured to the nearest 0.1 mm (using digital callipers or a microscope and graticule), parallel to the sulcus, from the anterior tip of the rostrum to the posterior edge. Otolith width should be measured perpendicular to the sulcus at the widest point of the otolith, especially for those species with fragile post-rostrums (Fig. 9.1). Measurement protocols for other fish structures vary, but selection is typically based on those structures that are most robust to digestive processes (e.g. jawbones and vertebrae) and on the availability of regressions to convert the size of the measured structure to an estimate of prey size eaten (Watt *et al.* 1997; Zeppelin *et al.* 2004).

Random sub-sampling of prey items can be used when there are large numbers of a particular prey type. Appropriate sub-sample size will depend on the variability in the size of the measured structure and the degree of precision required (see Bowen and Harrison 1994).

For cephalopods, the lower beak is generally used for identification and measurement, although upper beaks can also be used, but far fewer regressions are available. Rostral length is normally measured in squid and cuttlefish and hood length in octopods (Figs 9.2a, b; Clarke 1986). Rostral width has been recommended for *Loligo opalescens*. The size of krill can be estimated from measurements of body length (anterior edge of the eye to the tip of the telson, excluding setae) or carapace length (tip of the rostrum to the mid-dorsal posterior edge) (Croxall 1993).

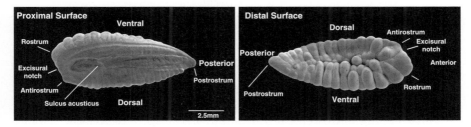

Fig. 9.1 Sagittal otolith from a teleost fish (family: Gadidae). (Photo courtesy of Dr Steven Campana, Bedford Institute of Oceanography, Canada.)

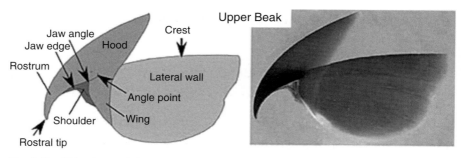

Fig. 9.2a Side views of the upper beak of *Stigmatoteuthis hoylei*. The rostral lengths should be measured from the tip of the rostrum to the jaw angle. Hood lengths should be measured from the tip of the rostrum to the tip of the hood. (Left, drawing courtesy of R. Young, University of Hawaii; Right, photo courtesy of R. Young.)

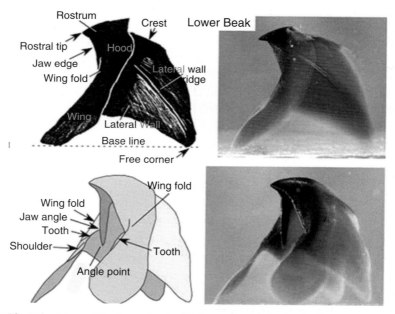

Fig. 9.2b Side views of the lower beak of *Stigmatoteuthis hoylei*. (Left, drawing courtesy of R. Young; Right, photo courtesy of R. Young.)

Separate regressions are often developed for each sex and reproductive status category, so these characteristics should be recorded when possible. Beaks can be measured using a microscope, digital camera and image measurement software (AxioVision 3.1) or using a binocular microscope equipped with an eyepiece graticule, or with Vernier calipers.

9.6 Quantification of diet composition using GI tract and faecal analyses

Biomass reconstruction using otoliths, beaks, and other hard part structures should account for structures that may be partially or completely eroded during digestion, as well as assessing levels of uncertainty (Bowen 2000; Hammond and Grellier 2005; Sigler *et al.* 2009).

9.6.1 Accounting for complete digestion of hard part structures

Numerical correction factors (NCFs) were introduced for pinniped faecal analysis (Harvey 1989) to take account of interspecific differences in otolith/beak recovery (the probability of passage), which had been shown to bias diet analyses in favour of species with large and robust hard parts (e.g. gadids, cephalopods) compared to smaller or more fragile prey with less robust structures (e.g. clupeids, scrombrids, salmonids, osmerids) (Prime 1979; Bigg and Fawcett 1985). NCFs are typically generated in captive feeding studies by comparing known numbers of prey consumed with estimates derived from reconstructing the number of prey, using MNI counts based on the structures that survive digestion (Harvey 1989; Bowen 2000; Tollit *et al.* 1997, 2007; Grellier and Hammond 2006). NCFs are applied to counts of prey before estimating prey mass. Fish otolith NCFs based on captive feeding studies typically range from 1.0 (where numbers fed are the same as the number estimated using MNI counts) to ~10.0 (i.e. only 10% of the numbers fed were recovered, as estimated from MNI counts). Relatively robust NCFs are available for key prey of just a few species of pinnipeds (notably grey seals, harbour seals, Antarctic fur seals (*Arctocephalus gazella*), Steller sea lions (*Eumatopias jubatus*), and California sea lions (*Zalophus californianus*). The application of NCFs is complicated by a range of extrinsic and intrinsic factors that may affect the values of these coefficients, such as meal size (Marcus *et al.* 1998), activity (Bowen 2000; Tollit *et al.* 2003), prey size (Tollit *et al.* 2007), study animal (Orr and Harvey 2001; Cottrell and Trites 2002), method of ingestion (Grellier and Hammond 2005) and the diet composition (Casper *et al.* 2006). Realistic NCFs are therefore hard to derive. Captive feeding studies using standard protocols are important to provide robust comparisons across prey species. Otolith robustness (pristine otolith length divided by otolith mass) is correlated with the probability of recovery in scats, but not in captive studies (see Harvey 1989; Tollit *et al.* 1997, 2007).

Differences in the recovery rates of prey species can also be reduced, but not eliminated, by using multiple structures rather than otoliths alone. The use of multiple structures is particularly useful for identifying and counting more fragile prey species (e.g. clupeids salmonids, and elasmobranchs). However, their use can also increase biases due to double counting prey across scats (Joy *et al.* 2006), unless counts for each species are based only on the skeletal element (of that species) which yields the highest count. Overall, while the application of NCFs (in captive studies of pinnipeds at least) clearly improves the accuracy of dietary biomass estimates (see Tollit *et al.* 2007; Philips and Harvey 2009), the effect of applying NCFs in general will depend on the relative proportion of prey species in the diet and the NCFs of these species. In one field study, as an example, applying NCFs reduced the apparent importance of gadids in the diet by about 30–40% (Tollit, unpubl. data).

NCFs are currently unavailable for cetaceans. Most dietary studies on cetaceans are based on stomach contents, for which the degree of digestion will vary according to the time since ingestion. Murie (1987) suggested that the proportion of fish otoliths recovered inside fish skulls could be used to indicate the time since ingestion, and similar indices based on, for example, otolith degradation could be constructed. However, calibration of these indices, and derivation of NCFs for a range of times since ingestion, would require extensive experiments.

9.6.2 Other factors affecting recovery of hard part structures

Most remains of fish hard parts pass through the digestive tract between 1 and 3 days after ingestion (minimum 2 h, maximum 142 h; Harvey 1989, Fea and Harcourt 1997; Orr and Harvey 2001; Staniland 2002; Tollit *et al.* 2003); but interspecific rates can vary by a factor of two, and scats can represent a composite of up to six past meals (Tollit *et al.* 2003). Cephalopod beaks are far more resistant to the digestive process (Bigg and Fawcett 1985) and can be retained in the GI tract for up to three weeks (Tollit *et al.* 2003). Small beak recovery is higher than for large beaks, which may break or be regurgitated (Yonezaki *et al.* 2005). Consequently, the frequency of large cephalopods may be underestimated.

Australian sea lions (*Neophoca cinerea*) grind most identifiable hard parts into a paste, making hard part analysis unreliable (Gales and Cheal 1992; Casper *et al.* 2006). Captive feeding studies with otariids have shown that the remnants of meals containing relatively large fish as well as cephalopods can be regurgitated, so that structures from these prey will not appear in the scats. The collection of regurgitates is therefore important (see Gudmundson *et al.* 2006). Some otariids will also ingest gastroliths (stomach rocks), which likely affect the digestive process, recovery rates, and the identification of hard parts (Needham 1997; Tollit *et al.* 2003).

9.6.3 Accounting for partial digestion of hard part structures

Estimating the size of fish from measurements of otoliths (or other structures) recovered from scats and GI tracts may be biased because of partial digestion. For example, sizes of fish eaten by captive harbour seals were underestimated by 0–76% when back-calculated from otolith measurements (Tollit *et al.* 1997). To compensate for this bias, digestion coefficients or correction factors (DCFs) can be applied to otolith measurements (e.g. North *et al.* 1983; Harvey *et al.* 1989; Reid 1995; Tollit *et al.* 1997), and also to measurements of various other fish skeletal structures (Tollit *et al.* 2004). Larger otoliths are digested proportionally more than smaller otoliths, meaning that care must be taken not to create bias by only selecting uneroded otoliths (Tollit *et al.* 1997; Grellier and Hammond 2006). Furthermore, the application of an average species-specific DCF may be unwise if there is high intraspecific variation in otolith digestion (Dellinger and Trillmich 1988; Tollit *et al.* 1997; Grellier and Hammond 2006).

Grade-specific DCFs are based on changes in the morphology and surface topography. Typically, structures are grouped into three or four different digestion grades (e.g. pristine, good, fair, poor), based on objective keys and photographic references (see Tollit *et al.* 1997; Leopold *et al.* 2001; Tollit *et al.* 2004; Grellier and Hammond 2006). Pristine graded structures are usually associated with regurgitations (or relatively undigested stomach contents) and therefore no correction is required. Structures classified in poor condition generally need not be measured (except perhaps to estimate a minimum size), as correction factors applied to otoliths in this condition result in wide confidence intervals around estimated prey lengths and weights (Tollit *et al.* 1997). Structures in 'good' condition are typically reduced 5–15% in size, while those in 'fair' condition are reduced 15–30% in size, though prey size is an additional variable (Tollit *et al.* 1997, 2004; Grellier and Hammond 2006). Grade-specific DCFs are best calculated from captive feeding studies and are applied before applying allometric body size regressions. In contrast to otoliths, in general there is little size reduction of cephalopod beaks (Harvey 1989) or crustacean carapaces (Staniland 2002).

9.6.4 Prey length and mass reconstruction

There are allometric relationships between otolith (or lower beak) size and fish (or cephalopod) size (see Casteel 1976; Härkönen 1986; Clarke 1986; Harvey *et al.* 2000). Fewer published relationships exist for other skeletal structures (but see Watt *et al.* 1997; Zeppelin *et al.* 2004). For most species, the relationship between otolith length and fish length can be described by a simple linear regression (sometimes with an inflection point), whereas conversion directly to mass is typically a power function.

Fish mass can be estimated directly from otolith length or by first estimating fish length. The two-stage procedure is particularly useful if information on seasonal variation in length–weight relationships is available, because the otolith–fish length relationship is more likely to be consistent across seasons. However, potentially, this approach leads to wider confidence limits (Casteel 1976). Most studies use linear regressions with log-transformation, but weighted non-linear regression is also used (Leopold *et al.* 2001).

Applying regressions introduces both random and systematic errors. The errors introduced by deriving fish weight from linear regression equations can readily be estimated using bootstrap techniques (Hammond and Rothery 1996; Pierce *et al.* 2007), provided that either the raw data or error parameters for the regression line are available. Investigators should be cautious about using regressions based on collections from outside the study region, small sample sizes, or fish sizes outside the size range observed by the study in question (Harvey *et al.* 2000).

9.6.5 Quantification methods

To allow across-study comparisons, we recommend that authors report the total numbers of samples collected that contained (1) no prey remains (empty), (2) at least some identifiable prey remains, and (3) only prey remains that were 'unidentifiable' (hard parts too eroded for identification) or 'unidentified'. Regurgitations, stomach contents, and scats should be reported separately as they provide different representations of the diet.

The relative importance of prey in diets can be expressed in a variety of ways. Numerical counts of prey (e.g. percentage of the number of individuals for each prey taxon as a percentage of the total number of individuals found in all samples) is a simple method that is susceptible to overestimating the importance of small prey, mainly due to differences in the number of prey consumed per meal. Percent frequency of occurrence (percentage of samples containing a given prey taxon) is the simplest method to represent diet. It describes the number of animals eating a prey type and is probably least affected by interspecific differences in prey recovery. In general, occurrence indices may overestimate the importance of small or trace prey (especially when eaten in small numbers), while biomass indices may overestimate the importance of large prey. Volumetric indices are most useful for fresh stomach samples, while composite indices (such as index of relative importance, Pinkas *et al.* 1971), dominance, abundance, and diversity indices can be useful for comparative purposes (see Lance *et al.* 2001). The energetic contribution of prey to the predator's diet is probably the currency of most relevance to foraging animals. Detailed discussions of the advantages and disadvantages of the various measures of diet have been presented in reviews by Berg (1979), Hyslop (1980), Pierce and Boyle (1991), and Laake *et al.* (2002).

Numerical importance (N_i) for prey category i can be calculated as:

$$N_i = \frac{\sum_{k=1}^{s} n_{i,k}}{\sum_{i=1}^{\omega}\sum_{k=1}^{s} n_{i,k}} \text{ or } N_i = \frac{\sum_{k=1}^{s}\left[\frac{n_{i,k}}{\sum_{i=1}^{\omega} n_{i,k}}\right]}{s}$$

where ω = number of prey categories, s = number of samples, and $n_{i,k}$ is the number of individuals of the ith prey category in the kth sample. The first of these estimators is the more usual, in which the distribution of each prey category between scats is ignored and a single scat with a high number of a particular prey category can thus have a strong influence on the estimated overall importance of that prey category. In the second estimator, the contribution of each scat is effectively weighted equally.

How averages should be calculated depends on the nature of the samples and on the question asked (see Pierce *et al.* 2007 and p. 208 below).

Percentage frequency of occurrence (%FO), can be re-scaled so that, summed across all prey types, the values total 100%, and is then termed modified FO (MFO; Bigg and Perez 1985). Olesiuk *et al.* (1990b) proposed a split-sample FO (SSFO) estimator which examines species occurrence in each scat sample individually, apportioning the contribution of each prey category to each scat depending on the number of other species present, and assuming that each prey category present in each scat has been consumed in equal proportions. These two variants of the percentage FO method differ only in that MFO takes an equal weighting approach and SSFO an unequal weighting approach (see equations below). Olesiuk (1993) showed that when using SSFO the diet composition percentages for the primary prey varied by a factor of two or three, depending on the assumed composition within each scat. It is this composition of prey within a sample that biomass models aim to estimate.

$$FO_i = \sum_{k=1}^{s} I_{i,k} \; ; \; \%FO_i = 100 \times \frac{\sum_{k=1}^{s} I_{i,k}}{s} \; ; \; MFO_i = \frac{\sum_{k=1}^{s} I_{i,k}}{\sum_{i=1}^{\omega}\sum_{k=1}^{s} I_{i,k}} \; ; \; SSFO_i$$

$$= \frac{\sum_{k=1}^{s}\left[\frac{I_{i,k}}{\sum_{i=1}^{\omega} I_{i,k}}\right]}{s}$$

where I = indicator function equal to 1 if the ith prey category is present in the kth sample, and 0 if it is absent.

Similarly, there are two approaches to estimating the weight (W) or the reconstructed biomass (BR) of prey consumed in the diet. Once again, each scat can be considered as contributing a variable amount (e.g. VBR) or a fixed (equal weighted) average amount (e.g. FBR).

$$W_i = \sum_{k=1}^{s}\sum_{j=1}^{n_{i,k}} w_{i,j,k} \text{ or } W_i = \sum_{k=1}^{s} n_{i,k} \times \bar{w}_{i,k} \; ;$$

$$VBR_i = \frac{\sum_{k=1}^{s} n_{i,k} \times \bar{w}_{i,k}}{\sum_{i=1}^{\omega}\sum_{k=1}^{s} n_{i,k} \times \bar{w}_{i,k}} \; ; \; FBR_i = \frac{\sum_{k=1}^{s}\left[\frac{n_{i,k}\times\bar{w}_{i,k}}{\sum_{i=1}^{\omega} n_{i,k}\times\bar{w}_{i,k}}\right]}{s}$$

where $n_{i,k}$ is the number of prey of category i in sample k, $w_{i,j,k}$ is the weight of the jth individual of prey category i in sample k, and $\bar{w}_{i,k}$ is the average weight of an individual of prey category i in sample k.

Most recent dietary studies have used the VBR (Hammond and Rothery 1996) rather than the FBR method, and the variable estimator also seems to perform marginally better in the analysis of captive feeding data (Tollit *et al.* 2007). Further work on foraging patterns is required to determine, in particular, if meal size actually varies systematically with prey type and availability (justifying VBR). Overall, given the assumption that a reconstructed biomass of scats does reflect variability in foraging success and meal size (consumption), we recommend applying the VBR method, coupled with the calculation of confidence intervals and an assessment of outliers. Choice of VBR assumes a negligible impact of other factors that likely influence reconstructed biomass, such as differences in digestion and subsequent deposition, and the inclusion of partial scat samples. Given these assumptions, comparison with FBR is considered worthwhile. The same weighting issue arises for pooling multiple collections.

Relatively little attention has been given to evaluating the uncertainty associated with estimates of the species composition of diets (Pierce and Boyle 1991; Tirasin and Jorgensen 1999; Hammond and Grellier 2005). To account for uncertainty due to sampling, non-parametric 95% confidence intervals (95% CI) for diet compositions can be generated by bootstrapping (Reynolds and Aebischer 1991). Bootstrap techniques can be also be applied to each stage of the quantification process, such as assessing errors around regressions or correction factors, provided that some information is available about the underlying uncertainty in each part of the calculations (Hammond and Rothery 1996; Tollit *et al.* 1997; Santos *et al.* 2001; Stenson and Hammill 2004; Pierce *et al.* 2007). Alternatively, variance estimators for diet composition can be constructed using finite population sampling methods (Cochran 1977) and delta method approximations based on the Taylor series (Seber 1973; Laake *et al.* 2002).

Given potential biases and often low precision, point values for prey proportions in a diet should be treated with appropriate caution (Hammond and Grellier 2005; Matthiopoulos *et al.* 2008). Diet studies need to address sampling and digestion-related biases, report confidence intervals, and highlight limitations. Consumption estimates are needed to estimate marine mammal predation of commercial fish species or of endangered prey (see Chapter 8). In addition to diet composition, accurate consumption estimates rely on good data on population size, age structure, and energetic requirements (Mohn and Bowen 1996; Shelton *et al.* 1997; Stenson *et al.* 1997; Winship and Trites 2003).

9.7 Molecular identification of prey remains

Protein electrophoresis and immunoassays using polyclonal antisera have been used with some success in detecting single species from GI tract and scat material. However, discerning mixtures of prey appears problematic and proteins degrade during digestion, reducing antigenicity (Pierce *et al.* 1990, 1993). An enzyme-linked immunosorbent assay (ELISA) enables rapid screening to obtain accurate data on gut contents (Sunderland 1988).

In contrast, there have been major advances in using DNA to study diets. This is based on the ability to identify unique pieces of DNA from either the predator (Reed *et al*. 1997) or the prey (Jarman *et al*. 2002; Casper *et al*. 2007a) species. There are two approaches to the identification of species-specific prey DNA sequences. The first technique involves polymerase chain reaction (PCR) amplification of prey DNA from tissue homogenates (e.g. stomach or scat samples) using group-specific primers (e.g. fish, cephalopods, krill; see Jarman *et al*. 2004; Deagle *et al*. 2005). To distinguish the different sequences (species) represented, the amplified DNA is then analysed using a technique such as high-resolution gel/capillary separation, DNA cloning, or restriction fragment length polymorphism. The DNA can then be sequenced. Sequences can then be matched to species using basic local alignment search tool (BLAST) database searches. The second approach involves the amplification of prey DNA using species-specific targets (Jarman *et al*. 2002; Casper *et al*. 2007a). This has become popular because it is relatively simple and inexpensive to design PCR primer sets that target organisms at various taxonomic levels. Group-specific DNA primers allow an even broader survey of prey types to be conducted (see King *et al*. 2008). These PCR-based techniques have been applied to DNA extracted from scat and stomach remains, as well as prey remnants collected after surface feeding events. Reed *et al*. (1997) were also able to identify individuals, species (pinniped), and the sex of seals at mixed-species haul-outs. PCR techniques were used to identify different species of salmon from bone fragments recovered from seal scats (Purcell *et al*. 2004; Parsons *et al*. 2005) and also using fish scales collected after killer whale feeding events (Ford and Ellis 2006).

Tollit *et al*. (2009) used group-specific nested primers followed by denaturing gradient gel electrophoresis (DGGE), a technique which separates amplification products based on their melting behaviour as they denature, to detect >40 different species from the scats of Steller sea lions. DNA identification increased the number of prey species detections by 22% compared with conventional morphological identification. Captive feeding studies with sea lions have shown that the detection of prey in faecal matter is limited to those consumed during a 48-h period before defecation (Deagle *et al*. 2005), suggesting analysis of prey flesh may be more representative of recent feeding and not a composite of meals from many days.

To date, conventional PCR has provided only occurrence data rather than quantitative estimates of the proportion of each prey eaten. Consequently, quantitative real-time PCR (qPCR) methods, which measure the amount of DNA by fluorescence monitoring of PCR, have been performed with seal scats and these deserve further development. To date, qPCR studies highlight high detection sensitivity (0.01%) and the potential to estimate relative quantities of a target species, but there is a need to assess prey DNA degradation during digestion in mixed-prey species studies (Deagle and Tollit 2007; Matejusová *et al*. 2008).

Potentially, biases equivalent to those that affect quantitative estimates of diet from hard parts will have to be overcome. Due to development costs, qPCR methods are presently likely to be limited to very specific questions, such as the contribution of salmonids in a predator's diet. As DNA mass target detection systems improve and become less costly, multiplex PCR, microarrays, and, in particular, pyrosequencing appear to have great potential in the future (see review in King *et al.* 2008; Deagle *et al.* 2009; Dunshea 2009).

Finally, near infrared spectroscopy (NIRS) is relatively new. It estimates the composition of an organic sample when it is irradiated with light in the NIR spectrum. Organic material absorbs NIR light at wavelengths characteristic of particular bonds. Therefore, the amount of light reflected at a given wavelength is indicative of the concentration of compounds with that particular bond. Kaneko and Lawler (2006) tested the method using scats from captive otariids that were fed mixed-species meals, and this study represents the only ever test. In five of six cases, NIRS accurately and precisely quantified how much of a given diet component a seal had eaten the previous day. The authors suggest the technique may be especially useful in cases where there are particular, or few, prey species of interest.

9.8 Fatty acid (FA) signatures

Fatty acids (FAs) are the largest constituent of neutral lipids, such as triacylglycerols (TAG) and wax esters (WE), as well as of the polar phospholipids (PL). All FAs consist of carbon atom chains, which are most commonly even-numbered in length and straight, containing 14–24 carbons and 0–6 double bonds with a methyl (CH_3) terminal at one end and an acid (carboxyl, COOH) group at the other (see reviews of Dalsgaard *et al.* 2003; Budge *et al.* 2006; Iverson 2009b). FAs of carbon chain-length 14 or greater pass into the circulation intact and are generally taken up by tissues the same way. Although some metabolism of FAs occurs within the predator, such that the composition of predator tissue will never exactly match that of their prey, many FAs can be deposited in adipose tissue with relatively little modification and often in a predictable way (Iverson *et al.* 2004; Iverson 2009b).

Three characteristics of FAs and their storage permit them to be used as trophic tracers. First, since a relatively limited number of FAs can be biosynthesized by animals, especially at higher trophic levels (Cook 1996), it is possible to distinguish dietary versus non-dietary sources. Second, unlike proteins and carbohydrates, which are completely broken down during digestion, FAs are generally not degraded during digestion and are taken up by tissues in their original form. Third, fat is stored in animal bodies in reservoirs, which can be substantial. Thus, FAs accumulate in storage sites over time and represent an integration of dietary intake over days, weeks, or months, depending on the species and its energy intake and storage rates (Iverson 2009b).

It has been known for decades (e.g. Klem 1935) that FAs are transferred from prey to predator both at the bottom and top of marine food webs (reviewed in Dalsgaard *et al.* 2003; Budge *et al.* 2006; Iverson 2009b), permitting inferences about consumer diets. FAs can be used to study diets in marine mammals in three ways. First, by examining changes in FA distributions, or 'signatures' (Iverson 1993) of the predator alone, qualitative questions about spatial or temporal variations in diets can be addressed, both among and within individuals or populations (e.g. Iverson *et al.* 1997a, b; Walton *et al.* 2000; Beck *et al.* 2005). Second, the presence or abundance of unusual FAs in predator tissues can be traced to the consumption of a particular prey species or taxa (e.g. Pascal and Ackman 1976; Thiemann *et al.* 2007). Third, and requiring the most careful considerations, FAs can used to quantitatively estimate diet from the FA signatures of predators using quantitative FA signature analysis (QFASA, Iverson *et al.* 2004), which employs a statistical model to compute the likely combination of prey FA signatures that comes closest to matching that observed in the predator FA storage sites (Iverson *et al.* 2004, 2006, 2007).

9.8.1 Tissues for analysis

Predators

Adipose tissue (including blubber), milk, and blood, contain the most direct information about diet (Budge *et al.* 2006; Iverson 2009b). The more membrane-structured the tissue (e.g. structural blubber such as tailstocks in cetaceans and muscle), the greater the contribution from endogenously conserved FAs which may obscure dietary influences to lesser or greater extents. Selection of tissue type will depend on the research question and sampling limitations of the species. Both adipose tissue stores and non-structural blubber contain an integrated record of dietary intake over a period of weeks to months, and perhaps longer in some species. However, the time frame of integration has only been investigated in a few species and thus more research is needed. The FA composition of milk also reflects diet, but the temporal nature of this information depends on the reproductive strategy of the particular species. In capital breeders, such as phocid seals and many baleen whales, milk FAs will be derived from blubber mobilization and thus reflect diet over the months of fattening prior to lactation. Conversely, in income breeders (e.g. otariids, small odontocetes), milk FAs will reflect the most recent dietary intake (i.e. days), as well as some mobilization of FAs from fat depots. Quantitative diet estimates from milk FAs using QFASA have not yet been attempted or validated. Finally, FAs can also be isolated from blood in the form of chylomicrons, which are the lipoproteins that specifically carry FAs from recent digestion throughout the bloodstream. Chylomicrons only persist in the blood for 2–6 hours, but when correctly isolated from blood, have provided accurate estimates of the most recent meal in grey seals (Cooper *et al.* 2005).

Subcutaneous adipose tissue/blubber can be sampled easily from live animals using a medical biopsy punch (e.g. 6 mm diameter), normally inserted through a small incision made in the skin through the full depth of the blubber layer (e.g. Kirsch *et al.* 2000). In most situations, it is considered best to leave the incision open, rather than suturing it closed, for the best draining and healing. Samples are easily obtained from dead animals by incision, but should be taken as soon as possible after death and certainly within 24 h of death. The FA composition of true fat storage sites (i.e. subcutaneous adipose tissue, non-structural blubber) appears to be homogeneous over most of the main body in many species (Koopman *et al.* 1996; Layton *et al.* 2000; Cooper 2004; Thiemann *et al.* 2006; Iverson *et al.* 2007). Although this should be confirmed in each species, it appears that the best site for sampling subcutaneous adipose tissue is simply where most fat is normally and actively stored. Of greater importance for blubber is the depth sampled (e.g. Koopman *et al.* 1996; Arnould *et al.* 2005; Strandberg *et al.* 2008). In pinnipeds, blubber is not 'stratified' per se, but there exists a gradation in FA composition from the inner to outer portions. Studies have shown that the full-depth blubber layer provides information on longer term diet, while the inner half alone reflects a more recent diet (Cooper 2004; Iverson *et al.* 2004; Tollit and Iverson, unpubl. data). Small samples only taken near the skin are not appropriate for diet assessment (Thiemann *et al.* 2004, 2009). In contrast to pinnipeds, the blubber of cetaceans, particularly small odontocetes, can exhibit extensive mor-phological and FA stratification, but the degree to which this occurs is species-specific (Koopman 2007). In some cases, FA stratification is sufficiently extreme that only the innermost layer can be used to infer diet, as the FAs stored in the outer layers appear to be largely endogenously derived and conserved. However, FA stratification appears to be less pronounced in larger cetacean species such as the bottlenose whale (*Hyperoodon ampullatus*) (Hooker *et al.* 2001a) and sperm whale (Koopman 2007).

Milk for FA analysis can be obtained directly from evacuation of mammary glands or taken from the stomach contents of neonates up to 8 h following ingestion, as the FA composition remains unchanged within the gastric milk fat globule until it is disrupted in the intestine (Iverson 1988).

Blood may also be used for FA analysis, but because of the ephemeral nature of dietary lipid transport in blood, the time of collection with respect to feeding is critical. It is essential that the TAG-rich chylomicrons carrying dietary FAs are: (i) visibly present in samples (as indicated by a cloudy or milky hue) and (ii) isolated from other blood lipids and lipoproteins (Cooper *et al.* 2005), as analysis of whole blood, plasma, or serum will lead to highly erroneous diet inference. The isolation of chylomicrons from other lipoprotein classes is performed by ultra-centrifugation of freshly collected serum or plasma samples (both give identical results). Even short-term freezing may cause disruption of the chylomicron lipids.

Prey

If FAs are to be used to estimate diet quantitatively, then prey must be sampled as has been extensively reviewed elsewhere (Iverson *et al.* 2004; Budge *et al.* 2006; Iverson 2009b). In brief, the onus is on the investigator to reasonably sample the prey field of the predator, to sample prey in the same manner in which the predator consumes these prey (i.e. for most marine mammals prey is consumed whole), and to sample a sufficient number of individuals of each prey species to allow assessment of within-species variability and between-species differences or overlap (Budge *et al.* 2002; Iverson *et al.* 2002, 2004). Prior to chemical analysis, each individual should be measured (mass, length), homogenized, and a weighed aliquot taken for analysis of fat content and FA composition. A thorough quantitative evaluation of within- and between-species variability to confirm the ability to reliably differentiate prey species in the estimation procedures is required.

9.8.2 Sample storage and chemical analysis

Guidelines for the optimal storage of tissues for lipid and FA analysis are summarized in Budge *et al.* (2006). Exposure to air will oxidize FAs in the sample, with a loss of especially highly unsaturated FAs. Freezing at −80 °C in an airtight container is one recommendation for long-term storage, but the lipids in the sample can be preserved indefinitely by immediately immersing the sample in chloroform ($CHCl_3$) containing 0.01% butylated hydroxytoluene (BHT) as antioxidant (however, note the biohazard issue with using chloroform). If frozen at −20 °C, especially under a nitrogen atmosphere, such samples may be safely stored for years. Only glass vials with Teflon-lined caps can be used for FA storage in solvents, as $CHCl_3$ will extract the plasticizers in other types of containers and contaminate the isolated lipids.

Methods for extraction from tissues, preparation, and analysis of FAs have been extensively reviewed (Christie 1982; Ackman 1986, 2002; Parrish 1999; Iverson *et al.* 2001; Budge *et al.* 2006). Briefly, lipids are best extracted with a modified Folch *et al.* (1957) procedure employing $CHCl_3$ and methanol (MeOH) (Iverson *et al.* 2001), which also allows quantification of fat content (see 9.8.4). Once extracted, acyl lipids are *trans*-esterified (i.e. converted to FA methyl esters, FAME, or FA butyl esters, FABE), ideally using an acidic catalyst, such as sulphuric acid (H_2SO_4) or hydrochloric acid (HCl) in methanol or butynol, respectively. Comparable results can be obtained using fresh anhydrous BF_3 and H_2SO_4 catalysts (Iverson *et al.* 1997a; Thiemann *et al.* 2004). However, anhydrous BF_3 in MeOH is no longer guaranteed by chemical suppliers, therefore we now recommend the use of the H_2SO_4 catalyst. FABE are only required in the case of some odontocetes that produce very short-chain volatile FAs, which are otherwise lost when using FAME.

Analyses of FA composition are performed using temperature-programmed gas chromatography (see methods in Iverson *et al.* 1997b, 2002; Budge *et al.* 2002). It is crucial to select a polar capillary column that allows adequate separation of all peaks of interest, such as the many isomers of long-chain poly-unsaturated fatty acids (PUFA) and some mono-unsaturated fatty acids (MUFA) (e.g. *n*-11 and *n*-9 isomers of 20:1 and 22:1). Excellent separations of marine FAs are achieved using a very polar column coated with 50% cyanopropyl-methylpolysiloxane, specifically the DB-23 model from Agilent Technologies, but other columns can be used. FAs are identified and integrated (i.e. quantified) using manufacturers' software. However, blindly accepting the FAME data generated by the computer software is a dangerous practice, given the complexity and number of marine FAs; therefore, chromatogram IDs and separations should be rigorously checked and reintegrated where necessary (Budge *et al.* 2006).

9.8.3 Using predator FAs to qualitatively infer diet

Evaluating variation in the full array of FAs (FA signatures) among individuals and populations of marine mammals is a promising, qualitative, way to look at trophic interactions and to detect dietary differences (S. Smith *et al.* 1997; Iverson *et al.* 1997a, b; Dahl *et al.* 2000; Thiemann *et al.* 2008; Tucker *et al.* 2009). Because over 70 different FAs are routinely identified in marine lipids, multivariate statistical techniques are generally required to best use the information contained in the data, although visual graphical inspection allows assessment of absolute differences. That is, finding a 'significant' difference in the levels of a specific FA among groups does not indicate whether this difference is biologically meaningful or whether the overall FA signature differs between groups. Multivariate analyses, which also allow pattern recognition, are generally the most powerful as they use the maximum number of FAs (depending on sample size) for differentiating predators and resolving trophic interactions (reviewed in Budge *et al.* 2006; Iverson 2009b).

The use of a FA 'biomarker approach' (i.e. using a single or group of unusual FAs or an unusual abundance of a FA to infer a predator diet item) will, in principle, rarely be possible in marine mammals given that they generally occupy the top of the food chain, and thus integrate consumption over several trophic levels. Although this has been done with some degree of success in several instances within simple systems, since it is generally a risky practice to infer diets directly from one or a few FAs, we point to reviews of this subject (Budge *et al.* 2006; Iverson 2009b).

9.8.4 Using predator and prey FAs to quantitatively estimate diet

QFASA is a first-generation statistical tool designed to quantitatively estimate predator diet using FA signatures of predator and prey. The basic approach of QFASA is to determine the mixture of those prey species FA signatures that most closely resembles that of the predator's FA stores, after accounting for the effects of predator metabolism, to thereby infer its diet. Details of the initial QFASA approach are provided by Iverson *et al.* (2004) and further discussed in subsequent studies and reviews (Budge *et al.* 2006; Hoberecht 2006; Iverson *et al.* 2006, 2007;

Beck *et al.* 2007; Nordstrom *et al.* 2008; Thiemann *et al.* 2008; Iverson 2009b; Tucker *et al.* 2009).

Briefly, QFASA proceeds by applying experimentally derived weighting factors ('calibration coefficients', CCs) to individual predator FAs to account for the effects of predator metabolism on FA deposition. It then takes the average FA signature of each prey species (or group), and estimates the mixture of prey signatures that comes closest to matching that of the weighted predator's FA stores by minimizing the statistical distance between that prey species mixture and the weighted predator FA profile. Lastly, this proportional mixture is weighted by the fat content (i.e. relative FA contribution) of each prey species to estimate the proportions in the predator's diet.

A number of important issues must be recognized in order to predict diet using QFASA (see Iverson *et al.* 2004; Budge *et al.* 2006; Iverson 2009b for detailed discussions). Perhaps the most important issue is that of accounting for predator metabolism (e.g. Nordstrom *et al.* 2008; Meynier 2009). At present, calibration coefficients are used. These are simple ratios for each FA in the predator stores divided by the level of that FA in the diet consumed over a period long enough for complete FA turnover. In principle, the FA signature of the predator's lipid stores should resemble this diet as much as possible and any differences would be attributable to metabolic processing of individual FAs. Although CCs are currently the only method put forward to account for predator metabolism, and they have been shown to result in reasonably accurate estimates of core diet, they remain a simple mathematical attempt to describe potentially complex biochemistry. Individual FA deposition could be affected by a number of factors, including physiological status (e.g. lactation, starvation) or external factors (e.g. prey fat content and possibly FA composition).

CCs have been estimated for a handful of pinniped species fed long-term on herring and, while many CCs are similar among these studies, recent validation studies suggest that predator species-specific CCs should be used if possible (e.g. Hoberecht 2006; Nordstrom *et al.* 2008). Whether these are truly species effects or diet/study effects are not yet known. Cetacean CCs have been harder to generate, mainly because of the requirement to sample full-depth blubber in captive animals that are on public display. In captive Steller sea lions, some dietary-related CCs varied with diet composition (Tollit, unpublished data). The application of these CCs in the QFASA model has been shown to be critical to optimal diet estimates (Iverson *et al.* 2004) and further studies of CCs are critical to further development.

Another issue in QFASA is which FA subset to use. Not all FAs provide information on diet (e.g. some arise in predators solely from biosynthesis), and those FAs that do not should be removed from analysis. Other FAs may be affected by the reliability of the CC calculated for them. Furthermore, not all FAs identified as useful and of dietary origin in the original model (Iverson *et al.* 2004), can be consistently and reliably detected by some laboratories, depending on the equipment used (e.g. Hoberecht 2006). The choice of FAs may also affect how well each subset selection discriminates between different prey species. Subset selection should be thoroughly tested, and using captive validation studies when possible. Similarly, further studies are required to ascertain FA turnover time (normally approx. 1–6 months) and the potential interaction with life history events.

Simulation is a useful way to explore prey differentiation using QFASA (e.g. Iverson *et al.* 2004, 2006; Tucker *et al.* 2009). Understanding the detection limits of prey, the fitting procedures of statistical models (Stewart 2005), and inclusion of within-species variability in prey FA and fat content in estimates require further research.

To date, QFASA has been able to provide reasonably accurate estimates of simple diets and diet switches in captive pinnipeds (e.g. Cooper 2004; Iverson *et al.* 2004; Nordstrom *et al.* 2008). However, there may be more inconsistent estimates and increased rates of misclassifications if the diet is diverse, measured over a very short time (<10 days) or if it consists of multiple species with similar FA signatures (e.g. Hoberecht 2006; Tollit and Iverson, unpubl. data). Nevertheless, natural relatively complex diets have been estimated using QFASA for free-ranging seals and seabirds, which were corroborated using either animal-borne video or other methods of diet analysis (e.g. Iverson *et al.* 2004, 2007; Tucker *et al.* 2008).

9.9 Stable isotopes and other markers

Elements in nature typically occur in more than one stable form due to differences in atomic mass. Small differences in the mass of these stable isotopes mean they behave differently in biogeochemical reactions. In kinetic and other rate-limiting processes, the relative abundance of heavier to lighter isotopes can change among reservoirs within the body of an animal. Isotopic distributions in nature provide the basis for tracing the origins of elements and molecules spatially and through trophic interactions. The most common elements used in the study of marine mammals have been H, C, N, O, and S. These light elements are typically fixed during primary production and consist of 2–3 stable isotope forms: ^1H, ^2H, ^{12}C, ^{13}C, ^{14}N, ^{15}N, ^{16}O, ^{17}O, ^{18}O, ^{32}S, ^{34}S.

Measurements of slight differences in the relative abundance of one isotope over the other are possible through established mass spectrometric techniques. These usually involve the comparison of the relative abundance of the heavier to the lighter isotope of an element in unknown samples to those in international standards. This is the basis of the delta notation in isotopic measurements that represents a ratio of ratios:

$$\delta X = [R_{sample}/R_{standard} - 1] \times 1000$$

where X is the heavy isotope of interest (e.g. ^{13}C, ^{15}N, ^{34}S) and R is the ratio of the heavy to light isotope (e.g. $^{12}C/^{13}C$, $^{14}N/^{15}N$, $^{32}S/^{34}S$). The international standards are relatively arbitrary and so delta values can be negative (e.g. $\delta^{13}C$ values) or positive (e.g. $\delta^{34}S$ values). Units of measurement are parts per thousand ‰. Measurement error typically ranges from ± 0.1 ‰ for $\delta^{13}C$ to ± 2 for $\delta^{2}H$.

There are three primary concepts in the application of stable isotope measurements to marine mammal dietary studies. The first is that stable isotope values in food webs are largely determined by those values in inorganic substrates. The process of fixation of elements during primary production involves isotopic discrimination, whereby one isotope is differentially incorporated. Once fixed, elements move through food webs, and the measurement of stable isotope ratios in consumers can give some idea of the source of primary production in cases where sources may differ in their stable isotope values. The second concept is that of isotopic discrimination between prey and predator, with more or less consistent changes in isotope ratios as one ascends the food chain. However, not all tissues are created equally, and isotopic discrimination between diet and predator tissues can vary according to tissue type, diet quality, and differential metabolic routing of macromolecules within the organism. Finally, the turnover of elements can differ between tissues and the half-life ranges—from a matter of days in the case of blood plasma to months in the case of muscle, and even years in the case of bone collagen. Dietary information will thus be integrated over different time spans depending on the tissue used.

Dietary reconstructions in marine mammals require fundamental knowledge of the baseline isotope values in the food webs of interest, the isotopic discrimination factor between diet and a specific tissue, likely metabolic routing of macromolecules, and the turnover rate associated with the tissue sampled. There are few cases where such knowledge will be complete. The main limitation of stable isotope analysis is that it provides relatively coarse information on diet, since the contributions of all prey are reduced to a single ratio for each isotope and the small number of variables inevitably limits discrimination of diet composition. Nonetheless, there are some advantages to using stable isotopes in trophic studies involving marine mammals, and also in conjunction with other methods such as FA and QFASA (Budge *et al.* 2008; Tucker *et al.* 2008). Both methods provide information on food assimilated and integrated over a relatively long time span.

9.9.1 Tissues for analysis

Skin, blubber, teeth, baleen, and internal organs have been used from stranded animals, and whiskers, muscle, and blood components from sedated animals. At sea, remotely sampled biopsy plugs of skin and blubber are possible, as is the use of sloughed skin (Todd *et al.* 1997; Ruiz-Cooley *et al.* 2004).

Although diet-tissue discrimination factors are still poorly understood, they do allow dietary isotopic values to be predicted and to understand the period of dietary integration. However, some controlled captive studies of phocid (Hobson *et al.* 1996) and otariid (Kurle 2002) seals have helped derive approximations for various blood components and whiskers (Hall-Aspland *et al.* 2005). The average transit time of dermal cells from the basal lamina to the skin surface, where they are sloughed, corresponds to about 75 days in the bottlenose dolphin (*Tursiops truncatus*) and beluga (*Delphinapterus leucas*) (Hicks *et al.* 1985; St. Aubin *et al.* 1990), but is probably longer for larger cetaceans. So, skin biopsies probably integrate isotopic information on diet for at least that period. For metabolically inactive tissues like baleen, nail, or whiskers, stable isotope values are typically unchanged following formation and so provide a temporal record of past diets (Schell *et al.* 1989).

Teeth contain both organic (i.e. collagen) and inorganic fractions in the dentine and enamel (Clementz and Koch 2004) and can be sub-sampled using the internal growth layers, or growth layer groups (GLG, see Chapter 5). The organic fraction can be analysed for all the light elements, and the inorganic fraction for $\delta^{13}C$ and $\delta^{18}O$ values. Age-specific trophic level estimates can be based on analysis of individual annuli (Hobson and Sease 1998; Hobson 2004a; Hanson *et al.* 2009). Mendes *et al.* (2007) inferred changes in trophic level and migration patterns in sperm whales from isotope measurements of tooth annuli.

9.9.2 Trophic modelling

Stable nitrogen isotope values ($\delta^{15}N$) in marine consumers show a step-wise increase with trophic level. Thus, by knowing the magnitude of this trophic enrichment corresponding to the tissue sampled, it is possible to predict (by subtraction) the average $\delta^{15}N$ value of the prey or, if the baseline food web $\delta^{15}N$ value is known, to model the trophic position of the organism of interest (e.g. Hobson and Welch 1992). Meta-analyses have suggested a $\delta^{15}N$ trophic enrichment factor for marine mammals of about 3.4 ‰ (Post 2002).

In many cases, the appropriate stable isotope values for primary production are poorly known, or are known to differ seasonally or over the large spatial areas used by many marine mammals. In such cases, it may be better to use $\delta^{15}N$ measurements of higher trophic-level (TL) primary herbivores (i.e. assumed to be at TL 2) as a baseline (Hobson *et al.* 2002; Hooker *et al.* 2002c; Ruiz-Cooley *et al.* 2004). Another consideration in applying $\delta^{15}N$ measurements to derive estimates of trophic level is that the values are averages. So, if two trophic sources were consumed in equal proportions by a consumer, then the intermediate trophic level predicted by the use of this single isotope measurement (i.e. [TL1 + TL2]/2) would be identical to that of another consumer feeding exclusively at that intermediate trophic level. Thus, it is essential to be aware of the inherent ambiguity in trophic models.

Theoretically, the transfer of maternal-based nutrients to offspring via suckling represents a trophic increase in the position of the neonate relative to the mother. Such a trophic increase should be reflected in higher $\delta^{15}N$ values in the offspring compared to the nursing parent (but see Jenkins *et al.* 2001). Indeed, this $\delta^{15}N$ enrichment effect has been observed in the first annulus of seal teeth (Hobson and

Sease 1998; Hobson *et al.* 2004a; Newsome *et al.* 2006, Hanson *et al.* 2009). A complementary decrease in dentine $\delta^{13}C$ values corresponding to the pre-weaning period also corresponds well with a neonate diet rich in ^{13}C-depleted lipids, since lipid carbon in milk contributes to protein synthesis in offspring. York *et al.* (2008) used $\delta^{15}N$ and $\delta^{13}C$ measurements of archived tooth annuli of Steller sea lions to infer how age of weaning was influenced by large-scale oceanic change in the Gulf of Alaska during the 1970s.

9.9.3 Source of feeding and marine isoscapes

Stable isotope analyses also provide the possibility of delineating spatial informa-tion on where marine mammals fed. Stable carbon isotope measurements are known to provide spatial information by latitude (Rau *et al.* 1982; Hobson *et al.* 1997a; Cherel *et al.* 2005). Off-shelf, pelagic food may also be more depleted in ^{13}C compared to on-shelf or benthic food, probably partly because of the depletion of $\delta^{13}C$ values of primary production in low nutrient conditions (Laws *et al.* 1995; France 1995). There is a longitudinal gradient in $\delta^{13}C$ values in the marine food webs along the northern Gulf of Alaska which can be detected in the tissues of Steller sea lions (Kurle and Gudmundson 2007). In addition to those of C and N, stable isotopes of other elements including S, O, and H have considerable potential to provide useful marine isoscapes. For example, $\delta^{34}S$ values may also be more enriched in benthic or inshore vs. pelagic food webs, and both δD (deuterium) and $\delta^{18}O$ are sensitive to salinity and ocean temperature (Hobson *et al.* 2010).

9.9.4 Other elements and compounds

Concentrations of other elements and compounds, especially inorganic and organic contaminants, in the tissues of marine mammals may provide information on diet and feeding provenance. Numerous studies have found good correlation between $\delta^{15}N$ values and contaminant loads in marine organisms (Atwell *et al.* 1998; Das *et al.* 2000; Fisk *et al.* 2001). As such, contaminant measurements themselves can be used to infer trophic level as well as geographical segregation in some systems (Shao *et al.* 2004). Recent advances in our understanding of strontium isotope ratios, and how these can influence tissue values in animals influenced by estuaries or terrestrial runoff, suggest this heavier isotope will be of use (Hobson *et al.* 2010).

9.9.5 Field and laboratory methods and data analysis

Tissues for stable isotope measurements should be frozen following collection. However, short-term storage of soft tissues in 70% ethanol has negligible effects on stable isotope measurements (Hobson *et al.* 1997b). Oven-drying may be used, but temperatures should not exceed 60 °C. Hard tissues like fur, whiskers, claws, teeth, and baleen can be stored dry. Hobson and Sease (1998) and Newsome *et al.* (2007) describe methods for isotopic analysis of the organic and inorganic fractions of teeth. Bone collagen extraction methods are found in Newsome *et al.* (2006).

Tissues are typically cleaned of surface contaminants before drying and homogenizing to a fine powder. For hard keratinous tissues, oils can be removed using a variety of solvents, but a 2 : 1 chloroform : methanol solution works well. Sonication can assist in surface cleansing. For soft tissues, following thawing or decanting of ethanol, freeze-drying is the preferred next step as this renders materials easy to powder by mortar and pestle or by mechanical grinders or mills. It is essential that the material analysed is homogenous.

Lipids are considerably more depleted in ^{13}C than most other animal tissues and, because they occur in varying concentrations depending on nutritional considerations, are best removed from bulk tissues such as muscle, liver, and blood in order to remove this source of variation on the tissue δ^{13}C value. This can be achieved using various techniques including soxhlet extractions or quicker solvent rinses on small powdered materials (Hobson *et al.* 2002). However, lipid extraction has also been shown to have a small but measurable effect on tissue δ^{15}N values. A good discussion on how to approach marine food web tissue analyses is provided in Søreide *et al.* (2007).

About 1 mg is required for most powdered samples of δ^{13}C, δ^{15}N and δ^{34}S, but this can vary depending on the elemental concentrations, and all samples must be weighed precisely (± 0.1mg). Continuous-flow, isotope-ratio, mass spectrometry (CFIRMS) is used almost universally these days. Researchers should contact appropriate labs to determine precise protocols and seek out labs that use *organic* laboratory standards (that are of a similar C : N ratio to their unknowns) when measuring organic materials. Researchers are also encouraged to provide blind replicates of their samples, as it is the replicate measurement of lab standards run with the samples or on a series of the same unknown that provides the measurement error that should be quoted (Jardine and Cunjak 2005).

It is common to portray results as biplots using two isotope measurements: most typically with δ^{13}C values on the x-axis and δ^{15}N values on the y-axis. Because different tissues can involve different isotopic discrimination values for a consumer, authors should either only include plots with the same consumer tissues represented or normalize all tissues to their diet equivalents by applying appropriate tissue-specific discrimination values. Other common uses of stable isotope data are to reconstruct dietary inputs to consumers using mixing models. The rule-of-thumb is that inputs from *n* sources of isotopically distinct foods can be uniquely resolved using *n*-1 stable isotopes (Phillips 2001). In cases of too many sources, ranges of inputs from specific sources can use probabilistic models (Phillips and Gregg 2003).

The successful application of stable isotope techniques for inferring the diet and source of feeding in marine mammals will depend on how well the researcher is able to characterize isotopically the food web being used. As such, an isotopic assay of prey and consumer tissues is encouraged over measurements of the consumer alone. Choice of tissue and knowledge of the period of dietary integration represented by the isotope values is also critical. Researchers are encouraged to consider the use of more than one tissue in order to gain insight into diet over different periods for the same animal. This is a rapidly advancing field, and we anticipate more refined analyses being possible through the description of marine isoscapes on the one hand and the careful use of controlled dietary studies on captive animals on the other. Finally, stable isotope methods will typically augment, but not replace, the other tools we have to investigate marine mammal diet.

9.10 Summary

The emergence of various new techniques described in this chapter underlines the great potential to use multiple dietary assays. Ultimately, no single technique will provide all the answers, and researchers should aim to use as many lines of evidence as possible when weighing the evidence for marine mammal dietary compositions, especially if the species under study is considered a wide-ranging generalist consuming many different taxa. Stomach, regurgitate, or scat hard part analysis provides vital definitive species identification (as well as size) and should therefore form the baseline of new dietary studies, assuming samples are readily available. Many of the key limitations of hard part analysis are now well understood. The concurrent use of DNA methods on soft and hard prey remains in scats and GI tracts shows great promise and the field is developing quickly. FAs and isotopes typically provide less direct evidence of diet (and lower resolution in the case of stable isotopes - i.e. identifying trophic level) but importantly, over longer periods and samples can be collected directly from individual animals. This makes them attractive alternatives for many species of cetacean and wide-ranging pinnipeds.

10
Telemetry

Bernie McConnell, Mike Fedak, Sascha Hooker,
and Toby Patterson

10.1 Introduction

Marine mammals can be obscure and cryptic. Some travel to remote regions of the world's oceans and most spend a high proportion of their time underwater, often at great depth. This means there are limited opportunities to directly observe important features of marine mammal biology and this problem has been the main motivation for the development of marine mammal telemetry systems. Perhaps the first tangible telemetry system deployed on a marine mammal was a clockwork time-depth recorder attached to a Weddell seal in 1964 (Kooyman 2007).

The word telemetry derives from the Greek *tele* meaning far and *metros* meaning measurement. Telemetry is a diverse set of methods, all of which monitor one or more aspects of an animal's behaviour, physiology, or environment more or less remotely from the observer in order to answer many different biological questions. Each species presents its own challenges and opportunities and so there are many different technical solutions—each optimized to answer a given biological question. Studies may scale from individual physiology (Ponganis 2007) and behaviour through to population parameters such as survival (Murray 2006), the probability of hauling out (Simpkins *et al.* 2003), abundance (Huber *et al.* 2001; Ries *et al.* 1998), and distribution (Matthiopoulos *et al.* 2004). In this chapter we consider *active* telemetry, where an animal is instrumented with an electronic device that uses energy to collect information which may be recovered later or relayed by radio or sound. This includes time-depth recorders (TDRs) which have been a workhorse for marine mammal researchers for many years. *Passive* remote sensing, the detection of sound generated by animals, is covered in Chapters 3, 12, and 13.

We do not attempt to provide a complete catalogue of existing systems, which would be a book in its own right and would soon be out of date. This chapter provides a general background to the fundamentals of telemetry systems and provides examples of various tag configurations and studies. Our (undoubtedly biased) bibliography is chosen to assist a newcomer's entry into the vast telemetry literature, rather than to do adequate justice to historical developments and the many individuals who have contributed greatly to the development of the field. Indeed, every citation should be prefixed with the phrase 'for example'.

We first consider general tag design and attachment constraints imposed by the obscure and diverse lifestyles of marine mammals. We then consider the choice of various systems used to determine geographical locations. We discuss the wide range of sensors currently available to tell us more about the animal and its environment. Many of these can aid location determination. Having collected the data on the tag the next challenge is to get it into the hands of the researcher, either by recovering the tag or by using some form of data relay. The researcher then has to visualize the information obtained and place it within an environmental frame-work. Finally we deal with the quantitative aspects of data analysis.

10.2 Design considerations

From both ethical and scientific viewpoints, a telemetry tag should have a minimal effect on an animal (Wilson and McMahon 2006). However, there will always be some physical effect due to its size, mass, or drag, and so the tag should be as small as possible to minimize its biological effect. In addition to minimizing size per se, good design can reduce the hydrodynamic drag of a tag (Pavlov et al. 2007). Good design can also reduce biofouling which can cause additional drag as well as sensor malfunction (Whelan and Regan 2006).

Since the battery is often the largest part of a tag there is a need to maximize energy storage density as well as its efficient use (Fedak et al. 2002). Lithium thionyl chloride is the most popular energy storage medium for telemetry tags. Batteries of this type are extremely energy dense (about 7 MJ/L, similar to trinitrotoluene), have wide temperature tolerances ($-60\,°C$ to $+85\,°C$), and a long shelf life (>10 years). However, their high energy density means that they must be treated with caution, disposed of appropriately, and may be the subject of transportation restrictions. Whilst secondary (rechargeable) batteries are generally less energy dense, they do offer the opportunity to incorporate energy scavenging systems (Paradiso and Starner 2005)—the principal candidate being solar power. The prospective energy budgets of solar scavenging must take into account winter periods, time spent underwater, biofouling, and any diel behavioural patterns.

Energy efficiency is achieved through the use of ultra low-power components and extreme duty cycling. The latter means that the tag is effectively switched off for most of its operation, briefly switching on selected components when there is a need to collect or relay data. The intelligent coding, compression, and transmission of data can also result in large energy savings.

Whilst it may seem obvious that the purchase and operating cost of a telemetry tag should be minimized, the unit cost has a bearing on good science. Most research grants are capped and the understandable outcome is that the sample size is the maximum that the remaining budget will permit. However, if the aim of the study is to make an inference about a *population* process, an a priori statistical power analysis will assist the ethical and scientific decision of whether the sample size is, indeed, sufficient (Lindberg and Walker 2007).

10.3 Attachment

Ethical concerns are equally important when considering tag attachment (see Chapter 1). Any procedure should minimize disturbance and discomfort to the animal. Attachment methods for large whales are necessarily remote, but some cetacean and most seal deployments require the animal to be captured and anaesthetized or restrained. These issues are covered in Chapter 2.

There are three main approaches to tagging seals: gluing to fur, attaching to flipper tags, and implantation. For radio tags, the prime position in pinnipeds is on or near the back of the head to take advantage of surfacing periods; but placing an instrument in the middle of the back may be appropriate if information is only transmitted when the animals are hauled out. Rapid setting, two-part epoxy resins are the most popular approach for attaching devices to the fur, albeit the tag will certainly detach at the annual moult. An epoxy-hardener mix can generate a high exothermic reaction and excessive amounts carry the risk of burning the seal. Cyanoacrylate glues have the advantage (and disadvantage to witless researchers) of setting almost immediately, and some are sufficiently waterproof to be suitable candidates. Whilst being inherently weaker, silicones have the advantage of avoiding a hard edge that can result in local abrasion. Recent advances in dental glues also warrant attention.

In many cases tags remain on pinnipeds until moult and are then usually lost at sea. Animals may be recaptured when they predictably return to a location; in which case, the use of instruments attached with cable ties to a mesh glued to the animal's fur allows the rapid recovery of the instruments and leaves only the mesh attached until moult (Fig. 10.1a). For animals that cannot be recaptured, radio-release mechanisms may be used in certain circumstances (Andrews 1998; Watanabe 2006a).

To overcome the problem of fur-glued tags detaching at the annual moult, miniature telemetry tags may be incorporated into flipper tags (Ries *et al.* 1998; Huber *et al.* 2001; Simpkins *et al.* 2003). Since flipper tags preclude tracking whilst the animal is at sea they are used primarily to model haul-out patterns of animals detected during aerial surveys. Their small size limits their range and longevity.

Fully implantable tags have the advantage of creating no drag and of indefinite attachment (Lander *et al.* 2005). In fact, the only time that an implantable seal tag will detach is after death and carcass decay. Horning *et al.* (2005) discuss this idea for their 'Life History pop-up Argos tag' whose operation signifies seal death. However, the implantation operation is time-consuming and the risk of infection and rejection of the tag needs to be tested in each species because different species respond differently to implantation.

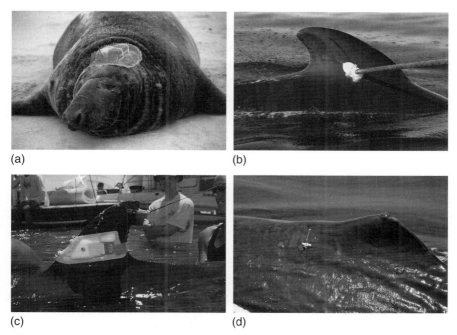

Fig. 10.1 Different forms of tag attachment are used depending on the species and study duration. (a) tag attached with cable ties to mesh glued to the pelage of a grey seal facilitating tag recovery upon seal recapture; (b) suction-cup attached D-tag deployed by pole onto the dorsal surface of a pilot whale; (c) suction-cup attached Trac Pac and pinned roto-radio on the dorsal fin of a bottlenose dolphin; (d) satellite tag fired by crossbow or air-rifle into the skin and blubber of a sperm whale reduces drag and prolongs the tag attachment duration. (Photos courtesy of: Robin Baird, Sanna Kuningas, Chicago Zoological Society - Sarasota Dolphin Research Program, Bruce Mate.)

In tusked species (walrus and narwhal) the tusk can be used for tag attachment (Jay and Garner 2002; Laidre *et al.* 2003). For walrus, remotely attached dart-tags have also been applied into their tough hide using a gas-powered rifle or crossbow (Jay *et al.* 2006). Dugongs and manatees have a tailstock to which a collar can be fitted. A line with a weak link then connects to a transmitter which trails at the surface (Deutsch *et al.* 1998; Sheppard *et al.* 2006).

Attaching tags to cetaceans is more difficult because they lack hair and do not haul-out (Hooker and Baird 2001; Lander *et al.* 2001). Animals must either be captured (see Chapter 2), or tags deployed remotely. Tags can be attached either by penetrating the skin surface (pinning through the dorsal fin or by using a barb/hook attachment to the blubber), or without penetrating the skin surface (using a suction-cup attachment).

For short-term deployments (usually less than a day) where animals will be tracked continuously using VHF telemetry, tags can be attached using either single or multiple (usually 2–4) suction cups (Fig. 10.1b). The tag can be attached to the suction cup either directly or via a tether. This package can then be deployed remotely at short-range (2–3 m) on bow-riding species using a hand-held pole (Baird *et al.* 2001), or at greater distances (5–10 m) using a cantilevered pole (M.P. Johnson and Tyack 2003). A crossbow or airgun may need to be deployed for species which are less predictable in their surfacing pattern (Hooker and Baird 2001). Tags are preferably placed on the flank around the dorsal fin in order that radio-signals are most likely to be detected during surfacing. To ensure recovery, such packages require a buoyant housing. The duration of short-term tag attachment is very unpredictable, and therefore a timed-release system may be used (M.P. Johnson and Tyack 2003). An alternative design for small cetaceans uses a series of smaller suction cups attached to a hinged thermoplastic saddle conformed to the shape of a dolphin's dorsal fin (Westgate *et al.* 2007). The suction cups hold the pack to the dorsal fin, and a dissolvable link causes the pack to jettison after several hours. A limitation of this system is the need for capture in order to attach the package.

For captured animals, long deployments (several months) may be achieved by pinning the tag to the dorsal fin (Fig. 10.1c; Lander *et al.* 2001), or dorsal ridge (Westgate *et al.* 2007). Biocompatibility of pin material and a bimetallic combination of nuts and washers is crucial to avoid pressure necrosis and ensure corroding to avoid damage caused by migration of the tag out of the tissue (Irvine *et al.* 1982; Lander *et al.* 2001). Long-term attachments on the dorsal fin also need to consider both thermoregulatory and hydrodynamic function. The vasculature of the fin many vary between species and the quantity, size, and placement of pins needs consideration (Lander *et al.* 2001). Therefore, advice should be sought prior to undertaking pinned dorsal-fin attachment due to the high risk of tissue damage. There is an urgent need for the design of a more benign long-term attachment system (Pavlov *et al.* 2007). For short-term VHF tracking (25–30 days), a roto-radio (combining a small, epoxy-encapsulated VHF transmitter with a cattle ear roto-tag) can be attached to the trailing edge of the dorsal fin (Scott *et al.* 1990a).

For larger species which cannot be captured, long-term attachments (>100 days) require penetrating implantable tags. Satellite tag attachment via crossbow or air-rifle has become much refined since the first deployments in the mid-1980s (Heide-Jorgensen *et al.* 2001; Mate *et al.* 2007). A slim cylindrical tag is now fired directly into the blubber leaving only the antenna protruding. However, for small- and medium-sized species, such as killer whales and melon-headed whales, with thinner blubber layers, completely implantable tags are not feasible. In such cases, remotely attached tags can be deployed by crossbow or air-rifle using a design which leaves the main unit above the skin surface with one or two penetrating barbs holding this in place (Andrews *et al.* 2008).

The importance of quantifying the biological effect of such procedures on each study species has been widely noted (Boyd *et al.* 1997; Geertsen *et al.* 2004; Martin *et al.* 2006; McCafferty *et al.* 2007). In most cases the immediate behavioural impact of these attachment and deployment methods appears to be low-level (Mate *et al.* 2007; Hooker *et al.* 2001; Andrews *et al.* 2008), although adverse reactions to suction-cup tagging were observed for bottlenose dolphins in New Zealand (Schneider *et al.* 1998). Some practitioners advise discarding the first dive prior to analysis (Miller *et al.* 2004a). Longer term reactions are much more difficult to quantify, although data suggests these are also minor (Geertsen *et al.* 2004).

10.4 Location determination

Most telemetry systems provide some sort of 2D positional information (Millspaugh and Marzluff 2001). The frequency of these position fixes can vary from daily to sub-minute intervals, and their accuracy can vary from hundreds of kilometres to tens of metres. However, as in all data collection systems, animal behaviour will influence the quality and quantity of the location data collected (Rettie and McLoughlin 1999; Aarts *et al.* 2008).

The most primitive location fixing method is to use the reception of a VHF tag to indicate its proximity, or to use two or more receivers with directional antennae to triangulate a fix (Kenward 2000). VHF signals can only be received when in 'line of sight' of the transmitter. Thus a receiver mounted in a plane has a greater range than one that is handheld. Simple VHF tags have the benefit of being cheap and thus may be deployed in sufficiently large numbers to infer *population* haul-out behaviour (Simpkins *et al.* 2003) or size (Ries *et al.* 1998). If animals are predictably distributed, automatic terrestrial or ship-borne (Hooker *et al.* 2002b) stations can be established. Otherwise the logistic cost of transporting receivers over the target species' range can be prohibitive.

Acoustic tags have the advantage that, unlike VHF, they can communicate underwater. However, care must be taken that their frequency is outside the detection range of the study species and its prey. Since the range is usually less than 1 km, acoustic tag studies are usually restricted to sites where the target animals are predictably close. Wright *et al.* (2007) monitored the presence of tagged harbour seals in an estuary using a fixed array of 15 acoustic receivers. Wartzok *et al.* (1992) pioneered depth-modulated acoustic tags that can be tracked in three dimensions using the time delay to an array of fixed receivers. This system has been used by many workers to study the movement of polar seals in relation to breathing holes in ice (Harcourt *et al.* 2000; Simpkins *et al.* 2001).

Light-based geolocation (Hill 1994; Hill and Braun 2001) depends on the tag recording the perceived times of sunrise and sunset each day. Positions can be estimated after the data have been retrieved using navigation equations. However, there may be latitudinal errors up to hundreds of kilometres, depending upon the time of year and latitude (Musyl *et al.* 2001). Sea surface temperature or dive depth (and thus minimum depth of sea) may be used as a covariate with day length to assist latitude estimation (Beck *et al.* 2002; Teo *et al.* 2004; Nielsen *et al.* 2006).

The Argos satellite system (Argos 2008) has been the workhorse for many hundreds of marine mammal studies. Location is determined using Doppler shifts in the frequency of the transmissions perceived by one or more of the polar-orbiting spacecraft. The minimum transmission (uplink) interval is 40 s and the average visibility of a pass is 10 minutes. The frequency and quality of the location fixes depend mainly on the number of uplinks received per pass, and this depends on the behaviour of the tagged animal. In addition, time of day, latitude, transmitter stability, and proximity to radio interference also affect results. Each location fix is assigned a *categorical* Location Class (LC), but since 2008 Argos have also supplied error ellipsis estimates as a continuous variable. The best LC (3) means that the 68% of latitudinal or longitudinal errors are less than 150 m. However, few locations obtained from an animal at sea achieve such LCs. Vincent *et al.* (2002) provide realistic confidence intervals to *all* LCs provided by Argos. They also advise that individual tags should be similarly calibrated before deployment (Soutullo *et al.* 2007). Another reason for not relying upon published calibrations is that Argos improves their location estimation algorithms from time to time.

The global positioning system (GPS) can provide locations to an accuracy of a few metres. But its cold-boot requirement to have an uninterrupted 30 s (or more) view of the sky to obtain the critical ephemeris data does limit its use in marine mammals. One exception is the successful tracking of dugongs (Sheppard *et al.* 2006). However, the Fastloc® system overcomes this problem by separating two processes that are normal carried out on a GPS receiver. A snapshot (<100 ms) of satellite transmission is captured, from which the identities and pseudoranges of the available satellites are calculated and then compressed into 32 bytes. Although this calculation may take up to 20 s, it can be carried out when the tag is underwater. When a pseudorange record is relayed ashore it is combined with archived ephemeris data, available online as RINEX files, and a GPS quality location is calculated. A similar approach is used in the Navsys TrackTag GPS system (MacLean 2009), except that the captured satellite data are stored with no further processing. Since the raw data required for each fix is about 32 Kbytes, the TrackTag technology is best suited to retrieval tags. An additional advantage of GPS-based systems is that the user can control the rate of attempted location fixes and the associated energy drain. Also, more frequent locations reduce the uncertainty of estimating animal movement in between fixes (Lonergan *et al.* 2009).

Dead reckoning (see Wilson *et al.* 2007 for a review) primarily uses depth and velocity to determine relative movements at, potentially, very high temporal resolution. When complemented with a geographical location system, such as GPS, dead reckoning can provide detailed absolute location fixes.

10.5 Sensors

Sensors can provide information about tag function, the animal, and its environment (Hooker and Boyd 2003). Measures of tag function include battery voltage, antenna tuning, and drift in the wet/dry sensor. Such sensor data and software diagnostic data provide feedback not only to the design engineer but also to the biologist attempting to untangle the probability of tag failure/loss from animal death. For population studies tag failure/loss can also be estimated by double tagging, although the failure/loss of each tag must be assumed to be independent (Bradshaw *et al.* 2000).

Telemetry can be used to test hypotheses relating animal movement and behaviour to feeding. Ingestion of cold food results in the stomach undergoing a 'precipitous temperature drop followed by an approximately exponential rise' (a PDER [predischarge event recording], Catry *et al.* 2004), and these can be detected by a stomach temperature tag (Andrews 1998). These tags must be small enough to insert via a stomach tube, but large enough to be retained within the stomach. Retention time can be increased by attaching the tag to biodegradable ethnafoam (Austin *et al.* 2006a). The ingestion of water can be distinguished by its increased rate of warming (Catry *et al.* 2004; Kuhn and Costa 2006). However, whether the meal size can be predicted from the shape and magnitude of the PDER is a matter of some debate (Hedd *et al.* 1996; Bekkby and Bjorge 1998; Kuhn and Costa 2006). Nevertheless, stomach temperature tags inform where feeding is taking place, shed light on the relationship between feeding and movement and dive type (Austin *et al.* 2006a, b; Horsburgh *et al.* 2008), and complement indirect evidence of feeding (Robinson *et al.* 2007).

Feeding is also associated with jaw movement, which can be detected by a Hall effect sensor. Liebsch *et al.* (2007) tested such a device on both captive and free-ranging seals and concluded that the success of the technique depends upon the range of sizes of prey taken by individuals. Their attempts to infer meal size were equivocal.

The incorporation of a still or video camera (e.g. Marshall *et al.* 2007; Moll *et al.* 2007) provides direct evidence of feeding and prey type as well as the available prey field (Davis *et al.* 1999; Bowen *et al.* 2002; Hooker *et al.* 2002a; Mitani *et al.* 2004; Mori *et al.* 2005; Parrish *et al.* 2005; Fulman *et al.* 2007). However, the amount of data collected is too large to transmit using current telemetry systems so it is necessary to physically retrieve the instrument.

Whilst feeding is associated with energy gain, metabolic expenditure can be estimated using heart rate (see Chapter 8). Heart rate can be measured using implanted or surface-mounted ECG electrodes (Fedak *et al.* 1988; Boyd *et al.* 1999b; Greaves *et al.* 2004) and the data can either be stored for retrieval or relayed in real time via an acoustic link. Fahlman *et al.* (2008) used overall dynamic body acceleration (ODBA) derived from a 3-axis accelerometer to estimate field metabolic rate in Steller sea lions.

An animal's buoyancy affects its swimming energetics and also reflects changes in condition (fat : lean ratio). Biuw *et al.* (2003) inferred changes in buoyancy in southern elephant seals from their rate of drift in drift dives. Watanabe (2006a) used 3-axis accelerometers to investigate stroke pattern in relation to buoyancy changes in Baikal seals. M. P. Johnson *et al.*'s (2003) D-tag incorporates 3-axis accelerometers, as well as magnetometers that monitor 3-axis attitude. These tags have been used to determine buoyancy from stroke patterns in sperm whales (Miller *et al.* 2004b) and North Atlantic right whales (Nowacek *et al.* 2001a). Energy budget in northern fur seals has been inferred using the acoustic detection of flipper stroke rate (Insley *et al.* 2007).

The D-tag also records sound and has been used extensively to detect and characterize echolocation pulses from toothed whales (Miller *et al.* 2004a; Johnson *et al.* 2006). Such information can then be used to calibrate passive acoustic monitoring (PAM) surveys (Tyack *et al.* 2006). In a stereo configuration the D-tag has also been used to characterize the echo characteristics of the prey field (Stanton *et al.* 2008). Perhaps the most significant application of the D-tag is to monitor and record the effect of anthropogenic sound on behaviour (Nowacek *et al.* 2004; Madsen *et al.* 2006).

Understanding the physical structure of the ocean, identifying its water masses, and predicting its behaviour depend on measuring temperature, salinity, and density to high accuracy. Archival tags been be used in some circumstances to measure temperature, salinity, and light (Boyd *et al.* 2001; Hooker and Boyd 2003; McCafferty *et al.* 2004), but miniature low–power, conductivity–temperature–depth (CTD) sensors have been specially developed (Valeport Ltd, Totnes UK) and incorporated into CTD-Satellite Relay Data Loggers (SRDLs; Sea Mammal Research Unit, Scotland). They measure temperature to an accuracy of $\pm 0.005\,^{\circ}$C with a resolution of $0.001\,^{\circ}$C and salinity to an accuracy of ± 0.01 mS/cm with a resolution 0.002 mS/cm. Such accuracies and precisions allow these instruments to be used for oceanographic measurement, which has resulted in conductivity and temperature profiles of the water column being obtained from several marine mammal species. Biuw *et al.* (2007) used these data to relate the behaviour and condition of southern elephant seals to their *in situ* oceanographic conditions. Since marine mammals often frequent remote polar regions that have been sampled only to a limited extent, these data add to our knowledge base (Boehme *et al.* 2008; Lydersen *et al.* 2002; Charrassin *et al.* 2008). In addition, the data have been distributed by the World Meteorological Organization (WMO) Global Telecommunication System (GTS) to operational forecasting centres where they are assimilated into models providing ocean forecasts and long-range seasonal and climate predictions.

10.6 Information relay

Recovering information from a tag is equivalent to transmitting data over a noisy communication channel (MacKay 2003). We compare the properties of different transmission channels used in marine mammal telemetry in Table 10.1.

Table 10.1 *Properties of different communication channels used in marine mammal studies. The values in this table are somewhat arbitrary and will vary from study to study. The aim is rather to focus on the relative properties of different transmission modes within a common framework and terminology.*

Channel	Capacity	User latency	Setup latency	Spatial coverage	Receiver availability
Acoustic mobile	High	Immediate	Immediate	Local 1–5 km	Largely continuous
Acoustic automatic	High	Immediate	Immediate	Local 1–5 km	Continuous
VHF mobile	Low	Immediate	Immediate	Local 20 km	Largely continuous
VHF automatic[1]	Low	Slow	Immediate	Local 20 km	Continuous
GSM SMS	Medium	Minutes	10 s	Coastal < 35 km	Continuous
GSM—GPRS/FTP	High	10 s	10 s	Coastal < 35 km	Continuous
Argos	Low	Hours	Immediate	Global	Periodic
Iridium SBD service	High	Minutes	20 s	Global	Continuous
Data sharing	Low	Weeks/months	Immediate	Depends on number of tagged animals and their social patterns	Very intermittent
Argos Life History tag	Low	Years	Immediate upon post-mortem release	Global	Periodic
Retrieval	High	Months	NA	Logistic availability	Periodic

[1] *Automatic* refers to a receiver that automatically logs data. GSM (Global System for Mobile Communications) is a mobile (cell) phone network. Data can be relayed via text messages (SM's) or via a GPRS/FTP internet session. SBD is iridium's Short Burst Data (SBD) service. *Data sharing* over a network of tagged individuals is referenced in the text. The *Life History* tag is described by Horning and Hill (2005). *Retrieval* tags are physically retrieved to obtain stored data.

This abstraction of all the various means of getting data from a marine mammal into the framework of communication channels is intended to encourage the reader to appreciate the fundamental problems involved—and so be in a better position to invent better solutions.

Two data (signal) carriers are commonly used in marine mammal telemetry—acoustic and radio frequency (RF) waves. At the receiver end the signal to noise ratio is crucial in determining the rate at which data can be relayed. The signal from both types of carrier diminishes because of the spreading of the wave front, described by the inverse square law, and because of absorption by the medium through which the signal is moving. At the molecular level, RF energy at certain frequencies can resonate (and thus be absorbed by) oscillations of molecular bonds. In addition, free ions (i.e. salts in the sea) have relatively large masses and can therefore absorb larger amounts of energy directly, transforming it directly into heat. Thus the RF absorption of seawater is considerably greater than that of fresh water.

The high degree of RF attenuation through water is the reason that sound may be chosen as a data carrier. However, sound is still absorbed by water to some extent, and it may also be absorbed and scattered by the air–sea and sea–seabed interface. Furthermore, it may be trapped within, or reflected off, temperature–salinity discontinuities within the water body itself (Lurton 2002).

Noise is present everywhere, including that generated within a receiver. RF noise may also be generated by powerful neighbouring frequencies or other tags competing for the attention of the receiver. In water, acoustic noise can be produced by currents or wave action, marine life, and anthropogenic sources.

The *capacity* of a channel is the maximum rate at which information can be relayed, and is determined by its bandwidth and noise. All other things being equal (which they seldom are), higher carrier frequencies have a greater bandwidth to transfer information as well as requiring smaller radiating antenna or transducers. However, absorption generally increases (though seldom monotonically) at higher frequencies. With acoustic carriers it is essential that the carrier frequency is above the audible threshold of the study species. Note that pulsed operation of a high-frequency carrier may result in lower frequency audible clicks, and many acoustic transducers have lower frequency side-lobes that may be within the hearing range of the study species. It can be easy to be lulled into imagining that a tag's operational frequency range is well outside the hearing range of the study species.

For marine mammal studies measures of telemetry capacity are scale-dependent. For example, capacity can be high (e.g. $\sim$125 bits per second, assuming that noise corrupts an average 50% of received uplinks) over the duration of a successful one-second uplink to an Argos satellite. However, over a day there may be only 100 such uplinks and so the daily rate is much lower (in this case c. 0.14 bits/s). Ultimately, the researcher is most interested in channel capacity at the scale of the duration of the tagging study. That is, how much information in total is collected from an individual animal.

The *transmission latency* of a channel is the delay between transmission and safe delivery to the recipient. With VHF and acoustic tracking this latency can be negligible. At the other extreme, the transmission latency for the retrieval of data from a data storage tag (the tag itself is the physical communication channel in this case) may be many months—if ever. This latency is of operational interest to the user. Are the data needed immediately, perhaps in some experimental study or one where real-time decisions are essential for the conduct of the study—or can data be retrieved after the event with no loss to the objectives of the study?

Set-up latency is the time between a cold start and the establishment of a channel. For example, it takes about 10–15 s to establish a GPRS-GSM (mobile phone) connection before data transmission can start (McConnell *et al.* 2004). The concept of a cold start is important since a transmitter unit is usually powered down between transmission sessions to minimize long-term energy consumption. Set-up latency must be considerably less than a normal surfacing period for cetaceans or seals that seldom haul-out. Moreover, a surfacing period may be interrupted with wave wash, thus reducing the probability of a successful channel set-up. A major advantage of the Argos system is that set-up latency is, for practical purposes, zero.

If the data channel is two-way there is the potential to acknowledge (handshake) that data records have been successfully received. An example is the acknowledgement of a GSM SMS text message that implies it has been successfully delivered to the service provider. There may be a further delay (perhaps several hours) until the final delivery to the intended recipient. This data acknowledgement is important in avoiding energetically expensive repeat transmissions of the same data. The current situation with Argos is that there is no handshake, so Argos incurs the added cost of repeat transmissions to try and ensure most data are received, but, even then, gaps in the data are inevitable. This is another example of a trade-off, in this case between saving power and saving time spent at the surface interacting with the receiving station. Future Argos enhancements may permit some form of uplink data acknowledgement. A novel method of relaying and acknowledging data is by *data sharing* over a network of tagged animals, a few of which have the capability or opportunity to relay the data ashore (Small *et al.* 2005; Lindgren *et al.* 2008). Once ashore, data acknowledgements slowly back percolate throughout the network.

For data to be relayed, both transmitter and receiver must be *available* and within range at the same time. Using Argos as an example, a satellite must be available above the horizon, the tag must be out of the water, and at least 40 s (for marine animals) must have elapsed since the previous transmission. There is the potential for on-board calculation of satellite availability, at the scales of individual passes or diurnal patterns, so that transmissions may be inhibited at times of low satellite availability. However, the energetic savings of this approach must be balanced against the dangers of losing satellite synchrony. This may be caused by the animal swimming large distances from the location for which the predictions are made and the possibility of an operational change in the satellites themselves. Whilst the Iridium (http://www.iridium.com) Short Burst Data service (SBD) has the advantage of continuous, global availability and data handshaking, its current start-up latency is in excess of 15 s, thus limiting its applicability for data relay at sea. Getting data back from retrieval tags depends upon the probability of animal/tag recovery and for many species this is not a practical option. Recovery probability can be increased by translocating an animal from its rookery/resting site to a distant site and awaiting its return (Oliver *et al.* 1998). Tag retrieval from only live animals means that this method is of limited use in survival studies.

10.7 Data modelling and compression

Relaying data is energetically expensive and the channel used and the behaviour of the animal may limit the relay rate. Therefore data compression and modelling of stored data may be used. This is not applicable to tags that relay data in real time and do not store data (e.g. some VHF and acoustic tags) or to retrieved archival tags with large memory capacity. Fedak *et al.* (2002) discuss various approaches to overcome channel capacity limitations. In one example they state that 'of the approximately 500,000 bits collected each day, 50,000 bits are sent but typically only 5,000 bits are received'. Yet this is not necessarily a handicap. For example, the researcher may be content with an unbiased *sample* of all dives recorded by the tag. Similarly, of these dives the researcher may be content with a dive modelled by a series of significant inflection points rather than raw depths collected every 4 s.

10.8 Visualization and analysis of individual paths

The track and behavioural data obtained from telemetry devices have a time stamp, but these are often complex with respect to their interpretation. Ideally, the data should be considered alongside environmental information to provide a context for the behaviour being measured (see Chapter 4). The environmental data may be time-invariant (e.g. coastlines and bathymetry), but most (e.g. oceanographic and prey abundance) varies with time. Such data, as well as the output of oceanographic models, come from a multitude of sources and, at the time of writing, a comprehensive catalogue is available at http://www.justmagic.com/GM-GE.html.

Visualizing such complex datasets is challenging. MAMVIS (Fedak *et al.* 1996) visualizes animated track, dive, and *in-situ* oceanographic data within their temporal oceanographic context. As well as being a necessary tool for the researcher, visualization also provides an opportunity for outreach to other researchers and the public. An example is the Ocean Biogeographic Information System's SEAMAP initiative (Halpin *et al.* 2006) that archives and displays track data for many marine species. Satellite Tracking and Analysis Tool (STAT) (Coyne and Godley 2005) provides an integrated system for archiving, analysing, and mapping animal tracking data. At a smaller scale GeoZui4D (Ware *et al.* 2006) visualizes the intensive and complex type of data obtained from the D-tag (M. P. Johnson *et al.* 2003).

Movements are indicative of biological processes operating on individuals (Turchin 1998). Therefore most analyses of movement data ultimately seek to discover something about the reason for the observed movements and relate them to life history, behavioural ecology, or relationships to habitat (Patterson *et al.* 2008). Marine mammal telemetry data sets can be described as irregular time series of noisy observations (Vincent *et al.* 2002; Pepin *et al.* 2004), of positions (Fig. 10.2). The nature and degree of the errors dictate what sort of biological insight can be obtained from telemetry data. For instance, large-scale migrations can often be inferred despite large noise, but the fine-scale nature of the movement may remain elusive. The scale of the error and the scale and function of the movement combine to constrain our biological inference. Therefore, analysis of individual path data has often been broken into two stages: (1) estimating the location, and (2) estimating

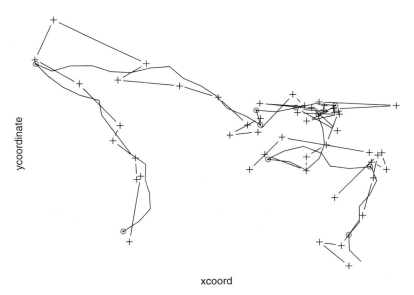

Fig. 10.2 Simulation of animal movement and observations. The black line represents a path simulated from a CRW with behavioural switching used in Patterson *et al.* (2008). The dots ⊙ represent 'foraging' bouts. The grey line and crosses mimic the telemetry observation process. Random *t*-distributed error is added (Vincent *et al.* 2002; Flemming *et al.* 2006) and the resulting path is sub-sampled to give 50% of the original positions. The result is a noisy and temporally, irregular sequence of positions that restrict our view of the underlying behaviour.

what the animal may be doing given its location, the habitat occupied, and its life history status. The crucial point is that (1) and (2) are fundamentally inseparable. Analyses that do not consider both can lead to erroneous conclusions (Bradshaw *et al.* 2007).

The problem of determining the most likely location from noisy telemetry data applies to both terrestrial and marine applications, across many species. Therefore, we include examples relevant to marine mammal researchers, regardless of the species under consideration. We group various analysis methods into two broad sections: (1) heuristic methods that aim to pick out the most accurate points given assumptions about the feasible speed of travel or other ancillary data, and (2) statistical methods that aim to construct a more likely path given noise in the data.

10.8.1 Heuristic approaches to position filtering

Variation in the magnitude of position error means that some points from a tag will be very bad while others may be quite accurate. Argos data are a good example of this, with errors spanning at least an order of magnitude in scale (Vincent *et al.* 2002). Therefore, a common approach to handling errors in location has been to attempt to determine which points have the lowest error and discard the remainder. We term these heuristic methods. One of the more widely used for Argos satellite data is that of McConnell *et al.* (1992). This assumes a known release position and uses the linear speed of travel between consecutive points to determine if a position is retained. Because removal of a point along the path affects the calculated speed between the remaining adjacent points, the algorithm operates forwards and backwards in time to determine which points should be removed. The maximum linear speed of travel must be assumed but is typically given as between 1 and 3 ms^{-1}. Austin *et al.* (2003) used a 'three-stage' heuristic filtering process to determine the movements of grey seals. The first stage calculated four rates of travel between the two previous and two subsequent locations, removing any locations implying a travel speed exceeding 2 ms^{-1}. The second stage applied the speed filter of McConnell *et al.* (1992). The third and final stage removed points above the 99th percentile of distances travelled over a 7-day period. These three methods removed approximately 30% of the data collected. Another heuristic approach applied to Argos data has been that of Hays *et al.* (2001). This aims to improve the accuracy of calculations of the travel speed in turtles. Successive points are not used to calculate speed; rather locations separated by several days are used. This tends to lead to more realistic estimates of speed of travel.

Limitations of heuristic approaches

Heuristic approaches are limited because all estimates of position, even from GPS data, contain some error. Even when points are discarded the remaining data are still subject to uncertainty and the extent to which any subsequent analyses are influenced by remaining error or artefacts is never quantified. This sort of problem has been discussed in Lévy flight analysis (Bartumeus *et al.* 2005; Benhamou 2004) by Bradshaw *et al.* (2007). Other approaches such as first-passage time (FPT) methods (Fauchald and Tveraa 2003) applied in marine contexts by Pinaud and Weimerskirch (2005, 2007) are likely to be similarly vulnerable to the same issues. However, all these approaches have the fundamental limitation that they exclude data which, even though it may carry a high level of uncertainty, also contains useful information. In principle, we need to avoid doing this. On the other hand, telemetry positions can often be so poor that removal of the worst data serves only to enhance our view of the true movements. Therefore, correction and analysis methods must strike a balance between using the largest possible proportion of the data and also down-weight the influence of potentially aberrant data.

10.8.2 Statistical approaches to path analysis

Statistical methods that avoid most of the disadvantages of the heuristic approach seek to achieve at least one of the following goals: (a) correction of position estimates, (b) quantification of uncertainty, (c) integrated analysis of position and behaviour.

Curvilinear smoothing

One way to estimate a signal from noisy time series is to fit a non-parametric curve, such as a spline, to data. Tremblay *et al.* (2006) compared linear interpolation of positions to several spline-fitting methods for simulated data. They found that some curvilinear interpolation techniques improved the accuracy of reconstructed tracks in a variety of marine birds and mammals.

Splines may interpolate data, i.e. intersect with available data and construct a smooth path between them, or smooth noisy data by producing a path that is a compromise between fitting the points exactly and the smoothness of the generated path (Venables and Ripley 2003). Essentially, splines use a local approximation of the data via some simpler function such as a polynomial. While splines are convenient—in that they are simple, computationally fast, and easy to apply—they have some drawbacks. First, while a curvilinear path may be a more realistic description of animal movement than a straight line (Anderson-Sprecher and Lenth 1996; Tremblay *et al.* 2006), the curve is nevertheless generated from a biologically arbitrary formula and thus the parameters of the spline have no biological interpretation. In this sense they are a purely empirical function. Furthermore, curvilinear approaches involve purely spatial and geometric solutions to a problem that involves temporal as well as spatial dimensions. The time between points cannot easily be accommodated into the reconstruction of an animal path using curvilinear methods. Additionally, information on the quality of the locations (e.g. Argos location quality categories) cannot be included, so all points are treated equally, despite the fact that some may be known to be more accurate than others.

However, some non-parametric statistical approaches may be more flexible and address some of the shortcomings identified above. Generalized additive models (GAM) and generalized additive mixed models (GAMM) (Wood 2006), can use smoothing splines within a linear modelling setting. These have been applied to Argos data from pinnipeds (Lonergan,[1] unpublished). They accommodate non-constant weights and variances to be attributed to particular points, allowing, for example, service Argos quality information to be used to weight the influence of particular points. Additionally, other variables such as locations of haul-out sites or colonies can be included within the estimation.

State–space methods

The problem of estimating location from noisy data fits neatly into a statistical approach known as state–space modelling (SSM) (Jonsen et al. 2003, 2005; Patterson et al. 2008). State–space models have recently been applied to marine mammal telemetry data (Johnson et al. 2005, 2008), but have also been used in terrestrial applications using radio-telemetry data (Anderson-Sprecher 1994; Anderson-Sprecher and Ledolter 1991) and for tracking pelagic fishes from light-based geolocation data (Sibert et al. 2003).

Before describing the structure of SSM it is informative to consider what a state is in the SSM sense and how it is represented. The state is represented by the state vector, $\mathbf{x}_t$, which contains the variables required to describe the state of the animal. The state vector might contain only location data or other variables such as the current behaviour. This might look like:

$$\mathbf{x}_t = \begin{pmatrix} Longitude(t) \\ Latitude(t) \\ Behaviour(t) \end{pmatrix}$$

Note that it is indexed by the time t. The SSM actually combines two sub-models. The first is known as a process model which updates the state vector:

$$\mathbf{x}_t = f(\mathbf{x}_{t-1}, \theta) \tag{10.1}$$

where f denotes a movement model such as a correlated random walk with parameters θ. Note that in (eqn 10.1) the current state at t is a function of the previous state at $t-1$.

A variety of process models may be used. For instance, a random walk with a drift term allow for the representation of diffusive movement as well as advective, directed movement (Nielsen et al. 2006; Sibert et al. 2003, 2006). The model of Jonsen et al. (2005) is a differenced random walk, with parameters governing the turning angle and the degree of autocorrelation in the headings of successive movements.

---

[1] Mike Lonergan, NERC Sea Mammal Research Unit, University of St Andrews, Scotland.

The second component to the SSM is known as the observation model which describes the likelihood of obtaining a particular observation $\mathbf{y}_t$ given the true, but unobserved state. This is often described in the following form:

$$\mathbf{y}_t = g(\mathbf{x}_t, \phi) \tag{10.2}$$

Here $g(.)$ describes the observation model, often a probability distribution, with parameters ϕ. For example, to model the distribution of errors in Argos positions, (Jonsen et al. 2005) used several Student's t-distributions estimated from the data collected by Vincent et al. (2002).

The combination of the process and observation models allows for a realistic model of animal movement to be linked to a model of the telemetry observation and data collection process. States/locations are proposed according to eqn 10.1 and then updated by the observation model shown by eqn 10.2 (see Fig. 10.2). The resulting inferred path (or state sequence) is then a mixture of the predictions of the process model and the uncertainty in the locations. Therefore, SSM allow for statistically robust prediction that embraces, rather than hides, the inherent uncertainty in the position.

While conceptually elegant and well suited to telemetry applications, SSMs require complicated statistical methods for estimating the most likely path (or state sequence) and associated parameters. For what are known as dynamic linear models with a Gaussian observation model, a filtering technique known as Kalman filtering and smoothing may be used to predict the most likely path and calculate the likelihood of the model given data. This has recently been applied to Argos data by D. S. Johnson et al. (2008) and has been used in the estimation of paths from radio tracking (Anderson-Sprecher and Ledolter 1991; Anderson-Sprecher 1994) and geolocation data (Nielsen et al. 2006; Sibert et al. 2003, 2006). While Kalman filtering assumes Gaussian errors and a linear movement, the model is fast to compute as a precise analytical prediction of the states and likelihood is available. Additionally, variants on the Kalman filter can deal with non-linear models or non-Gaussian errors (Royer and Lutcavage 2008). Moreover, recent work has combined heuristic methods and Kalman filtering. Patterson et al. (2010) used data from tags which transmitted Fastloc GPS data via service Argos. This gave the usual Argos data set and also more accurate GPS data from the same animals. Kalman filtering of speed-filtered tracks led to average errors of between 6 and 13km (Fig. 10.3). Additionally, they found approximately correct confidence intervals for locations were estimated at typically reported values (between 80% and 95%).

While useful for position estimation and interpolation, Kalman filtering is impractical for models that also aim to estimate behavioural modes such as categorization of foraging bouts from movement data. In these models, animals are allowed to switch their state between bouts of short, non-directed movements and periods of sustained faster and directed movements. This is intended to mimic animals intensively foraging on productive patches and rapid directed movements between productive areas. However, Bayesian estimation methods can cope with

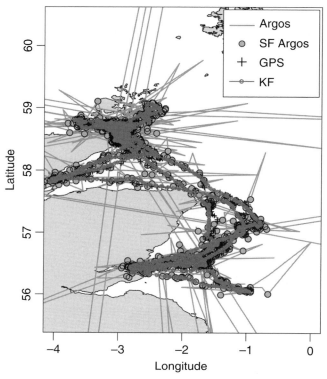

Fig. 10.3 An example of state space modelling from Patterson *et al.* (in press). The grey lines show raw Argos positions from a grey seal. Of these some are removed by the speed filtering method of McConnell *et al.* (1992). The remaining positions (grey dots) were smoothed to construct a path using continuous-time Kalman filtering (magenta). Comparisons of these to GPS paths gave root-mean-square errors of 6.29 km. The root-mean-square error described the average expected error. The figure was prepared by Lea Crosswell.

these sorts of models. Markov chain Monte Carlo (MCMC (Gilks *et al.* 1995)) have been used to estimate models with behavioural switching (Jonsen *et al.* 2003, 2005, 2006, 2007; Morales *et al.* 2004). These methods were applied to grey seals to determine sex-linked differences in foraging locations (Breed *et al.* 2006). Another Bayesian method, particle filtering (Liu 2008), has also been demonstrated on simulated data (Royer *et al.* 2005; K. H. Andersen *et al.* 2007; Patterson *et al.* 2008). Both MCMC and particle filtering are computationally demanding. The freely available software WinBUGS (Spiegelhalter *et al.* 1999) protects the user from some complications of the estimation process. However, this software has limitations and often researchers must write their own code.

While the biological realism of SSMs is arguably greater than the other methods discussed here, the underlying description of the biology associated with movement remains simplistic. However, the SSM is capable of more intricate and realistic process models that can incorporate the influence of physiology, ontogenetic development, and habitat. This indicates that much more work is required in these areas. Nevertheless, of the main methods considered here, the SSM is by far the most statistically rigorous analysis method for marine mammal tracking data.

We have considered here three approaches that have been applied to marine mammal telemetry data. *Heuristic methods*, although fundamentally simplistic, are useful as a pragmatic approach for pre-analysis and possibly to process data prior to the application of statistical methods (Patterson *et al.* in press). They have the advantage of being simple to automate and therefore apply to routine exploration and visualization tasks. However, we advocate the application of methods that are as simple as possible to minimize the opportunity for spurious hypothesis generation. We also note that they may be simple in removing the worst outliers from a data set before more detailed and rigorous analysis is applied (e.g. Johnson *et al.* 2008). *Spline- and other curve-fitting methods* attempt to correct data but are likely to be subject to over-fitting[2] and are limited in their biological interpretation. More advanced non-parametric statistical models such as generalized additive models are useful for handling telemetry data as they can deal with time, as well as space, and varying degrees of error. *State–space methods* represent an appealing approach to the analysis of path data. They can fit a model with biologically meaningful parameters and incorporate behavioural aspects. However, they are statistically and mathematically complicated and often computationally demanding. Much further research is required in this area to make SSM more usable and applicable to marine mammal research.

10.9 Concluding remarks

Telemetry is multidisciplinary, involving biologists with field skills, electronic and software engineers, material scientists, information theoreticians, and often statistical and ocean modellers. The art of successful telemetry design and deployment depends crucially on close interaction between these workers and a good understanding of each other's aims and constraints. There is more than an ounce of truth in the old adage that 'you shouldn't design a tag for a species you haven't been bitten by'! The point here is that not only is it fun to jump out of your own discipline, it is essential. In addition, techniques and ideas in terrestrial ecology, medicine, and pervasive computing are ripe for exploitation in the field of marine mammal biology. For the field of telemetry to progress, lateral and high-risk ideas must be encouraged (e.g. Wikelski *et al.* 2007).

---

[2] Over-fitting is the situation where the data model is too flexible. The data is modelled so well that the model loses general explanatory power and simply mimics the variability in the data. This means that such a model has a very low error for the data set from which it was generated but fails to predict new data.

Yet telemetry is just one part of a biologist's toolkit, and its full potential is realized when complementing other research such as diet, prey field abundance, longitudinal reproductive studies, and oceanography. Such integrated and large-scale studies (e.g. Block *et al.* 2002; Welch *et al.* 2002) require a step change in our ability to analyse and interpret multidimensional data sets if we are to capture the potential of current techniques in telemetry and remote observation. The hardware and software of data capture and transmission have moved ahead of our ability to analyse the resulting data.

11

Foraging behaviour

Mark A. Hindell, Dan Crocker, Yoshihisa Mori, and Peter Tyack

11.1 Introduction

Marine mammals are both taxonomically and ecologically diverse, and this diversity is reflected in their range of foraging behaviours. They include herbivorous grazing by dugongs through mid-trophic level grazing on zooplankton, higher level predation of fish and squid, to the predation of apex predators by killer whales. But marine mammal foraging is difficult to study because it usually takes place when the animals are submerged, making direct observation of hunting and feeding generally impossible. Nevertheless, studying foraging in a three-dimensional aquatic environment presents researchers with advantages not available for terrestrial species because relatively large aquatic animals can be equipped with electronic tags to sample critical information about behaviour (see Chapter 10). The vertical dimension (depth) can be simply measured by recording pressure. Distance (and therefore swim speed) can be measured simply through the use of turbines or accelerometers. When combined together, these simple measurements can provide an enormous amount of behavioural information, and provide the basis of many of the techniques used to construct our interpretation of the foraging behaviour of marine mammals.

Given these challenges, marine mammal biologists have used a variety of techniques which increasingly depend on sophisticated technology and complex statistical analyses. In this chapter we review these methods, from simple direct observations of animals feeding through to complex devices that simultaneously record 3D tracks and vocalizations. We also examine how these data contribute to our growing understanding of the foraging ecology of this diverse group of species.

11.2 Observation of foraging

11.2.1 Following focal animals

Direct observations are of little use for the majority of marine mammal species that spend most of their lives far from land and feed too deep for aerial observation of

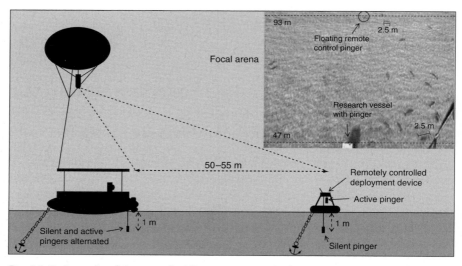

Fig. 11.1 Example of the use of a blimp-cam to study dugong movements. In this study the reaction of dugongs to acoustic pingers was being monitored. The width of the field of view was estimated using dugong lengths as a reference. The inset shows the dugongs in the focal arena as seen from the video footage taken by the blimp-cam. (From Hodgson *et al.* 2007 with permission from Wiley Blackwell.)

foraging. However, there are exceptions. The goal when collecting data on foraging behaviour, as well as other behaviour, is to have a continuous and unbiased sample. Standard methods of following marine mammals from small boats and observing them when they surface has been used effectively to quantify aspects of foraging behaviour in small or surface-feeding cetaceans (Mann 1999; Gibson and Mann 2008). This approach can be problematic for other taxa, where one seldom sees more than the animal surfacing to breathe or surfacing with a fish in its mouth. Overhead video from a towed aerostat has provided continuous observation of prey and the foraging behaviour of coastal dolphins in shallow water (Nowacek *et al.* 2001a). Use of this system has uncovered previously poorly described 'pinwheel' feeding, and enabled analysis of sequences of foraging behaviour (Nowacek 2002). Similarly, a 'blimp cam' (Fig. 11.1) has been used to monitor the feeding behaviour of dugongs in response to passing boat traffic (Hodgson and Marsh 2007; Hodgson *et al.* 2007).

11.2.2 Passive acoustic observations

Odontocetes use echolocation to find and locate prey, and several studies have used passive acoustic techniques to investigate the movements of foraging odontocetes. For example, Wahlberg (2002) tracked the clicks of foraging sperm whales to study their movement and use of bio-sonar (see Section 11.6 and Chapters 3 and 12 for further details).

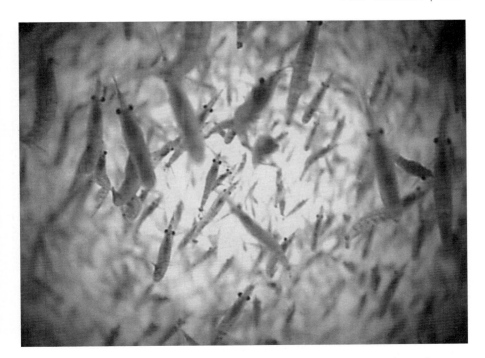

Fig. 11.2 Example digital pictures of krill prey recovered from Antarctic fur seal deployments showing krill swarm with strong down-welling light. (Photo courtesy of Sascha Hooker.)

11.3 Cameras

Another approach to visualizing foraging behaviour is the use of animal-borne cameras which record the animal's perspective and enables direct quantification of prey encounters and foraging success (Parrish *et al.* 2005). Cameras can be used to take movies or still photographs, and the choice of format depends largely on the nature of the questions being asked and duration of the deployment. A major limitation with video cameras has been the duration of footage that can be collected. Early versions were restricted to 180 min of videotape recording. More recent devices that use flash memory still have memory constraints that require the cameras to be programmed to turn on at specific times, or at specific depths, but we should expect this limitation to lessen with improved technology. Sampling is therefore a crucial component of any study using cameras. Some studies have opted to take a sequence of still photographs (Hooker *et al.* 2002a; Watanabe *et al.* 2006b). This is useful when trying to quantify prey encounters and the distribution and abundance of prey (Fig. 11.2), but it provides no information on prey capture. This can be achieved more easily with movies, which require far more memory.

Nonetheless, estimates of prey capture success have been obtained for several species (Parrish *et al.* 2000, 2005; Bowen *et al.* 2002; Sato *et al.* 2002).

The most sophisticated camera systems developed so far collect a suite of behavioural and environmental data simultaneously with video, which allow powerful insights into fine-scale foraging strategies (Davis *et al.* 1999, 2003). Integrating data from three-axis accelerometers, compass bearings, and video footage have allowed the 3D movements of Weddell seals to be reconstructed, and to identify where in the path prey have been encountered, as well as the species of prey captured (Davis *et al.* 1999).

To date, the large size and high cost of cameras has precluded their use on many species or for long deployments. The large size of most cameras (e.g. Crittercam is housed in a 10×25 cm aluminium housing, Marshall 1998) has the potential to influence behaviour of the animal carrying them, most likely through increased drag, but for short deployments of several days, no discernible effects in key foraging variables have been detected (Littnan *et al.* 2004). Recent advances in camera miniaturization and memory capacity (Hooker *et al.* 2002a) suggest that animal-borne video will become more widely used.

11.4 Reconstruction of foraging behaviour

In the absence of direct observation of foraging, investigators most commonly resort to reconstructing an animal's behaviour from 3D movement data. The ability to reconstruct the foraging behaviour of oceanic animals has been greatly improved in recent decades by the miniaturization of electronic components and sensors and with the introduction of high sensor sampling frequencies that allow the detailed measurement of acceleration (Sato *et al.* 2003).

11.4.1 Time–depth recorders

From the first data-loggers that recorded pressure traces on smoked paper or film (Kooyman *et al.* 1976), the application of advances in electronics to tag development has resulted in a continuing progression of increased capacity, accuracy, and precision of tags, along with a dramatic increase in the number and kinds of sensors available to researchers (Chapter 10). Time–depth recorders (TDRs) measure pressure to calculate depth, with sensor resolution depending on tag specifics including depth capacity and the number of bits used to store data. At present, commercially available tags typically have a depth resolution of 0.5 m and $\pm 1\%$ accuracy. These allow the reconstruction of dive and surface times along with 2D (i.e. time and depth) profiles of dives. In addition to measuring changes in position in the water column, tags can contain a variety of sensors that measure other behavioural components (e.g. swim-speed, acceleration, geomagnetic heading), physiology (e.g. heart rate, stomach temperature, muscle temperature), and the external environment (e.g. water temperature, illuminance, conductivity).

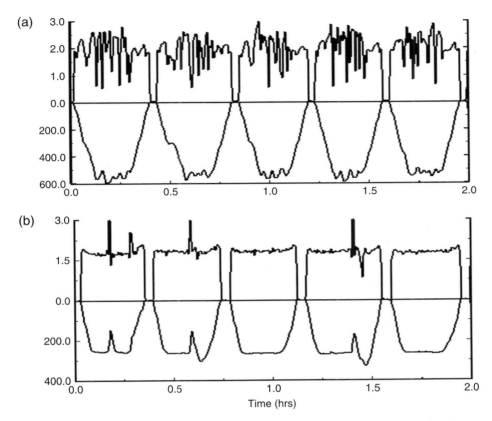

Fig. 11.3 Swimming speed and diving depth from presumed foraging dives in a female (a) and male (b) northern elephant seal. Dramatic changes in swimming speed are associated with vertical excursions at the bottom of dives.

Data acquired by tags can be logged for recovery or telemetered to satellites for remote access (Fedak *et al.* 2002; see Chapter 10).

One of the limitations of reconstructing foraging behaviour from depth–time series is that these data only provide information about the vertical component of an animal's movement. While rates of depth change provide some indication of swimming effort, these interpretations can be confounded by changes in dive angles and the linearity of the animal's track through the water. The addition of data on swimming speed allows diving angles and 2D dive shape to be estimated, in addition to providing better information on swimming effort (Fig. 11.3) (Le Boeuf *et al.* 1992). Within the limits of the sampling interval and depth resolution, data on swim speed can be used to calculate acceleration. Since swim speed sensors measure rotations of an impeller, these instruments require calibration. Sensors can be calibrated on life-size models in swim flumes that emulate water flow over the body surface (Ponganis *et al.* 1990). However, stroking behaviour from a real animal can create turbulence that affects the sensor (Ropert-Coudert and Wilson 2005) and water flow can vary with instrument placement and animal girth. One approach that has received considerable use is to calibrate the sensor based on comparison to the concurrently sampled depth changes. Assuming that some portion of the animal's descent or ascent movement is vertical, a plot of depth change vs. impeller rotations should have a distinct edge that can be used to develop an equation relating speed to sensor rotations (Blackwell *et al.* 1999). Biased or inaccurate swim speeds may be generated if diving is never vertical, the animal's body position alters during swimming so that the instrument is not parallel to the vector of the animal's movement, or the impeller is fouled with sand, algae, or ice. The recent development of 2D accelerometers that sample at high frequencies has given additional insight into swimming effort and body position during foraging (Sato *et al.* 2003). In addition, these sensors allow an assessment of body angle for use in calibrating swim speed.

Although potentially more informative, 3D reconstructions of an animal's trajectory underwater is technically much more challenging and is much less common. The earliest successful 3D data were obtained using acoustic telemetry, where acoustic transmitters attached to Weddell seals were tracked by a hydrophone array mounted in the surrounding sea-ice (Wartzok *et al.* 1992). The technique has been used successfully on several species (Harcourt *et al.* 2000; Simpkins *et al.* 2001; Hindell *et al.* 2002), but its use has usually been restricted to pinnipeds within fast-ice environments, as the hydrophone arrays must be fixed. Baumgartner and Mate (2003) also used hydrophones suspended from buoys at known locations. The positions of animals within the arrays can be precise to several metres, but, once outside, the array precision declines exponentially. Recent development of geomagnetic sensors allows an animal's heading to be estimated and when combined with distance travelled, dive profiles can be reconstructed in three dimensions (Davis *et al.* 1999; Mitani *et al.* 2004). This technique is sensitive to large-scale movement of the water column in the form of currents, so for long-term, open-water measurements, occasional known surface locations are required to correct for extrinsic drift (see Chapter 10).

Passive acoustics can also be used to localize the movements and foraging behaviour of cetaceans that are vocalizing during foraging. One of the best examples of this is from bottom-mounted grids of hydrophones operated as submarine test ranges, mainly by the US Navy (Ward *et al.* 2008). At the Atlantic Underwater Test and Evaluation Center in the Tongue of the Ocean within the Bahamas, a grid of 82 hydrophones spaced at intervals of 2 nautical miles allows the detection of foraging vocalizations by cetaceans occupying the underwater range. Other military arrays, such as the SOSUS arrays, within various parts of the world's oceans have also proved to be useful for following the behaviour of large whales at long range (e.g. Clark and Gagnon 2004). However, because these arrays are both fixed in specific locations and access to them is restricted to various degrees by military security, new and more flexible passive acoustic monitoring systems are being used to monitor the behaviour of vocalizing cetaceans (e.g. Croll *et al.* 2001).

11.4.2 Sampling strategies

Perhaps the most important factor in reconstructing foraging behaviour using data-loggers is the temporal resolution of the data obtained. Sampling interval can create artefacts or bias when inappropriately matched to event duration (Boyd 1993; R.P. Wilson *et al.* 1995; Hindell *et al.* 2000). The sampling interval must be shorter than the event duration to reliably detect an event, and significantly shorter to accurately measure its duration. Sampling theory generally suggests that the sample interval needs to be half the duration of the duration of the shortest event to be resolved. R.P. Wilson *et al.* (1995) recommended that a minimum sampling rate for depth is 10% of the dive duration. Ultimately, sampling frequency for a recoverable tag is balanced against the memory capacity of tags and anticipated deployment duration. The recent development of large memory capacities in tags has facilitated dramatic increases in sampling frequencies, despite the addition of new sensors. For example, depending on how data are stored, a 16-MB capacity instrument can record depth every 1 s for 185 days. Sampling the frequency of depth profoundly influences estimates of dive shape, duration, and surface interval and their use for inferring foraging behaviour. Some sensors require high sampling frequencies. A good example is accelerometers, which typically sample at frequencies of 16–32 samples per second (Watanabe *et al.* 2006a).

Historically, these memory constraints sometimes required the use of bins to summarize stored data. This approach records the frequency of diving behaviour in user-defined depth and duration bins while sacrificing the actual time–depth data. While increases in memory capacity have made this approach largely unnecessary for the use of recoverable data-loggers, it is still frequently used when satellite linked time–depth recorders (SLTDR) are used to transmit behavioural data. For numerous species where instrument recovery is difficult or impossible, a successful approach has been to use the Argos satellite system to receive data transmitted when the animal is at the surface. Comparisons of TDRs and SLTDRs have suggested that while satellite transmitted data is usually only a fraction of the data collected by the TDRs, depth and duration bins remain accurate representations of actual dive behaviour (Burns and Castellini 1998). Recent advances in tag design have successfully maximized use of the Argos bandwidth to allow transmission of the shapes of individual dives and detailed thermal and conductivity profiles collected by animals (Fedak *et al.* 2002).

11.4.3 Interpretation of 'foraging' dives

Although time–depth data can be used to identify different types of dive behaviours, there has been relatively little validation of dive type functionality, making it difficult to refer to dives as serving a particular purpose. However, depth–time series data are often used to *infer* foraging behaviour at a variety of spatial and temporal scales. For example, changes in behaviour within individual dives may indicate foraging. Also, changes in the dive cycle and bout structures of diving and surfacing can provide information on both prey distribution and foraging decisions. The maximum depth of dives can yield important information about search strategies and potential prey species. For example, some marine mammal species exhibit strong diel patterns to diving depth, suggesting the use of vertically migrating prey.

In species that exhibit a variety of diving behaviours or that dive without extended surface intervals while at sea, one common approach is the classification of dives based on their 2D profile. Manual classification of dives using visual inspection has been used, but the repeatability and subjectivity of this classification scheme may limit the value of this approach. High inter-observer agreement has been demonstrated in northern elephant seals (e.g. Le Boeuf *et al.* 2000). Although manual classification encourages researchers to look closely at individual diving behaviour, automated computer algorithms are often needed to process the tens of thousands of dives that are now routinely collected (Thums *et al.* 2008b).

Several quantitative approaches have been used for classifying dives into types. Discriminant function analyses (DFA) are built using a sub-sample of dives classified by visual inspection and a variety of descriptive dive variables (Schreer and Testa 1995; Krafft 1999; Beck *et al.* 2003a; Hassrick *et al.* 2007). The function is then applied across all dives and this approach has often yielded high reclassification success (>95%). However, discriminant functions built for this purpose can be highly sensitive to sampling frequency, with dramatic reductions in the reclassification success of data sets with differing sampling rates (Crocker, unpubl. data). This is particularly true of variables that are represented by counts, such as the numbers of vertical excursions or 'wiggles' at the bottom of dives. Other studies have used various forms of hierarchical and non-hierarchical cluster analysis, particularly k-means clustering, to classify dives (Schreer and Testa 1995; Lesage *et al.* 1999; Robinson *et al.* 2007). Cluster analysis has been successful in objectively grouping dive behaviours, but is potentially problematic due to sensitivity of the analysis to the co-linearity of input variables (a common situation with diving data). Also, since cluster characteristics are based on the input data and can vary between datasets, these groups may not be as useful when comparing across studies. Currently, a wide variety of multivariate techniques are being explored for their use in dive classification, including: variants of factor analysis; non-metric, multidimensional scaling; random forests' classification trees; and pattern recognition analysis.

Inherent in the desire to classify dives by shape is the assumption that those shapes represent biologically interesting differences in behaviour. However, putative functions of recurring dive shapes have only been supported by correlation with other relevant measurement variables for a few species (e.g. Baechler *et al.*

2002; Kuhn and Costa 2006; Horsburgh *et al.* 2008). In the absence of direct validation of dive shape and function, dives with distinct bottom times are often used as proxies for foraging behaviour. In some cases these putative functions are supported by comparison to tracking data (e.g. animals reduce movement rates while exhibiting putative foraging dives and speed up when exhibiting putative transit dives). The addition of swim speed or acceleration data has also been used to support foraging proxies based on dive shape. For example, in some species the vertical excursions at the bottom of dives are associated with rapid changes in swim speed (Amano and Yoshioka 2003; Hassrick *et al.* 2007), while deep 'V-shaped' dives are sometimes associated with the extensive use of gliding behaviour (Sato *et al.* 2003). Swim-speed data have also revealed that similar dive shapes can mask very different behaviours. For example, male northern elephant seals exhibit flat bottom dives where the animals swim continuously, and similarly shaped dives where swimming is absent and the animal is apparently stationary on the bottom (Hassrick *et al.* 2007). Identical behaviours can also result in different dive shapes as the buoyancy of an animal changes (Crocker *et al.* 1997; Biuw *et al.* 2003). Despite these limitations, inferences about behaviour from dive shapes have shown value in identifying important variation in behaviours with respect to sex (Hindell *et al.* 1991; C. Campagna *et al.* 1999; Le Boeuf *et al.* 2000; Beck *et al.* 2003a; Baird *et al.* 2005; Page *et al.* 2005a) and development (Le Boeuf *et al.* 1996; Horning and Trillmich 1997; McCafferty *et al.* 1998; Burns 1999; Pitcher *et al.* 2005).

Groups or clusters of dives, called bouts, can also be thought of as the behavioural unit for analysis. This analysis depends on quantitative estimates of bout criteria, or on the lengths of the shortest intervals between bouts. For animals that dive intermittently, an iterative statistical method has been widely used to define bouts (Boyd *et al.* 1994). A minimum foraging bout is defined based on the distribution of surface intervals between dives. The end of each bout is found by comparing the mean surface interval within a bout to that of the next dive. Bout ending criteria can also be used to define transitions between bouts of dive shapes. Log-survivorship (Slater and Lester 1982), log-frequency (Sibly *et al.* 1990), and log-normal (Tolkamp and Kyriazakis 1999) models have been used in some species and contexts. These approaches are based on the assumption that the mean dive interval within a bout is consistent among bouts and ignores other variables of potential importance to a diving predator, such as depth. Since the depth, duration, and interval between dives may all be adapted for optimal foraging, Mori *et al.* (2001) proposed the use of sequential differences analysis, which uses the frequency of differences in several dive characteristics between successive dives, to determine bout ending criteria.

Analysis of behaviour at the level of bouts has provided insight into the basis for foraging decisions in marine predators (Mori and Boyd 2004). Bout-level analysis also allows insight into the patch characteristics of prey resources used by marine predators. Temporal variation in predator behaviour can provide insight into the spatial distribution and quality of patchy and ephemeral prey resources, how these features change annually or seasonally, and how predators respond to these changes. Behavioural bouts can be analysed to provide insight into prey distribution at large (hundreds of km), meso (tens of km), and fine (hundreds of metres) spatial scales (Boyd 1996) and integrated with tracking data. Changes in the bout structure of diving or dive shape have been used to address changes in prey distribution with geographical (McConnell *et al.* 1992; Lea *et al.* 2002), seasonal (Mattlin *et al.* 1998; J.-Y. Georges *et al.* 2000; Harcourt *et al.* 2002; Burns *et al.* 2004), and annual variation (Boyd *et al.* 1994; Boyd 1996, 1999). Behavioural proxies for foraging can also enable the identification of environmental features that are possible foraging habitats, or are important to search strategies (Guinet *et al.* 2001; Lea and Dubroca 2003; Bradshaw *et al.* 2004).

11.5 Feeding success

While feeding behaviour can be directly measured by observing prey consumption for relatively few species, foraging success is a fundamentally different variable and can be more difficult to measure. For species that haul-out on land, changes in mass across a time interval can be used as a measure of foraging success, and this can be applied at the level of individuals or using population samples. Body composition measurements from isotopic dilution (Iverson *et al.* 1993) or ultra-sound measurements coupled with morphometrics (Gales and Burton 1987) can be used to estimate energy acquisition during foraging trips. Proxies such as transit rate, dive shape, and bout structure have shown direct relationships to measures of foraging success (Le Boeuf *et al.* 2000; Mori and Boyd 2004; Crocker *et al.* 2006). An advantage of this approach is that it measures foraging success at the level of assimilation, taking into account the energy expenditure associated with various foraging strategies. A limitation is its synoptic nature, providing information about foraging success over entire trips, but this may be an appropriate scale at which to work with some species, especially lactating otariid pinnipeds because of the information this contains about the rate at which mothers are able to provision their offspring (Boyd *et al.* 1997; Boyd 1999).

To place successful foraging behaviour in the context of habitat features, a technique that allows an assessment of fine-scale energy assimilation is needed. One recent approach uses changes in the diving behaviour of animals to assess changes in buoyancy (Crocker *et al.* 1997; Webb *et al.* 1998; Beck *et al.* 2000; Biuw *et al.* 2003). In deep-diving animals, buoyancy is determined largely by body composition and the ratio of body lipids to lean tissue. Several species have been shown to exhibit 'drift' dives where active swimming behaviour is absent for large portions of dives (Crocker *et al.* 1997; Biuw *et al.* 2003; Page *et al.* 2005b). Changes in this 'drift rate' throughout can provide an estimate of relative changes in body condition that can be mapped spatially or examined relative to changes in diving behaviour (Thums *et al.* 2008a). While this technique provides an exciting window into real-time foraging success, it is applicable to a relatively small number of species and only detects relative changes in body condition that do not directly equate to foraging. For example, recovery of lean tissue losses after reproduction, and growth of a fetus, which is largely lean tissue, can superimpose decreases in buoyancy over increases caused by the deposition of fat. In future, the use of instruments with accurate accelerometers may allow this method to be applied to a broader range of species that do not drift passively during diving. The objective of the method would be to resolve the buoyancy force vector from the propulsive force vectors detected by the accelerometers.

One approach to validating behavioural proxies for foraging is to place the behaviour in the context of hydroacoustic visualization of prey fields (Winn *et al.* 1995; Croll *et al.* 1998). This approach is impractical for many species because we are unable to sample prey at the relevant spatial scales for the predator. Another approach to directly visualizing prey during foraging behaviour involves the use of animal mounted cameras (see p. 245).

Recent developments in the telemetry of stomach temperature have enabled feeding frequency to be studied directly in some species. This technique is based on the principle that the body temperature of most prey is significantly colder than that of endothermic predators, so that stomach temperature drops measurably after prey ingestion (Fig. 11.4). While used with considerable success in free-ranging seabirds, its application to marine mammal studies have been more problematic. Captive studies with harp seals, harbour seals, Steller sea lions, California sea lions, and elephant seals have shown that stomach temperature can be used to reliably detect feeding events (R. Gales and Renouf 1993; Hedd *et al.* 1996; Andrews 1998; Bekkby and Bjorge 1998; Kuhn and Costa 2006) and that food can usually be differentiated from sea-water ingestion (Hedd *et al.* 1996; Kuhn and Costa 2006). However, these studies have also suggested that reliability decreases with the rates of prey ingestion, and that estimation of meal size is more difficult due to variability in changes in stomach temperature and a tendency to overestimate small meal sizes (Kuhn and Costa 2006). Recent data from stomach temperature telemeters in free-ranging animals show that feeding occurs in bouts, but animals also display single feeding events (Kuhn and Costa 2006) and in this case the technique may be capable of estimating meal size.

The primary factor limiting the use of stomach temperature telemetry in free-ranging animals appears to be the ability to keep the transmitter in the stomach during foraging (Lesage *et al.* 1999; Horsburgh *et al.* 2008). This problem has been addressed in phocid seals by increasing the size of stomach transmitters using biodegradable ethafoam (Austin *et al.* 2006a; Kuhn and Costa 2006). This technique has increased record duration to as much as 40 days, though transmitter residence times were highly variable among individuals. However, the approach has also yielded several important insights into foraging behaviour.

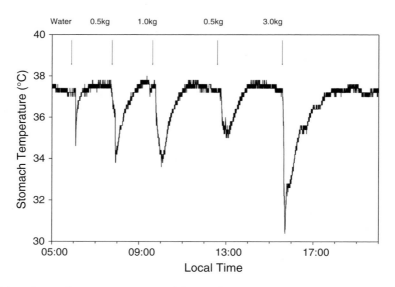

Fig. 11.4 Stomach temperature record for a sub-adult male northern elephant seal fed four meals (0.5 kg, 1.0 kg, 0.5 kg, and 3.0 kg, denoted by arrows) of whole herring. First decline in stomach temperature was a result of seawater ingestion. (From Kuhn and Costa 2006.)

Despite short record durations, stomach temperature provided support for several key foraging proxies in northern elephant seals (Kuhn *et al.* 2009) and grey seals (Austin *et al.* 2006b). Feeding events were inversely correlated with transit rates and strongly associated with the putative functions of dive shapes (Horsburgh *et al.* 2008).

The development of high sampling rate and multiple sensors has provided another approach to obtaining information on the fine-scale behaviour of marine mammals. Johnson and Tyack (2003) describe a tag (known as a D-tag) designed to sample orientation, acceleration, and depth at 50 Hz. This rate is sufficient to sample movement in enough detail to measure behaviours such as fluking, turning, and movements related to foraging. Detailed analyses of the vertical acceleration of sperm whales during the gliding portions of their ascent and descent enable estimation of the tissue density of individual whales, which likely corresponds with body fat (Miller *et al.* 2004b).

11.6 Acoustics and echolocation

Monitoring an animal's movement underwater is only one way of investigating foraging. Many species of odontocetes use sound as a fundamental component of their foraging, and this can be recorded by hydrophones deployed on boats or buoys (e.g. Benoit-Bird and Au 2003) or on tags attached to the animals.

For example, Akamatsu *et al.* (2000) report using a tag that was designed to detect feeding clicks from *Neophocoena*, along with depth, swim speed, and acceleration. Madsen *et al.* (2002) report an ultrasonic recording tag that was used to record the full waveforms of echolocation clicks from sperm whales. The D-tag (Tyack and Recchia 1991) also records sound at rates of up to 192 kHz, along with depth, temperature, and orientation. Use of these tags on odontocetes such as sperm and beaked whales, which use echolocation to forage at depth, have revealed remarkable details about how these animals use bio-sonar in foraging.

The D-tag records the orientation of a clicking whale, but the acoustic record of the clicks from the tagged whale contains artefacts as a consequence of being off-axis (as the hydrophone is in fixed orientation on the moving whale) and sometimes in the near-field. The acoustic structure of the clicks can be obtained by linking far-field recordings with data on the location and orientation of the clicking whale with respect to the recording hydrophone. For example (Zimmer *et al.* (2005b) recorded the clicks of a tagged sperm whale, *Physeter macrocephalus*, from a quiet research vessel towing an array of hydrophones. They were able to track the whale using sightings and by listening to the clicks, and used the tag to provide data on the orientation of the whale with respect to the hydrophones. There was a relatively omni-directional, low-frequency component to the clicks, with a source level of about 190 dB$_{peak}$ re 1 μPa at 1 m (Watkins 1980; Watkins and Daher 1984), and also a highly directional higher frequency component with an on-axis source level estimated to be 230 dB$_{peak}$ re 1 μPa at 1 m (Mohl *et al.* 2000, 2003). Zimmer *et al.* (2005a) used data from two simultaneously tagged Cuvier's beaked whales, *Ziphius cavirostris*, to obtain the source level and directionality of far-field clicks from one whale recorded on the other tagged whale.

Acoustic detection of feeding in odontocetes relies on the echolocation clicks being associated with prey detection and capture, and there are several lines of evidence for this. Data from sperm whales indicated that as they descend, they start producing clicks (Watwood *et al.* 2006). After these initial clicks, which may occur at intervals of >2 s, sperm whales produce clicks at regular intervals of 0.5–2 s. At depth, these regular clicks are interspersed by accelerations of clicks at intervals typically from 0.2 to 0.02 s. These 'buzzes' tend to occur at inflections in the depth profile of the diving whale, and the rate of change of roll and orientation increases during the 3 sec at the end of the buzz, suggesting that the whale is manoeuvring to capture prey (Miller *et al.* 2004a). Two species of beaked whales (*Ziphius cavirostris* and *Mesoplodon densirostris*) have patterns of search and buzz clicks during deep foraging dives similar to that of sperm whales, and the echoes from prey are clearly visible on audio recordings from tagged *Mesoplodon* spp. (M. Johnson *et al.* 2004).

These results suggest that, for both sperm and beaked whales, the buzzes represent attempts to capture prey. Some echograms for *Mesoplodon* spp. show that prey can occasionally escape the whale, but most prey appear to be relatively sessile, and most capture attempts appear to be successful. A critical area for future research will involve study of this deep ecosystem. Sperm whales may take about the same biomass from the ocean as all human fisheries (Whitehead 2003), suggesting that this deep prey resource is much more significant than is generally appreciated.

11.7 Foraging tactics

The foraging tactics of an individual (how it searches for and acquires its food) is determined by extrinsic (e.g. prey characteristics) and intrinsic factors (e.g. sex, body size). A variety of techniques can be used to study marine mammal foraging tactics, and these are, to a large extent, dictated by the spatial and temporal scale of the questions being asked. These can be broadly characterized as fine-scale investigations which rely on high–resolution, sub-surface behavioural data of the type provided by time–depth recorders (TDRs, see Section 11.4) and animal-borne video, or coarser scale 2D interpretations of movement (see Chapter 10). There are also encouraging attempts to integrate the two types of data.

11.7.1 Central place foraging

Central place foraging can be modelled as a special case of the marginal value theorem, and occurs when an animal has a fixed position that it must return to after each foraging excursion. For example, adult female otariids are classic central place foragers during the breeding season. Females suckle their pups onshore, but need to make regular trips to sea to feed to support their own metabolic requirements and the cost of milk production (J. -Y. Georges *et al.* 2000; D. Thompson *et al.* 2003; Lea *et al.* 2006). The characteristics of central place foraging also make fur seals sensitive to local environmental variations, and this can be detected in an number of foraging behaviours as well as in parameters such as mass gain and pup growth. Some species, such as Antarctic fur seals are therefore regarded as indicators of marine resources (Hindell *et al.* 2003).

All diving predators can be regarded central place foragers because they need to return to the surface (the central place) to breathe (Kramer 1988; Houston and Carbone 1992; Mori 1998a). The requirement to return to the surface to replenish oxygen stores is a major constraint on foraging shared by all marine mammals, and has been the source of a large body of theoretical studies and practical research (see Section 11.8).

There are relatively few fine-scale investigations of foraging behaviour because of the relatively coarse resolution (greater than 1 km and temporal resolution of several hours) of movement and diving derived from Argos satellite tags. Tags that provide frequent and accurate GPS data are becoming more common, and will enable finer scale investigation of foraging (Tremblay *et al.* 2006; see Chapter 10).

11.7.2 Search tactics and characterizing the prey field

The study of search tactics in marine mammals is still in its infancy, partly due to the technical difficulties associated with obtaining data of sufficient quality, and partly due to the lack of underlying theoretical models for 3D systems. Search strategy models are well developed in 2D systems and have been applied to a wide variety of organisms (Viswanathan *et al.* 1999). Most search strategy models are modifications of simple random-walk models, where an animal moves at random through the environment in search of patchy and randomly distributed resources (Bovet and Benhamou 1988). The nature of the search strategy adopted by an individual is often regarded as providing information on the prey field with which it is interacting (Hindell 2008).

Correlated random walks (CRW) are models of random movements, but they are used to build up independent probability distributions of the two components of movement: step distance and turning angle. Inferences about an animal's search strategies are drawn by comparing the actual movements of an individual, typically derived from a satellite telemetry path, with theoretical random distributions generated by the CRW. This approach has been applied to the movements of grey seals and has identified three types of movement: directed movement with long distance steps that were under-predicted by the CRW; resident movement, with animals remaining in a small region that were over-predicted by the model; and animals studied which were found to display a CRW, which included almost half of the individuals studied (Austin *et al.* 2004). These latter seals were searching in a way that suggested that their prey was unpredictably distributed.

Levy flight models (Viswanathan *et al.* 1999; Hays *et al.* 2006) are another special case of random walks, where step lengths are not constant, but drawn from a probability distribution with a power-law tail. Theoretical studies indicate that Levy flight strategies are an optimal way to search for randomly distributed objects. Such strategies have been demonstrated for some marine groups (Sims *et al.* 2007), but not others (Edwards 2008). There are few examples of Levy flight behaviours in marine mammals (Austin *et al.* 2004; Edwards 2008).

A more statistically rigorous approach incorporates CRW into state–space models (SSMs). Animal paths are time-series data, with a set of locations in sequential (although rarely equal) time steps, and can therefore be analysed using SSMs (Jonsen *et al.* 2003; Patterson *et al.* 2008). These are time-series models that allow unobservable behavioural states to be inferred from observed data, and can therefore be used to identify transitions from 'transit' behaviours to 'foraging' behaviours. A useful feature of these methods is that the two principal sources of uncertainty in location data, errors in the measurement process and the inherent stochasticity in the movements, can be accounted for separately. However, the true value of SSMs is in their ability to model movement behaviour in a flexible and reliable manner, with robust methods for dealing with the error structure of the data (Jonsen *et al.* 2005). Most applications of SSMs use Bayesian modelling approaches, which make them technically difficult. This may account for their limited use in marine mammal studies to date (Breed *et al.* 2006, 2009), although they have been applied to other marine taxa, particularly turtles (Jonsen *et al.* 2006)

The path of an animal through space and time will contain periods with differing rates of travel and direction, which will reflect different behavioural modes. The simplest examples of this are the rapid, direct movements that occur during transit between food patches, and the slower, less directed movements within a patch, which is the area restricted search (ARS) behaviour (Fauchald *et al.* 2000; Fauchald and Tveraa 2006). Early attempts to distinguish between these two behavioural modes measured changes in speed, angularity, or sinuosity along a track (Robinson *et al.* 2007), but lacked the ability to quantify the scale of the ARS. The scale at which an animal exhibits ARS contains fundamental information about the nature of the distribution of the prey, and is of considerable utility in broader ecological studies.

First-passage time (FPT) is one approach that identifies both the location and scale of ARS. The amount of time that an animal spends within any given area along its path will depend on the speed and sinuosity of its path within that area. FPT is defined as the time required for an animal to cross a particular area (conventionally a circle) of a given size. In these analyses a series of increasingly large circles are applied to every point in the path. If the path is highly sinuous, the rate of increase in mean FPT with increasing area is larger than when the path is linear. The circle radius with the highest variance is taken to define the animal's operational spatial scale (Tremblay *et al.* 2007). Regions along the path where movements match these scales can then be mapped. This approach has been successfully used for a number of marine predators (Pinaud and Weimerskirch 2002; Bailey and Thompson 2006).

Animal paths can also be considered as fractal objects (Laidre *et al.* 2004b). More sinuous paths have a higher fractal dimension than linear ones, so fractal analysis is an alternative way of identifying ARS behaviours (Robinson *et al.* 2007). One shortcoming of both fractal and FPT techniques is that they tend to produce a uniform patch size along a path. Futhermore, while within a species these analyses may be useful they often do not scale independently and are therefore not true fractals (Turchin 1996). A recently described variant on this, fractal landscape analysis (Tremblay *et al.* 2007), offers a potential improvement.

All the methods outlined above are based on 2D path data, and do not incorporate the all-important third dimension that characterizes most marine mammal foraging. The potential for combining these types of spatially explicit models with diving behaviour data is considerable, and is a rich area for future research.

11.7.3 Functional response

A predator's functional response, which describes the relationship between an individual's rate of consumption and the density of its prey, is an important component of population regulation; therefore quantifying the nature of the functional response of a predator is an important first step in modelling future population trajectories. Functional responses are typically non-linear and asymptotic because constraints exist on the ability of individuals to find food when it is scarce and to consume more at high prey densities. However, functional responses are invariably difficult to quantify in field situations, particularly for marine mammals. This is because data are needed on both the amount of food consumed by the predators (see Section 11.5), and a synoptic estimate of the prey abundance. Furthermore, the data are required over a range of values for prey abundance. For long-lived species, this requires a time-series of data, encompassing a wide range of environmental conditions requiring studies to span many years. A functional response has not been empirically described for any species of marine mammal, although the long-term studies of the Antarctic fur seals at South Georgia have been able to demonstrate a non-linear relationship between prey abundance and an index of predator consumption (Boyd and Murray 2001).

However, as the general form of the functional response is well known, it can still be used as a component in broader ecosystem models (Assenburg et al. 2006). A generalized Type 2 functional response for grey seal consumption of cod was a key component in a more complex model that estimated predator-induced mortality (Trzcinski et al. 2006). Similarly, ecological models incorporating generalized functional responses were used to demonstrate the adverse outcomes of commercially harvesting minke whale prey (Mackinson et al. 2003). The nature of the functional response can also be used in species-specific foraging models (Mori and Boyd 2004), and these are discussed in more detail below.

11.8 Foraging models

11.8.1 Optimal foraging models

Optimal foraging models are, in general, made up of three components: decisions, currencies, and constraints (Stephens and Krebs 1986). The question in optimal foraging studies is how an animal should behave (i.e. make decisions) under the constraints to maximize the currency (which is some measure that is representative of fitness). In the case of foraging in marine mammals, the most important constraint is physiological. As they cannot breathe while submerged, the oxygen used for activities under water has to be stored before diving, at the cost of time at the water surface (and therefore reduced foraging opportunities). For pinnipeds, lactation on land can be also be a constraint, particularly for otariids because females regularly have to return to sea during lactation to renew their energy reserves. Also important are ecological factors, such as the vertical, horizontal, and temporal distribution of prey in the water column. Predation risk can also be a critical determinant of behaviour in diving animals (Heithaus and Frid 2003). Foraging decisions made by marine mammals, which need to take all these factors into account, can be reflected in their diving behaviours. These behaviours can be characterized by parameters such as dive duration, dive depth, number of dives, and the movement of body and flippers (see Section 11.5). Therefore, optimal foraging models in marine mammals consider how they find food under diverse and dynamic physiological and ecological circumstances, and are developed using empirical diving behaviour data. It should be noted that these approaches have rarely been applied to cetaceans, mainly due to the lack of available diving data (but see Acevedo-Gutierrez *et al.* 2002).

The currency maximized in many optimal foraging models is the net rate of energy intake (i.e. net energy intake divided by the time spent taken to obtaining it). An alternative currency that is sometimes used is foraging efficiency (energy intake divided by the energy spent obtaining it). In diving animals, with their complex 3D environment, energy intake through foraging is taken to be positively correlated with the time spent at a foraging patch at a depth within the water column. Thus, the energy intake during foraging can be expressed as a function of time spent at a depth, $g(t)$, where t is a time spent at a foraging patch. For some species of cetacean, counts of acoustic signatures for prey capture may provide more accurate estimates of prey intake rates than time at depth (see Section 11.6). Due to the physiological constraint of finite oxygen stores, and the consequent need to return to the surface, the time taken to obtain energy consists of both dive time (the time foraging at depth plus the time travelling between the depth and the surface) and the time spent on the water surface for preparing and/or recovering from the dive (i.e. the full dive cycle). Longer or more energetic dives consume more oxygen (Hindell *et al.* 2000), this means the surface interval for preparing and/or recovering is longer as more oxygen needs to be replaced. For species which display such a relationship (not all do), dive time is a function of the surface time and can be expressed as $f(s)$, where s is the time spent on the surface. Therefore, the time taken for foraging is dive time plus surface time, or $f(s) + s$, and energy intake during the time is $g(t)$, indicating that the rate of energy intake, G, in diving animals can be expressed as:

$$G = g(t)/[f(s) + s] \qquad (11.1)$$

This is the basic equation of a static optimal foraging model in diving animals, and can be used to investigate the behavioural options available to animals in the context of ecological constraints such as the depth of a foraging patch. The optimal behaviour for a given environment can be found by solving for the variables that maximize G. For example, the simplest model predicts that the optimal dive time for a given depth corresponds to the travel time between a foraging patch and the surface (x), assuming that the energy intake is proportional to the time spent in the foraging patch, t. Since this time is the difference between dive time and travel time, the basic equation can be re-expressed as:

$$G = (f(s) - x)/[f(s) + s] \qquad (11.2)$$

Defining the function of $f(s)$, the optimal surface time (s^*), and optimal dive time, $f(s^*)$, which maximize this expression can be found for a given x which corresponds to dive depth.

More complex models can be used to investigate other aspects of foraging behaviour, such as optimal dive depth in relation to body size, optimal use of anaerobic diving, and the effect of patch quality on optimal dive time, or the optimal dive time in relation to predation risk. These models have indicated that: (1) optimal dive time should increase with increasing foraging depth (Kramer 1988; Houston and Carbone 1992), (2) optimal diving depth is shallower than the depth at which prey density is the highest (Mori 1998a), (3) large animals should make longer and deeper dives than a smaller animals (Mori 2002), (4) anaerobic diving is profitable under some conditions (Carbone and Houston 1996; Mori 1999), (5) prey patch quality affects optimal dive time, making it possible to estimate prey patch quality from diving behaviour, assuming the animal makes dives that are optimal for a given prey patch quality (Mori et al. 2002), (6) optimal dive time should be shorter if the predation risk is high (Heithaus and Frid 2003).

These models consider foraging behaviour at the temporal scale of a single dive cycle, but alternative temporal scales such as dive bouts and even the entire foraging trip (if the animal returns regularly to the land) are also important. Considering the scale of dive bouts, it is possible to predict how many times an animal should dive in a patch before leaving for a new patch (Mori and Boyd 2004). At the scale of a foraging trip, it is possible to estimate how long and how often an animal should remain at sea, or ashore, based on resource availability (Boyd 1999). In these cases, the basic equation (11.1) is still applicable, although with some appropriate expansion. For example, other measures should be incorporated such as travelling time between prey patches, travelling time from land to ashore, and time spent at land. There is also a model which describes optimal dive bout organization, or the combination of optimal dive time and number of dives repeated for a given patch characterized by depth and quality (Mori 1998b). This model is an optimal patch-use model, which predicts when an animal should leave the patch. These types of models can also be used to predict other parameters such as the optimal number of dives during a bout or the optimal number and duration of foraging trips during a breeding season. The best studied marine mammal in this regard is the Antarctic fur seal, which has been shown to maximize its rate of gain at all three of these temporal scales: the scale of dive cycle, dive bout, and foraging trip (Boyd et al. 1995b; Boyd 1999; Mori and Boyd 2004).

11.8.2 Stochastic models

The models described above are static models, or 'function maximization' models, which assume static environmental conditions. In these models, animals have complete information on the environmental conditions. However, in reality no animal will have perfect knowledge of foraging conditions. For example, an animal does not always know whether prey will be found before diving, or how much prey will be found in the future, since both prey (and prey patch) distribution are often unpredictable. To determine the optimal foraging behaviour under these uncertain conditions, the alternative approach of simulation models is often used. For example, Thompson and Fedak (2001) consider the situation in which dive durations are influenced by a seal's assessment of patch quality which varies stochastically. They found that for shallow dives, there should always be a net benefit from terminating dives early if no prey is encountered early in the dive, and the magnitude of the benefit was highest at low patch densities. This model is specific to diving animals such as seals, but there have been many optimal foraging models under a stochastic environment developed for non-marine species that may be useful (e.g. Iwasa *et al.* (1981).

Bayesian approaches can also be applied to foraging models, particularly where there is incomplete information. Bayesian models have the advantage that they can update their expectation of the state of the environment by combining prior knowledge with sampling information to make foraging decisions (McNamara *et al.* 2006; Valone 2006). Although these Bayesian foraging models have not yet been applied to foraging in diving animals, they should be very powerful for understanding the foraging behaviour of marine mammals.

Another approach is the use of dynamic programming models, or state–space models (Mangel and Clark 1986). These models consider the 'state' of an animal at a particular place or time. It seems obvious that a different optimal behaviour would be predicted for animals in different physiological conditions, such as pregnancy, and that different stages in an animal's life history are likely to require different optimal foraging decisions. For example, whether an animal is nearly full or starving will influence its choice between a safe but poor patch vs. a rich but dangerous patch. Moreover, even if the states of two animals are similar, younger individuals who have a chance to breed next season should make different decision from that made by older individuals. The state–space models can show sequences of optimal decision-making for a given state and stage. In the context of foraging in diving animals, Ydenberg and Clark (1989) presented a dynamic programming model which predicted the optimal usage of anaerobic metabolism for foraging dives. They demonstrated that anaerobic diving is profitable when prey are aggregated, mobile, and hard-to-find. Since state–space models consider an optimal sequence of decision-making under given conditions, they are a powerful tool for investigating the life history of animals.

12

Studying marine mammal social systems

Hal Whitehead and Sofie Van Parijs

12.1 Introduction

12.1.1 The definition of social structure

Pinnipeds and cetaceans often aggregate, sometimes in their thousands, and on occasion in exceptional densities. Many of these gatherings are not obviously the result of attraction to environmental features. Thus, these are actively maintained groups, not passive aggregations. Groups imply social structure, and social structure can affect ecology, genetics, population biology, and thus issues of conservation and management (Wilson 1975; Sutherland 1998). Hence, both for interest in the social lives of marine mammals, as well as a general understanding of the biology of the animals, their places in ecosystems, and the effects of human activities on them, we need to study social structure. Additionally, because the habitat of marine mammals is so different from that of terrestrial species, an understanding of the social structure of marine mammals provides an important comparative perspective on the forces of mammalian social evolution (Connor *et al.* 1998).

Those who study the social structures of non-humans often use the framework of Hinde (1976) as their conceptual basis. From this perspective, social structure is fundamentally about interactions between individuals. A relationship between two individuals is the content, quality and patterning of their interactions. The social structure of a population is the nature, quality, and patterning of the relationships among its members.

A glossary of many of the terms used herein is given at the end of this chapter.

12.1.2 How do we study social structure?

Many marine mammalogists who adopt Hinde's (1976) conceptual framework for the study of social structure face an immediate problem. Except when pinnipeds are hauled out on land or ice, most behavioural interactions, the basis of the framework, between aquatic mammals are unobservable. Cetologists have long circumvented this roadblock, and have made considerable progress in understanding the social structure of whales and dolphins, by observing 'associations' rather than interactions (Whitehead *et al*. 2000). The ability to use visual observations as a tool for studying pinnipeds at sea is very limited. Most information gathered in the past has been restricted by the available technology, which has limited information gathering to the interactions of small numbers of individuals. However, advances in and novel uses of existing technologies have enabled pinniped scientists to begin to build a more complete picture within which associations can begin to be assessed at sea.

Formally, two animals are associated if they are in circumstances in which inter- actions usually take place (Whitehead 2008b). As interactions are generally mediated by communication, a definition of association should be based upon the circumstan- ces under which individuals communicate, and thus ideally studies of communication underlie those of social structure. Using associations rather than interactions as the basis of a study of social structure might be considered to be making the best of a bad job. However, some relationships among animals are not expressed by overt inter- actions (Whitehead 2008b). Synchronous movement is an example of when associ- ations may be better measures of relationship strength than observable interactions.

When collecting behavioural data, an important distinction is between events and states (Altmann 1974). Events occur virtually instantaneously, whereas states are continuous. Breaches, lobtails, and upsweep vocalizations are events, feeding and travelling are states. From a social perspective, interactions are events and associations states. It is common to define several states, such as milling, travelling, resting, and socializing (Mann 2000), so that at any instant an individual or group is in one state (or perhaps more than one if they are not mutually exclusive).

Those studying the societies of marine mammals often define associations using spatio-temporal groups: animals are considered associated if they are members of the same group. This makes sense if it can be assumed that interactions generally take place within groups (Whitehead 2008b). Once again, studies of communica- tion can help buttress this assumption.

Observations of interactions, associations, or groups are then used to calculate relationship measures, which quantify the second level of Hinde's framework, the relationships between dyads. Relationship measures for marine mammals usually take the form of interaction rates or association indices. These can then be synthesized into descriptions and models of social structure, Hinde's third level, using ordinations, cluster analyses, network analyses, lagged association rates, and other univariate and multivariate techniques.

Social structures are influenced by, and influence, demography, genetic popula- tion structure, population biology, culture, patterns of kinship, and fitness. Tech- niques, such as Mantel tests, allow data and models of social structure to be linked to data on age, sex, genetics, range use, movement, and diet allowing hypotheses about the drivers and effects of social structure to be investigated (Whitehead 2008b). Almost none of this is feasible without knowledge of the identity of the interacting, associating, or grouped animals. Thus, a prerequisite for moving beyond a most basic analysis of social structure is some method of distinguishing individuals. Available techniques are summarized in Chapter 2.

So to study marine mammal social systems we usually need:

- a method of identifying individuals;
- observations, or other records (such as those from acoustics or telemetry), of interactions, associations, or groups;
- if using associations or groups, a basic understanding of the communication system of the species;
- statistical methods for producing relationship measures between individuals from records of interactions, associations, or groups;
- displays and models of social structure produced from the relationship measures; and
- supplemental data in areas such as sex, age, kinship, ranging, diet, non-social behaviour, and statistical methods for integrating these with the social data and results.

Marine mammalogists have quite frequently used ingenuity to bypass behavioural observations. Indirect but revealing inferences have been made about social behaviour and social structure from genetic data (Amos *et al.* 1993), parasite fauna (Best 1979), anatomy (Brownell and Ralls 1986), scarring (MacLeod 1998), and life history data (Kasuya *et al.* 1988).

12.1.3 Styles of studying social structure

The social systems of species of the two principal orders of marine mammals have been studied in contrasting manners. Studies of pinniped social structure have largely been based on understanding mating systems and mother–pup relationships while the animals are hauled out on land or ice (Stirling 1983; Boness 1991; Bowen 1991; Le Boeuf 1991; Insley *et al.* 2003). In contrast, cetacean social structures have usually been approached from the perspective of a fission–fusion society in which groups form and break-up. Animals are identified photographically, associations are recorded either directly or using group memberships, association indices are calculated between pairs of animals and used as relationship measures to construct models of social structure (Whitehead *et al.* 2000). A rarer, but particularly revealing, type of research is the detailed study of relationships using focal animal follow data (Mann and Smuts 1998; Mann 2000). There has been much less study of social structure in sirenians (see P.K. Anderson (2004) for a summary).

12.2 Field research

12.2.1 Identifying individuals

Studies of social structure generally require many identifications (usually, at least thousands) of many individuals, typically many more than are required in studies of population size using mark–recapture methods (see Chapter 3). Quantities of data required to produce useful descriptions of social structure are estimated by Whitehead (2008a). For instance, for an association index (which estimates the proportion of time that a pair of animals spend together) to have a coefficient of variation of less than 20%, requires about 15 observations of the pair associated; and, when the population is fairly homogeneous socially (coefficient of variation of true association indices of 20%) for the estimated association indices in a population to have a correlation greater than 0.4 with their true values needs an average of at least five observed associations of each dyad in the population (Whitehead 2008a). In practice this means that among the methods used to identify individuals (see Chapter 2), those that are cheap and simple to employ in the field, and cheap and straightforward to analyse in the laboratory are most suitable. Currently, photo-identification best fulfils these requirements, and has been the most important field method for studying the social structures of marine mammals, especially those of cetaceans. For pinnipeds, visual observations of individual markings at haul-out sites as well as tagging (i.e. flipper, radio, satellite, or other) have served as the primary forms for identifying individuals. However, photo-identification is rapidly becoming more widespread as technology facilitating pelage recognition improves (e.g. Karlsson *et al.* 2005). Acoustic techniques are also becoming increasingly useful as a method for individual recognition (e.g. Van Parijs and Clark 2006).

12.2.2 Collecting interaction, association, and group data

Observational field data can be collected using several protocols. The collection of observational data is more difficult at sea than on land, and protocols for doing this more crucial (Mann 1999; Whitehead 2008b). A primary distinction is whether to follow or survey. In a survey, when animals are encountered, their identities, associations, and perhaps behavioural states are recorded, and then attention moves towards finding another individual or group. In a follow, the individual or group is tracked and remains the focus of attention. It is usual in a group follow to collect information, as far as possible, on all members of the group, usually their identities, but perhaps including their behavioural states, fine-scale associations within the group, or interactions. In an individual follow, interactions and perhaps associations between the focal animal and any others are usually recorded. The protocols can be nested within one another. For instance, Gero and Whitehead (2007) conducted focal follows of individual calves within group follows of sperm whale groups.

Once a follow protocol is decided, a second, related choice is the sampling protocol. These are described and discussed in some detail by Altmann (1974), Lehner (1998), and Mann (1999). Options include '*ad libitum*' (all behaviour recorded as far as possible), 'focal-animal' (the behaviour of the focal animal and those it interacts with are recorded), 'all-event', 'predominant activity' (the state of the majority of the members of a group), 'point' (the state of an animal

Table 12.1 *Using interactions or association data: some guidelines. (Adapted from Whitehead 2008b.)*

	Interactions	Associations
Dyadic measure:	Usually counts of interactions	Usually 1:0 (associated : not associated)
Applicable when:	Interactions reliably and frequently observable	Interactions not reliably or frequently observable; coordinated behaviour predominates
Follow protocol:	Usually individual follow is best	Usually survey or group follow is best
Sampling protocol:	All interactions involving focal animal with times and interactant identities is best	Associations are noted at regular times or when they change

or animals at a particular time), or 'scan' (the state of an animal or animals when observed).

The major styles of collecting social data on marine mammals can be classified by the follow protocol (survey, individual follow, or group follow), and whether social structure is being studied using interactions, associations, or groups. Combinations of these suggest optimal sampling protocols (Whitehead 2008b) for the study of marine mammals in different circumstances. In general, it is more efficient to use group follows or surveys when collecting association or group data, whereas, except in unusual circumstances, interaction data can only be collected reliably during individual follows. Recommended field methodologies for collecting observational social data on marine mammals are given in Table 12.1 (also see Whitehead 2008b).

Group follows, recording interactions—Recording all interactions taking place within a group of animals requires high visibility, small group size, and low interaction rates. However, this is sometimes possible with hauled-out pinnipeds (e.g. Kovacs 1989), in captive situations (e.g. Davies *et al.* 2006), or ideal field situations and with prominent behaviours (see Connor 2001). In such cases, interactions are recorded when observed, together with the identities of the animals, the type of interaction, and the time. Additional information that may be useful includes the locations of the members of the group (so, who could interact with whom) and non-social behaviour, either in state (e.g. resting, scanning) or event (e.g. a movement, echolocation signal) form.

Individual follows, recording interactions—This is the methodology which produces the finest scale portrait of the social behaviour of marine mammals at sea (e.g. mother–calf relationships in bottlenose dolphins; Mann and Smuts 1998). Time-referenced interactions involving the focal animal are recorded along with the identity of the individuals involved. Additional information that should generally be recorded at regular intervals includes geographical positions, the behavioural state of the focal animal, its non-social behaviour (e.g. feeding attempts), and which animals are in a location to interact with the focal animal. This style of study requires excellent observation conditions, so that the focal animal can be followed and interactions observed consistently.

Group follows, recording associations or groups—This is the methodology often used with larger odontocetes and sometimes with baleen whales. Usually, group membership is noted at regularly spaced sampling times or whenever it changes, or photographs are taken opportunistically, with times recorded, to document group membership in retrospect. The location and predominant behavioural state of the group is also recorded. Additionally, fine-scale associations such as synchronous movement, or behavioural events such as breaches of vocalizations, are noted, preferably with the identity of the performer.

Surveys, recording associations or groups—When animals are encountered, their identities and associations or groups are noted, and then the survey moves on to find more animals. This method is often used with small-boat surveys of cetaceans, with the identities of animals in well-defined groups being determined on encounter, generally using photo-identification. It is usual, and recommended, also to record geographical location and behavioural state, as well as additional social measures, such as distinct subgroups, when feasible.

Most observational studies of marine mammals are made from 3–20-m long motor vessels or auxiliary sailing vessels. Vessels larger than about 20 m are generally too high and unmanoeuvrable to collect good observational and identification data. In all boat-based studies, it is important to minimize the effects of the boat on the animals' behaviour, both for ethical reasons and so that recorded social behaviour is representative. Studies of captive animals or hauled-out pinnipeds do not need boats, and in some cases inshore or freshwater animals can be watched and individually identified from land. Aerial platforms are also theoretically feasible. However, fixed winged aircraft are not stable, and it is hard to observe animals for any time, while helicopters are noisy and expensive. Powered or tethered lighter-than-air balloons or blimps are in many ways ideal, providing stable, quiet platforms which can be manoeuvred into optimal locations (Hain 1991; Nowacek *et al.* 2001a). However, they are currently expensive and often difficult to operate out at sea.

12.2.3 Collecting social data without observing animals

Identifying individuals other than visually has usually proved challenging. While marine mammals can be individually identified using genetic profiles (Palsbøll *et al.* 1997), given the number of identifications needed for useful studies of social structure, expense, and the ethical and logistic difficulties of collecting many samples from the same animals usually precludes this method. There have been successes in acoustic individual identification (e.g. Campbell *et al.* 2002), and as marine mammals use the acoustics so prominently for communication (Tyack 1999), if individual acoustic identification is feasible then studies of vocalizations have the potential to provide the raw material for models of social structure, although they will be more powerful if tied to observational data. For example, Van Parijs and Clark (2006) have been able to study the mating tactics of acoustically identified individual bearded seals, *Erignathus barbatus*, over 16 years.

Tags identify animals, and modern tags have the ability to collect interaction or association data (see Chapter 10). For instance, 'D-tags' record sounds and so can identify vocal exchanges which could be assigned to the tag-holders (Johnson and Tyack 2003). Unfortunately, though, such tags are unlikely to be deployed in sufficient numbers at the same time to give a comprehensive model of social structure on their own.

Quite simple tags may be more effective. If the tags exchange information with each other when they come within proximity, either acoustically or by radio, then they effectively record associations (see Chapter 10). With a moderate number of these types of tags deployed, a large body of association data might be collected.

12.3 Relationship measures

The second stage of Hinde's (1976) framework for the study of social structure is the analysis of relationships between animals. So records of interactions or associations need to be transformed into relationship measures.

12.3.1 Interaction rates

Perhaps the simplest relationship measure is the interaction rate, the rate at which a pair interacts per unit time, so we might have a touching rate, or a fighting rate. The 'per unit time' can refer to continuous time within a study or just periods during which the pair were in circumstances under which they could interact (e.g. both may be hauled out at the same site). For either concept, we need measures of effort: during how much time could we have observed interactions? Some interaction measures are asymmetric, so that if A interacts with B, this does not necessarily imply that B interacts with A. Examples include winners of fights, leadership in dives, or beak-to-genital touching. With asymmetric interactions, a number of additional relationship measures are feasible. These include Van Hooff and Wensing's (1987) directional consistency index and de Vries' *et al.*'s (2006) dyadic dominance index which vary between 0, indicating equal interaction rates from A to B and B to A, and 1 when A is always the actor and B the receiver, or vice versa.

12.3.2 Association indices

When association data are collected, the usual relationship measure is an association index (Cairns and Schwager 1987) which is an estimate of the proportion of time that a pair is in association. Thus association indices are naturally symmetric (although in some circumstances it may be possible to identify active and passive participants in an association). To construct association indices, the study needs to be divided into sampling periods, which could be hours, days, surveys, or encountered groups. Then, for each period and dyad, say A and B, we note whether A and B were observed, and if both were observed, whether they were observed in association with one another. Counts are made of the number of sampling periods in which A and B were observed associated, x, those when both were observed but not associated, y_{AB}, those when A was observed but not B, y_A, and those when B was observed but not A, y_B. The two most popular association indices in marine mammal work are:

$$\text{Simple ratio} : x/(x + y_{AB} + y_A + y_B)$$
$$\text{Half-weight} : x/(x + y_{AB} + (y_A + y_B)/2)$$

If individuals are equally likely to be identified during any sampling period and all associates are identified, then the simple ratio is an unbiased estimate of the proportion of time individuals spend together (Ginsberg and Young 1992). However, if individuals are more likely to be identified when not associated, or not all associates are identified, then the half-weight is less biased. As associates are often missed in photo-identification studies of marine mammals, and at least one member is more likely to be identified when they are in separate groups, the half-weight is generally preferred. For more information on these and other association indices and their biases, see Cairns and Schwager (1987) and Whitehead (2008b).

12.3.3 Temporal measures

One element of Hinde's (1976) definition of relationship that is rarely considered in practice (but see Mann and Smuts 1998) is temporal patterning. There are ways to quantify the temporal patterning of a dyadic relationship (see Whitehead 2008b), for instance using dyadic variants of the lagged association rate (described below).

12.3.4 Matrices of relationship measures

Using the methods described above, or others, we can produce one or more relationship measures (usually interaction rates or association indices) between each pair of animals in the population. These can be arranged into square matrices, such as that shown in Table 12.2. If there is interest in particular elements in this matrix (relationships between particular dyads), then standard errors, or other measures of confidence, should be attached to each (see Whitehead 2008a). More usually, however, we consider the general patterns in these tables as indicators of the social structure of the population.

Table 12.2 *Half-weight association indices among nine female sperm whales observed off the Galapagos Islands on 19 days between 12 April 1998 and 12 April 1999. Sampling periods are days, and two whales are associated if they dived within 5 min of one another.*

Sperm whale identification number	#3700	1.00								
	#3701	0.32	1.00							
	#3702	0.32	0.46	1.00						
	#3703	0.31	0.22	0.22	1.00					
	#3704	0.34	0.27	0.33	0.32	1.00				
	#3705	0.38	0.30	0.22	0.21	0.19	1.00			
	#3706	0.19	0.08	0.23	0.22	0.20	0.37	1.00		
	#3707	0.32	0.08	0.38	0.07	0.27	0.30	0.46	1.00	
	#3708	0.12	0.07	0.36	0.34	0.31	0.21	0.36	0.29	1.00
		#3700	#3701	#3702	#3703	#3704	#3705	#3706	#3707	#3708
					Sperm whale identification number					

12.4 Describing and modelling social structure

Matrices of interaction rates or association indices (e.g. Table 12.2) are not easily assimilated by the human brain, especially when there are more than a handful of individuals, or more than one relationship measure. Thus visual displays, model-fitting techniques, and statistical measures are used to analyse such tables. As marine mammal study populations are generally larger than those of most large terrestrial mammals and the observational data sparser (Whitehead and Dufault 1999), these integrative methods have become disproportionately important. Displays and models of social systems are richer if information is available to allocate individuals into classes, usually age/sex classes, but there are other ways of classifying animals, for instance by mitochondrial haplotype, or using the results of a social clustering method (see below).

12.4.1 Visual displays

Of the methods commonly used to display matrices of association indices or interaction rates, four are illustrated for the data in Table 12.2 in Fig. 12.1 (for further information on these and other display techniques, see Whitehead 2008b):

Histograms of association indices or interaction rates indicate the variation among relationships. For instance, in Fig. 12.1a, a rather homogeneous social system is indicated by rather little variation in association indices. Note that quite different social systems could produce similar patterns in histograms. For instance, a range-based social structure where individuals interact often with their neighbours and weakly with non-neighbours would show a bimodal pattern of a few strong relationships and many weak ones, as would a social structure

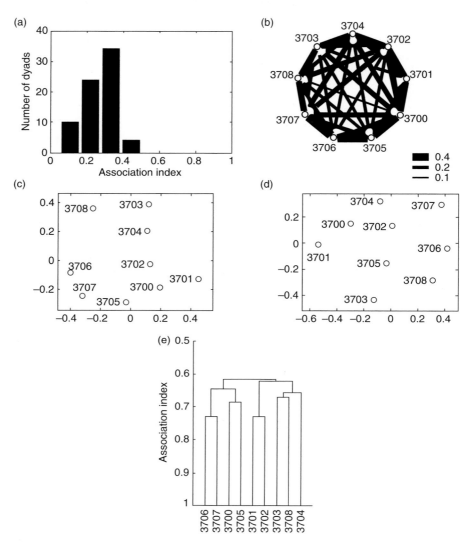

Fig. 12.1 Visual displays of association data for nine sperm whales shown in Table 12.2: (a) histogram of association indices; (b) sociogram; (c) principal coordinates analysis (first two axes explain 39% of variance); (d) non-metric multidimensional scaling (stress 0.19); and (e) average linkage cluster analysis (cophenetic correlation coefficient 0.58). All indicate a relatively homogeneous social system.

consisting of small permanent social units, with many interactions within units and few between.

Sociograms are displays in which individuals are represented by nodes (Fig. 12.1b). These are linked by lines whose widths indicate the strength of the relationship measure. These useful displays become unwieldy when many individuals are studied, although the sophisticated network drawing displays of programs like NetDraw can help (see below).

Principal coordinates analysis and *non-metric multidimensional scaling* display individuals as points with the distances between the pairs of points, ideally, being inversely related to the dyad's association index or interaction rate (Figs 12.1c and 12.1d). Ideally, in the principal coordinates method the measure of relationship is linearly related to the distance, whereas in non-metric multidimensional scaling it is monotonic. These methods can use any number of dimensions but, given human perceptual abilities, it is normal to display just two. They can indicate general attributes of social systems, such as social units, as well as how particular individuals fit into the social structure. The success of the principal coordinates method for displaying a matrix of relationship measures in a given number of dimensions is indicated by the proportion of variation accounted for (with perhaps >40% indicating a useful display). 'Stress' (varying between 0 and 1) assesses the success of non-metric multidimensional scaling, with values less than about 0.1 indicating satisfactory ordinations. The criteria of non-metric multidimensional scaling are less stringent than those of principal coordinates, so it generally produces a more satisfactory arrangement in a given number of dimensions. However, non-metric multidimensional scaling is an iterative technique and may not work with more than about 50 individuals.

Hierarchical cluster analysis produces a dendrogram (tree diagram; Fig. 12.1e) which seems to give a very useful display of the relationship measures. However, it presumes a hierarchically arranged social structure in which the clusters at one level are the elements of higher level clustering. This is true of some social systems, such as those of resident killer whales (Bigg *et al.* 1990), but if it is not the case then cluster analysis can be misleading. The cophenetic correlation coefficient, which measures the fit between the relationship measures of dyads and their level of clustering in the dendrogram, is a good way of assessing the value of a cluster analysis. Values greater than 0.8 are often taken to indicate that a dendrogram is a reasonable display of the matrix of relationship measures. There are several types of hierarchical cluster analysis, of which average linkage is generally recommended for social analysis (Whitehead 2008b). Hierarchical cluster analyses, as well as non-hierarchical cluster analyses, can be used to split populations into social units or communities such that relationships are strong within units or communities and weak between them.

12.4.2 Testing for preferred/avoided companions

It is increasingly being realized that hypothesis testing is often a poor way to make inferences, especially when studying wild animals, partially because null hypotheses are unrealistic (Johnson 1999). However, in social analysis, there are circumstances in which realistic null hypotheses exist and should be tested. It is entirely reasonable that animals may see each other as 'equivalent' (Schusterman *et al.* 2000), and not

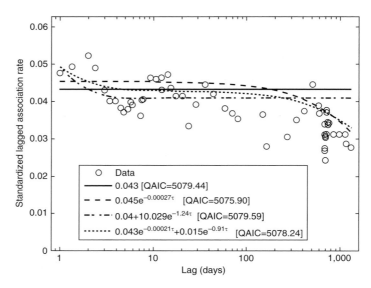

Fig. 12.2 Standardized lagged association rates plotted against time lag (τ in days) of female sperm whales and immature sperm whales from data collected off the Galapagos Islands between 1985 and 1995. Several models of the exponential family are fitted to the data. That with the lowest QAIC could be considered 'best'.

distinguish between social partners. Bejder *et al.* (1998), working on Hector's dolphins (*Cephalorhynchus hectori*), developed a permutation test of the null hypothesis that individuals have no preferred or avoided companions. Extensions of this test control for differences in gregariousness among individuals and demography, as individuals within the study area at the same time are more likely to be observed in association with one another (Whitehead 2008b). These tests are often a prelude to further analysis of the social structure. If the null hypothesis is not rejected, then displays and analyses, such as dendrograms and network analyses, have little validity. For the data that produced the association indices in Table 12.2, the variant of the Bejder *et al.* test that permutes associations within sampling periods, while maintaining the total number of associations of each animal in each sampling period, did not reject the null hypothesis of no preferred or avoided associates at $P = 0.51$, confirming the homogeneous nature of this social structure indicated in the ordinations (Fig. 12.2).

12.4.3 Network analyses

The social system of a marine mammal population can be viewed as a network, with nodes (individuals) being connected by edges (relationships). The network can be described by a matrix of relationship measures (such as Table 12.2) and is then amenable to the large, and quickly growing, body of quantitative techniques known as network analysis. For introductions to the application of network techniques to

Table 12.3 *Weighted network measures that can be calculated for each individual (node) from matrices of association indices or interaction rates (a_{IJ} is the association between individuals I and J; $a_{II} = 0$ for all I). (Adapted from Whitehead 2008b.)*

Measure:	What it means:	Formula for weighted network
Strength	How connected to other individuals	$s_I = \sum_J a_{IJ}$
Eigenvalue centrality	How well connected, in terms of number and strength of connections, and to whom	$e_I = (1^{st}$ eigenvector of $a)_I$
Reach	Overall strength of neighbours	$r_I = \sum_J a_{IJ} \cdot s_J$
Affinity	Weighted mean strength of neighbours	$f_I = r_I / s_I$
Clustering coefficient	How well connected are neighbours to one another	$c_{IJ} = \dfrac{\sum_J \sum_K a_{IJ} \cdot a_{IK} \cdot a_{JK}}{\max(a_{JK}) \cdot \sum_J \sum_K a_{IJ} \cdot a_{JK}}$

non-human social systems see Croft *et al.* (2008) and Wey *et al.* (2008). Network analysis has been applied to marine mammal social systems (Lusseau 2003; Lusseau and Newman 2004). A drawback of almost all current applications of network analysis to non-human social systems has been that the network has been considered binary, with relationships either existing (1) or not existing (0). However, especially in rather labile marine mammal social systems, all pairs of animals possess a relationship, but they differ in strength. Thus attention is shifting to weighted networks in which edges are represented by continuous variables, for instance by association indices or interaction rates (Lusseau *et al.* 2008). Although less well developed than the analysis of binary networks, there is a growing body of techniques that can be applied to weighted networks (Newman 2004). Sifting out those that are useful for the study of marine mammal social systems can be a challenge, but here are some recommendations.

For each individual we can calculate a suite of nodal statistics (Table 12.3). The mean and standard deviation of these are then usually presented for the population, as well as for classes of animals (e.g. males and females), and in some cases between classes (for instance, the strength of relationships between males and males, males and females, females and females). Relationships between measures can also be examined. For instance, if individuals with high strength also have high affinity (the mean strength of an individual's neighbours), then there are preferential links among 'important' individuals. This situation is known as assortative mixing, and is often found in social networks (Newman 2003). Lusseau *et al.* (2008) showed how permutation tests, variants on those described earlier, can be used to test a variety of null hypotheses about weighted social networks, such as that relationships are similar within and between classes or that there is no assortative mixing. Network analysts have also developed a wide array of techniques for subdividing networks, some of which (such as Newman's 2006 eigenvector modularity) are very useful for finding social units and other social entities within study populations.

12.4.4 Lagged association rates

Temporal patterning is a part of Hinde's (1976) definition of social structure that is poorly covered by visual displays, tests for preferred companions, and most network analysis. Lagged association rates explicitly address temporal change in social relationships, and have been used to describe cetacean social systems (e.g. Karczmarski *et al.* 2005). The lagged association rate of lag τ is the probability that two individuals associated at time 0 are also associated τ time units later. Lagged association rates can be estimated from observations of the associations between identified individuals using formulae provided by Whitehead (2008b). However, these assume that all associates of an identified individual are identified, which is often not the case in studies of marine mammals. Instead, we use the standardized lagged association rate, which is the probability that given that B was associated with A at time 0, an identified associate of A τ time units later is B (estimation formulae in Whitehead 2008b).

The estimated lagged association rate, or standardized lagged association rate, is usually plotted against time lag, τ, often with time lag logged, as in Fig. 12.2. Lagged association rates usually decline in relation to the time lag because associations weaken or break down, but the pattern of decline tells us much about the temporal dynamics of a social system. For instance, in Fig. 12.2, standardized lagged association rates of sperm whales decline over periods of a few days, indicating disassociation of groups of social units, and a few years, indicating emigration from social units or mortality. A number of ancillary methods help us interpret lagged association rates and standardized lagged association rates (see Whitehead 2008b for details and Fig. 12.2 for an example).

Confidence intervals in lag duration can be estimated using the temporal jack-knife technique in which, for instance, month-long periods of data are omitted in turn. Null association rates, and standardized null association rates, give the expected values of the lagged rates if associations were random within the study population, and thus allow us to assess over what time scales individuals show preferences in their associations. Models, often of the exponential family, can be usefully fitted to lagged association rate data by maximizing the sum of logged likelihoods over different lags. The estimated parameters of these models (and their standard errors from the temporal jackknife procedure) quantify social processes, for instance the rate of disassociation per hour or day, or the typical group size. If several models are fitted, their fits can be compared using the quasi-Akaike Information Criterion, QAIC (Whitehead 2007 see Fig. 12.2).

Lagged association rates are particularly useful for studying the social systems of marine mammals because they integrate large amounts of quite sparse data (i.e. with little information on any particular individual or period, but with many individuals studied over a long time).

12.4.5 Describing mating systems

To describe mating systems, we need to know who mated with whom and, especially, which matings produced viable offspring. In marine mammals, the maternal side of the mating process is usually easily inferred, because in most species the mother–infant bond is sufficiently tight that successful mothers can be easily identified (an exception is the sperm whale, in which frequent babysitting and some allo-suckling makes maternal discrimination less straightforward; see Gordon 1987). Paternal input is harder to assess. In land-mating pinnipeds, copulations can be observed (e.g. Le Boeuf 1974), but may not always correlate with paternity (e.g. Worthington *et al.* 1999). For most marine mammals that mate in the water, copulation is rarely observed, and even more rarely can the identities of the individuals be ascertained. Thus most conclusions about marine mammal mating systems are based upon molecular genetic analysis (e.g. Hoffman *et al.* 2007).

To assign paternity to an individual with some certainty, genetic data are needed from the individual, as well as a substantial proportion of the potential fathers. The analysis is much more effective if genetic data are also collected from the mother of the focal individual, thus allowing the paternal genes to be determined by elimination. Currently, microsatellites are the preferred method of determining the paternities of marine mammals (see Selkoe and Toonen 2006), but other markers are being developed. Once paternities are determined, rates of reproductive success can be calculated for different ages and the behavioural strategies of males.

12.4.6 Other methods of social analysis

We have concentrated so far on the methods that have been most used, or we think have most potential, in the study of marine mammal social systems. There are other techniques which are important in particular circumstances, or are just beginning development:

- A well-developed methodology is available for examining dominance hierarchies (see de Vries 1998; Bayly *et al.* 2006). Individuals may be assigned dominance indices or dominance ranks, and the linearity of dominance hierarchies can be assessed using methods such as Landau's (1951) index. The analysis of dominance hierarchies is particularly important for land-breeding pinnipeds (e.g. Le Boeuf and Reiter 1988) and captive populations (e.g. Samuels and Gifford 1997).
- The analyses described to this point have assumed just one relationship measure, usually an interaction rate or association index. However, in many studies we can calculate two or more relationship measures. For instance,

Gero *et al.* (2005) measured association indices between bottlenose dolphin dyads in four different behavioural states: travelling, feeding, socializing, and resting. Whitehead (1997) suggests several techniques for displaying and analysing such multivariate social data.

- Especially among the larger odontocetes, there appear to be permanent, or nearly permanent, social units. However, these social units often group with one another, making their delineation less than straightforward. While cluster analyses may be used to distinguish members of social units, they may produce ambiguous or incorrect clusters. Lagged association rates and other analyses can suggest the development of more appropriate techniques in particular circumstances. For instance, for long-finned pilot whales (*Globicephala melas*), 'key' individuals were defined as those identified on at least 4 days at least 30 days apart (Ottensmeyer and Whitehead 2003). Social units were formed from key individuals, and animals identified with key individuals on at least 3 days, separated from one another by at least 30 days.
- Although informative mathematical models can be fitted to lagged association rates and network statistics, they are post hoc statistical models and are therefore descriptive rather than prescriptive. Ideally, one or more biological models of social structure would be translated into statistical terms, and then these fitted to the data. Unfortunately, with current software and hardware, fitting such theoretically initiated models to real social data is demanding.

12.4.7 Useful software

In the study of marine mammal social systems, computers have become essential tools. A list of some of specialized software that can be useful for analysing data using methods such as lagged association rates and network analyses is given in Table 12.4.

12.5 Broader issues

We have highlighted methods for describing and modelling the social structures of marine mammals. Having produced displays, statistics, and models of social structure, most scientists wish to place their results in a broader perspective to make comparisons among social structures, to investigate the evolutionary forces in social evolution, and to find out how social structure may affect other areas of biology. Here is a summary of a few of these issues (for more details see Whitehead 2008b).

12.5.1 Evolutionary forces behind marine mammal social structures

Experiments are almost never practical to examine the function and evolution of behaviour in marine mammals. So we are left with comparisons between individuals,

Table 12.4 *Software that may be useful for social analysis. (Adapted from Whitehead 2008b.)*

Name	URL	Free?	Notes
MatMan	www.noldus.com/site/doc200401030	No	Manipulates matrices, good for analyses of dominance and reciprocity
SOCPROG	myweb.dal.ca/~hwhitehe/social.htm	Yes*	Wide range of social analyses
UCINET	www.analytictech.com/ucinet_5_description.htm	No [+]	Range of network and other analyses
Pajek	vlado.fmf.uni-lj.si/pub/networks/pajek	Yes	Range of network analyses
NetDraw	www.analytictech.com/netdraw.htm	Yes	Visualizes networks
GraphViz	www.graphviz.org	Yes	Visualizes networks

* To run SOCPROG in its original form you need to have installed MATLAB plus the Statistics toolbox, which are not free. However, there is a compiled version of SOCPROG available for which MATLAB is unnecessary.

[+] Free evaluation version available.

groups, populations, and species (see Box 12.1). Many of the statistics and measures discussed earlier can be used in such studies, often by relating them to non-social attributes. For instance, the functions of mating strategies can be examined by relating male reproductive success, usually obtained through genetic paternity analysis, with attributes such as age and mass, or the function of grouping by comparing group size with attributes of prey distribution. Kinship is believed to be one of the major forces for sociality, with Hamilton's (1964) rules predicting a relationship between genetic relatedness and affiliative interactions. This is sometimes tested in marine mammals (e.g. Möller *et al.* 2006) by using a Mantel test to compare dyadic association indices with estimates of kinship derived from analyses of microsatellites.

If comparisons are made between species, for instance group size versus brain size across species (Marino 1996), there is a potential problem of independence. Related species may tend to have similar brain sizes because of common descent. The method of independent contrasts (Felsenstein 1985) restores independence to such analyses.

12.5.2 How can we study culture in marine mammals?

It is becoming increasingly clear that the standard paradigms of behavioural ecology, including Hamilton's (1964) rules, are insufficient to explain the patterns of diversity of cetacean behaviour. We need to evoke culture, defined as the

> **Box 12.1   Comparing and classifying social structures**
>
> We may wish to compare social structures between populations or species, between pinnipeds and cetaceans, between marine and terrestrial mammals, or with other animals. Some measures which can be used in such comparisons include:
>
> - Skew in reproductive success, especially among males (e.g. Worthington *et al.* 1999)
> - Group size (although it is important that groups are defined in comparable ways)
> - Sexual and age segregation. To what extent do animals associate with their own sex, and how does social behaviour change with age?
> - Presence and size of social units in females, or coalitions in males
> - Rates of disassociation from groups and units
> - 'Social differentiation': an estimate of the coefficient of variation among association indices between members of a community (Whitehead 2008b).
>
> These, or other, measures can potentially be used to classify social systems, although classification should only be used when clear classes exist.

transmission of information or behaviour through social learning (Rendell and Whitehead 2001). The study of culture in marine mammals, and other wild populations, has generally used an exclusionary protocol, in which patterns of behaviour that cannot be explained by genetic differences, ontological change, or environmental variation are ascribed to culture (e.g. Krützen *et al.* 2005).

This methodology has been criticized on several grounds. For example, it falls back on culture only as a last resort when no other explanations can be found, even though it may not have been possible to exhaustively investigate those alternatives; or that many important aspects of culture, such as foraging techniques, covary with ecology or genetics, and that the boundary line for exclusion is unclear (Laland and Janik 2006). New methods are being developed, including a dyadic multiple regression technique that apportions behavioural variation into genetic, ecological, and cultural sources using extensions of the Mantel test (Whitehead 2008b).

12.6 Acknowledgements

Many thanks to Don Bowen, Ian Boyd, Richard Connor, and John Durban for useful comments.

Glossary (extracts from Appendix in Whitehead 2008b)

Aggregation: Spatio-temporal cluster of individuals that is entirely the result of some non-social forcing factor.

Association: Two animals are associated if their circumstances (spatial ranges, behaviour states, etc.) are those in which interactions usually take place.

Association index: An estimate of the proportion of time that a pair of animals [is] in association.

Assortativity: The extent to which nodes in a network are connected to nodes that are similar to themselves.

Asymmetric relationship: A dyadic relationship in which the members interact with one another at significantly different rates.

Clustering coefficient: In a network, the extent to which the nodes connected to a focal node are themselves connected.

Culture: Information or behaviour shared by a population or subpopulation that is acquired from conspecifics through some form of social learning.

Dendrogram: Tree-diagram, in which individuals are represented by nodes and the branching pattern indicates degrees of association, the results of a hierarchical cluster analysis.

Dominance: 'An attribute of the pattern of repeated, agonistic interactions between two individuals, characterized by a consistent outcome in favor of the same dyad member and a default yielding response of its opponent rather than escalation' (Drews 1993).

Dominance hierarchy: An ordering of individuals such that more highly ranked individuals generally win agonistic encounters over, or receive submissive behaviour from, those ranked lower.

Dominance index: A measure of an individual's ability to dominate others in its community.

Dominance rank: The ranking of an in individual, within its community, in its ability to consistently win repeated agonistic encounters with other members of the community.

Eigenvector centrality: In a network analysis, eigenvector centrality is a measure of how well connected an individual is. Mathematically, it is the first eigenvector of the matrix of edges or weights.

Equivalence: Things, including social partners, that become mutually interchangeable through common spatiotemporal or functional interactions.

Fission–fusion: 'A society consisting of casual groups of variable size and composition, which form, break-up and reform at frequent intervals' (Conradt and Roper 2005).

Follow: A research strategy in which the researcher's attention stays with an individual or group (as opposed to a survey).

Group: Sets of animals that actively achieve or maintain spatio-temporal proximity over any time scale and within which most interactions occur.

Interaction: An action of one animal directed towards another or affecting the behaviour of another.

Kinship: Genetic relatedness through common ancestry.

Lagged association rate: The probability that a dyad are associated at some time after a recorded association.

Mantel test: Permutation test of the significance of the relationship between the corresponding, non-diagonal, elements of two similarity or dissimilarity matrices indexed by the same individuals, with the null hypothesis being that there is no relationship.

Modularity: For some arrangement of individuals into clusters, the difference between the proportion of the total association within clusters and the expected proportion for randomly associated individuals.

Network: Pattern of connectedness among members of a population.

Null association rate: The expected probability that members of a dyad are associated at some time after a recorded association if association had no time dependency.

Ordination: Visual display in which points represent individuals and their proximity to one another indicates their degree of association.

Reach: A measure of indirect connectedness in a network such that nodes with high reach are connected indirectly to other nodes of high degree or strength (Flack *et al*. 2006).

Relationship: A synthesis of the content, quality and patterning of the interactions between two individuals, where patterning is both with respect to each others' behaviour and to time.

Relationship measures: Quantitative descriptors of the content, quality or temporal patterning of dyadic relationships.

Social differentiation: The degree to which the dyads within a population differ in their probability of association, measured using an estimate of the coefficient of variation of the true association index.

Social structure: A synthesis of the nature, quality, and patterning of the relationships among the members of a population.

Sociogram: Diagrammatic representation of social structure in which individuals are represented by nodes, and edges between nodes indicate the strength of the dyadic relationship.

Strength: In a weighted network analysis, the sum of the weights of the edges connected to a node.

Survey: A research strategy in which an individual or group is first encountered, then observed, and then the researcher moves on to another individual or group (as opposed to a follow).

Typical group size: The mean group size that an individual, or set of individuals, experiences.

13
Long-term studies

W. Don Bowen, Jason D. Baker, Don Siniff, Ian L. Boyd, Randall Wells,
John K.B. Ford, Scott D. Kraus, James A. Estes, and Ian Stirling

13.1 Introduction

Marine mammals are large species with longevities ranging from 9 to >100 years (Trites and Pauly 1998). They typically have low rates of population increase and relatively high adult survival. Thus, along the continuum of life histories, marine mammals are generally considered K-selected species that are expected to exhibit relatively low maximum theoretical annual rates of increase (~2–15%/year). Nevertheless, environmental variability acts directly on individuals, at a range of scales, to depress the growth rate. Factors affecting the latency to sexual maturity, including growth rate which is in turn affected by food availability, may be involved. Similarly, survival and reproductive rates are critical determinants of population growth rate and these are also affected by food availability. In addition to reasonably well-known seasonal and inter-annual variability in the environment, variability at decadal or longer scales can have important effects on behaviour and life history traits and thus the dynamics of populations. Therefore, characterizing variability in population trends and distribution and the mechanisms underlying those dynamics requires that studies be conducted at appropriate scales. Given their longevity, we should expect marine mammal life histories to have evolved adaptations to deal with environmental variation at a range of temporal and spatial scales ranging from months to decades and metres to hundreds of kilometres (e.g. Whitehead 1996). Thus, short-term studies of individuals cannot hope to understand how they cope with longer-term environmental variation.

The value of long-term ecological studies has been recognized for some time (Cody and Smallwood 1996). Fine examples of long-term studies can be found among birds (Weimerskirch 1992; Cooke *et al.* 1995; Grant and Grant 2000), small mammals (Broussard *et al.* 2005), ungulates (Clutton-Brock *et al.* 1982; Clutton-Brock and Pemberton 2004), and carnivores (Packer *et al.* 1998). There are also a growing number of long-term studies on marine mammals, as the examples in this chapter speak of so well. The chemistry of what makes a long-term study work involves foresight mixed with robust short-term goals of high science value in their own right—because we need to keep funding going—but also in the context of understanding the underlying mechanisms involved in the long-term processes. Thus, it is difficult to start out with the objective of creating a long-term study. In this chapter, we illustrate both the value and difficulties associated with long-term studies, how these studies often make use of multiple tools and approaches, and how they may have changed as the nature of the questions asked and our ability to make measurements have changed as new technologies have become available.

13.2 Grey seal (*Halichoerus grypus*)
W. Don Bowen

13.2.1 Motivation

One of the longest running pinniped ecological studies has been conducted on grey seals at Sable Island, off the east coast of Canada. Since the early 1960s, the number of pups born annually has been monitored, providing a demographic context for a wide range of shorter term studies on the behaviour, physiology, and ecology of this species. Research on this population had the rather modest initial goal of monitoring the distribution of the sealworm parasite (*Pseudoterranova decipiens*) in grey seals. The larval stages of the parasite devalue harvested groundfish, such as Atlantic cod (*Gadus morhua*). However, as the seal population grew, the potential impact of grey seals as a vector of the sealworm parasite and as a predator of commercially harvested fish stocks became the primary motivations for continuing research funded by the Department of Fisheries and Oceans (DFO) (Bowen 1990). This resulted in a broader range of studies, including those on diet and reproduction.

A primary goal of the life history research was to understand sources of variation (i.e. age, sex, and year) in survival rates and fecundity using mark–recapture statistical models. Because grey seals are the longest lived pinniped species with a maximum observed age on the breeding colony of 44 and 35 years for females and males, respectively, a long-term view was needed to attempt to understand sources of variation over a significant fraction of an individual's lifespan. This realization was the primary motivation for the standardization of re-sighting branded individuals in the breeding colony. However, a second and perhaps more important motivation, was the desire to contribute to a broader understanding of how life history traits of individuals vary in response to both intrinsic and extrinsic factors (Bowen *et al.* 2006). One extrinsic factor was clearly evident early on. Population size was increasing exponentially and, by the early 1980s, this had been occurring for two decades (Bowen *et al.* 2007). Although it was not possible to estimate when density-dependent changes in life history traits would occur, there was a presumption that they would and there was a need to study those changes. In fact, it was the desire to study changes in vital rates that motivated the branding of a sample of weaned pups each year from 1998 to 2002 (Bowen *et al.* 2007).

13.2.2 Nesting of short-term objectives within long-term objectives and change in focus through time

Although the potential of Sable Island as a site for long-term studies was realized in the mid-1980s, taking full advantage of this potential only became possible in about 1990 through collaboration with Daryl Boness, at the Smithsonian Institution, Washington, DC and Sara Iverson at Dalhousie University. The decision to combine resources and research questions permitted a wide range of short-term physiological and behavioural studies, conducted by graduate students and post-doctoral fellows, to be nested within our long-term objectives. Pooled resources and the increase in intellectual capital of multidisciplinary collaborations of physiologists, biochemists, geneticists, behavioural, and population ecologists was essential to the maintenance of the underlying long-term research effort. To improve our knowledge of population dynamics, there was a need to investigate the energetic cost of reproduction (Iverson *et al.* 1993), the factors affecting offspring size and growth rate (Boness *et al.* 1995), the relationships between offspring size, survival probability and age and size at recruitment, and the effects of age and condition on the probability of reproduction and survival. Continually improving technology and new methods, such as fatty acid signature analysis, meant that we could also study the foraging ecology of this size-dimorphic species in greater detail to determine the relationships between habitat use, diet, and reproductive success in both sexes. Although we had reason to expect sex differences, the nature and extent of those differences came as a surprise (Beck *et al.* 2003a; Austin *et al.* 2006a, b; Breed *et al.* 2006).

13.2.3 Standardization of methods and effects of changing technology and techniques

A critical element to the success of our long-term research was the decision by Arthur Mansfield and Brian Beck in 1963 to begin a hot-iron branding programme to permanently mark recently weaned pups. In most years, a sample of seals was marked with unique brands, permitting individuals to be studied over time. Intermittently between 1963 and 2002, 8850 individuals were branded, providing a sample of 7200 uniquely marked individuals of known-age. The core of our demographic research from 1983 through 2010 was a series of standardized re-sighting censuses of branded adults conducted weekly throughout the breeding season to estimate patterns of survival and fecundity. Although the fundamental goal of these censuses remained constant over time, the development of affordable GPS units permitted the accurate location of each sighted animal to be determined, and the collection of other data allowed us to address questions relating to habitat use, site fidelity, and reproductive success.

286 | **Long-term studies**

13.2.4 Challenges

The advantages of long-term studies are clear enough; however, there are significant challenges. Live capture of adults at a remote field site requires a team of qualified people and rather expensive helicopter transportation to and from the Island. Thus securing sufficient funding is a recurring problem. Another potential problem of long-term studies is that the types of data collected become routine, but may not be the most informative with respect to the questions asked. To some extent this is inevitable, but it does mean there is a need to evaluate which long-term measures should be taken. Despite those recurring challenges, most interesting questions about population biology can only be addressed by taking a long-term approach. In the case of grey seals on Sable Island, after 40 years of exponential population growth, we are just now seeing changes in vital rates that suggest resource limitation. Without our long-term life history data we would have little hope of understanding either the nature of those changes or their underlying causes.

13.3 Hawaiian monk seal (*Monachus schauinslandi*)
Jason D. Baker

13.3.1 Motivation

The Hawaiian monk seal is among the rarest of marine mammals with only ~1000 remaining individuals scattered throughout the entire Hawaiian archipelago. In 1976, the monk seal was listed as endangered under the US Endangered Species Act. Before that, monitoring of the species consisted primarily of occasional counts. By the early 1980s, a research programme was established at each of the North western Hawaiian Islands (NWHI) subpopulations to understand better why the species was in trouble and to devise recovery strategies.

The Hawaiian monk seal has the distinction of being the subject of a long-term and thorough demographic study, on a par with that undertaken for any large, free-ranging mammal in the world. A suite of demographic parameter estimates have been updated annually for each NWHI subpopulation within a few months after the annual field seasons have ended (Harting 2000; Baker 2004; Baker and Thompson 2007). This near real-time information is scrutinized and considered in the context of evaluating past recovery efforts and designing future conservation measures.

13.3.2 Nesting of short-term objectives within long-term objectives and change in focus through time

The fundamental long-term objectives of the population assessment programme have remained constant, which is one reason the accumulated data are so valuable. Because the NWHI are so remote and largely lack a permanent-support infrastructure, by far the largest cost and difficulty of conducting research there is getting the necessary people and supplies in place to support the fieldwork. Conducting the monk seal population annual assessment field camps has meant that special short-term studies could be added at relatively low cost. In this way, studies on foraging ecology, disease, and behaviour have been facilitated. However, it has proven far easier to *add* data collection tasks than to *remove* them. Data collection incorporated into the monitoring programme, even if it is originally intended as a means of obtaining data for a short-term ancillary project, tends to be over-valued as a permanent component of the long-term objectives. People conducting long-term research tend to be conservative by nature and are loathe to stop the collection of any data that have become part of protocols. Over time, the data collection burden becomes ever larger, compromising the efficiency of achieving core long-term objectives. Eventually, there is a risk that some data collection protocols are carried out annually even when no one can remember why! It is essential to regularly re-evaluate protocols and winnow them when appropriate.

13.3.3 Standardization of methods and effects of changing technology and techniques

When monk seal population assessment began in earnest in the early 1980s, Thea Johanos and others showed remarkable foresight as they developed data collection and management protocols. To this day, the fundamentals that they established remain. Advancements in technology and techniques have primarily been applied to make the process of data collection, management, storage, and analysis more efficient and accurate. In the first year of the population assessment programme, field staff recorded diverse information on everything they thought might be important and did not use standardized formats. Subsequently, key data elements were identified and data collection was condensed onto relatively few forms with standardized coding. Another important advancement has been entering data shortly after collection on laptop computers in the field into purpose-designed databases which incorporate error checking and editing features. This greatly facilitates rapid post-field data analysis. Further, a suite of computer programs has been developed which operate on standardized output files to generate sum-mary population analyses (e.g. abundance and vital rates estimates). A recent advance was the replacement of hand-drawn 'scar cards' and film photography to identify individuals using natural marks with a searchable digital photographic identification system.

13.3.4 Challenges

Monk seals in the NWHI are in many ways ideally suited for long-term monitoring studies. They are non-migratory and reliably spend time on shore on relatively few tiny islands where they can be readily approached without disturbance and observed for individual identification. Their regrettably low numbers mean that populations can be censused and individual re-sighting probabilities are high. On the other hand, the challenges are considerable. The monk seal metapopulation is dispersed throughout the most isolated archipelago on earth. Out of six subpopulations, four can be accessed only by ship, and air transport to the remaining two is sporadic and expensive. Because monk seals have an asynchronous breeding system (i.e. females do not all give birth within a short time), in order to evaluate pup production and reproductive rates, field seasons must span many months, whereas some other pinnipeds have most of their pups within a few weeks of each other (Johanos *et al.* 1994; Harting *et al.* 2007). Another considerable challenge is finding enough qualified field technicians. Field camps are staffed primarily by people hired seasonally, who, in addition to having mastery of complex scientific and data management protocols, must also be able to safely operate and maintain small boats, capture and handle seals, and possess numerous remote camping skills.

Another challenge is less intuitive and unrelated to field work. Good population monitoring programmes have momentum. This helps maintain consistency in methods, comparability of data collected, and dogged commitment to the long-term time series so rarely available. Even so, the business of collecting new field data, maintaining the integrity of databases, and generating the latest up-to-date analyses, can be all-consuming. It then becomes difficult to step back and evaluate field and analysis methods, research priorities, and to recognize emergent patterns in the accumulated data. Such activities are paramount to ensuring the success of long-term research programmes.

13.4 Weddell seal (*Leptonychotes weddellii*)
Don Siniff

13.4.1 Motivation

What turned out to be a long-term study on Weddell seals at McMurdo Station, Antarctica, began in 1968. Weddell seals were known to return to traditional pupping colonies each year to give birth and breed and were highly approachable. Our initial motivation was to use newly developed radio transmitters to gain insights into haul-out patterns, movements, and mother–pup interactions. It was obvious during the first season that observations under the sea ice were needed if any understanding of behaviour was to be accomplished. Given the time limitations on observations imposed by SCUBA gear, in 1969 an underwater TV camera and video recorder were used. This was a considerable technical challenge in those days. With the help of very good electrical engineers, we obtained recordings of mother–pup behaviour, male territorial behaviour, and underwater breeding. As this work moved into the third field season, it became evident that to investigate such areas as homing tendencies, territorial behaviour, and territorial longevity, observations of the same individuals for more than one reproductive season were required. Thus, interest in longer term observations began to develop.

The funding prospects available at the beginning determined, in large part, the strategy of the study. In 1968 the US National Science Foundation (NSF) funded all Antarctic science and this continues today. The emphasis within NSF, at that time, was not focused on long-term efforts. Research focused on experimental designs. Thus, any suggestion of investigating questions that could only be answered by long-term data sets, were looked upon as 'monitoring' and were generally not favoured during the review process. Continuation with the Weddell seal studies required the development of 3-year projects with specific hypotheses that could be pursued within that time frame. Thus, for the first 10 years or so of this study, the major motivation was to fulfil these short-term objectives and publish the results, with the general view that the more we knew about the individuals we used as experimental units the better our results would be.

13.4.2 Nesting of short-term objectives within long-term objectives and change in focus through time

The research using underwater television worked well and opened the door for behavioural research. Also, because Weddell seals were relative easy to handle using the bagging technique developed by Ian Stirling (see Chapter 2), it was easy to obtain tissue samples. The initial efforts were focused on studies of behaviour, with supplemental work considering the analysis of blood samples and karyotyping. But it quickly became obvious how much more valuable the studies would be if there existed life history data such as age, location of birth, and identification of the birth mother. Thus, even though we had tagged many adults and pups during our first few years, we began in 1973 to attempt to tag all the pups born into the McMurdo population as well as a significant number of adult females. From this point on, every three years, short-term projects that would give publishable results were established within the framework of the long-term study involving the tagging of pups and adults. By 1976 we had accumulated enough population data to complete an analysis of the population dynamics (Siniff *et al.* 1977). In 1988, the database and the continuing work to maintain the database was moved from the University of Minnesota to the University of Alaska with Ward Testa and Mike Castellini. They continued with the population work, but also used the known-aged animals in reproductive and physiological studies. Again, the long-term work was nested within short-term projects, focusing on physiology, movement patterns, and foraging studies, using satellite tags (Testa 1994; Burns *et al.* 1997). In both of these efforts the use of known-aged individuals was extremely important. Because of other commitments for the Alaskan investigators, the project returned to the University of Minnesota in 1993. The long-term data about individuals was becoming valuable to many different projects. For example, in the early 2000s, these data were used to look at genetic relationships (Gelatt *et al.* 2000). In 2002 the project moved to Montana State University with Bob Garrott and Jay Rotella as principal investigators. Recently, these longer-term data have been used to explore linkages with oceanography, trends in age at first reproduction (Hadley *et al.* 2006; Proffitt *et al.* 2007), and for documenting events credited to climate change (Siniff *et al.* 2008).

13.4.3 Standardization of methods and effects of changing technology and techniques

The development of an Access database that could hold all the tagging and recapture data, as well as ancillary information, was a tale of trial and error. Alongside tagging, about seven censuses of tagged seals were carried out each year. These censuses became the central part of the database. Initially, censuses were recorded in the standard 'field notebook' and information was then transferred to summary log books. These methods introduced errors into the data. In the mid-1980s, small hand-held computers began to be used to record tag numbers and other data in the field. This allowed real-time data checking. In spite of these efforts errors in the database still occurred and required further development and standardization of data entry protocols after 2001 (Cameron and Siniff 2004).

13.4.4 Challenges

From the beginning there had been no specific intention of establishing a long-term data set and was certainly not the vision shared by reviewers or the funding agency. If the development of a long-term data set was something supported by all from the onset, we would have done many things differently in the beginning. Unless an investigator has some such promise of long-term funding from the onset, there is a need to plan short-term efforts that yield publishable results. When a long-term study comes to an end questions arise about curatorial responsibilities, and how access to the data should be managed. The final repository, meta-data formats, access control, and data ownership all need careful attention.

13.5 Antarctic fur seals (*Arctocephalus gazella*)
Ian L. Boyd

13.5.1 Motivation

The studies of Antarctic fur seals at Bird Island, South Georgia, developed from the overarching question concerning the *krill surplus hypothesis* (R.M. Laws 1977). This stated that the reduction of baleen whale populations in the Southern Ocean through the first half of the twentieth century had led to an abundance of krill. The hypothesis also led to the prediction that this had been exploited by other krill predators which had shown compensatory population growth. Bird Island, is situated at the west end of the island of South Georgia and lies at a critical location in the path of the Antarctic Circumpolar Current that transports krill from upstream regions of high krill production. Therefore, in theory the long-term monitoring of Antarctic fur seals at Bird Island, amongst other krill-eating species such as penguins and albatrosses, could reveal the dynamics of Southern Ocean krill stocks and allow testing of the krill surplus hypothesis (Croxall *et al.* 1988).

A principal motivation, in the context of applied science, for developing long-term monitoring of Antarctic fur seals was the establishment of the Convention for the Conservation of Antarctic Marine Living Resources (CCAMLR) in 1982. This is the vehicle for the management of fisheries in the Southern Ocean, and it was one of the first management systems to be established based on principles of an ecosystem approach to management. There was a determination not to repeat the over-exploitation of whales, and then fish, which had taken place until that time (Boyd 2002d). Since economic indicators suggested that krill could have been the subject of the next fisheries boom in the Southern Ocean, there was recognition that appropriate regulatory mechanisms needed to be established in a manner that would prevent levels of exploitation that could further exacerbate the ecosystem perturbations already under way. The CCAMLR established the principle that any exploitation should not adversely affect populations of predators, including seals and whales, which also relied upon the exploited resource. Consequently, there was a need to establish monitoring to assess the extent to which this was happening. In subsequent years, the research carried out in conjunction with monitoring allowed the reproductive and population dynamics of fur seals to be used as a biological indicator of krill populations (Reid *et al.* 1999; Boyd and Murray 2001).

13.5.2 Nesting of short-term objectives within long-term objectives and change in focus through time

In addition to providing support for fisheries management in the Southern Ocean, monitoring provided a foundation for a wide range of other types of studies of the demographics, population dynamics, and behavioural ecology of Antarctic fur seals. These studies were a vital component of sustaining both the flow of funding and the scientific integrity of the monitoring. The value of monitoring—involving the collection of long-term data about reproductive rate, maternal foraging during lactation, survival, and growth rate—would have been greatly reduced without research built around those data to underpin their interpretation. Moreover, the studies of fur seals have been a component of a wider monitoring programme at Bird Island that has included two species of penguins and three species of albatross. Together, they formed a formidable array of long-term data sets from a range of species that exploited the resources of the Southern Ocean at a broad range of spatial scales. Individually, these would have been difficult to fund and sustain, especially given the costs of operating in a remote location like South Georgia, but as a group they have sustained the critical mass necessary for continuity over several decades.

Several important studies with general implications in biology could not have been undertaken without the long-term studies. Examples include the studies of the importance of sexual selection as a driver behind social structure (Hoffman *et al.* 2007), the testing of behavioural models (Boyd 1999), and understanding the basis of non-linear functional responses (Mori and Boyd 2004).

13.5.3 Standardization of methods and effects of changing technology and techniques

The studies of fur seals at Bird Island developed in parallel with other studies elsewhere in the Antarctic (Reid *et al.* 2004), stimulated by a coordinated approach developed under the CCAMLR Ecosystem Monitoring Programme. The long-term studies at Bird Island provided a template for the standard methods used within this formal programme. Only data gathered using these standard methods can be submitted to CCAMLR and included formally within the ecosystem assessment process that is then used to set krill fishing levels. This has been a strong stimulus to maintain standardization through time, and also to allow data collected from fur seals at Bird Island to be directly comparable with data collected elsewhere by other researchers.

The techniques used have relied greatly upon gaining easy access to animals without causing disturbance, and the sites (see Fig. 2.4, Chapter 2) used for long-term studies were chosen carefully at the start to give them the maximum chance of fulfilling the long-term vision. A significant factor in the long-term success of the research at Bird Island was also the selection of a range of simple, easily measured variables to be used in monitoring the population and behavioural dynamics of the seals. Cheap, reliable, low-technology solutions were vital and these have been sustained and supplemented, but not replaced, by new methods involving PIT tagging and various forms of telemetry. However, in general, since the objective was to sustain long-term data sets with consistent properties it was essential to find solutions that were low cost. For example, studies that have documented the demographics of Antarctic fur seals (Boyd *et al.* 1995c) and how these vary in relation to physical oceanography (Forcada *et al.* 2005) have relied upon consistent, simple sets of measurements from known-age seals identified using flipper tags (see Chapter 2).

13.6 Bottlenose dolphin (*Tursiops truncatus*)
Randall Wells

13.6.1 Motivation

In 1970, little was known about the ranging patterns of dolphins. A few reports suggested that some dolphins in the south-eastern USA might be residents, but these were the result of casual sightings of distinctively marked individuals. Blair Irvine, at Mote Marine Laboratory in Sarasota, Florida, developed a curiosity about dolphin movements. Having worked previously with Bill Evans, Sam Ridgway, and Steve Leatherwood for the Navy at Pt Mugu, California, Irvine borrowed some of the tags they were developing for pelagic dolphins. The original idea was to test the pelagic tags and the use of freeze brands as part of a pilot tagging study of bottlenose dolphins in the Sarasota Bay area. The initial question posed by Irvine and his assistant (RSW) was very simple: Do the same dolphins live here year-round? The initial findings from tagging work in 1970–1971 indicated that bottle-nose dolphins were year-round residents in local waters. Multiple sightings of some of the same animals during that period suggested that they might live in relatively stable social groups. This set the stage for long-term studies and the development of Sarasota Bay as a 'natural laboratory' for dolphin research. The sheltered waters of the bay, along with the ability to observe and safely capture and release individuals on a relatively predictable basis, made this a valuable field site.

Over time, we collected a suite of information on most of the identifiable resident dolphins, including age, sex, genetics, maternal lineage, birth order, health and body condition, signature whistles, reproductive success, ranging patterns, and social association patterns. Access to this cast of local characters, including some recognizable individuals that we have never managed to capture, has motivated a wide variety of collaborators from around the world to study the Sarasota Bay dolphins.

13.6.2 Nesting of short-term objectives within long-term objectives and change in focus through time

Our long-term research objectives have evolved over time. Confirmation of the multigenerational residency of dolphins in the Sarasota Bay area allowed us to define core objectives, including: (1) monitoring the dynamics of the local dolphin community; (2) determining life history parameters; (3) identifying long-term ranging and habitat use patterns; (4) describing social structure features and patterns; (5) identifying and mitigating anthropogenic threats; and (6) developing and testing field research techniques such as tags, data-loggers, or aerostats. The research has been adapted as new field techniques become available, and as new conservation issues and management needs emerge.

In the mid-1970s, the US Marine Mammal Commission needed information on ranging patterns to begin to develop management plans for bottlenose dolphins, leading to an emphasis on research on tagging and tracking to explore a geographical basis to stock definition (Irvine *et al.* 1981). This work led to photographic identifi-cation studies (with more than 4000 identifiable dolphins currently catalogued), genetic research, and expanded surveys into adjacent areas for comparisons (Duffield and Wells 2002). Beginning in the late 1980s, large-scale unusual mortality events involving bottlenose dolphins in the south-eastern USA led to the expansion of nascent health assessment studies in Sarasota Bay, in an attempt to understand dolphin health before carcasses wash up on the beach (Wells *et al.* 2004). Capture–release efforts for life history studies were expanded in scope to measure health and body condition parameters, metabolic rates, and thermoregulatory capabilities, and to coordinate with local stranding networks. Further refinement of the health assess-ment approach included sampling for existing and emerging organic environmental contaminants and trace elements in order to relate their concentrations to health, immunology, and reproductive success (Wells *et al.* 2005). In recent years, increasing interest in the effects of harmful algal blooms on dolphin health has led to the measurement of red-tide toxins in dolphins during health assessments, and focused studies on the direct and ecological impacts of red tides on dolphins. The ability to briefly and safely handle wild dolphins during health assessments has facilitated the compilation of the first records of hearing abilities in a wild dolphin population, in response to increasing concerns about the impacts of ocean noise. Dolphin distribution surveys initiated in the 1970s, and continuing today, have led to and support detailed quantitative ecological studies relating habitat use to prey distribution and abundance, which in turn are conducted in conjunction with trophic research involving stable isotopes, stomach contents, and fatty acids. Early recordings of individually specific 'signature whistles' for Sarasota Bay dolphins set the stage for subsequent studies over the next three decades of the development, stability, and function of such whistles, including novel play-back experiments during capture–release sessions that resulted in recognition of the use of these whistles by the dolphins as abstract 'names' (Janik *et al.* 2006). Monitoring over long periods is providing the basis for understanding the impacts of longer term environmental changes, including natural events such as harmful algal blooms, and anthropogenic changes such as commercial net fishing bans and coastal development.

13.6.3 Standardization of methods and effects of changing technology and techniques

While exploring, testing, and embracing new techniques and technology as appropriate, consistency has been maintained in the most basic aspects of our data collection protocols, in order to facilitate long-term comparisons. The most important component of the research is the compilation of longitudinal records for identifiable dolphins, building robust dossiers on each individual (Wells 2003). This information is used in a wide variety of descriptive and hypothesis-testing research projects.

One of the unique aspects of dolphin research in Sarasota Bay is the ability to safely capture and release individual dolphins, and to be able to observe and monitor individuals over time. This has facilitated the development and testing of tags, where re-captures for tag removal can occur if warranted. As a programme of the Chicago Zoological Society, the Sarasota Dolphin Research Program has, as a primary purpose, to train cetacean researchers in the use of standardized field techniques, which has proved useful when developing new studies, such as tagging studies of Franciscana dolphins in Argentina.

13.6.4 Challenges

The largest challenge is maintaining continuity of staff, archives, and infrastructure over decades.

13.7 Killer whale (*Orcinus orca*)
John K.B. Ford

13.7.1 Motivation

Killer whales living in the coastal waters of British Columbia have been the subject of long-term field studies since the early 1970s. There was concern about the rapidly expanding live-capture of killer whales in the waters of British Columbia and Washington State in the mid-1960s for aquaria throughout the world. It was unclear whether this removal was sustainable as population structure and abundance, calving intervals, and survival rates were unknown. In response to this concern, the Canadian Department of Fisheries and Oceans tasked Michael Bigg with providing the urgently needed biological data and management advice.

Mike Bigg's initial approach to the problem in 1971 was to launch a coast-wide public sighting and questionnaire programme to determine the approximate numbers and distribution of killer whales. This survey suggested that the numbers of killer whales in British Columbia were on the order of a few hundred, and revealed that killer whales could be found on a daily basis in summer in Johnstone Strait, off north-eastern Vancouver Island. In 1972, Mike Bigg, Graeme Ellis, and Ian MacAskie began observations of the whales in Johnstone Strait and discovered that many individuals bore nicks on their dorsal fin and scars on the grey-coloured 'saddle patch' at the base of the fin. Realizing that these could serve as 'natural tags', the focus shifted to photographically identifying and tracking individual whales. This novel field technique was considered quite radical at the time, but Bigg and co-workers persisted despite scepticism. By 1976, Bigg *et al.* had confirmed that the killer whale population in nearshore waters of British Columbia was less than 300 whales and live-captures ended.

Although the management urgency had passed, Mike Bigg and colleagues realized that an excellent opportunity existed to learn much more about killer whale biology. Since every individual was, or could be, individually identified, over time it would be possible to understand many facets of the animals' life history and ecology. What began as a short-term management question evolved into a multidisciplinary study that is now well into its fourth decade. Mike Bigg died in 1990, but his long-term colleagues and dozens of younger researchers and students have ensured that this important study continues.

13.7.2 Nesting of short-term objectives within long-term objectives and change in focus through time

Central to the long-term study of killer whale demography is the annual census of individuals by photo-identification undertaken mainly by Graeme Ellis and myself in British Columbia, and Ken Balcomb in the adjacent waters of Washington State, with contributions by numerous colleagues. These demographic data have provided the foundation for many other short- and long-term studies.

Our research has provided insight into the lives of these apex predators, as well as many surprises. Instead of a single population of killer whales off the British Columbia coast, there are three sympatric, but non-associating, genetically distinct ecotypes—residents, transients, and offshores (Bigg *et al.* 1990; Ford *et al.* 2000). These ecotypes differ in appearance, social structure, movement patterns, vocalizations, and, most importantly, in their foraging ecology. Resident killer whales feed on fish, particularly salmon (Ford and Ellis 2006), while transients prey on marine mammals and occasionally seabirds (Ford and Ellis 1998). The recently discovered offshore ecotype is still poorly known but appears to feed on fish. Resident killer whales live in multigeneration matrilineal groups that are as stable as that of any mammal—individuals of both sexes stay in their natal group throughout their lives. As many as five matrilineal generations can coexist in one group. Resident killer whales have what appears to be a unique outbreeding mechanism—each matriline has a distinctive dialect that encodes its genealogy and relatedness to other groups in the community, which may inhibit mating among close relatives (Ford 1991). These findings were only possible due to the long-term nature of this study.

13.7.3 Standardization of methods and effects of changing technology and techniques

The success of our study owes much to the establishment of standardized protocols for identifying and naming individual whales and their social groups. The alpha-numeric system for naming whales developed in the early 1970s by Mike Bigg is still used today, while the naming system for groups has evolved slightly to better reflect their matrilineal structure. Updated catalogues with each whale's name, identification photo, and genealogy are published regularly. Study of the whales' acoustic behaviour and dialects has also benefited from a standardized naming system for their distinctive and highly stable call types.

New technologies have enabled us to address questions that in the past have gone unanswered. DNA analyses of skin from biopsy samples confirmed our suspicion that sympatric ecotypes are genetically distinct, and that breeding takes place within the population but outside the matriline, and is influenced by dialect differences (Barrett-Lennard 2000). Contaminant analyses of blubber from the same biopsies revealed extremely high levels of PCBs in many individuals, which may put the animals at risk of a variety of health problems (Ross *et al.* 2000). Advanced hydrophone arrays have allowed the study of vocal exchanges within and between groups, and autonomous underwater acoustic recording devices located in remote parts of the coast are helping to fill in gaps in our understanding of the whales' seasonal movement patterns through identification of group-specific dialects.

13.7.4 Challenges

As with most long-term studies, finding sufficient and reliable funding has, at times, been a struggle. Each year, many weeks of on-the-water field work are needed. Analysis of identification photos and updating of databases requires many more weeks of work in the laboratory. The annual photo-identification effort can appear to be rather mundane to funding agencies, but it is critical to the success of our studies as well as to the conservation of these populations. For example, long-term unbroken effort to maintain a precise annual registry of individuals in the population enabled us to detect subtle year-to-year changes in age-specific survival rates in resident killer whales, and to link these to varying prey abundance. Although not considered a funding priority by federal agencies on either side of the border, a sharp decline in the abundance of resident killer whales in the late 1990s resulted in a listing under the Canadian Species at Risk Act and the US Endangered Species Act. With those listings has come more secure funding to identify limiting factors and threats facing west coast killer whales and to promote their recovery.

13.8 North Atlantic right whale (*Eubalaena glacialis*)
Scott D. Kraus

13.8.1 Motivation

The North Atlantic right whale was hunted nearly to extinction, and despite protection for over 70 years, it still numbers under 450 animals. In the 1950s most scientists thought the species was nearly extinct, but the University of Rhode Island's Cetacean and Turtle Assessment Program (1979–1982) first comprehensive survey for marine mammals and turtles in the western North Atlantic (CeTAP 1982), found that more than 200 right whales had survived. In 1980, the New England Aquarium started surveys in Canada. The goals of both the CeTAP and Aquarium surveys were to evaluate the potential effects of oil exploration and refinery development on marine mammals. Both programmes identified areas where relatively large number of right whales aggregated seasonally. This work was followed by studies on the feeding behaviour of right whales around Cape Cod, aerial surveys in the winter off the south-eastern United States, and surveys off the southern tip of Nova Scotia. By 1985 these efforts had identified the major coastal Western North Atlantic right whale habitats (M.W. Brown *et al.* 2007).

However, right whales live a long time and inter-birth intervals are long (3 plus years), and it dawned on researchers that collecting baseline data on reproduction and population dynamics was going to take at least another decade. Then in 1990, data emerged demonstrating the leading causes of death in right whales were collisions with ships and entanglements in fishing gear. Following this, right whale reproductive rates declined throughout the 1990s with two periods characterized by very low calf production (Kraus *et al.* 2007). By the mid-1990s marine biologists were widely aware that right whales were in trouble—primarily because of human activities conducting business as usual—and this prompted an increase in research activity on this species.

In spite of the increase in knowledge about North Atlantic right whales and active conservation efforts, 50% of all confirmed right whale deaths are still due to collisions with ships or entanglements in fishing gear (Kraus and Rolland 2007). In addition, 75% of all right whales display scars indicative of fishing entanglements.

13.8.2 Advances in methods and changing objectives through time

Early research on right whales consisted of surveys aimed at assessing population size, and some opportunistic studies of behaviour. Right whale studies changed with the development of individual identification methods in the early 1980s. The right whale catalogue and sightings database provided information on the age and sex of individuals, movements and habitat use patterns, reproduction, mortality, and the impacts of human activities on the population (Hamilton *et al.* 2007). Advances in genetics supplemented the catalogue, with information on sex, genetic migratory patterns, habitat use, inbreeding, paternity, and the genealogy of the entire population (Frasier *et al.* 2007).

To assess right whale reproduction, researchers pioneered non-invasive methods by measuring steroid hormone metabolites in faecal samples to identify the sex of individuals, pregnancy and lactation, and to evaluate levels of stress hormones (Rolland *et al.* 2007). Faecal analyses also provided data on right whale energetics, marine biotoxins, and parasites (Rolland *et al.* 2007).

Potential effects of ecosystem changes on whales were studied using ultrasound to measure the blubber thickness of free-swimming right whales, photoanalysis to assess health and body condition, and advanced forensic methods to determine causes of death. Researchers, using suction cup tags with acoustic, movement, and depth recorders, studied underwater right whale behaviour, including subsurface feeding, responses to ships, other whales, and potential alarm calls. Another acoustic programme deploys passive listening, recording, and transmitting devices to locate right whales if surveys cannot be conducted due to weather conditions or poor visibility. Finally, to answer right whale biology and management questions, models have been developed to describe trends in population growth and demographics, and to help monitor human sources of mortality (see Kraus and Rolland 2007).

One unusual feature in the history of right whale research is the North Atlantic Right Whale Consortium. Originally formed in the mid-1980s by five research groups, its purpose was to coordinate research, bring attention to the plight of the right whale, and elicit federal support for research and management efforts. The Consortium re-organized in 1996 with an open membership, a commitment to the open exchange of information, and a mandate to ensure the survival of right whales in the western North Atlantic. Sightings and photographic data are now made publicly available following a review process.

13.8.3 Challenges

Right whales live a very long time. Females require 10 years to reach sexual maturity, and may have a reproductive lifespan longer than a researcher's career. Since most of these whales will outlive the scientists studying them, a researcher may only expect to see a portion of a right whale's life, which can limit the scope of scientific studies and their conclusions. For example, a 10-year analysis of right whales in the 1980s would have shown population growth—analysis of data from the 1990s would have shown a decline—and averaged over 25 years, there is slight (ca 1%) growth. Herein lies another difficulty inherent in working with small populations—poor statistical power for assessing trends. When population assessments are subdivided by age, sex, and habitat-use patterns, sample sizes are not statistically robust.

The overarching challenge to long-term studies is the sustained fundraising and dedicated fieldwork needed. The scientific rewards are large, but they are slow to arrive, making them unpopular with foundations and agencies. And because these whales may outlive the researchers, it may be that the best security for the species and the continuity of the science and management efforts will be in communal ownership of the data.

13.9 Sea otters (*Enhydra lutris*) and kelp forests
James A. Estes

13.9.1 Motivation

Sea otters have been studied in the coastal marine ecosystems of the central and western Aleutian Islands for 38 years. While most ecological studies of large vertebrates (including marine mammals) have taken the approach of trying to understand environmental influences on behaviour and population biology, the overarching goal of this research has been to understand the ecological influences of sea otters on their environment.

13.9.2 Approaches

We have used theory and natural history to formulate our questions, and we have used field observations to answer them. Our questions were motivated by two pioneering developments in ecology: Hairston, Smith, and Slobodkin's (1960) famous 'Green world hypothesis', now commonly known as HSS—the idea that autotrophs are abundant (i.e. the world is green) in large part because predators limit their herbivores; and early experimental studies by R.T. Paine (Paine 1976) and others demonstrating the ecological importance of predation. Recognizing that sea otters eat sea urchins and that sea urchins eat kelp, HSS led us to predict that sea otters help maintain kelp forests by limiting the number of sea urchins.

History, geography, and time provided a means of testing this hypothesis. Sea otters were hunted to near extinction in the Pacific maritime fur trade. Recovery following the fur trade was geographically asynchronous, especially in the Aleutian archipelago where ocean passes limited inter-island dispersal. Spatial contrasts among islands with and without sea otters, and temporal contrasts following the re-colonization and growth of sea otter populations (and in more recent years, the unanticipated collapse of these same populations), provided a natural experiment for examining the ecological role of sea otter predation. Early contrasts of islands with and without sea otters supported the prediction that the kelp forest ecosystem was maintained through sea otter predation on sea urchins. Our subsequent work has addressed the strength and generality of this trophic cascade: indirect effects of the trophic cascade on other species and ecosystem processes, co-evolutionary consequences of sea otter predation on marine plants, and the rates and dynamics of ecosystem change associated with changes in sea otter density.

13.9.3 Methods

Nearly all of our conclusions emerge from contrasts in the distribution, abundance, and behaviour of species among islands with and without sea otters, or through time as sea otter abundance has varied. We have determined the generality of the otter–urchin–kelp trophic cascade by obtaining measures of the abundance and species' composition of kelps, and the density and population structure of sea urchins, from randomly selected sites in areas with and without sea otters. The data are obtained through a standardized process of randomly sampling the seafloor at any given site with 0.25 m^2 quadrats (Estes and Duggins 1995). Using the same methods and measurements, the changes in the rates and dynamics of ecosystems over time have been chronicled as sea otters reinvaded unoccupied islands and their populations expanded, and more recently following sea otter population collapses (ostensibly due to increased killer whale predation) at islands where they had previously been abundant. Other measures obtained over similar temporal and spatial scales have been used to explore the indirect effects of sea otter predation on other species and ecosystem processes. For instance, we have used growth rates of filter-feeding mussels and barnacles and stable carbon isotope ratios to measure differences in primary production and the relative importance of kelp between islands with and without sea otters. Similar contrasts have been used to document the effects of sea otters on kelp forest fishes, the diet and foraging behaviour of Glaucous winged gulls and Bald eagles, and the interaction strengths between predatory sea stars and their invertebrate prey (Estes *et al.* 2004). Using the knowledge obtained from these real-time ecological studies, we have expanded these contrasts to appropriate scales of space and time in an effort to understand the evolutionary effects of sea otter predation. For example, comparisons of secondary metabolites in marine plants, the abundance and spatial distributions of plants and herbivorous invertebrates, and the rates of plant tissue loss to herbivory between the Northeast and Southwest Pacific ocean (a physically similar system that has always lacked sea otters or predators of comparable influence) suggest that sea otters and their immediate ancestor decoupled the evolutionary arms race between plant defences and herbivore resistance in the Northeast Pacific Ocean, thus creating a high-quality food resource for other primary consumers (Steinberg *et al.* 1995). This process in turn appears to have influenced such seemingly unlikely events as the evolution and extinction of Steller's sea cow, and the evolution of large body size in North Pacific abalones (Estes *et al.* 2005).

13.9.4 Rewards

The most important reward from these long-term studies of sea otters and kelp forest ecosystems has been the excitement of discovery in an intellectual arena (the ecological roles of large, mobile, and often secretive species) that has been difficult for others to penetrate. Our excitement has been bolstered by the use of our findings in biology and ecology textbooks, frequent invitations to speak and to write, attention from time to time by the media, and a clear sense of interest and appreciation by students and members of the informed lay public.

13.9.5 Challenges

One important cost of conducting long-term research in the Aleutian Islands has been the isolation of the study location. A changing logistic infrastructure associated with the closure of military bases in the Aleutian Islands has necessitated new approaches to gaining access to the area and to obtaining on-site field support. Funding has never been a significant challenge as the US National Science Foundation, the US Department of the Interior, and various branches of the US military have generously supported new ideas and developing management problems.

13.9.6 Serendipity

Species interactions are dynamic processes that are difficult or impossible to understand from observations of static systems. For instance, we would have been unable to see with any clarity into the sea otter's ecological role in kelp forest ecosystems had the species been hunted to extinction, or conversely had sea otters never been depleted anywhere. Serendipitous opportunities provided by the fur trade and by the more recent, and wholly unexpected, entry of killer whales as new apex predators in the system have been essential to the design of our studies and in our resulting understanding of this system.

13.10 Polar bears (*Ursus maritimus*)
Ian Stirling

13.10.1 Motivation

In the autumn of 1970, I started a population study of polar bears in the eastern Beaufort Sea and their ecological relationships with seals and sea-ice conditions. Like most studies of its kind in the Arctic at the time, it was projected to last a maximum of about 5 years. Part of the project entailed gathering data on the species and ages of seals killed by polar bears, sampling Inuit harvests of ringed seals to assess reproductive success, and doing aerial surveys to begin to delineate geographical variation in seal distribution and abundance. This study also began at a time when the operative paradigm was that polar oceans were relatively stable ecologically. Thus, once the basic population processes of a marine-based species had been worked out, it was expected the information could be used to conserve or manage that population for many years.

It was serendipitous that several projects coincided with T.G. Smith's long-term studies of the population dynamics and ecology of ringed seals (T.G. Smith 1973) over many of the same areas at similar times. A pivotal result we obtained through collaborative studies in the mid-1970s was confirmation of a roughly 90% decline in ringed seal reproduction and pup survival in a single year. Although the underlying mechanisms remain speculative, the decline itself was clear and appeared to be correlated with unusually heavy ice conditions. It took about 3 years before ringed seal reproduction fully recovered. Because the bulk of the polar bear's diet is made up of ringed seals, especially young of the year, when seal reproduction and pup survival dropped, so did polar bear body weights, reproductive success, and cub survival. All these parameters recovered after the seals did, but the 1974–76 cohorts of both polar bears and seals remained under-represented in the population.

Because large-scale population assessments of polar bears are very expensive, most are budgeted for a 3–4 year period, but the results are expected to be adequate for determining sustainable harvests for Inuit hunters for 15 years or more before repeat surveys may be conducted. Similarly, 'baseline studies' undertaken to assess potential environmental impacts of offshore industrial activities are typically done over short time frames. The brevity of these studies precludes a possible understanding of whether fluctuations or trends in the environment might be affecting the results.

From these early studies, it was clear we needed to know a great deal more in two major areas if we hoped to be able to conserve and manage polar bears effectively over the longer term. First, we needed to learn enough about the scale, frequency of occurrence, and trends of natural ecological fluctuations in the marine environment in order to assess how they affected the bears. Second, we needed a much greater overall understanding of how interrelationships between polar bears, seals, and changes in ice conditions function over the longer term (Stirling *et al.* 1999).

It was also clear that only a team approach could accomplish such long-term goals since several areas of specialized expertise would be needed, most of the funding would have to be raised externally, and that long-term continuity would have to be built through coordination of a diversity of shorter term projects. Thus, collaboration with a network of dedicated colleagues and a series of productive graduate students coordinated around a focused theme appeared to be the only way forward.

13.10.2 Standardization of methods and sampling

To approach the question of understanding natural environmental variation, one needs continuity and the ability to repeat collection of the same data in the same way in the same seasons, in the same areas, and often with at least some of the same animals, every year or at some scientifically acceptable interval. Over the long term, different people will need to make the same measurements repeatedly. Thus, the basic measurement involved tagging with a permanent individual identifying number tattooed inside the upper lip. Sub-samples of bears were also weighed. Over the years, various other samples, such as fat, blood, and skin were collected for independent concurrent projects as well as being archived for future projects.

13.10.3 Two long-term approaches

Natural variation in the same study area

When polar bears are on the sea ice, they are widely distributed over huge areas and usually at relatively low densities, which makes trying to maintain a long-term quantitative sampling programme on a population scale in that habitat prohibitively expensive. With the benefit of hindsight, it would have been remarkably valuable to have been able to maintain a long-term study with annual sampling in the eastern Beaufort Sea, from the early 1970s through to the present (Stirling 2002). Unfortunately, that simply wasn't possible, mainly because of cost. Thus the primary focus for a long-term study of population ecology became the polar bears of western Hudson Bay while they were ashore along the Manitoba coast during the ice-free period in summer and autumn. This enabled us to sample from the whole population, but over a much smaller study area than that of the sea ice, and searching for white bears on a snow-free background was much easier than looking for them on the ice. These two factors made a long-term study on the western Hudson Bay polar bear population more cost-effective than would have been possible in any other area in Canada (Regehr *et al.* 2007).

We studied reproductive biology, physiology of fasting, behaviour, distribution and movements on land, terrestrial denning habitat and productivity of cubs in spring, population estimation, human–bear interactions, assessment of the effectiveness of different immobilizing drugs, the ecological significance of supplemental feeding, effects of forest fires on the denning areas, satellite tracking of bears on the sea ice, and the age and sex composition of the harvest from this population by Inuit hunters in Nunavut. However, all projects required the capture of polar bears and, hence, facilitated the accumulation of basic population monitoring data annually for over three decades. One problem was that because of unpredictable budget fluctuations, there was sometimes more inter-annual variability in sample sizes than was desirable. Regardless, carrying out back-to-back shorter term studies facilitated continued funding and ensured that at least the minimum core of population monitoring data could be collected annually. Eventually, because of the long-term data set, it was possible to quantitatively confirm the influence of events such as the eruption of Mt Pinatubo and, most importantly, the negative effects of long-term climate warming (Stirling and Parkinson 2006). Put simply, none of this would have been possible with a study done in the more usual time frame of 3–5 years, or even 10 years.

Long-term orientation toward ecological relationships

The second long-term approach involved studying aspects of the ecological inter-relationships between polar bears, seals, and sea ice, but did not always necessitate working in the same areas or for continuous periods (Stirling and Øritsland 1995). We studied polar bears and ringed seals over a period of 35 + years that were designed to address particular aspects of their ecology and, collectively, increase our basic understanding of their interrelationships (Thiemann *et al.* 2008). These studies included investigating the population ecology of polar bears and their use of habitat in the Beaufort Sea, seal reproductive success in a variety of areas, recordings of underwater vocalizations of seals and walruses, aerial surveys of seals in much of the Canadian Arctic, quantifying the behaviour of undisturbed free-ranging polar bears, physiological studies of the energetic requirements of polar bears, monitoring the age and sex composition of the Inuvialuit harvest of polar bears in the Beaufort Sea, and fatty acid analyses to quantify geographical variation in the diet of polar bears throughout the Canadian Arctic.

Through those studies, we began to understand the ecological significance of polynyas, shore leads, and different kinds of sea-ice habitats to both polar bears and pinnipeds. Of more specific importance, we were able to quantify aspects of the close relationship between the size of ringed seal and polar bear populations over large areas of the Canadian Arctic and the sensitivity of the dependence of polar bears on ringed seals, including the critical importance of young of the year. Fortuitously, we conducted a second population study of polar bears in the eastern Beaufort Sea in the mid-1980s that coincided with another major decline in ringed seal productivity and pup survival, which, though not as severe as that in the mid-1970s, had a marked negative effect on the polar bears. When evaluated along with less complete data from both the 1960s and the 1990s, it appears clear that, up to about the mid-1990s at least, periodic declines in the productivity of both ringed seals and polar bears, in association with heavy ice conditions, were a decadal scale event. These results underscore the risk involved with trying to sustain the maximum sustained yield of polar bears from a short-term study and the importance of taking a conservative approach to harvesting if financial constraints permit only periodic short-term re-assessments.

13.10.4 Challenges

To date, few agencies have been willing to fund long-term studies. However, addressing the effects of natural environmental fluctuations or climate warming on polar bears will only be possible using focused long-term studies on carefully chosen representative populations.

13.11 Conclusions

Several themes emerge from these narratives that are worth noting. First, few long-term studies set out with that goal initially, although we expect this is changing as the success of long-term studies becomes more apparent to both research and funding agencies. Second, each of these long-term studies have provided insights into species' biology and ecosystem structure and functioning that could not have been achieved otherwise. Third, establishing core measurements near the outset is a critical component of success, but being responsive to expanding routine measurements is also critical as new methods are developed and the questions asked change over time. Finally, sustaining funding for long-term studies can be challenging. In some cases, such as for the Northern right whale, the need for long-term research will be with us for decades and probably centuries to come, so new approaches to funding, such as through the establishment of 'trusts' or 'foundations', is probably required. Although asking interesting and important questions is critical, establishing multidisciplinary teams seems to provide one recipe for success.

14

Identifying units to conserve using genetic data

Barbara L. Taylor, Karen Martien, and Phillip Morin

14.1 The biology of structure and the role of genetics

Genetics has provided new tools to understand units to conserve (UTC) for marine mammals. Direct observations of processes important to the management and understanding of marine mammal biology are impossible or costly for most species. Although insights to marine mammal movements have been gained using photographic identification and tagging (see Chapter 2), particularly for pinnipeds, these tools are impractical for many species, particularly those with more pelagic ranges and/or higher abundance. The long generation times of marine mammals also present challenges because studies conducted over just a few generations may not be informative about questions involving longer time-scales. Marine mammal taxonomy lags behind work on terrestrial species because of issues with inaccessibility coupled with practical difficulties in maintaining collections of large skulls. Genetics, therefore, becomes a much more important tool for marine mammal scientists than for those studying terrestrial species, since genetic data often provide the only means to address pressing questions in a timely fashion.

Nevertheless, genetics are not always the right tool for the job. It is important to discriminate when genetic studies are most useful and when genetic studies have been applied properly. Like other fields, genetics has special terms that we indicate in bold defined in Box 14.1 at the end of this chapter. Although genetic technologies and analytical methods will continue to change, the foundation of the relationship between biology and genetic patterns and the basic principles of project design will not. Project design generally is optimized through the following steps: (1) frame the question; (2) choose the marker(s) with consideration of how life history affects the markers; (3) if possible, do power analysis to aid in estimating sample size and the number of markers required to address the question; (4) obtain/choose the samples; (5) conduct the laboratory analysis;

(6) rigorously assess data quality and return to step 5 as needed; (7) employ appropriate analytical methods; and (8) interpret and write-up results.

While each step is necessary, none is more critical than carefully framing the research question. This careful reflection on project design is especially important because samples often result as by-products of other research rather than as the primary goal of the research. We begin, therefore, by describing general properties of marine mammal biology and genetics that will provide the foundation for good genetic project design. Additional examples and interpretation of population genetic structure are given in Hoelzel *et al.* (2002).

Although genetic markers are covered in detail below, some basics are needed to understand how genetic patterns relate to marine mammal biology. There are two types of DNA that yield very different insights: mitochondrial DNA (mtDNA) and nuclear DNA (nDNA). mtDNA is uniparentally inherited from the mother and has some unique properties useful in marine mammal research. Marine mammals are all relatively long-lived with high survival rates and relatively low reproductive rates. Because a female produces few offspring, her fitness is increased by investing in maternal care. This care may extend over several years, allowing learned behaviours, ranging from feeding strategies to migratory routes, to be transferred to offspring (both male and female). These learned behaviours are reflected in the patterns of mtDNA. These mtDNA patterns are important for identifying some types of unit to conserve discussed in more detail in Section 14.2.3, Demographically independent populations.

Since nDNA is biparentally inherited, it therefore provides data on male and female **gene flow**, and thereby indicates the structure of the breeding population. Most traits subject to selection result from the expression of nDNA, making these markers appropriate for questions relating to population structure that is most important to evolutionary significance.

Inferences drawn from genetic data are strengthened by good sampling design, which in turn depends on considering the biology of the species in question. For example, obtaining samples during the mating season can greatly simplify interpretation of data for answering questions about population structure. Without such 'pure state' data, interpretation of data from samples obtained on a feeding ground, where individuals may be present from multiple breeding populations, may prove impossible. Even at a breeding site, careful consideration should be given to obtaining a random sample if strong site fidelity is present. For example, some otariid females may return to the beach of their birth. If the question is to compare population structure from one island to the next, then sampling should avoid obtaining all samples from a single beach at each island, as the level of relatedness on a single beach may be higher than between beaches within a single island. Because estimates of differences between islands are based on comparing within to between island frequencies, inappropriate sampling could produce biased results indicating stronger separation than is actually present.

14.2 Scale—units to conserve

Are the mammal-eating killer whales part of a single **lineage** or have they originated multiple times? Is the form of the Irrawaddy dolphin found off Australia a new species? How many demographically independent populations of coastal spotted dolphins are there along the central American coasts? Do all humpback whales in the North Atlantic go to a single breeding ground? Is it likely that declining Steller sea lion rookeries in the western Aleutians will be rescued by rookeries in Russia or the eastern Aleutians? All these questions have been addressed using genetic data. Most of these questions involve identifying units to conserve. However, an accurate understanding of population structure is critical to correctly estimate trends in abundance, to assess the impact of a particular mortality source like fisheries by-catch or whaling, or to model the role of marine mammals in an ecosystem. Framing the research question involves first identifying the scale, either spatial or temporal, that is of interest. We begin at the longest time and greatest spatial scale with taxonomic units and work toward the shortest time and smallest spatial scale with demographically independent populations.

14.2.1 Taxonomy

Laws and treaties, including the Convention on the International Trade of Endangered Species (CITES) and the red list—developed by the International Union for the Conservation of Nature (IUCN) and the US Endangered Species Act (ESA)—and similar domestic legislation, make an implicit assumption that scientists have already properly defined species and subspecies. However, new species of cetaceans are being named nearly every year, including the North Pacific right whale (Rosenbaum *et al.* 2000), Perrin's beaked whale (Dalebout *et al.* 2002), and the Australian snubfin dolphin (Beasley *et al.* 2005). These species' descriptions were based largely or solely on genetic data. This reliance on genetic data departs from previous practice in taxonomy (the naming of species and subspecies), which has primarily used morphology, especially skull measurements, to designate species.

Designation of a species without corroborating morphological data, as in the right whale case, remains controversial among taxonomists. This case demonstrates why traditional taxonomy lags behind and may not be appropriate for species' designations involving marine mammals: adequate skull collections are not, and often will not become, available. The precise definition of what constitutes a 'species' or 'sub-species' has long been a subject of debate. Two dozen 'species concepts' exist (Mayden 1997) that can result in widely varying numbers of species when applied to any given dataset (Sites Jr and Marshall 2003). This debate reflects the underlying reality that species (and subspecies) boundaries are inherently 'fuzzy' and are not amenable to a one-size-fits-all definition (Hey *et al.* 2003).

To promote more timely progress in cetacean taxonomy, a workshop was held in 2004 to develop and agree on new definitions and criteria for species and subspecies (Reeves *et al.* 2004). Workshop participants agreed that: (1) multiple species concepts should be acknowledged; (2) two congruent lines of evidence should be required to define a species, which would ideally incorporate both morphological and genetic data but could be based on two independent genetic markers; (3) only a single line of appropriate evidence should be required to define a subspecies and this may be a genetic marker; and (4) subspecies could be either a geographic form or an incipient species. Because these guidelines validate the use of genetic data in taxonomic designations, they pave the way for a more timely resolution of taxonomic uncertainty in many marine species for which substantial skull morphometric datasets are unlikely to be obtained in the foreseeable future.

14.2.2 Evolutionary significance

There are several types of units that are below the species level but are still likely to be following independent evolutionary paths: subspecies and units representing an important evolutionary component of the species. The latter units go by several names including evolutionarily significant units (ESUs) and distinct population segments (see USFWS 1996 for a full definition of DPS) under the US ESA and 'regional populations' by the IUCN (Gärdenfors *et al.* 2001). In this text we use the ESU term to broadly represent units needed to conserve the essential genetic variability for future evolutionary potential. Thus, the vision is to maintain sufficient evolutionary potential that species in the future can respond to environmental challenges through adaptation as well as they can today. The division between subspecies and the ESU-type unit is unclear given the new criteria for cetacean subspecies, as described above, but it seems likely that they differ by a matter of time elapsed since separation. For subspecies, the most closely related units have been separated long enough that morphological differences may accrue. ESUs, in contrast, are experiencing sufficiently low gene flow that local adaptation may occur. A rule of thumb often used to distinguish ESUs is that the level of gene flow between them is less than one disperser per generation (Gärdenfors *et al.* 2001).

14.2.3 Demographically independent populations (DIPs)

Demographically independent populations (DIPs) are units for which the internal dynamics within each group are more important to the group's maintenance than immigration from neighbouring groups. Units at this scale have been referred to by many names, including 'management units' (Moritz 1994), 'stocks', and 'population stocks' (Angliss and Wade 1997). DIPs are important for many management issues focused on shorter temporal scales, on the order of a human lifespan. Many marine mammals are highly social animals that form associations to exploit a marine habitat that varies over short time-scales of months to ones of decades. It is highly unlikely that individuals move about randomly like molecules and simply diffuse across space. Instead, a far more likely structure is that marine mammals operate like their terrestrial cousins and form herds that move according to their prey, predators, and needs to rear their young in a safe environment. For some species, maximizing fitness has resulted in concentrated breeding grounds thousands of miles from their feeding grounds, as is the case for North Atlantic humpback whales (Palsbøll *et al.* 1997). For other species, prey specialization has resulted in localized resident populations, as is the case for bottlenose dolphins in the Gulf of Mexico (Sellas *et al.* 2005).

Because structure at this level exists on a continuum, framing questions about DIPs is difficult for researchers. There is no one level of discreteness that applies to all situations. DIPs are relevant to both applied questions (e.g. are fishery kills too high for a local population (DIP) to sustain?) and academic questions (e.g. are Steller sea lion populations declining because of shifts in prey distribution?). It is rarely the case that researchers are interested in the question 'Is there population structure or not?', yet this is the question posed by many analytical methods. There are some general principles of how genetic differentiation relates to gene flow that assist researchers in framing questions relating to DIPs and assessing whether a genetics approach will provide useful insights.

DIPs are typified by a genetic pattern of differences in allele frequencies. The magnitude of genetic differentiation between two groups (F_{ST}) is often represented by the formula

$$F_{ST} = 1/(4N_e m + 1) = 1/(4N_e dT + 1) \text{ for nuclear makers, or:} \quad (14.1)$$

$$F_{ST} = 1/(2N_e m + 1) = 1/(2N_e dT + 1) \text{ for mitochondrial markers} \quad (14.2)$$

where N_e is the **effective population size**, d is the annual **dispersal** rate, T is **generation time**, and $m = d*T$. The value of F_{ST} ranges between zero, with no differentiation, to one, when there is no overlap between the strata compared.

This formula demonstrates how the large abundances and long generation times of many marine mammal species work to reduce the level of differentiation expected for a given dispersal rate. Consider, for example, comparing two strata (putative populations) with moderate abundances of 1000, an even sex ratio, and a generation time of 10 years with one individual dispersing per year. For nDNA, (eqn 14.1) $F_{ST} = 0.000025$ and for mtDNA (eqn 14.2) $F_{ST} = 0.0001$. The researcher who wants to detect population structure at this level needs to distinguish between $F_{ST} = 0$ (no population structure) and at best $F_{ST} = 0.0001$ (worthy of designation as a DIP). Both geneticists and managers need to be wary of studies attempting to detect population structure with small sample sizes because low statistical power may make interpretation difficult. In fact, if increasing sample sizes to appropriate levels are not possible, genetic studies may be inappropriate (see Section 14.4, Analytical methods).

Although F_{ST} is a useful parameter because it gives the researcher an ideal magnitude of differentiation, there are some important caveats. All models make assumptions that simplify from biological reality. One assumption that is particularly important to be aware of for marine mammals is the assumption that populations are in equilibrium. While all populations naturally fluctuate through time, most marine mammals have recently been hunted to small fractions of their normal range of abundances. There are two effects for which researchers should carry out sensitivity tests to guard against false conclusions. The first effect is a case for a population that is typically at high abundance, but has recently been severely reduced in number. This population will retain the genetic diversity from times of high abundance, and estimating statistical power to detect structure should use the historical effective population size. The second effect is a case for a population with a long generation time and sampling that is not random with respect to age. Rapidly growing populations will have cohorts that are more related to one another than would be the case in a stable population. The interpretation of genetic data when populations are out of equilibrium is difficult and analytical methods are still in their infancy. Since analytical methods are being rapidly developed, researchers should make a practice of seeking out new analytical methods that may be more appropriate for the case-specific conditions of their question.

In studies of marine mammals, researchers are often better able to detect fine-scale population structure with mitochondrial sequence data than with nuclear markers. There are several possible explanations for this pattern. First, because the mitochondrial **genome** is **haploid**, the rate of **genetic drift** in mitochondrial markers is double that of the **diploid** nuclear genome. This difference is reflected in the factor-of-two difference in the denominators of equations 14.1 and 14.2. Second, because only females pass their mitochondrial genome on to their offspring, the effective population size for the mitochondrial genome is half that of the nuclear genome (assuming an even sex ratio and random mating). Thus, the haploid, uniparentally inherited nature of the mitochondrial genome means that, for a given level of dispersal, the expected degree of differentiation for mitochondrial markers is four times greater than for nuclear markers.

Finally, male-biased dispersal or male-mediated gene flow can result in stronger differentiation at mitochondrial markers than nuclear markers. Due to its strictly maternal inheritance, the mitochondrial genome only reflects the population structure in the female component of the population. Thus, if females exhibit a high degree of site fidelity while males are more cosmopolitan, genetic structuring may be present in the mitochondrial genome but not in the nuclear genome. Similarly, the maternal transmission of mtDNA will affect patterns of differentiation in some highly **migratory** species. For instance, humpback whales exhibit strong, maternally transmitted fidelity to high-latitude feeding grounds, resulting in strong mitochondrial differentiation among those feeding grounds (Palsbøll *et al.* 1997). However, because whales from multiple feeding grounds converge on the same breeding ground, females are likely to breed with males from different feeding grounds, resulting in mixing of the nuclear genome (Palsbøll *et al.* 1997).

The North Atlantic humpback whale example nicely illustrates the utility of different markers to questions at different scales for units to conserve (UTCs). Consider a case where whales in the Gulf of Maine are being killed by entanglement in fishing gear and by ship strikes. In addressing the question of whether the local 'population' can sustain this mortality, it would be inappropriate to count whales from other discrete feeding grounds (like off Greenland) into mortality rate calculations because whales from other feeding grounds do not recruit into the Gulf of Maine feeding aggregation at demographically significant levels. Thus, mtDNA is both necessary and sufficient to address this question. Using nDNA would be neither necessary nor sufficient; and in this case would give an incorrect picture of structure, as whales from the demographically discrete feeding grounds could not be distinguished because of mixing of the nuclear genome on the breeding grounds. Thus, based on genetic criteria alone, whales from Greenland and the Gulf of Maine belong to different DIPs but the same ESU. The ESU question requires nDNA markers.

Some questions can only be addressed with nDNA. For example, consider a case where whaling is occurring on a feeding ground that has whales from multiple breeding grounds. If one breeding population is small, and hence their recovery could be delayed by excessive whaling, managers would need to know which breeding population the harvested whales were coming from. Such a determination would require the use of an assignment test, in which animals of unknown origin are assigned to populations on the basis of their genetic make-up. Because mtDNA is a single non-recombining unit with limited variability, most breeding populations have many shared mtDNA **haplotypes**. Consequently, it is generally not possible to identify the breeding population origin of samples using mtDNA. With nDNA, there is the possibility of using many independent markers to increase variability, thereby increasing the precision of population assignments.

14.3 Genetic markers

The toolbox of genetic markers continues to grow with the advances in genomic technologies and the availability of whole genome sequences. Some methods and markers (such as **allozymes** and **DNA fingerprinting**) have lost their appeal because of technical limitations or replacement by higher quality or more broadly applicable methods. We will focus on the markers that are currently most commonly used, or show the greatest promise for application, to detect genetic variation from the individual to the species level, all of which can be useful in defining units to conserve. Other reviews of genetic markers and applications contain additional information for application to molecular ecology questions that are not addressed here (e.g. Beaumont and Nichols 1996; Sunnucks 2000; Luikart *et al.* 2003). In the following sections, we will discuss the benefits and limitations of markers for each level of variation of interest for defining UTCs. Analytical methods to identify structure at various levels assume that markers are 'neutral'.

Thus, we do not discuss markers that code for actual proteins and therefore are subject to selection and focus on non-coding markers (though this does not ensure that they are selectively neutral). A summary of the applicability of different markers to each level of investigation is given in Table 14.1.

14.3.1 Mitochondrial DNA sequencing (mtDNA)

Mitochondrial DNA has been the primary molecular marker for **phylogenetics** and phylogeography for over 20 years. Several characteristics of mtDNA make it highly suitable for these types of studies:

- High copy number in cells, so that it can be easily amplified by **PCR** from even very poor samples.
- Maternal haploid inheritance with no recombination. Haploid status means that the effective population size of mtDNA is 25% the effective population size of nDNA, resulting in faster **lineage sorting** of haplotypes in populations. The smaller effective population size allows mtDNA to more rapidly resolve to **monophyly** after divergence of populations. Lineage sorting generally resolves to monophyly in $4N_e$ generations (Avise *et al.* 1987; Avise 1989), where N_e is the effective population size.
- Availability of PCR primers that are widely applicable across taxa for amplification of well-characterized portions of the mtDNA genome, including the non-coding control region (Ross *et al.* 2003), cytochrome b (Kocher *et al.* 1989), and the 'DNA Barcode' region of cytochrome c oxidase subunit I (CO-I) (Glaubitz *et al.* 2003; Waugh 2007).

Table 14.1 *Appropriate use, difficulty, and cost of molecular techniques. The number of Xs in each cell indicates the level of appropriateness.*

	mtDNA sequence	Microsatellite	SNP	AFLP	Nuclear locus sequence
Taxonomy	XXX	X	XX	XXX	XXX
ESUs	XXX	XXX	XXX	XXX	X
DIPs	XXX	XXX	XX	XXX	NA
Assignment tests	X	XXX	XX	XXX	NA
Individual ID, relatedness	NA	XXX	XX	XX	NA
Historical DNA	XXX	X	XXX	NA	NA
Difficulty	Low	Moderate	Low	*High	Low–moderate
Cost	Low	Moderate	**Low–moderate	Low	Moderate–high

Key: XXX, highly appropriate; XX, appropriate; NA, not appropriate.

* Initial methods development and optimization can be very labour-intensive, but subsequent typing can be rapid and easy.

** Cost depends on the number of SNPs used, **ascertainment,** and genotyping methods.

Mitochondrial DNA is especially useful for population structure at the species, ESU, and DIP levels, as the rate of mutation fixation and allele frequency differences, due to complete or partial reproductive isolation of populations resulting from genetic drift, are greater for mtDNA than for nuclear loci. For many mammalian species, sex-biased dispersal also favours differentiation of mtDNA among populations, when females are the non-dispersing sex.

Despite its wide use and utility, mtDNA also has some limitations. The relatively high mutation rate, especially in the non-coding control region, can lead to **homoplasy** that decreases the signal among more divergent taxa (Galtier *et al.* 2006). In addition, fragments of the mitochondrial genome frequently (in evolutionary time) become incorporated into the nuclear genome, creating copies that evolve at a different rate than the mtDNA, and which can confound analysis (Sorenson and Fleischer 1996; Bensasson *et al.* 2001). Finally, the fact that the whole mitochondrial genome still only represents a single non-recombining locus means that it represents only a single gene lineage, or **gene tree**, and may not represent the true phylogeny of the taxa. In some cases, lineage sorting of mtDNA can be at odds with the true taxonomy, especially with more widely divergent taxa (Murphy *et al.* 2001).

14.3.2 Microsatellites

Microsatellites are a class of highly repetitive and variable nDNA loci that are widely distributed in almost all genomes. They consist of tandemly repeated copies of one to six nucleotides, with the number of copies varying between alleles in the population. Most microsatellite loci used for population studies are di-, tri-, or tetra-nucleotide repeats, as these are the most variable and easiest to score (genotype). Several review papers describe the characteristics, strengths, and weaknesses of microsatellites (Jarne and Lagoda 1996; Schlötterer and Pemberton 1998).

The great benefit of microsatellites as nuclear markers is that they are typically highly **polymorphic** in populations, providing a high level of statistical power for individual identification (Herraeza *et al.* 2005; Seddon *et al.* 2005), paternity (Morin *et al.* 1994; Fung 2003), and population structure analyses (Luikart and England 1999) with relatively few loci, typically between 10 and 20. Microsatellite loci are often also conserved across taxa, so PCR primers designed to amplify a locus in one species will often work in closely related species (and sometimes across widely divergent taxa, e.g. S.S. Moore *et al.* 1991).

While by far the most widely used nuclear marker for analyses from the individual to the species level, microsatellites have several limitations that can be significant, and careful consideration should be made prior to starting a project. Microsatellite **genotypes** are obtained by PCR amplification followed by **electrophoresis** and **allele** size inference. Scoring of alleles can have many pitfalls, potentially resulting in population biases that can affect results (van Oosterhout *et al.* 2004; Hoffman and Amos 2005; Morin *et al.* 2009a). Comparison of genotypes across laboratories, technologies, or even just time can be difficult, as allele size inferences can vary with the physical and chemical conditions during electro-phoresis (LaHood *et al.* 2002; Davison and Chiba 2003). Finally, the mutation rates and patterns vary greatly among microsatellites, making application of an appro-priate model of evolution difficult (Luikart and England 1999).

14.3.3 Single nucleotide polymorphisms (SNP)

Single nucleotide polymorphisms (SNPs) have only very recently been introduced to the fields of **systematics**, population genetics, and molecular ecology, but show great promise for accurate, efficient nuclear genotype data that provide some technical and analytical benefits over microsatellites (Brumfield *et al.* 2003; Morin *et al.* 2004). SNPs are simply mutations at individual nucleotide positions in the DNA sequence, and are the most abundant type of variation in the genome, being found typically every few hundred nucleotides, on average, in mammalian genomes (Aitken *et al.* 2004; Morin *et al.* 2004, 2007). This makes SNP discovery relatively easy (Aitken *et al.* 2004; Morin *et al.* 2004). Higher numbers of SNPs are needed relative to microsatellites because statistical power is related to the number of independent alleles in the whole set of markers, and SNPs are most often bi-allelic (Chakraborty *et al.* 1999, Krawczak 1999; Kalinowski 2002). For this reason, estimating the number and types of SNPs needed is an important first step in any study (Ryman *et al.* 2006). Recent applications of SNPs have shown that they can be used for individual identification, parentage, and population assignment (Herraeza *et al.* 2005; Seddon *et al.* 2005; Rohrer *et al.* 2007; Morin *et al.* 2009b), population structure ('Turakulov and Easteal 2003; Seddon *et al.* 2005; C.T. Smith *et al.* 2005, 2007; Rosenblum and Novembre 2007; Narum *et al.* 2008), phyloge-netics (Shaffer and Thomson 2007), and studies of historical samples such as museum collections of bone and baleen (Morin and McCarthy 2007).

SNPs are still in their infancy relative to mtDNA and microsatellites, especially for applications to natural populations, so there is much to learn about their benefits and limitations. We know that individual SNPs provide far less informa-tion than individual microsatellite markers for most applications, and this may limit their applicability for fine-scale population structure analysis (Morin *et al.* 2009b) and relatedness.

Methods are still being developed for more rapid and efficient SNP discovery, genotyping, and data analysis. SNPs have become the marker of choice for studies of humans, medical genetics, and economically important or model organisms, so adaptation for their use in conservation research of natural populations will no doubt follow quickly.

14.3.4 Amplified fragment length polymorphisms (AFLP)

AFLP methods have been around for many years, and are widely used for genetic diversity studies of plants and agriculturally important animals. The AFLP method makes use of **restriction endonucleases** to digest genomic DNA into small fragments, and then preferentially amplifies a small portion of those fragments, some of which are variable within and among populations (Bensch and Akesson 2005). The great benefit of this method is that it is universally applicable to DNA from any organism without the need to design species-specific PCR **primers**. AFLPs are also able to generate a substantial number of variable loci with relatively few primer combinations. These characteristics have made AFLPs attractive for analysis of relatedness (J. Wang 2004), population structure, and phylogenetics (Giannasi *et al.* 2001; Kingston and Rosel 2004).

Despite the apparent benefits of AFLPs, they have not been widely used for animal studies (Bensch and Akesson 2005). One reason is the type of data generated; AFLP 'alleles' are **dominant**, so that the allelic state is unknown (i.e. each **locus** is scored as present or absent, so a **homozygote** for the present allele cannot be distinguished from a **heterozygote**). This means that, although the method produces data from many loci, the loci are less informative than **co-dominant** markers. The data are also nearly impossible to compare across studies, as all the data essentially need to be generated at the same time and with the same conditions to allow the many alleles to be compared and correctly sized. The quality of the data can also be highly dependent on DNA quality and strict adherence to laboratory and allele scoring protocols. While the methods are, in principle, straightforward, the robust application of them has often proven difficult.

14.3.5 Nuclear locus sequencing

To date, direct sequencing of multiple nuclear loci has been used primarily for phylogenetic studies (Palumbi and Baker 1994; Murphy *et al.* 2001). Sequencing of multiple individual loci from multiple individuals remains an expensive technology, both in terms of the laboratory work and the time it takes to analyse the data. Researchers must also know enough about the genes and genomes of the organisms studied to design PCR primers for each locus. One way to do this in the absence of whole genome projects for the target species is to use conserved primers designed based on the alignment of genes from multiple divergent species. This has been done for cetaceans and other mammals using the method of '**exon** priming, **intron** crossing' (EPIC; Palumbi and Baker 1994), also called 'comparative anchor tagged sequences' (CATS: Lyons *et al.* 1997; Aitken *et al.* 2004). The benefit of direct sequencing is that all variant sites can be detected for the phylogenetic information content of each locus (as opposed to genotyping one or a few SNPs in a locus). Most researchers have opted for less information from each locus (e.g. SNPs, AFLPs) while sampling more loci.

As with most genetic methods, however, sequencing technologies are changing rapidly. New 'next generation' methods are extremely high throughput, generating over a million base pairs of sequence from a single experiment (Margulies *et al.* 2005). Initial applications of this technology have been for rapid sequencing of whole genomes or transcriptomes (the transcribed genes), but it is also amenable to highly parallel sequencing of smaller sets of genes (Meyer *et al.* 2008). While it may be a while before the costs come down sufficiently to make this approach practical for population level studies, it is likely that phylogenetics will benefit quickly. The question will be, how many genes are needed to obtain the true phylogeny (Gatesy *et al.* 2007)?

14.4 Analytical methods

As outlined above, there are many types of units to conserve that are of interest to conservation geneticists. Because these different units differ greatly in the expected magnitude of genetic differentiation, it is important to choose an analytical method appropriate to the unit that you need to define. In this section, we review the genetic patterns that are expected for each of the UTCs described above, and the analytical methods that are useful for detecting those patterns. Genetic analytical methods are rapidly evolving, with new approaches and critiques of existing methods published frequently. Consequently, we focus on the general types of analyses that are most useful in identifying UTCs at different levels rather than attempting to review specific methods. We conclude by discussing the importance of assessing the strength of an analysis for inferring population structure.

14.4.1 Choosing methods to match questions

Taxonomy

Analytical approaches to identifying taxonomic units are reviewed in Sites and Marshall (2003). Approaches can generally be categorized as tree-based and non-tree-based. Tree-based approaches consist of using a dataset to construct a phylogenetic tree and making inferences based on the topology of the resulting tree. Non-tree-based approaches rely on patterns in the genetic distance within and among hypothesized taxa and the identification of diagnostic characters. Kingston and Rosel (2004) used a combination of tree-based and non-tree-based approaches to examine genetic differentiation in pairs of sister taxa in the genera *Delphinus* and *Tursiops*.

In the past decade, considerable attention has been devoted to the development of **Bayesian** approaches to the reconstruction of phylogenetic trees (reviewed in Ronquist 2004). These powerful methods allow explicit incorporation of uncertainty into phylogenetic analyses. The availability of increasingly large datasets from ever-increasing numbers of taxa has driven the development of new analytical methods capable of dealing with the size and complexity of datasets now available. The 'supertree' (Sanderson *et al.* 1998; Bininda-Emonds 2004) and 'supermatrix' (Gatesy *et al.* 2004; de Queiroz and Gatesy 2007) approaches allow diverse datasets to be integrated into a single phylogenetic analysis. These approaches may eventually be able to provide resolution in groups such as the *Stenella–Tursiops–Delphinus* complex, which has long been problematic (LeDuc *et al.* 1999).

ESUs

The requirement that ESUs be evolutionarily distinct means that they will be strongly genetically differentiated. Therefore, many of the tree-based methods that are used for identifying taxonomic units are often applied to the identification of ESUs. ESUs often display a **phylogeographic signal**. However, species with very high genetic diversity, and those in which population structure has changed in the recent evolutionary past, may not exhibit a strong phylogeographic signal even when multiple ESUs are contained in a sample set.

Moritz (1994) suggested that in order for two groups to be considered separate ESUs, their mitochondrial sequences should exhibit a pattern of **reciprocal monophyly**. Again, while the presence of reciprocal monophyly between geographically, morphologically, or ecologically distinct groups can provide strong support for the designation of ESUs, the lack of this pattern does not mean that management as separate ESUs is unwarranted. Reciprocal monophyly develops as a result of lineage-sorting between two reproductively isolated groups. The rate at which lineage-sorting occurs is affected by the effective population size and generation time (Avise *et al.* 1987; Avise 1989). Because many marine mammal populations have very large abundances and some, particularly the large whales, have long generation times, complete reciprocal monophyly may take millions of years to develop, even between groups that are completely reproductively isolated. In fact, some currently recognized species of cetaceans are not reciprocally monophyletic (e.g. *Delphinus delphis* and *D. capensis*; LeDuc *et al.* 1999; Kingston and Rosel 2004). Of course, complete reproductive isolation is not expected between ESUs. Even the very low levels of gene flow expected between these units can be sufficient to prevent reciprocal monophyly from developing.

Bayesian clustering algorithms can be very powerful tools for identifying ESUs. These algorithms typically use multilocus genotypes to cluster samples into groups. Because they work at the level of the individual, these methods do not require the a priori stratification of samples nor any a priori hypotheses regarding the locations of population boundaries. Thus, these methods can be particularly valuable in cases where a species is continuously distributed over a large geographical area with few obvious barriers to movement, as is the case for many marine mammals. The development of Bayesian clustering algorithms is a rapidly developing research area, with new methods being introduced every year. While older methods made very limited use of spatial data (e.g. 'BAPS' (Corander *et al.* 2003, 2004), 'PARTITION' (Dawson and Belkhir 2001), 'STRUCTURE' (Pritchard *et al.* 2000; Falush *et al.* 2003)), many newer 'geographically constrained' methods use information regarding the spatial location of samples to constrain the clustering algorithm so that the final groups exhibit some degree of geographic cohesion (e.g. 'TESS' (Chen *et al.* 2007), 'GENECLUST' (Francois *et al.* 2006), 'GENELAND' (Guillot *et al.* 2005)). Because these geographically constrained methods are more likely to produce groups that are geographically contiguous, they may be more easily applied to management questions.

Bayesian clustering algorithms represent a subset of methods from a rapidly evolving discipline known as 'landscape genetics' (Manel *et al.* 2003). One of the primary aims of landscape genetics is the detection of genetic discontinuities, making many of its methods well suited to the definition of ESUs (and, to a lesser extent, DIPs). Although new methods and software are being developed every year, many of the available landscape genetic methods are included in the software package Alleles in Space (Miller 2005).

DIPs

The lower levels of genetic differentiation expected for DIPs make them far more challenging to detect using currently available analytical methods. Bayesian clustering algorithms hold some appeal, but performance analyses indicate that they will often be unable to detect the low levels of genetic differentiation expected for some DIPs (Latch *et al.* 2006; Waples and Gaggiotti 2006; Chen *et al.* 2007). Other landscape genetic methods for detecting genetic discontinuities are also well-suited to the task of defining DIPs (e.g. Monmonier's algorithm (Monmonier 1973), 'Wombling' (Crida and Manel 2007)), but have not been adequately tested to see how they will perform in this context.

To date, the most common analytical approach to defining DIPs is traditional null hypothesis testing. Hypothesis testing has been subject to growing criticism in the last decade. It requires the researcher to define hypothesized units in advance, a step that is often highly subjective and can greatly influence results (Taylor and Dizon 1999; Martien and Taylor 2003). Statistical power to detect genetic differentiation is often low, but is rarely estimated (Avise 1995; Taylor and Dizon 1996; Taylor *et al.* 1997; Martien and Taylor 2003). There also remains considerable uncertainty and disagreement regarding how and when to account for the impact of conducting multiple hypothesis tests (Perneger 1998; Ryman and Jorde, 2001; Moran 2003; Nakagawa 2004). The choice of α (the value at which a p-value is considered statistically significant) is arbitrary, and represents a hidden policy decision regarding the relative importance of Type I errors (concluding that there is population structure when there really is not) and Type II errors (failing to detect population structure that really is present) (Taylor and Dizon 1999).

Despite these drawbacks, hypothesis testing remains the primary tool for identifying DIPs. Statistics for estimating genetic differentiation in a hypothesis test generally fall into two categories—frequency-based statistics and distance-based statistics. Frequency-based statistics rely exclusively on allele or haplotype frequencies to estimate genetic differentiation. Distance-based statistics take into account both allele/haplotype frequencies and the evolutionary relationships between alleles/haplotypes. Thus, with a distance-based statistic, two strata whose haplotypes differ by an average of only one or two nucleotides will be considered less differentiated than two strata whose haplotypes differ by an average of three to five nucleotides, even if the differences in haplotype frequencies are comparable between the two pairs of strata. Despite the fact that they seem to make better use of the available information (by taking account of both frequency data *and* evolutionary relationships), distance-based statistics tend to perform more poorly than frequency-based statistics for detecting low levels of genetic differentiation (Hudson *et al.* 1992; Goudet *et al.* 1996). An exception is when sample size is very small relative to genetic diversity (Chivers *et al.* 2002). In these cases, many alleles/haplotypes will be represented by a single individual and will therefore contribute no information to a frequency-based statistic. Reviews of the statistical power of different measures of differentiation under a variety of circumstances are provided by Hudson *et al.* (1992) for mtDNA and Goudet *et al.* (1996) for microsatellite markers.

The critical parameter for determining whether two putative populations are demographically independent is the annual rate of dispersal between them (d in equations 14.1 and 14.2 above). While a finding of statistically significant genetic differentiation can be used to infer a low dispersal rate, a more direct approach is to use genetic data to estimate dispersal rates. Estimates of dispersal rates, along with their uncertainties, are far more useful to managers and decision makers than the results of a hypothesis test (Taylor and Dizon 1999). They can easily be brought into a formal risk-analysis to evaluate the relative costs and benefits of various management actions. They can also be incorporated into population dynamics' and genetics' models, thereby informing other aspects of research on species that are of management concern.

Estimating dispersal rates remains an analytical challenge. Several Bayesian and maximum likelihood methods have been developed in recent years (Beerli and Felsenstein 1999, 2001; Fisher *et al.* 2002). However, they can be computationally prohibitive and have been found to perform poorly in some instances (Abdo *et al.* 2004). Nonetheless, as new methods are developed and computers continue to get faster, estimation of dispersal rates is likely to become the standard in defining DIPs.

14.4.2 Assessing the strength of inference

Attempts to define units to conserve are often plagued by a mis-communication between researchers and managers. All too often, the manager wants an answer to the question, 'Are there multiple units-to-conserve in this geographic region?', while the researcher sets off to address the question, 'Can I detect genetically distinct groups in this geographic region?' These two questions do not always have the same answer. There are many reasons why a geneticist might fail to accurately define UTCs. Awareness of these potential pitfalls can help researchers and managers alike to interpret the results of genetic studies better and use them to define UTCs.

Many studies of population structure are hampered by inadequate sample size or inadequate sample coverage. This can be particularly problematic in studies of marine mammals, due to the difficulty of obtaining samples in a marine environment. Studies based on opportunistically collected samples often end up with sample distributions that better reflect the distribution of people (e.g. fishermen collecting samples from by-caught animals, beachcombers happening across stranded animals, etc.) than the distribution of the study species (Chivers *et al.* 2002). In species that exhibit social structure, inclusion of multiple samples collected from a single group could result in disproportionate representation of related individuals in some strata, which could bias the results of tests of genetic differentiation. In these cases, it may be necessary to exclude samples from potential relatives from some strata, further reducing the sample size. For pelagic species that rarely come into contact with people, collecting adequate samples for even phylogenetic studies may be extremely challenging.

The probability of correctly identifying population structure with traditional hypothesis tests can be assessed through power analysis (Martien and Taylor 2003; Ryman and Palm 2006; Morin *et al.* 2009b). Estimating statistical power is computationally burdensome. Faster computers and the development of new, user-friendly programs for estimating power will likely result in power analyses becoming more commonplace. However, in order to calculate statistical power, the researcher must specify a specific alternate hypothesis. In other words, the researcher must specify the level of differentiation s/he is trying to detect. This remains an important stumbling block to the useful application of power analysis (Morin *et al.* 2009b).

Many of the analytical methods outlined above require the a priori stratification of samples. If samples can be stratified in advance into groups that accurately reflect the underlying population structure, the likelihood of detecting genetic structure between those groups is much higher than if the analysis has to rely on methods that do not require stratification. If samples are stratified incorrectly, however, that error will propagate through and affect the entire analysis (Martien and Taylor 2003). Such an error could result in the definition of UTCs that does not accurately reflect underlying population structure or, worse, failure to define UTCs at all. The problem of a priori stratification can be partly dealt with by examining the sensitivity of results to stratification, or by starting with small strata and relying on careful consideration of patterns of significant and non-significant test results, to guide the grouping of strata into UTCs (e.g. O'Corry-Crowe *et al.* 2006). The sensitivity of results to the inclusion of data from stranded animals should also be examined, as the stranding location of an animal may be poorly correlated with its home range while alive.

DIPs remain the most challenging UTC to detect using genetic data. Most genetic analytical methods are developed by academic scientists for evolutionary studies. While these methods are likely to prove very useful for clarifying taxonomy and identifying ESUs, most are unlikely to detect the relatively high (from an evolutionary perspective) levels of gene flow typical of DIPs (Waples and Gaggiotti 2006). Because of its focus at the ecological rather than evolutionary scale, the emerging field of landscape genetics holds promise for the development of analytical tools appropriate for the definition of DIPs. It is critical, however, that new methods be subject to rigorous performance testing prior to their application in a management setting. Although such performance testing is becoming more common (e.g. Abdo *et al.* 2004; Latch *et al.* 2006; Waples and Gaggiotti 2006; Chen *et al.* 2007), methods are rarely tested at the levels of differentiation typical of DIPs. A new project, known as TOSSM (testing of spatial structure methods) aims to fill this gap by testing the performance of methods in a management context (IWC, 2004; Martien *et al.* 2009). Results of the TOSSM project and other studies like it will be critical in the development of analytical tools that enable us to make full use of genetic data to define UTCs.

Box 14.1 Genetics' definitions

Allele: One of two copies of a gene or locus in a diploid genome. One allele is inherited from each parent.

Allozymes: Protein variants that cause them to migrate differently in gel electrophoresis, or to stain differently based on enzymatic activity.

Ascertainment: The process of finding and selecting polymorphic loci for use in population analysis. Ascertainment bias results from use of a limited number of genetically restricted populations or samples, and can bias the outcome of genetic studies.

Bayesian: A theory of statistics that estimates the probability of events. A genetics example would be using genetic data from putative populations to estimate the probability of different rates of dispersal (usually represented by a probability distribution over the range of plausible dispersal rates).

Co-dominant: Genetic markers for which both alleles in a sample can be detected, resulting in genotypes that may be homozygous for either allele, or heterozygous.

Diploid: Having two sets of chromosomes, one inherited from each parent.

Dispersal: Emigration and reproduction of individuals outside of the population in which they were born.

DNA fingerprinting: The use of combined genotypes from many loci to generate a multilocus genotype that is highly likely to be unique to an individual.

Dominant: Genetic markers for which only one allele can be determined, the other being recessive or undetected, though present. Genotyping of dominant markers results in only presence/absence genotypes.

Effective population size (N_e): For nuclear DNA, roughly the number of reproducing adults in the population, but is reduced if the sex ratio is unequal. For mtDNA, N_e is roughly the number of reproducing females in the population. Both are affected by variance in individual reproduction. A genetics text will give several formulae that may be applied depending on available data, including formulae to apply if temporal data are available.

Electrophoresis: The separation of DNA or proteins by size and electrical charge. The DNA or protein is placed at one end of a matrix or polymer and exposed to an electrical current that moves it through the matrix, with smaller fragments moving faster than larger ones.

Exon: Any region of DNA within a gene that is transcribed to the final messenger RNA (mRNA), which is, in turn, translated into a protein. Exons are often, but not always, interspersed with introns within genes.

Gene flow: The movement of genetic material among populations through dispersal of individuals.

Gene tree: A tree-like figure consisting of branches that connect individuals, showing the evolutionary relationships or similarity among those individuals. Basal branches represent more ancestral relationships.

Generation time: In the context of genetic drift, generation time is the average age of parents.

Genetic drift: The process of changing allele frequencies due to random inheritance and loss of alleles in a finite population. The rate of genetic drift between populations is affected by the rate of mutation, the effective population size, and time since divergence.

Genome: The collective genetic material cells or mitochondrial organelles of an individual. The nuclear genome consists of the diploid chromosomes inherited from both parents in sexually reproducing organisms, and the mitochondrial genome consists of the circular chromosome found in the mitochondria within each cell.

Genotype: The designation of alleles for a locus. This usually refers to the combination of two alleles at each locus in an individual, but can also refer to the presence or absence of an allele for dominant loci.

Haploid: Having only one copy of a genome. The nuclear genome is only haploid in gametes, but the mitochondrial genome is always haploid.

Haplotypes: An identity given to the sequence or genotype of a haploid genome (an allele of a locus in a diploid genome).

Heterozygote: An individual locus containing copies of two different alleles.

Homoplasy: Repeated mutation at a site in DNA resulting in alleles that are identical in state but not identical by descent.

Homozygote: An individual locus containing two copies of the same allele.

Intron: Portions of genes that do not code for part of the gene product (protein), and that are not present in the final messenger RNA (mRNA) after the exon segments have been spliced together.

Lineage: For mtDNA, this refers to the lineal transmission of an mtDNA haplotype through a group of organisms. It can refer to any lineal descent of individuals or taxa.

Lineage sorting: The process by which reciprocal monophyly develops between two reproductively isolated groups. Bias in estimating dates of separation can occur when small populations are founded from a large diverse population and subsequently lose diversity and become fixed for the marker being used. Without knowing the population history, the genetic differences between the small populations will be falsely interpreted to represent the time taken to evolve such lineages.

Locus: A region of DNA defined by characteristics of the region (e.g. a gene), or by molecular markers (e.g. PCR primers for a microsatellite or SNP).

Migratory/migration: Seasonal movement between geographic areas used for breeding and feeding, or sometimes using different feeding grounds in different seasons. Geneticists sometimes incorrectly use 'migration' to mean 'dispersal' or 'gene flow' and therefore use the symbol 'm' in formulae (see equations 14.1 and 14.2).

Monophyly: All members of a taxa share a common ancestor to the exclusion of other taxa. This can apply to any character being used for phylogenetic analysis.

PCR: Polymerase chain reaction: a laboratory method for replicating a specific segment of DNA *in vitro* by the use of two PCR primers, DNA polymerase, and the building blocks of DNA, to cause an exponential increase in the number of copies of the DNA segment.

Phylogenetics: The study of the evolutionary relatedness among taxa, using genetic characters.

Phylogeographic signal: A pattern in which samples that are collected from the same geographical region also appear close to each other on a phylogenetic tree.

Polymorphic: Having two or more alleles in a population.

Primers: Synthetic short fragments of DNA used in PCR to specify the segment of a genome to be replicated.

Reciprocal monophyly: A group of taxa (species, subspecies, or populations) is considered reciprocally monophyletic when all members of each taxon share a most common recent ancestor that is not ancestral to any other taxa.

Restriction endonucleases: Enzymes commonly found in all organisms that function to splice or degrade DNA by cutting it into fragments at specific DNA sequence recognition sites. They are extracted from bacteria for use in laboratory methods that require DNA fragmentation.

Systematics: The study of the relationships among taxa through evolutionary time.

15

Approaches to management

John Harwood

15.1 Introduction

One of the definitions of management provided by the *Oxford English Dictionary* is 'the conservation and encouragement of natural resources such as game, fish, wildlife'. This captures the fact that the management of any living resource involves a compromise between ensuring the continued existence of the resource (the 'conservation' aspect) and obtaining an economic benefit from it (the 'encouragement' aspect). Historically, the economic benefits from marine mammal populations were usually realised by killing animals and processing their carcasses. These benefits rely on the fact that a population which has been reduced from its pristine, unexploited abundance will have a positive growth rate, and this growth can be harvested. The theory behind this is described in more detail in Section 15.2. More recently, economic benefits have been obtained by charging tourists to view marine mammals at close quarters. Unlike the benefits from harvesting, the benefits from this ecotourism are not directly related to the size of the population. In addition, indirect economic benefits may be obtained as a result of a change in the abundance of a marine mammal if, for example, this results in an increase in the abundance of another species that is commercially exploited.

The existence of a number of marine mammal populations has been jeopardized either as a direct consequence of some form of deliberate or incidental overkill, such as the vaquita, *Phocoena sinus*, in the Gulf of California (D'Agrosa *et al.* 2000), or as a result of environmental change (possibly the western stock of the Steller sea lion, *Eumetopias jubatus*, in the Bering Sea and North Pacific Ocean, Benson and Trites 2002). In such cases, the focus of management is almost entirely on ensuring the population's recovery; any economic benefits that result from management are of minor concern. Chapter 16 describes the techniques that can be used to manage these populations. In this chapter, I will focus on the management of populations that are still sufficiently large that they can provide economic benefits. It begins with a description of the theory that underpins the sustainable use of living resources, and then proceeds to a brief history of the commercial harvesting of marine mammals. Most of this harvesting was not managed in any conventional sense, but the success, and more frequent failures, of different industries provide valuable insights into the ways in which exploitation can place populations in jeopardy. Subsequent sections describe these insights and the attempts that have been made to develop approaches that can achieve a lasting balance between the need to conserve marine mammal populations and the desire to obtain economic benefits from them.

15.2 Sustainable use and the importance of economic factors

The theory underpinning the sustainable use of a marine mammal population is straightforward: any reduction in the size of a population should make more resources available to the surviving individuals, and this should be reflected in increased growth, reproduction, and survival. The reduced population should therefore tend to increase in size, and this potential growth could, in principle, be harvested on a continuing basis without jeopardizing the population.

The magnitude of this sustainable harvest, or yield, will be determined by the potential growth rate of the population and the number of individuals in the population. It will have its maximum at a population size intermediate between the pre-exploitation level and extinction. The actual harvest will be determined by the amount of effort put into harvesting the resource. In the case of marine mammals, appropriate measures of effort include the number of catcher boats involved in commercial whaling, or the number of people taking part in a seal hunt. Figure 15.1a shows how equilibrium population size and harvest vary with the level of effort for a hypothetical population in which the maximum sustainable yield (MSY) occurs at a population that is half of the pre-exploitation level. Taylor and DeMaster (1993) suggest that, for most marine mammal species, MSY is likely to occur at 50–85% of the initial population size.

However, yield is not the only component of sustainable use. Clark (1976) has shown that we can only properly understand the process of exploitation if we consider economic factors as well as biological ones. The profitability of an industry is determined not only by the value of the resource it exploits but the costs of harvesting and processing that resource. These costs will inevitably rise as effort increases. Figure 15.1b shows how the profitability of an industry varies if there is a simple linear relationship between effort and cost. The maximum profit occurs at a lower level of effort, and therefore a higher population size, than that providing the MSY. Unfortunately, this rather desirable outcome is only likely to occur if there is sole ownership of the resource. If access to the resource is not controlled (a situation known as open access), total levels of effort will inevitably rise well above those that provide maximum profits. New companies can, and will, enter the industry because they can still make a profit. Eventually total effort reaches the 'bionomic equilibrium' (Clark 1976), at which point

(a)

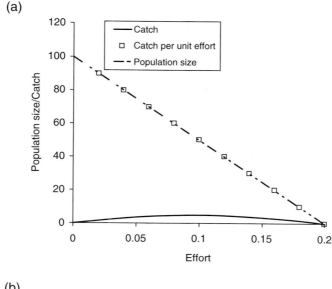

(b)

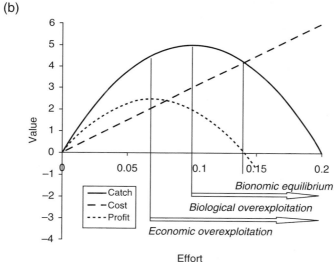

Fig. 15.1 (a) The relationship between the effort expended by an industry in catching marine mammals, the actual catch, and the equilibrium size of the marine mammal population. (b) An expanded version of the lower part of Fig. 15.1a showing the implications of the cost of exploitation. 'Profit' is the difference between the value of the catch and the cost (which is assumed to be proportional to effort) associated with a particular level of effort. Increasing effort above the level at which profit is maximized leads to economic overexploitation. Increasing effort above the level at which catches are maximized results in biological overexploitation. The bionomic equilibrium occurs when the value of the catch and the cost of the effort involved are exactly equal (i.e. profit is zero).

average profits are zero, the sustainable harvest is well below the MSY, and the equilibrium population size is only a small fraction of its pre-exploitation level (three-tenths, for the example in Fig. 15.1).

These arguments only apply if the primary aim of management is to maximize the economic income from a marine mammal population. However, some marine mammal populations, particularly those within the Arctic, are exploited by local communities not by commercial companies. Although goods produced from the harvested animals may be sold on the open market, the aim of management is to ensure the sustainability of the local community rather than to maximize its income from these goods. I will return to this topic in Section 15.6.

15.3 A brief history of marine mammal exploitation

Bonner (1989) provided convincing evidence that seals have been exploited for at least 30 000 years. He also suggests that the persistence of seal colonies in close proximity to human settlements for thousands of years implies that this exploitation was sustainable. Written records indicate that for hundreds, and probably thousands, of years the hunting of seals on the island of Anholt in the Kattegat, which links the Baltic to the North Sea, and on the Scottish island of Causamul was restricted to only a few days (one, in the case of Causamul) during the breeding season, and that only pups and large males were taken. The nature of seal exploitation changed dramatically in the late eighteenth century, largely because of the global expansion of commercial whaling.

Local populations of large whales were exploited by coastal communities in Europe at least as far back as AD 800–1000 (Allen 1980), but the technologies available were only effective for whales that came close inshore near human habitation. However, Basque whalers developed techniques for catching and processing whales over much larger geographic areas. By the end of the sixteenth century, they had expanded their operations to the coasts of North America and Spitzbergen, where they were soon joined by Dutch and British vessels (Macintosh 1965; Allen 1980). About the same time, the Japanese developed techniques for catching whales using large nets (Allen 1980). At the beginning of the eighteenth century American whalers modified the basic Basque whaling methods to allow them to take sperm whales, *Physeter catodon*, in deep water. This made possible a global expansion of whaling. As each local stock was depleted, the whalers moved further and further afield until, by the middle of the nineteenth century, whaling had spread throughout the Atlantic and Pacific Oceans and catch rates had declined alarmingly.

These voyages also brought the whalers into contact with the enormous fur seal colonies in the sub-Antarctic and Bering Sea. The whaling ships were perfectly equipped to extract oil from marine mammal carcasses, and it was easy to herd the breeding seals together in one convenient location and slaughter them. By the middle of the nineteenth century, over four million fur seals had been killed in this way (Busch 1985), and many colonies were almost or completely annihilated. Although the nineteenth-century harvests of fur seals were huge, they were dwarfed by the catches of harp seals, *Pagophilus groenlandicus*, on the Atlantic coast of Canada. Between 1800 and 1850 more than 12 million harp seals, mostly pups, were killed (Sergeant 1991).

Until the mid-nineteenth century, whalers were only able to catch those species that could be approached closely using a combination of oars and sail, and that could be secured with a hand-thrown harpoon. However, in 1864 the Norwegian Svend Foyn developed a gun capable of firing harpoons and mounted it on a steam-powered catcher boat that could hunt down the fast-swimming rorquals, such as the blue, *Balaenoptera musculus*, and fin whale, *B. physalus*, that had been inaccessible to whalers using the techniques developed in the eighteenth and nineteenth centuries. This technological breakthrough made possible a new era in whaling that followed exactly the same course as the previous one, but on a much shorter time scale.

By the early years of the twentieth century, most of the stocks of whales around the coast of northern Europe (Macintosh 1965) and in the North Pacific (Tønnessen and Johnsen 1982) had been reduced to commercial extinction, Whalers began to look further afield for other stocks to exploit. They found them in the Antarctic, where the same process was repeated. Attempts to regulate shore-based whaling were soon thwarted by the development of factory ships capable of processing whales at sea. In total nearly 1.5 million whales were caught over the first 75 years of the twentieth century (Tønnessen and Johnsen 1982).

The International Whaling Commission (IWC) was established in 1946, with the objective of 'conservation, development, and optimum utilization of the whale resources'. However, it failed to achieve any significant conservation during the first 30 years of its life, and instead it oversaw the almost complete destruction of whale stocks in the Southern Ocean.

Although there has been no commercial whaling under the management of the IWC for more than 20 years, the harvesting of marine mammals has not ceased. We will probably never again see 456 000 harp seals or 63 000 large whales killed in one year, as happened in 1951 and 1963, respectively. However, a number of members of the IWC have continued to catch whales for scientific purposes or because they lodged objections to the original decision on a pause in commercial whaling, and a number of seal hunts still continues. The combined catches of seals and whales in 2003 were just under 500 000—a reduction of only 20% from the average of 625 000 during the 1950s, when commercial whaling in the Antarctic was still in full swing (Table 15.1).

Table 15.1 *A comparison of the reported commercial catches of marine mammals in 2003 and the 1950s.*

Year	Species	Location	Number	Source
2006	Cape fur seal	Namibia	66,000	Associated Press
2003	Northern fur seal	Pribiloffs	2,000	Angliss & Outlaw 2006
2003	Hooded seal	West Ice	5,295	ICES 2006
2003	Harp seals	West Ice	2,277	ICES 2006
2003	Harp seals	East Ice	43,234	ICES 2006
2003	Hooded seal	Greenland	6,317	ICES 2006
2003	Harp seals	Greenland	67,522	ICES 2006
2003	Harp seals	NW Atlantic	289,512	ICES 2004
2003	Minke whale	Norway	647	IWC 2009
2003	Fin whale	Greenland	9	IWC 2009
2003	Humpback whale	Greenland & St Vincent	2	IWC 2009
2003	Grey whale	Russia	128	IWC 2009
2003	Minke whale	Greenland	199	IWC 2009
2003	Bowhead whale	Russia & USA	48	IWC 2009
2003	Minke whale	Iceland	37	IWC 2009
2003	Minke whale	N Pacific	151	IWC 2009
2003	Minke whale	Antarctic	443	IWC 2009
2003	Brydes whale	N Pacific	50	IWC 2009
2003	Fin whale	N Pacific	10	IWC 2009
2003	Sperm whale	N Pacific	50	IWC 2009
Total			**483,931**	
1950s	Northern fur seal	Pribiloffs	40,000	Angliss & Outlaw 2006
1950s	Hooded seals	West Ice	54,000	ICES 2006
1950s	Harp seals	West Ice	39,600	ICES 2006
1950s	Harp seals	White Sea	144,000	ICES 2006
1950s	Harp seals	Greenland	20,000	ICES 2006
1950s	Harp seals	NW Atlantic	326,700	ICES 2004
1950s	Large whales		55,000	Tønnessen & Johnsen 1982
Total			**679,300**	

15.4 Lessons from whaling and sealing

15.4.1 The International Whaling Commission

The IWC's attempts to regulate whaling from its inception until 1968 are described in great detail by Tønnessen and Johnsen (1982). The IWC's first actions were to set a limit on total catches and to agree on an opening and closing date to the whaling season. However, the total catch limit for the Antarctic was not divided between the various nations. This ensured open access (as defined in Section 15.2) to the resource and resulted in intense competition amongst the whaling nations and companies to obtain the largest share of the limited catch before the end of the ever-shortening season.

In response to a United Nations resolution calling for a moratorium on commercial whaling, the IWC's scientific committee developed the New Management Procedure or NMP (Punt and Donovan 2007), which was adopted by the Commission in 1974. This required the scientific committee to classify each whale stock into one of three categories depending on its current and pre-exploitation population size. In addition, continuing exploitation of stocks (such as many in the North Atlantic) for which there was a long catch history was permitted. Unfortunately, the NMP proved extremely difficult to implement, because small changes in the estimates of current or historical population size could result in large, abrupt changes in quotas.

In 1982 the Commission agreed to a pause in commercial whaling, which took effect in 1986, to allow for a comprehensive assessment of all whale stocks and the development of a more robust management approach, later referred to as the Revised Management Procedure (RMP). The scientific basis for the RMP, which is one of the first examples of a management approach based on 'operating models' described in detail in Section 15.7, was accepted by the Commission in 1994. However, despite repeated special meetings and workshops, the Commission had still not agreed on a formulation for the implementation of RMP at the time this chapter was written (i.e. 2009).

The lessons from the history of commercial whaling are clear. It provides one of the best examples of how open access to a resource will automatically result in both economic and biological overexploitation. In the case of whaling, attempts to protect some species by shortening the whaling season simply intensified competition between whaling nations and companies and hastened the industry's decline. The increasing efficiency of whalers during the period of intense competition also undermined the relationship between the measure of abundance (catch per unit effort or cpue) that was used to assess the impact of whaling on the stocks and the actual size of the stocks. In fact, the relationship between cpue and whale abundance is strongly non-linear even if efficiency does not change, because there is an upper limit to the number of whales that can be caught and processed in a working day. Provided the whale population is large enough to allow this number of whales to be caught, cpue will not change over a wide range of population sizes. Again, this is exactly what was observed in the period immediately after the formation of the IWC.

Although the NMP addressed some of these problems by relying primarily on estimates of whale abundance obtained independently of the whaling industry, it also relied upon a complex model of the biological processes that determine whale population dynamics. Recent experience in fisheries management has shown that management procedures that rely on complex models are outperformed by those that use simple rules (Punt and Hilborn 1997).

15.4.2 Northern fur seals

Although northern fur seals, *Callorhinus ursinus*, suffered from the same kind of overkill as large whales and many other fur seal populations in the first half of the nineteenth century, both Russia and the United States took action in the mid-nineteenth century to limit exploitation, particularly of adult females, on the breeding grounds. As was the case with whaling in the Antarctic, these catch regulations were circumvented by the development of a pelagic industry, which reduced the population to a few hundred thousand individuals by the early twentieth century. Pelagic sealing was ended in 1911, when Canada, Japan, Russia, and the United States signed the North Pacific Fur Seal Convention. Even though up to 96 000 sub-adult males were taken in some years after the signature of this Convention, the population increased steadily throughout the first half of the twentieth century (Angliss and Outlaw 2006). Major declines only occurred during periods of unregulated pelagic sealing, and when an attempt was made during the 1950s to increase juvenile survival by reducing the number of breeding females (T. Smith and Polachek 1981). As with the NMP, this decline demonstrates what can happen when a population is managed on the basis of an assumed biological mechanism that is not supported by empirical data. In this case, it was assumed, largely on the basis of evidence from the management of fish stocks, that juvenile survival would increase if the density of breeding females was reduced. It did not.

The successful management of the exploitation of northern fur seals was primarily a consequence of two things: the ownership of the islands where the seals bred was always clear, so it was possible for national authorities to limit access to the resource, and harvesting focused on only one component of the population (sub-adult males). Because the breeding system of northern fur seals is highly polygynous, there is generally a surplus of sexually mature males and changes in the size of this component of a population will have little effect on its dynamics.

15.4.3 Harp seals

As noted in Section 15.3, worldwide catches of marine mammals have, for the last 250 years, been dominated, at least numerically, by the various fisheries for harp seals. The longest running fishery has been the one associated with the Northwest Atlantic population of harp seals that breeds on ice off the coast of Newfoundland and in the Gulf of St Lawrence. Although extremely large catches were taken from this population in the nineteenth century, with a peak of 744 000 in 1832 (Sergeant 1991), it did not show the precipitous declines that have been observed in other exploited marine mammal populations. There appear to be three reasons for this: the breeding aggregations were largely within Canadian national waters and therefore access to the resource could be regulated; hunting focused almost entirely on pups, which suffer from high natural levels of mortality; and the entire resource was not available to the sealing industry every year because of weather factors and the distribution of ice. The last factor meant that, at least in some years, a fraction of the population had a refuge from exploitation.

However, the three factors that had allowed sustainable exploitation of the Northwest Atlantic population ceased to operate in the latter part of the nineteenth century and the early part of the twentieth. Steam-powered, metal-hulled ships were introduced, and this allowed sealers to access those parts of the breeding herd that had previously been inaccessible in poor ice years. Catches increased sharply after the Second World War when Norway joined the seal hunt, making regulation more difficult. The effect of this increase was exacerbated because large numbers of adult animals, which—unlike pups—suffer very low natural mortality, were taken for their skins. By 1971, the population size had fallen by around 50% from its post-war level (Sergeant 1991) and quota regulation was finally introduced. This regulation, and a collapse in the world market for seal products following the introduction of a ban on the import of pup skins by the European Union in 1983, halted the decline and allowed the population to recover.

15.5 Ecotourism

As noted in Section 15.1, it is also possible to obtain an economic benefit from a marine mammal population by charging tourists to see them. Such ecotourism has long been championed as an alternative to harvesting (Hoyt 2001) because much of the income is retained within local communities.

However, several studies (summarized in Lusseau and Bejder 2007) have shown short-term changes in the behaviour of marine mammals, and particularly cetaceans, that are approached by tourist boats. The biological significance of these changes has not been clear until recently, when Bejder *et al.* (2006) demonstrated that a population of bottlenose dolphins (*Tursiops* sp.) in Shark Bay, Australia that was repeatedly visited by tourist boats had declined over three consecutive 4–5-year periods, whereas a neighbouring less visited population had not. Similarly, R. Williams *et al.* (2006b) have shown that the energetic costs of the changes in behaviour observed in killer whales (*Orcinus orca*) that are regularly approached by tourist boats are relatively high.

These results do not, of course, mean that ecotourism is a bad thing. In general, the income associated with ecotourism demonstrates the value of marine mammals to local communities, and can offset the perceived costs associated with their consumption of commercial fish (see Section 15.6). Nevertheless, they suggest that simple rules of thumb for the management of ecotourism—such as forbidding close approaches—may be insufficient to prevent negative population consequences. More detailed studies of the responses of marine mammals to disturbance of this kind are needed before ecotourism can be considered as an entirely sustainable way of deriving economic benefit from marine mammal populations.

15.6 Obtaining indirect benefits from the management of marine mammals

In Section 15.1, I introduced the tantalizing prospect that indirect economic benefits could be obtained by managing a marine mammal population in such a way that the abundance of one of its commercially important prey species or competitors was increased. An obvious example would be a management programme that aimed to benefit a commercial fishery on one, or more, of the prey species of a particular marine mammal species by reducing the abundance of the marine mammal. But the perceived benefits of a change in marine mammal abundance may not be purely economic. For example, killer whale predation has been implicated in the decline of sea otters (*Enhydra lutra*) and Steller sea lions in the Bering Sea and North Pacific (Estes *et al.* 1998; Springer *et al.* 2003; T.M. Williams *et al.* 2004b), and harbour seals (*Phoca vitulina*) in Scotland. In these cases, it might be argued that conservation benefits could be obtained if killer whale numbers were reduced.

However, our knowledge of the likely ecosystem consequences of a change in predator abundance is currently very limited. Yodzis (2001) and Heithaus *et al.* (2008) have pointed out the many factors that render these consequences unpredictable, and Matthiopoulos *et al.* (2008) describe the extensive amounts of data on the interactions between the predator and all its prey species that are required to understand them. Other problems associated with ecosystem-based management of this kind are discussed in more detail in Section 15.8.

15.7 Defining and achieving the objectives of management

As we have seen, the history of exploitation for many marine mammal species has not been a particularly happy one, and even apparently benign forms of exploitation, such as ecotourism, can have unanticipated negative consequences. One cause of this unhappy history is that the aims of management do not necessarily coincide with those of the industries exploiting the resource. For example, most management agreements are intended to ensure the long-term persistence of the resource, but it is standard industry practice to discount the value of future harvests because of the uncertainty associated with these (Henderson and Sutherland 1996). As C.W. Clark (1976) has shown, if the discount rate is higher than the potential growth rate of the resource being harvested, it makes economic sense to harvest the resource to extinction as quickly as possible. Whaling and sealing are risky endeavours and therefore attract high discount rates, and most marine mammal populations have relatively low growth rates. So, overexploitation may well be in the financial interest of the industry.

When the aims of management are contradictory, it is important that they are prioritized. Thus, even though most developed and developing nations are committed to the precautionary principle (also known as the precautionary approach), which recognizes that there is a trade-off between the risk of serious or irreversible environmental degradation and economic benefit (Harwood and Stokes 2003), the nature of this trade-off needs to be stated explicitly.

The need for clear priorities was accepted by the IWC when it instructed its scientific committee to develop the RMP. Although there were three objectives (stable catches, highest continuing yields, and no serious increase in the risk of extinction as a result of exploitation), the Commission made it clear that avoiding an increase in the risk of extinction had priority.

The procedures used to manage a natural resource tend, these days, to be encapsulated into Harvest Control Rules (HCRs). These define what action should be taken, depending on the status of the resource being exploited. However, neither the environmental risks nor the economic benefits associated with a particular HCR can be evaluated with certainty. The uncertainty associated with these evaluations can conveniently be classified into four categories (Francis and Shooton 1997). Process variation is caused by the fact that no biological system, including those involving people, is perfectly predictable. Observation error is a consequence of the inevitable error associated with measurement, and the statistical uncertainty associated with estimates of key parameters derived from these measurements. Model uncertainty results from the fact that the mathematical models scientists choose to represent the real world will always be approximations, and may be completely inappropriate. Implementation error is a recognition that management procedures rarely have the precise effects that managers anticipate. It is a special case of the law of unintended consequences (Merton 1936). However, even a fully specified HCR that accounts for uncertainty and which is fed with appropriate data will not necessarily achieve the goals of management (Punt and Donovan 2007). For this reason, the IWC's scientific committee used an operating model framework to evaluate the effectiveness of the various rival HCRs that were proposed for the RMP.

Figure 15.2 illustrates how this framework operates. At its core is a process model, which attempts to summarize what is known about the biology of the target species that is relevant to exploitation and to account for process variation. The observation model mimics the way data on the system will be collected and analysed, and therefore accounts for observation error. The results of these analyses are then fed into the management model, which implements the HCR (but subject to implementation error) and applies it to the modelled population.

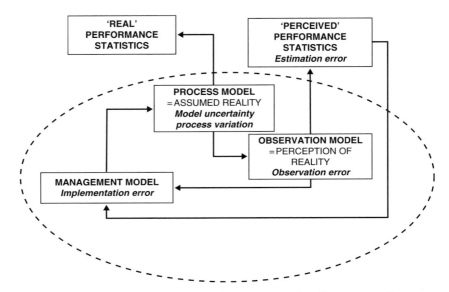

Fig. 15.2 An operating model framework for evaluating the effectiveness of a Harvest Control Rule—HCR (modified from Harwood and Stokes 2003). The process model attempts to capture what is understood about the dynamics of the exploited population and to model the effects of management. Different process models may be used to account for model uncertainty. The observation model mimics the way in which data on the population is collected and analysed. The results of these analyses inform the management model (which attempt to model the way in which the HCR is implemented) and are used to calculate the 'perceived' performance statistics. The latter can be compared with the 'real' performance statistics, calculated directly from the process model rather than via the observation model.

It is then possible to track the behaviour of the system using a set of performance statistics that reflect the objectives of management. By comparing the 'real' statistics, derived directly from the simulated population, and the 'perceived' ones, derived from the observation model, it is possible to compare the effectiveness of different HCRs. Model uncertainty is accounted for by using a wide range of process models to ensure that the performance of the HCRs is robust to this uncertainty. For example, the IWC scientific committee found that the performance of many HCRs was sensitive to the underlying structure of the stocks that were being exploited (Punt and Donovan 2007): harvesting a number of small stocks as if they were one large one could easily lead to severe depletion of some of them. It was therefore necessary to develop an HCR that could cope with uncertainty in stock structure.

As noted in Section 15.4, the RMP has yet to be implemented, but another HCR developed using the operating model framework is used to manage marine mammal populations. This is the potential biological removal (PBR) that may be allowed under the US Marine Mammal Protection Act (MMPA), and which is described in more detail in Chapter 16. The development of the PBR formula was relatively simple because the MMPA has a single clear objective: to prevent populations from 'depletion'. A population is considered to be depleted if is below the maximum net productivity level, or 50–70% of a historic population size that is thought to represent the carrying capacity of the environment (Wade 1998). However, this objective, on its own, is insufficient to allow the evaluation of rival HCRs because it does not specify a time frame. Clearly, an HCR that never allows any animals to be killed will achieve the objective of the MMPA, but there are many other HCRs that would allow some animals to be taken yet prevent depletion, or allow depleted populations to recover rapidly. Wade (1998) evaluated rival HCRs using two criteria: there should be a 95% probability that any population, regardless of its current size, will no longer be depleted after 100 years; and there should be a similar probability that a population that is not currently depleted will remain so for at least 20 years.

The PBR or the RMP are likely to permit the removal of only a small fraction (probably less than 2%) of the population. Some authors (Johnston, Meisenheimer, and Lavigne 2000) have suggested that the PBR algorithm provides a suitable precautionary HCR for species, like harp seals, that are still subject to commercial exploitation, and the joint ICES/NAFO Working Group on Harp and Hooded Seals (ICES 2006) has proposed that PBR should be used to set quotas for 'data-poor' populations of hooded seals (*Cystophora cristata*). However, these suggestions ignore the fact that PBR was developed with a single objective (preventing depletion), and that it assumes that the same proportion of animals is removed from all age-classes. The harp seal hunt is directed primarily at animals less than 1-year old, and it is well established (Beddington and Taylor 1973; Rorres and Fair 1975) that higher sustainable yields are possible if harvesting is directed at a single age-class. The HCR currently used to set quotas for the exploitation of harp seals in Canadian waters (Hammill and Stenson 2007) has many of the features of the PBR approach. It specifies that 'more stringent management measures' will be introduced if the population falls below a reference level set at '70% of its highest known abundance'. The aim of these stringent measures is to have an 80% chance of bringing the population back above the reference level within 10 years, although the measures to be used are not specified. The effectiveness of this HCR could be evaluated by using an operating model approach to compare the performance of a number of different management strategies. This evaluation would have to account for the uncertainties associated with the current monitoring strategy, which involves surveying the number of pups born every five years, and use a range of underlying process models.

Although the operating model approach provides a rigorous framework for evaluating the effectiveness of any fully specified HCR, there are two major problems associated with its use. First, it can only be applied when the objectives of management are clearly defined. In practice, there are likely to be many contradictory objectives and it is difficult to persuade all stakeholders in the resource to agree on how these should be prioritized. As a result, developing a set of operating models for any system will take a long time (although, hopefully, not as long as the implementation of the RMP). Second, implementing the computer models that are fundamental to this approach requires skills that are in short supply (Punt and Donovan 2007). These two problems mean that, for many years to come, management that has been evaluated using the operating model approach is likely to remain the exception rather than the rule.

15.8 The future of management

As we have seen, large numbers of marine mammals are still harvested each year and usually their populations are managed as a single resource. However, there is a growing acceptance that marine resources should be managed using an ecosystem-based approach, in which management priorities should 'start with the ecosystem rather than the target species' (Pikitch *et al.* 2004a). Such an approach would clearly encompass all the management issues that involve marine mammals, from direct exploitation to the indirect effects that may result from changes in behaviour as a result of disturbance. In particular, it acknowledges the fact that people are a key component of the system that is being managed, and recognizes that the success of management depends on good governance that provides incentives for all the stakeholders in the system to behave in a sustainable way (Hilborn *et al.* 2005). Garcia and Cochrane (2005) describe the approach that has been developed by the Food and Agriculture Organization of the United Nations, which defines the main objectives of an ecosystem approach to fisheries in terms of ecosystem well-being, human well-being, and the capacity of governance to achieve these. Such an approach is clearly as appropriate for the subsistence harvest of a marine mammal population by a local community as it is for a commercial fishery.

However, we know very little about the identity or the 'behaviour' of marine ecosystems, and this ignorance makes it difficult to manage them as discrete entities. Instead, ecosystem-based management has focused on the identification of ecosystem indicators that can provide reference points for management action. Rice and Rochet (2005) have outlined a protocol that can be used to select such indicators. The first step involves specifying the objectives of management, and the second involves determining whether the candidate indicators 'truly measure ecosystem status relative to the objectives'. Any attempt to make the first step is likely to encounter exactly the same problems that were experienced in the development of HCRs for single species management. The specified objectives of ecosystem-based management are often contradictory, and they are usually defined conceptually rather than quantitatively. This makes it difficult to know what aspects of ecosystem status need to be measured by the ecosystem indicators. The second step is also difficult to make because the performance of few, if any, of the candidate indicators will have been evaluated across a wide range of ecosystem states.

For example, three of the twelve ecosystem indicators (or 'Ecological Quality Objectives') used by the OSPAR Commission to monitor the impact of human activities on the 'health' of the North Sea involve marine mammals (trends in harbour seal and grey seal, *Halichoerus grypus*, numbers, and harbour porpoise, *Phocoena phocoena*, by-catch). However, it is not at all clear what information these indicators provide about the North Sea ecosystem. Nor is it clear what action should be taken if they change. In principle, these issues can be addressed by extending the operating model framework described in Section 15.7, so that the process models attempt to capture the behaviour of entire ecosystems rather than just individual populations. Some progress has been made in this area (Fulton *et al.* 2003), but there is an enormous amount of work still to do.

In addition, there is still considerable uncertainty about the role ecological analysis can play in ecosystem-based management (Harwood 2007). Corkeron (2004, 2006) has argued that this approach, when applied to fisheries, could allow some nations to justify the culling of top predators in order to protect high-value fisheries. As noted in Section 15.6, a reduction in marine mammal numbers might benefit commercial fisheries directed at their prey, but the scientific case for these inferred benefits is still unclear (Yodzis 2001). However, what is clear is that the original proponents of the ecosystem approach did not have this kind of tinkering with trophic interactions in mind (Pikitch *et al.* 2004b). Rather, their concern was that management of a particular species should take account of the consequences not just for the target species, but for all of the species with which it interacts, including the different sectors of society. For example, there is considerable evidence (Heithaus *et al.* 2008) that the effects of changes in the abundance of top predators, like marine mammals, on prey behaviour may be more important than the direct effects of predator-induced mortality. These behavioural effects also need to be accounted for in ecosystem-based management.

16

Conservation biology

Andrew J. Read

16.1 Introduction

Like many other large, long-lived marine vertebrates, marine mammals are vulnerable to the effects of a variety of human activities. Populations of cetaceans, pinnipeds, and sirenians may be threatened by excessive harvest, mortality in fisheries, boat strikes, disease, or habitat loss. These animals have important inherent and economic value and, as a result, their conservation is of considerable societal interest.

In this chapter I provide a road map for biologists working on marine mammal conservation, describing some of the problems and pitfalls likely to be encountered along the way. It comes at the end of this book because it relies, in many ways, on all the methods that have been described in the preceding chapters, whether they are: using genetics to define populations (Chapter 14), conducting surveys to estimate abundance (Chapter 3), assessing various approaches to management (Chapter 15), or even the ethics of undertaking research of different types on endangered species (Chapter 1). I have written the chapter that I needed, but couldn't find, as a graduate student. Conservation biology is a relatively new field; the Society for Conservation Biology was founded in 1985 and the journal *Conservation Biology* was first published in 1987. The field has now matured to the point where there are now university courses on marine mammal conservation and we can provide today's students with sound theoretical and practical skills to underpin the field.

Biological diversity exists at various scales of organization: genetic; population; species; communities; and ecosystems. Many threatened species exist as a series of relatively isolated populations. Thus, in this chapter I focus on the level of the population, assuming that much, but by no means all, genetic diversity can be conserved by working at this level of organization. The scientific community is focusing more of its conservation efforts at levels of organization below the species (Taylor 2005; Chapter 14).

In general, our understanding of marine biodiversity, and its conservation, has lagged behind that of terrestrial systems (Irish and Norse 1996). Marine ecosystems are difficult to study and threats to these systems are less obvious than comparable threats on land. Fortunately for us, however, marine mammals have received more scientific attention than many other marine species.

We have lost relatively little of the total diversity of marine mammals, especially when compared with losses in terrestrial ecosystems. To date, five species of marine mammals are known to have become extinct in recorded history: Steller's sea cow (*Hydrodamalis gigas*), Caribbean monk seal (*Monachus tropicalis*), Japanese sea lion (*Zalophus japonicus*), sea mink (*Mustela macrodon*), and baiji (*Lipotes vexillifer*) (Rice 1998; Mead *et al.* 2000; Turvey *et al.* 2007). The recent extinction of the baiji is of particular note because it represents the loss of an entire family (Lipotidae) of marine mammals. As described below, many populations and several species of marine mammals are currently at risk of extinction.

Most past extinctions of marine mammals were the result of uncontrolled harvest. Few modern conservation problems are caused by harvest practices, although many marine mammal populations are still depleted from past hunting. Many of today's conservation problems are more complex, involving multifaceted and poorly understood threats. Mitigation of these threats is the challenge that lies before us.

Due to the limitations of space, this is by necessity an idiosyncratic review; much more could be said about conservation methods for marine mammals. I direct the interested reader to excellent volumes by Twiss and Reeves (1999), N. Gales *et al.* (2003b), Reynolds *et al.* (2005), and the Marine Mammal Commission (2008).

16.2 What is conservation biology?

The job of conservation biologists is to determine which populations are at risk, identify the causes of decline, develop potential mitigation measures, assess their effectiveness, and, in an ideal world, document recovery. An unusual aspect of conservation biology is the normative nature of the discipline. Unlike most other physical and natural sciences, which are supposed to be value-free, the field of conservation biology is predicated on a value statement—that the loss of biological diversity caused by human activities is undesirable and should be prevented whenever possible. This does not mean, however, that the science practised by conservation biologists is not objective. Indeed, conservation biologists have a fundamental responsibility to practise good, objective science that informs policy.

One of the challenges for conservation biologists is to separate the normative aspect of their work (preventing extinction and conserving diversity) from the objective aspect (conducting good science). Another challenge for conservation biologists is to separate concern for the welfare of their study animals with an objective evaluation of the conservation status of a population. This is particularly true for marine mammals which often engender strong feelings. Nevertheless, conservation and animal welfare are not one and the same. This is not to downplay issues of welfare, which should be central to any study of animals, but it is important not to conflate issues of animal welfare with those of conservation. At the same time, however, it is important to understand that populations comprise individuals, and that in many cases, particularly those involving endangered populations, the well-being of individuals is an important consideration

It is also important to note that biologists play only a part, and sometimes a small one at that, in the conservation of marine mammal populations. In most cases, effective conservation requires a modification to human behaviour rather than a change in the biology of the threatened population. Thus, conservation biologists work alongside many other professionals, including lawyers, policy analysts, economists, educators, and social scientists. Humans are an integral part of almost all ecosystems and the most interested human participants, sometimes referred to as 'stakeholders', should be included in all aspects of conservation planning. As a result, conservation biologists often spend more time in meeting rooms than studying animals in the field.

16.3 The road map

Many conservation problems follow a similar course of development, from initial detection to implementation of potential solutions. It may be helpful to lay out a brief road map to help guide conservation efforts.

The first step in any conservation issue is initial *detection* of a problem. Some conservation issues are identified through scientific monitoring—surveys determine that a population is declining, or fisheries' observers document an unsustainable by-catch of marine mammals. In other cases, problems are detected by accident—an unusual number of marine mammals strand or anecdotal reports of aberrant behaviour are reported.

After initial detection of a potential problem, the next step is to conduct a scientific *assessment*. In this phase, it is essential to determine the conservation status of the population in question. What population is involved? Is this population isolated or a sub-population linked to others by emigration and immigration? What is its current abundance? Is it declining and, if so, at what rate? What demographic factors (increased mortality or reduced fecundity) are responsible for the decline? How many animals are being removed from the population and by what process? In this stage, quantitative demographic models provide insight into the status and trend of the population. Detailed demographic information is most useful at this stage, but even simple models can be helpful, as described below. This step can take years, and sometimes even decades, to complete.

Once a conservation problem has been documented and a potential anthropogenic cause identified, one should prepare for the *response* from representatives of whatever human activity has been identified as responsible for the problem. This response is often an assertion that no conservation problem exists (i.e. the population is not threatened) or that some other factor is responsible for the problem. This can be especially tricky if multiple factors are involved, especially if some potential causative factors are not caused by human actions. Sometimes, this response may bring into question the integrity of the scientists responsible for detection or assessment. Scientists can be placed in a particularly vulnerable position if they are responsible for both detection and assessment; this is another reason to ensure that our scientific methods are as objective and robust as possible.

Once it has been agreed that a problem exists and the cause has been identified, the search for *solutions* begins. This is when input from stakeholders can be particularly useful, because these individuals are the most knowledgeable about their interactions with threatened species. At this stage, it is often useful to have some formalized decision-making process to ensure that all potential solutions are evaluated fairly and that the way forward will actual meet conservation objectives. At this stage, population models are useful to evaluate the potential demographic effects of various mitigation measures. This is the point at which trade-offs between efficacy and cost must be evaluated; here again stakeholder involvement is essential. It is desirable to test potential solutions in field experiments or implementation trials, although this may be difficult or impossible with highly endangered populations.

Once an effective solution has been identified and deemed acceptable by stakeholders and policy makers, the next phase is *implementation*. In this stage it is critical to monitor the population for signs of recovery, such as positive population growth, reduced mortality, or increased fecundity. Such monitoring systems are vital to determine the success or failure of a particular conservation initiative, and should be accompanied by mechanisms that allow a measure to be modified or abandoned if it fails. Some conservation measures are temporary fixes, others are permanent. The latter requires institutionalization, so that they become part of normal practice. Successful measures will often include rewards or incentives to ensure that conservation practices are carried out, or disincentives or penalties if objectives are not met.

16.4 Which populations are at risk?

Conservation biology is typically a crisis discipline, in which we must determine how best to allocate scarce resources to the most pressing problems. A critical first step in this process is determining which populations are at greatest risk of extinction. The next step is setting conservation priorities—deciding which threatened or endangered populations should receive attention. It is not always the case, nor is it necessarily appropriate, that the most endangered species receive the most attention. Political, societal, and legal factors often intervene so that we devote more resources to species at lower risk. In addition, it may be unwise to expend a large proportion of available resources on a species that is doomed to extinction.

In an ideal world, resources should be allocated to recovery actions in a manner that maximizes the protection and recovery of the largest number of taxa (Possingham *et al.* 2001).

There have been many attempts to create a system of evaluating extinction risk, including quantitative analyses of the probability of extinction (see Section 16.5), rule-based systems, and the use of expert opinion. These approaches are reviewed in DeMaster *et al.* (2004). The most widely accepted international system is the Red List of Threatened Species, maintained by the International Union for Conservation of Nature (IUCN). The objective of the Red List process is 'to provide an explicit, objective framework for the classification of the broadest range of species according to their extinction risk' (IUCN 2008).

Red List assessments are compiled by species experts and reviewed by the relevant Species Specialist Groups. The IUCN Species Specialist Groups comprise experts who volunteer their time to these conservation efforts. Of these groups, four focus on marine mammals: cetaceans, sirenians, pinnipeds, and polar bears.

The Red List assessments use a set of quantitative criteria to determine the status of a species. The criteria include the following factors: reduction in population size, reduction in geographic range, small current population size, and quantitative analyses of extinction probability (IUCN 2008). Based on these criteria, a species is placed in one of eight categories, from *Extinct* to *Least Concern* (Fig. 16.1). The full set of criteria and rules are too detailed to describe here, but reflect the increasing risk of extinction caused by reductions in population size and geographical range. Species for which insufficient information exists to make an assessment are listed as *Data Deficient*.

For example, the beluga or white whale (*Delphinapterus leucas*) is designated as *Near Threatened*, meaning that the species is not immediately threatened with extinction, but that it could become so if existing management measures were withdrawn. In addition, as described below, some beluga populations are in critical need of conservation attention. To date, most Red List assessments have been conducted at the species level, but as these are completed there has been increasing emphasis on population-level assessments (Gärdenfors *et al.* 2001). The beluga population in Cook Inlet, Alaska has been listed as *Critically Endangered* by the IUCN (Lowry *et al.* 2006), the highest level of risk for a wild population. This population is of particular conservation concern because of its very small size (fewer than 250 mature individuals) and negative population trend, even though the factor responsible for its dire status—excessive harvest—was controlled in the late 1990s. Clearly, under any system of ranking, Cook Inlet belugas deserve attention from conservation biologists and managers.

The Red List is not without its critics (see Mrosovsky 1997), but it remains the primary international standard for assessing conservation status (Rodrigues *et al.* 2006) and performs reasonably well in empirical assessment of extinction outcome (Keith *et al.* 2004). In recent years, marine mammal biologists have been active in

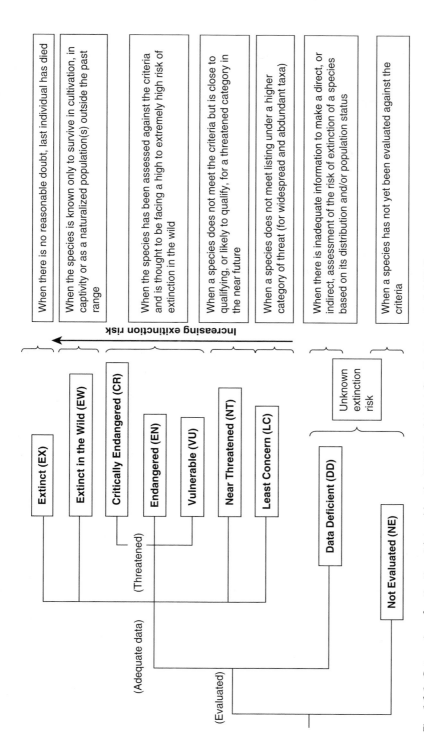

Fig. 16.1 Categories of extinction risk used by the IUCN Red List (after Rodrigues *et al.* 2006).

When there is no reasonable doubt, last individual has died

When the species is known only to survive in cultivation, in captivity or as a naturalized population(s) outside the past range

When the species has been assessed against the criteria and is thought to be facing a high to extremely high risk of extinction in the wild

When a species does not meet the criteria but is close to qualifying, or likely to qualify, for a threatened category in the near future

When a species does not meet listing under a higher category of threat (for widespread and abundant taxa)

When there is inadequate information to make a direct, or indirect, assessment of the risk of extinction of a species based on its distribution and/or population status

When a species has not yet been evaluated against the criteria

Increasing extinction risk

Extinct (EX)

Extinct in the Wild (EW)

Critically Endangered (CR)

Endangered (EN)

Vulnerable (VU)

Near Threatened (NT)

Least Concern (LC)

(Threatened)

(Adequate data)

Data Deficient (DD)

Unknown extinction risk

Not Evaluated (NE)

(Evaluated)

updating many species accounts as part of the Global Mammal Assessment (Schipper *et al.* 2008), although a great deal of work remains to complete these assessments at the population level.

Many countries have their own national system for classifying species at risk of extinction. In the United States, for example, species are categorized into one of two tiers, *Threatened* or *Endangered*, under the Endangered Species Act (ESA) of 1973 (Lowry *et al.* 2007). Two agencies are responsible for implementing the ESA with regard to marine mammals: the US Fish and Wildlife Service (USFWS) and the US National Marine Fisheries Service (NMFS). The USFWS has responsibility for polar bears, sea otters, walruses, and manatees, while NMFS has responsibility for cetaceans, fur seals, sea lions, and phocid seals.

In the United States, a species can be listed as threatened or endangered due to any of the five following factors: (1) destruction, modification, or curtailment of habitat; (2) over-utilization for commercial, recreational, scientific, or educational purposes; (3) disease or predation; (4) inadequacy of existing regulatory mechanisms; and (5) other natural or man-made factors (Fig. 16.2; Nicholopoulos 1999). Populations of vertebrate species, referred to as Distinct Population Segments, can be listed if they are: discrete from the rest of the species; significant to the existence or genetic composition of the species; and threatened or endangered, based on the five criteria described above (Federal Register 1996). Currently, 18 marine mammal taxa found in US waters are listed as threatened or endangered under the ESA (Table 16.1).

Thus, in 2007 the NMFS proposed listing the Cook Inlet beluga population as *Endangered* under the ESA (Federal Register 2007). The population qualified as a Distinct Population Segment because it is discrete (genetically and physically isolated from other populations) and significant to the species (no similar beluga habitat exists elsewhere). The NMFS concluded that the population warranted listing as endangered because of threats to its habitat, the inadequacy of existing regulation, predation by killer whales, and other factors (the unknown effects of past removals). The dire status of the population was confirmed by a population viability analysis (see Section 16.5). The NMFS proposal to list the Cook Inlet population cited the IUCN assessment (Lowry *et al.* 2006), demonstrating the utility of the international approach in national policy making.

In contrast to the IUCN Red List, the US ESA contains no quantitative criteria to guide listing decisions. Thus, it is often difficult to determine whether a population should be listed as threatened or endangered, or not listed at all. Conversely, there are no criteria that allow determination of when a species has recovered to the point where it should be down-listed or removed from the list completely. The lack of quantitative criteria can lead to inconsistencies in listing decisions, although all but two of the 18 marine mammal taxa listed as endangered or threatened under the ESA are also considered critically endangered, endangered, or vulnerable by the IUCN (Table 16.1; Lowry *et al.* 2007).

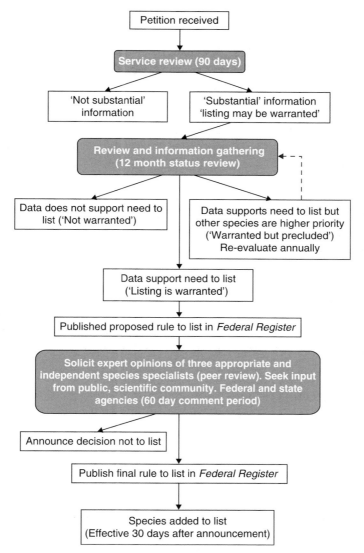

Fig. 16.2 The process for listing species as 'threatened' or 'endangered' under the US Endangered Species Act (ESA, after Nicholopoulos 1999).

Fortunately, there have been recent attempts to develop quantitative definitions of the terms *Endangered* and *Threatened* that relate directly to extinction risk, and to frame an accompanying set of decision rules for cases in which data are limited (DeMaster *et al.* 2004). This analytical approach has been used to assess the conservation status of several species of marine mammals in the United States (e.g. Gerber *et al.* 2007), although it has not yet been incorporated into general policy directives implementing the ESA or to amendment of the statute itself.

Table 6.1 *Domestic marine mammal species categorized as 'threatened' or 'endangered' under the United States Endangered Species Act (ESA) and their respective IUCN Red List status (from Lowry et al. 2007; see Fig. 6.2).*

Taxon	ESA status	IUCN status
West Indian manatee—Florida Population	Endangered	Vulnerable
West Indian manatee—Antillean Population	Endangered	Vulnerable
Southern sea otter	Threatened	Endangered[1]
Northern sea otter—SW Alaska population	Threatened	Endangered[1]
Caribbean monk seal	Endangered	Extinct
Hawaiian monk seal	Endangered	Endangered
Guadalupe fur seal	Threatened	Vulnerable
Steller sea lion—Eastern population	Threatened	Endangered[1]
Steller sea lion—Western population	Endangered	Endangered[1]
Blue whale	Endangered	Endangered
Bowhead whale—Western Arctic population	Endangered	Lower risk
Fin whale	Endangered	Endangered
Humpback whale	Endangered	Vulnerable
North Atlantic right whale	Endangered	Endangered
North Pacific right whale	Endangered	Endangered
Sei whale	Endangered	Endangered
Sperm whale	Endangered	Vulnerable
Killer whale—Southern resident population	Endangered	Lower risk[1]

[1] Listing status refers to entire species.

Many other countries have legislation intended to prevent the extinction of threatened and endangered species. In Canada, the relevant legislation is the Species at Risk Act (SARA) (Vanderzwaag and Hutchings 2005). Species designations under SARA are made by the Committee on the Status of Endangered Wildlife in Canada (COSEWIC), the national advisory body responsible for assessing the risk of extinction. Assessments of species status made by COSEWIC are based on guidelines adapted from the Red List criteria. To date, the COSEWIC Marine Mammals Sub-Committee has assessed the status of approximately 30 populations of marine mammals in Canada but, significantly, not all of those assessments have been translated into formal listings that would lead to tangible conservation action.

In Australia, the legislative mandate for protection of threatened species is the Environment Protection and Biodiversity Conservation Act (EPBC Act) of 1999. Nominations for inclusion of species on the list of threatened species are assessed by the Threatened Species Scientific Committee (TSSC), which prepares a Proposed Priority Assessment List of nominations for consideration by the Minister of the Environment. The categories of threat are similar to those used by the IUCN, but as for the ESA, there are no quantitative criteria or rules set out in the EPBC Act to guide the work of the TSSC or the final decisions made by the Minister.

16.5 Quantitative assessment of extinction threat

A range of population dynamics models can be used to predict the future status of a population (or populations), including quantifying the probability of extinction. When used in this fashion, such models are often described as *population viability analysis* (PVA), and have become a critical component of our conservation toolbox (Morris and Doak 2002; Marine Mammal Commission 2007). PVAs can be of varying complexity, but each contains the following elements: demographic features of the population, important factors of the animals' environment, and human impacts on the population. In addition, most PVAs include stochastic factors (those due to chance) that may influence the likelihood of extinction, such as those due to demographic, genetic, and environmental factors.

PVAs are used in several ways. The first is to assess the risk of extinction or of decline to such a small size that extinction is likely, often referred to as quasi-extinction. In this manner, a PVA can help to decide which populations are at most risk of extinction. For example, Runge *et al.* (2004) used a stage-based population model to assess the growth rates of four populations of Florida manatees (*Trichechus manatus*). This matrix model examined the survival and fecundity of seven stages of female manatees, ranging from first-year calves to breeding females. Model results indicated that three of the populations were increasing, but the population in south-western Florida was declining at about 1.1% per year. This analysis helped to focus management efforts on the population of manatees most in need of attention.

PVAs are also used to identify those life stages most sensitive to the effects of human activities by evaluating the relative contribution of demographic parameters to population growth. Put another way, this type of analysis examines the *elasticity* of population growth to the relative contributions of its component processes, such as juvenile survival, adult survival, and fecundity (Eberhardt and Siniff 1977; de Kroon *et al.* 1986). For example, Fujiwara and Caswell (2001) constructed a matrix population model of North Atlantic right whale (*Eubalaena glacialis*) demography. The model estimated a mean time to extinction of 208 years and identified the importance of adult female survival in determining the future status of this population. Management efforts that could prevent the deaths of two female whales each year would halt the population decline.

Finally, PVAs can be used to model the likely outcome of a particular management decision or to contrast the relative effects of competing alternatives. In this approach, we can either manipulate the model to simulate the anticipated effects of particular actions on population growth, or examine the sensitivity of population growth to particular interventions. For example, Gerber *et al.* (2004) constructed a stochastic age-structured model of California sea otters (*Enhydra lutris*) to examine the sensitivity of the population to different sources of mortality. The results of this exercise indicated that management actions that would reduce mortality due to disease by half, by controlling terrestrial inputs of pathogens, for example, would provide more conservation benefit than eliminating mortality due to gunshot wounds.

PVAs vary widely in their complexity, from simple count-based models to complex spatially structured, individual-based approaches. Morris and Doak (2006) provide two useful rules of thumb to consider when using population models in conservation. First, let the available data tell you which type of PVA to conduct. The level of model complexity should reflect the data available to parameterize the model. The most complex models are very data hungry and require detailed information on the demography and behaviour of a population. One of the great difficulties of working with very rare species is that data are often sparse or missing—we should not attempt to construct elaborate models for such data-poor species. The second rule is to understand what your model is doing. Several software packages offer 'off-the-shelf' PVAs. These programs can provide useful outputs, so long as the user understands the model assumptions and does not include risk factors for which no information exists. Do not use such software unless you have a good understanding of the underlying model assumptions and outputs. It is greatly preferable to construct a model from scratch, to ensure that the assumptions built into the mathematics are clearly identified and evaluated.

Finally, it is important to remember that any PVA is only a model, one of many possible representations of reality that reflects both the underlying mathematical structure and the data used to parameterize the model. PVAs can be updated and refined as more information becomes available and different questions are asked.

16.6 Experimentation

Conservation biology is a science and, as for all scientific endeavours, highly useful inferences can be drawn from carefully designed and controlled experiments. Experimentation with threatened and endangered species, however, poses a series of challenges and responsibilities for the investigator. First, ethical considerations must be addressed when conducting any experimentation with animals (see Chapter 1). The ethical responsibilities of a scientist are magnified when working with an endangered species—will the experiment place individual animals at risk? Will this risk affect the viability of a threatened population? As noted above, even the loss of a few individuals can affect the status of some highly endangered species.

Next, the experimenter must ensure that all appropriate regulatory requirements are met. In most cases, this will entail authorization from institutional animal care committees, compliance with codes of best practice, and authorization from national or regional permitting agencies. In many countries, experimentation with endangered species requires an extra layer of governmental approval, designed to ensure that science does not hasten the demise of a species. This is a necessary step, but one that can mire the scientific process in a bureaucratic quagmire, particularly in highly regulated countries, such as the United States.

Finally, it is necessary to design and conduct experiments with rare animals. It may be difficult to find animals in the field, making it difficult to obtain an adequate sample size of observations. Thus, it is critical to design experiments carefully, optimize sample sizes, and consider the statistical power required to adequately test a hypothesis (Taylor and Gerrodette 1993).

Despite these hurdles, there are many examples of field experiments that have contributed significantly to the conservation of marine mammals. Nowacek *et al.* (2004) conducted a field experiment to examine the response of North Atlantic right whales to controlled sound exposures. One of the two primary factors inhibiting the recovery of this species is mortality caused by ship strikes (Knowlton and Brown 2007). Therefore, this field experiment was designed to test the hypothesis that an alerting sound might be used to warn whales of the presence of oncoming ships. Whales were tagged with archival digital acoustic recording tags (D-tags), attached with suction cups, and then exposed to a variety of experimental and control acoustic stimuli. The digital acoustic tags allowed precise monitoring of the sounds heard by the whales and of their response to these stimuli. Out of six whales, five exposed to the alert signal responded strongly, but in an unexpected fashion. These whales abandoned their foraging dives and swam rapidly to and remained at the surface, thus likely increasing, rather than decreasing, their vulnerability to passing vessels. This unanticipated result indicated that alerting stimuli would not solve the problem of ship strikes and reinforced the need to separate whales and ships, or slow vessel traffic to safe speeds.

In another experiment, Hodgson *et al.* (2007) examined the response of dugongs (*Dugong dugon*) to acoustic alarms designed to reduce their accidental capture in coastal gill nets in Queensland, Australia. The actual by-catch rate of dugongs is relatively low, so an experiment with sufficient power to test the efficacy of alarms would require a large number of observations. Such an experiment would be expensive and likely result in the deaths of many dugongs. Rather than employ alarms in a full field trial with gill nets, Hodgson *et al.* (2007) examined the response of dugongs to active and silent alarms using a camera suspended from a small blimp. Dugongs did not respond to active alarms by moving away from, or orienting towards, the sound source, indicating that the alarms were unlikely to be effective in reducing by-catches of this species. Other management measures, such as the use of time-area fishery restrictions, must be employed to reduce the mortality of dugongs in Australian gill net fisheries.

Not all experiments are controlled by the experimenter; many scientists take advantage of natural contrasts to gain insights into demography or ecosystem function. Perhaps the most famous example of this approach in our field is the study of sea otters (*Enhydra lutris*) and kelp ecosystems in the North Pacific. The role of the sea otter as a keystone species was discovered by the pioneering work of Jim Estes, who contrasted the status of kelp forest ecosystems in areas where otters were abundant and those where they had been extirpated (Estes and Palmisano 1974). Throughout his career, Dr Estes has continued to take advantage of such natural contrasts to understand the dynamics of these marine mammal predators. Most recently, Estes and his co-authors (1998) documented that predation by killer whales (*Orcinus orca*) was responsible for the rapid decline of sea otters in western Alaska by contrasting population trends and survival rates of otters in areas where killer whales were present and absent (Estes *et al.* 1998). There are many insights to be drawn from such contrasts, but it takes a clever investigator to recognize their existence and potential.

16.7 Direct intervention

Most conservation solutions require a modification to human behaviour, rather than a change in the behaviour or ecology of the threatened species. In a few cases, however, scientists attempt to improve the conservation status of a population through direct intervention. These attempts are difficult, expensive, prone to failure, and certainly not for the faint-hearted. Nevertheless, under the right circumstances, and with the full support of government and the relevant stake-holders, they can be a critical tool with which to help avert extinction.

Direct intervention can take several forms: captive breeding, head-starting, re-introduction, translocation, and, in the cases of some marine mammals, disentanglement. Captive breeding has been a valuable tool for some endangered terrestrial species and the IUCN has constituted a Conservation Breeding Specialist Group (IUCN 2008). I am not aware of any actual attempts to improve the conservation status of a marine mammal species through captive breeding, although this approach has been discussed for several species and was suggested as a means to avert the extinction of the baiji (D. Wang *et al.* 2006).

Head starting, translocation, and disentanglement are all tools in the marine mammal conservation toolbox; I will illustrate their use with a single endangered species, the Hawaiian monk seal *Monachus schauinslandi* (Antonelis *et al.* 2006). Monk seals are distributed throughout the Hawaiian archipelago, with the largest sub-populations found in the remote islands and atolls of the North-western Hawaiian Islands (NWHI). Approximately 1200 individuals remain and the population is declining by about 4% per year.

In past decades, the sex ratio of monk seals at some breeding sites was highly skewed towards males; an unbalanced sex ratio is not uncommon with very small populations. As a result, many adult female seals were badly injured or killed by aggressive males during mating attempts, a behaviour referred to as *mobbing*. An effective response to this problem was the translocation of adult males from breeding sites in the NWHI, where mobbing was a problem, to the main Hawaiian Islands. The efficacy of male translocation as a conservation strategy was confirmed by population models (Starfield *et al.* 1995). The removal of these males successfully reduced the number of female deaths caused by male aggression. Of course, translocation can also be used as a means of establishing new populations of endangered species, to spread the risk of extinction, as has been done success-fully with California sea otters (Jameson *et al.* 1982).

The continuing decline of Hawaiian monk seals has been attributed, in large part, to the poor survival of juvenile animals caused by food limitation. Thus, there have been several attempts to improve the survival probabilities of young seals by giving them a head start. Weaned seals are brought into a captive setting where they are fed by hand and then released back into the wild. These activities have had limited success, at best. Of 104 juvenile seals head-started or rehabilitated since 1981, 68 were released into the wild, 22 died in captivity, and the remaining 14 animals were not healthy enough to be released and so were placed in aquariums for public display and research (Antonelis *et al.* 2006).

Finally, entanglement in marine debris is a continuing problem for Hawaiian monk seals, due to the frequent deposition of debris throughout the NWHI. To reduce the risk of entanglement of monk seals and other ecosystem components, the NMFS has been removing debris and disentangling seals. Between 1982 and 2003, 238 monk seals were disentangled from marine debris (Antonelis *et al.* 2006).

Entanglement in fishing gear is also a primary conservation threat to many marine mammals (see Sections 16.8 and 16.9) and, under some circumstances, disentanglement can be used to reduce the lethality of this threat. For example, right whales often encounter and carry off fixed fishing gear, typically strings of gill nets or lobster pots (A.J. Johnson *et al.* 2007). Without intervention, an entangled whale may die a long, slow, and painful death (Moore *et al.* 2007). Until effective methods of reducing entanglement are found, biologists will continue the expensive and dangerous task of attempting to disentangle these animals. Due to the highly endangered status of this species, it is necessary to continue these efforts, but they should not be viewed as a substitute for the changes to fishing practices required to eliminate entanglement in the first place (Reeves *et al.* 2007).

Critical to evaluating the success of any intervention, and to most conservation initiatives in general, is monitoring to determine the efficacy of the action. Do translocated animals return to their site of capture or do they stay in the area where they were released? Do head-started juveniles have an increased survival rate and, if so, is this increase enough to meet conservation objectives? Do disentangled animals survive after the intervention? Are such interventions sustainable over the long run? The answers to such questions are critical to the evaluation of conservation activities.

16.8 Fisheries by-catch

As noted above, many factors can threaten a population of marine mammals with extinction. One of the most ubiquitous threats, however, is by-catch in fishing gear (Read 2008), so I will briefly review some relevant strategies for dealing with this issue.

The existence of by-catch is often detected through anecdotal observations or strandings of by-caught animals. If a conservation problem is suspected, it is necessary to estimate the total mortality experienced by a population and determine whether this level of mortality is sustainable. Typically an observer scheme, in which independent observers are placed on fishing vessels, is used to document the by-catch *rate* (e.g. the number of marine mammals killed per haul). This observed by-catch rate is then extrapolated to an estimate of *total by-catch* by using some metric of fleet-wide fishing effort (e.g. total number of hauls).

Observers are commonly used to monitor catch and by-catch rates of commercial species, but there is often opposition to implementing this technique to monitor the by-catch of marine mammals. This opposition can arise due to the expense incurred, or because fisheries managers do not want to divert resources to a problem that they view outside their purview. Voluntary reporting schemes are often offered as an alternative, but such schemes significantly underestimate total mortality, particularly in areas where fishermen worry about management repercussions (Read 2005). There is no substitute for independent observers when estimating the by-catch of marine mammals.

Protocols for using independent observers should incorporate a statistical sampling design, considering the sample size required to generate estimates of sufficient precision, and taking into account the distribution of fishing effort in time and space. Observers can also provide important information on variation in by-catch rates due to fishing practices, season, or location; these observations can be used to design mitigation strategies to reduce by-catch. Observers should also be used to monitor the efficacy of conservation strategies, once they have been implemented in a fishery.

Once total mortality has been estimated, this information can be incorporated into a quantitative population model to determine the effect of by-catch on population viability. Often, however, it is not possible to construct a detailed demographic model due to the limitations of existing data. In such cases, simple models are available to assess the sustainability of by-catch mortality. One of these simple models is the *potential biological removal* (PBR) approach used in the United States (Wade 1998). The approach requires only estimates of population size and by-catch to determine whether or not fisheries removals will deplete a population of marine mammals; taxon-specific default values can be used for the other parameters used in the model. The PBR model explicitly considers uncertainty in parameter estimation in its performance (Taylor *et al.* 2000). The simplicity of the model is one of its most attractive features. Stakeholders may not agree with the management goals, in this case set by the Marine Mammal Protection Act (MMPA) and ESA, but they can readily understand the formulation and performance of this simple model.

Finally, if by-catch mortality exceeds a conservation threshold, it is necessary to devise a strategy that will reduce by-catches to sustainable levels. In the United

States, the approach has been to formalize stakeholder involvement in the 'take reduction process'. In this process, representatives of fisheries and environmental groups meet with scientists, managers, and other interested parties in a negotiated rule-making exercise. A professional facilitator helps the team work towards a consensus on changes to fisheries practices that will reduce by-catches to below PBR. The approach has been successful in reducing the by-catch of several populations of marine mammals to below PBR (Young 2001; Read 2005), although not in cases where by-catches must be reduced to zero, as in the case of the North Atlantic right whale (Johnson *et al.* 2007; Reeves *et al.* 2007). This model of stakeholder involvement could be adapted for use in many other countries after modification to reflect the political and social realities in each area.

16.9 Case study: by-catch of harbour porpoises in the Gulf of Maine

In 1982 Jim Gilbert and Kate Wynne began a study of the by-catch of harbour porpoises (*Phocoena phocoena*) in bottom-set gill nets along the coast of Maine. From interviews, logbooks, and a small number of observed trips they estimated that each of the fishermen they worked with caught about five porpoises annually. At about the same time, David Gaskin and I began an investigation into the by-catch of harbour porpoises in the gill net fishery in the Bay of Fundy. In 1986 we conducted interviews with gill net fishermen, who also reported catching about five porpoises each year. Despite our similar findings, we had no way of knowing whether they were typical of all 300 gill net fishermen in the Gulf of Maine.

To provide a statistically robust estimate of by-catch for the entire Gulf of Maine, the NMFS placed observers aboard a random sample of gill net vessels (started in 1989). The observers recorded the number of entangled porpoises as well as information on location, net configuration, and fish catches. The observer programme continues today, with 2–6% of fishing trips being monitored each year. The NMFS estimated that 2900 porpoises (CV = 0.32) were killed in 1990 and 2000 in 1991 (CV = 0.35) (Table 16.2). The findings of such a large by-catch led a coalition of environmental groups to sue the NMFS to list the Gulf of Maine population of harbour porpoises as 'Threatened' under the ESA in 1991. The NMFS proposed listing the population as Threatened in 1993, although it took no action on this proposal until 2001. The documentation of large by-catches, and the proposed listing, led to a flurry of work to assess the status of this harbour porpoise population in the mid-1990s.

To understand whether or not the by-catches of porpoises were sustainable, three pieces of information were required: population structure, abundance, and potential population growth. The first question was whether porpoises in the Gulf

Table 16.2 *By-catches of harbour porpoises (Phocoena phocoena) in the Gulf of Maine gill net fishery.*

Year	Observer coverage	Observed mortality	Estimated by-catch	CV	PBR
1990	0.01	17	2900	0.32	403
1991	0.06	50	2000	0.35	403
1992	0.07	51	1200	0.21	403
1993	0.05	53	1400	0.18	403
1994	0.07	99	2100	0.18	403
1995	0.05	43	1400	0.27	483
1996	0.04	52	1200	0.25	483
1997	0.06	47	782	0.22	483
1998	0.05	12	332	0.46	483
1999	0.06	14	270	0.28	747
2000	0.06	15	507	0.37	747
2001	0.04	4	53	0.97	747
2002	0.02	10	444	0.37	747
2003	0.03	12	592	0.33	747
2004	0.06	27	654	0.36	747
2005	0.07	51	630	0.23	610

Data are from National Marine Fisheries Service Stock Assessment Reports (NMFS 2008).

Observer coverage refers to the proportion of fishing effort observed each year.

Observed mortality refers to the number of porpoises observed killed—this is then extrapolated to *estimated by-catch* (and its associated coefficient of variation—CV) by scaling to total fishing effort.

PBR refers to the potential biological removal.

of Maine comprised a discrete population that should be managed separately from others in the North Atlantic. The primary tool used to assess population structure was molecular genetics. In an analysis of variation in the control region of mitochondrial DNA, Patricia Rosel found evidence to support the existence of a discrete population in the Gulf of Maine, although microsatellite markers exhibited little differentiation (Rosel *et al.* 1999).

The first comprehensive surveys of harbour porpoises in the Gulf of Maine were conducted during the summers of 1991 and 1992. These surveys used standard line transect techniques to generate estimates of 37 500 (CV = 0.29) in 1991 and 67 500 (CV = 0.23) in 1992 (Palka 1995a). The difference between the two estimates was likely due to inter-annual variation in the distribution of porpoises. A weighted average of the two estimates, 47 200 (CV = 0.19) was used for management purposes.

Considering the abundance and by-catch estimates together, it appeared that approximately 5% of the population was being removed each year as by-catch. Was this level of mortality sustainable? A PVA conducted by the NMFS suggested that the by-catch mortality experienced by this population in the early 1990s was sufficient to pose a plausible risk of extinction within 100 years, but the analysis was severely limited by the information available at that time. In particular, critical information required to parameterize the model, such as the age structure and potential rate of increase, were not available.

In lieu of a detailed PVA, therefore, sustainable removal levels were determined through the PBR approach. In 1995, the PBR was estimated at 403 porpoises per year and by-catches were approximately five times this level (Table 16.2). Thus, it was clear that immediate management action was required.

The initial response of the fishing industry was predictable—to criticize the estimates. Prior to the establishment of the NMFS observer programme, representatives of both Canadian and US gill net fisheries suggested that our results and those of Jim Gilbert and Kate Wynne were flawed because they were derived from interviews. The objections of the fishing industry became more muted after the NMFS studies independently confirmed the existence of a large by-catch.

In 1990 gill net fishermen from New England began meeting informally with fisheries managers, scientists, and representatives of environmental organizations. These meetings led to the formation of the New England Harbor Porpoise Working Group, which played a pivotal role in the development of conservation measures to address the by-catch of harbour porpoises in the Gulf of Maine. The Working Group met until 1996, when it was replaced by the Gulf of Maine Harbour Porpoise Take Reduction Team (TRT).

The TRT considered two measures to reduce the by-catch of porpoises in the Gulf of Maine: time-area fishery closures; and the use of acoustic alarms, widely known as 'pingers.' The first measure was unpopular with fishermen because it restricted their access to preferred fishing grounds. Scientists, including myself, were sceptical about alarms, but agreed to conduct an experiment to test their efficacy in 1994. The experiment involved 15 commercial fishermen who agreed to conform to a strict experimental protocol. The results of this double-blind field experiment demonstrated that alarms reduced the by-catch of porpoises by 92%, without affecting the catch of target species (Kraus *et al.* 1997). The TRT developed a complex plan in which some areas of the Gulf of Maine were closed to gill net fishing and others required the use of pingers. This plan was submitted to the NMFS in 1996, after six months of intensive negotiation.

NMFS failed to act on the recommendations, resulting in a second lawsuit settled in 1998 which required the NMFS to implement the plan. The regulations codified the use of alarms and time–area fisheries closures in the Gulf of Maine fishery. The presence of observers allowed continuous monitoring of the plan's efficacy. Initially, the strategy worked and estimates of porpoise by-catches dropped from 1200 in 1996 to 53 in 2001 (Table 16.2). Additional surveys were also conducted, revising the estimate of abundance upward to 89 054 (CV = 0.47) changing the value of the PBR, which was 610 in 2005 (Table 16.2). Unfortunately, after the nadir of 2002, by-catches crept back up to once again exceed the PBR in 2005. This increased by-catch was due to a lack of compliance with the regulations: 173 of 217 fishing trips observed in 2003 were not in compliance with area closure or pinger requirements (Cox *et al.* 2007). The NMFS reconvened the TRT in 2007 to consider a new set of measures and ensure that by-catches did not exceed PBR.

The most recent IUCN assessment of the species, conducted in 1996, lists harbour porpoises as *Vulnerable*, due in large part to by-catch mortality (IUCN 2008). No Red List assessment has been conducted specifically for the Gulf of Maine population. The four populations of the species in the Northwest Atlantic, including that in the Gulf of Maine, are considered of *Special Concern* by COSEWIC, due to the effects of by-catch. In 2001, NMFS removed the Gulf of Maine harbour porpoise population from the candidate list of species under the ESA because of the reductions in by-catch achieved at that time.

This case study illustrates several important points. First, the presence of fisheries observers was critical to providing robust estimates of the magnitude of by-catch mortality. These observations were necessary to persuade the fishing industry that a problem did, in fact, exist. Second, the involvement of the fishing industry and conservation groups early in the process allowed for a full and open assessment of potential conservation options. Discussions held within the New England Harbour Porpoise Working group and later at TRT meetings provided a forum in which options were proposed, evaluated, and accepted or rejected. Fishermen were present throughout the negotiation process and ensured that the outcome was both equitable and effective (at least in principle). Third, it was most helpful to have a clear and measurable conservation objective (PBR). This goal was very useful in evaluating potential mitigation approaches—we were able to ask whether the application of each measure, or combination of measures, would meet the conservation goal of reducing by-catches to below the PBR level. Finally, monitoring the fishery with the observer programme after the regulations were implemented allowed us to document the initial success, and subsequent waning effectiveness, of these conservation measures.

16.10 Future directions

Several topics of research are likely to be of great importance to the future conservation of marine mammal populations. First, it is clear that the effects of anthropogenic sound will become increasingly important in coming years (Hildebrand 2005). Dealing with this threat will require a new set of tools, including controlled exposure experiments and monitoring of behavioural and physiological responses (Nowacek *et al.* 2007).

Second, as human populations continue to expand in coastal areas, the habitats of many marine mammal populations have become urbanized. Many populations of small cetaceans, pinnipeds, sirenians, and even baleen whales (Kraus and Rolland 2007) now inhabit coastal waters adjacent to large urban centres. These marine mammals are exposed to a wide range of human activities including noise, non-acoustic disturbance, pathogens, and habitat modification. Teasing apart the effects of these individual factors and assessing their cumulative impact will require a new approach to population assessment and conservation. The same can be said for assessing the multiple effects of long-term climate change on marine mammal populations (Moore 2005).

As the number of populations of threatened and endangered marine mammal populations grows, we must learn to use our scarce conservation resources more effectively. Recently, the Marine Mammal Commission (2008) attempted to estimate the costs of marine mammal conservation efforts in the United States. The effort was hampered because there is no centralized system for tracking expenditures or determining the effectiveness of particular conservation actions. From this initial review it is clear, however, that a large expenditure does not necessarily translate into effective conservation action.

For example, Steller sea lions (*Eumetopias jubatus*) in the Aleutian Islands declined by more than 80% during the 1980s but, despite $US190 million being spent on research, the factors responsible for this decline are still unclear (Morell 2008). And $US45 million was somehow insufficient to reduce the number of right whales killed by entanglement and ship strikes between 2003 and 2006 (Reeves *et al.* 2007). In the future we will have to make better use of such resources to conserve marine mammal biodiversity.

16.11 Conclusions

Much of the work in conserving marine mammal populations is done by lawyers, economists, policy analysts, and other professionals who do not study marine mammals for a living. Thus, there is ample room to work in marine mammal conservation without a background in biology. It is also true that effective conservation practitioners are able to work across disciplinary boundaries. If you are a natural scientist interested in marine mammal conservation, you should learn how policy is formulated and implemented and, at least in the United States, make sure you understand the powerful role of litigation in conservation.

Finally, there are some cases in which the natural sciences have little to contribute to the conservation of marine mammals. For example, the critically endangered vaquita (*Phocoena sinus*) numbers only a few hundred individuals, yet is still taken as by-catch in a variety of gill net fisheries in the northern Gulf of California, Mexico. Biology has no role left to contribute to the conservation of this species (Jaramillo-Legoretta *et al.* 2007) other than to document its demise, if the actions needed to eliminate by-catch are not taken in the very near future. Scientists, interested in the conservation of this species do not need to conduct more surveys, create more models, or analyse more data. Instead, we are just another group of stakeholders engaged in a societal dialogue about the value of this species and the costs and benefits associated with its conservation.

References

Aarts, G., Mackenzie, M., McConnell, B., Fedak, M., and Matthiopoulos, J. (2008). Estimating space-use and habitat preference from wildlife telemetry data. *Ecography*, **31**, 140–160.

Abdo, Z., Crandall, K.A., and Joyce, P. (2004). Evaluating the performance of likelihood methods for detecting population structure and migration. *Molecular Ecology*, **13**, 837–851.

Acevedo-Gutierrez, A., Croll, D.A., and Tershy, B.R. (2002). High feeding costs limit dive time in the largest whales. *Journal of Experimental Biology*, **205**, 1747–1753.

Ackman, R.G. (1986). WCOT (capillary) gas–liquid chromatography. In R.J. Hamilton and J.B. Rossell (eds), *Analysis of Oils and Fats*, pp. 137–206. Elsevier, London.

Ackman, R.G. (2002). The gas chromatograph in practical analysis of common and uncommon fatty acids for the 21st century. *Analytica Chimica Acta*, **465**, 175–192.

Acquarone, M. and Born, E.W. (2007). Estimation of water pool size, turnover rate and body composition of free-ranging Atlantic walruses (*Odobenus rosmarus rosmarus*) studied by isotope dilution. *Journal of the Marine Biological Association of the United Kingdom*, **87**, 77–84.

Adams, G.P., Testa, J.W., Goertz, C.E.C., Ream, R.R., and Sterling, J.T. (2007). Ultrasonographic characterization of reproductive anatomy and early embryonic detection in the northern fur seal (*Callorhinus ursinus*) in the field. *Marine Mammal Science*, **23**, 445–452.

Adams, J.D., Speakman, T., Zolman, E., and Schwacke, L.E. (2006). Automating image matching, cataloging, and analysis for photo-identification research. *Aquatic Mammals*, **32**, 374–384.

Aitken, N., Smith, S., Schwarz, C., and Morin, P.A. (2004). Single nucleotide polymorphism (SNP) discovery in mammals: A targeted-gene approach. *Molecular Ecology*, **13**, 1423–1431.

Akamatsu, T., Wang, D., Wang, K.X., and Naito, Y. (2000). A method for individual identification of echolocation signals in free-ranging finless porpoises carrying data loggers. *Journal of the Acoustical Society of America*, **108**, 1353–1356.

Allen, K. (1980). *Conservation and Management of Whales*. Washington Sea Grant Publications, Washington.

Altmann, J. (1974). Observational study of behavior: sampling methods. *Behaviour*, **49**, 227–267.

Amano, M. and Yoshioka, M. (2003). Sperm whale diving behavior monitored using a suction-cup-attached TDR tag. *Marine Ecology Progress Series*, **258**, 291–295.

Amos, B., Schlötterer, C., and Tautz, D. (1993). Social structure of pilot whales revealed by analytical DNA profiling. *Science*, **260**, 670–672.

Anas, R.E. (1970). Accuracy in assigning ages to fur seals. *Journal of Wildlife Management*, **34**, 844–852.

Andersen, K.H., Nielsen, A., Thygesen, U.H., Hinrichsen, H.H., and Neuenfeldt, S. (2007). Using the particle filter to geolocate Atlantic cod (*Gadus morhua*) in the Baltic Sea, with special emphasis on determining uncertainty. *Canadian Journal of Fisheries and Aquatic Sciences*, **64**, 618–627.

Anderson, D.J. (1982). The home range: A new nonparametric estimation method. *Ecology*, **63**, 103–112.

Anderson D.R. and Burnham K.P. (1999). Understanding information criterion for selection among capture–recapture or ring recovery models. *Bird Study*, **46**, 14–21.

Anderson D.R., Burnham K.P., and Thompson W.L. (2000). Null hypothesis testing: problems, prevalence, and an alternative. *Journal of Wildlife Management*, **64**, 912–923.

Anderson, P.K. (2004). Habitat niche and evolution of sirenian mating systems. *Journal of Mammalian Evolution*, **9**, 55–98.

Anderson-Sprecher, R. (1994). Robust estimates of wildlife location using telemetry data. *Biometrics*, **50**, 406–416.

Anderson-Sprecher, R. and Ledolter, J. (1991). State–space analysis of wildlife telemetry data. *Journal of the American Statistical Association*, **86**, 596–602.

Anderson-Sprecher, R. and Lenth, R.V. (1996). Spline estimation of paths using bearings-only tracking data. *Journal of the American Statistical Association*, **91**, 276–283.

Andreassen, H.P., Ims, R.A., Stenseth, N.C., and Yoccoz, N. (1993). Investigating space use by means of radiotelemetry and other methods: a methodological guide. In N.C. Stenseth and R.A. Ims (eds), *The Biology of Lemmings*. The Linnean Society of London, vol. 63, pp. 103–112.

Andrews, R.D. (1998). Remotely releasable instruments for monitoring the foraging behaviour of pinnipeds. *Marine Ecology Progress Series*, **175**, 289–294.

Andrews, R.D., Pitman, R.L., and Ballance, L.T. (2008). Satellite tracking reveals distinct movement patterns for Type B and Type C killer whales in the southern Ross Sea, Antarctica. *Polar Biology*, **31**, 1461–1468.

Angliss, R.P. and Outlaw, R.B. (2006). Northern Fur Seal (*Callorhinus ursinus*): Eastern Pacific Stock. Rep. No. NOAA-TM-AFSC-168. Seattle, Washington.

Angliss, R. and Wade, P.R. (1997). Guidelines for assessing marine mammals stocks. *Report of the Gamms Workshop*, April 3–5, 1996. NOAA Technical Memorandum Nmfs-opr-12. Seattle, Washington.

Anon. (1967). Standard measurements of seals. *Journal of Mammalogy*, **48**, 459–462.

Anon. (1974). *Report of the Meeting on Age Determination in Baleen Whales*, Reykjavik, 5–12 April, 1973. Twenty Fourth Report of the Commission, International Commission on Whaling, Annex E, pp. 62–68. London.

Antonelis, G.A., Lowry, M.S., DeMaster, D.P., and Fiscus C.H. (1987). Assessing northern elephant seal feeding habits by stomach lavage. *Marine Mammal Science*, **3**, 308–322.

Antonelis G.A., Baker, J.D., Johanos, T.C., Braun, R.C., and Harting, A.L. (2006). Hawaiian monk seal (*Monachus schauinslandi*): status and conservation issues. *Atoll Research Bulletin,* **543**, 75–101.

AOAC. (2000). Chapter 33. In W. Horwitz (ed.), *Official Methods of Analysis*, p 18. AOAC International, Gaithersburg, MD.

Araabi, B.N., Kehtarnavaz, N., McKinney, T., Hillman, G., and Würsig, B. (2000). A string matching computer-assisted system for dolphin photoidentification. *Annals of Biomedical Engineering*, **28**, 1269–1279.

Argos. (2008). *Argos User's Manual*. CLS Argos, Toulouse, France.

Arim, M. and Naya, D.E. (2003). Pinniped diets inferred from scats: analysis of biases in prey occurrence. *Canadian Journal of Zoology*, **81**, 67–73.

Arnbom, T.A., Lunn, N.J., Boyd, I.L., and Barton, T. (1992). Aging live Antarctic fur seals and southern elephant seals. *Marine Mammal Science*, **8**, 37–43.

Arnould, J.P.Y. (1995). Indices of body condition and body composition in female Antarctic fur seals (*Arctocephalus gazella*). *Marine Mammal Science*, **11**, 301–313.

Arnould, J.P.Y., Boyd, I.L., and Speakman, J.R. (1996). Measuring the body composition of Antarctic fur seals (*Arctocephalus gazella*) by hydrogen isotope dilution. *Physiological Zoology*, **69**, 93–116.

Arnould, J., Nelson, M., Nichols, P., and Oosthuizen, W. (2005). Variation in the fatty acid composition of the blubber in Cape fur seals (*Arctocephalus pusillus pusillus*) and the implications for dietary interpretation. *Journal of Comparative Physiology B*, **175**, 285–295.

Arthur, S.M., Manly, B.F.J., McDonald, L.L., and Garner, G.W. (1996). Assessing habitat selection when availability changes. *Ecology*, **77**, 215–227.

Asper, E.D. (1975). Techniques of live capture of smaller Cetacea. *Journal of the Fisheries Research Board of Canada*, **32**, 1191–1196.

Assenburg, C., Harwood, J., Matthiopoulos, J., and Smout, D. (2006). The functional response of generalist predators and its implications for the monitoring of marine ecosystems. In I.L. Boyd and S. Wanless (eds), *Management of Marine Ecosystems: Monitoring Changes in Upper Trophic Levels*, pp. 262–274. Zoological Society of London, London.

Atwell, L., Hobson, K.A., and Welch, H.E. (1998). Biomagnification and bioaccumulation of mercury in an Arctic marine food web: Insights from stable-nitrogen isotope analysis. *Canadian Journal of Fisheries and Aquatic Science*, **55**, 1114–1121.

Augustin, N.H., Mugglestone, M.A., and Buckland, S.T. (1996). An autologistic model for the spatial distribution of wildlife. *Journal of Applied Ecology*, **33**, 339–347.

Austin, D., McMillan, J.I., and Bowen, W.D. (2003). A three-stage algorithm for filtering erroneous Argos satellite locations. *Marine Mammal Science*, **19**, 371–383.

Austin, D., Bowen, W.D., and McMillan, J.I. (2004). Intraspecific variation in movement patterns: modeling individual behaviour in a large marine predator. *Oikos*, **105**, 15–30.

Austin, D., Bowen, W.D., McMillan, J.I., and Boness, D.J. (2006a). Stomach temperature telemetry reveals temporal patterns of foraging success in a free-ranging marine mammal. *Journal of Animal Ecology*, **75**, 408–420.

Austin, D., Bowen, W.D., McMillan, J.I., and Iverson, S.J. (2006b). Linking movement, diving, and habitat to foraging success in a large marine predator. *Ecology*, **87**, 3095–3108.

Austin, M.P. (2002). Spatial prediction of species distribution: an interface between ecological theory and statistical modelling. *Ecological Modelling*, **157**, 101–118.

Avise, J.C. (1989). Gene trees and organismal histories: a phylogenetic approach to population biology. *Evolution*, **43**, 1192–1208.

Avise, J.C. (1995). Mitochondrial DNA polymorphism and a connection between genetics and demography of relevance to conservation. *Conservation Biology*, **9**, 686–690.

Avise, J.C., Arnold, J., Ball, R.M., Bermingham, E., Lamb, T., Neigel, J.E., Reeb, C.A., and Saunders, N.C. (1987). Intraspecific phylogeography: The mitochondrial DNA bridge between population genetics and systematics. *Annual Reviews of Ecology and Systematics*, **18**, 489–522.

Azevedo, A.F., Oliveira, A.M., Viana, S.C., and VanSluys, M. (2007). Habitat use by marine tucuxis (*Sotalia guianensis*). (Cetacea: Delphinidae) in Guanabara Bay, south-eastern Brazil. *Journal of the Marine Biological Association of the U.K.*, **87**, 201–205.

Bada, J.L., Brown, S., and Masters, P.M. (1980). Age determination of marine mammals based on aspartic acid racemization in the teeth and lens nucleus. In W.F. Perrin and A.C. Myrick Jr (eds), *Age Determination of Tooth Whales and Sirenians*, pp. 113–118. International Whaling Commission, London.

Baechler, J., Beck, C.A., and Bowen, W.D. (2002). Dive shapes reveal temporal changes in the foraging behaviour of different age and sex classes of harbour seals (Phoca vitulina). *Canadian Journal of Zoology*, **80**, 1569–1577.

Bailey, H. and Thompson, P. (2006). Quantitative analysis of bottlenose dolphin movement patterns and their relationship with foraging. *Journal of Animal Ecology*, **75**, 456–465.

Baird, R.W., Ligon, A.D., Hooker, S.K., and Gorgone, A.M. (2001). Subsurface and nighttime behaviour of pantropical spotted dolphins in Hawaii. *Canadian Journal of Zoology* **79**, 988–996.

Baird, R.W., Hanson, M.B., and Dill, L.M. (2005). Factors influencing the diving behaviour of fish-eating killer whales: sex differences and diel and interannual variation in diving rates. *Canadian Journal of Zoology*, **83**, 257–267.

Baker, A.S., Ruoff, K.L., and Madoff, S. (1998). Isolation of *Mycoplasma* species from a patient with seal finger. *Clinical Infectious Diseases*, **27**, 1168–1170.

Baker, J.D. (2004). Evaluation of closed capture–recapture methods to estimate abundance of Hawaiian monk seals, Monachus schauinslandi. *Ecological Applications*, **14**, 987–998.

Baker, J.D. and Thompson, P.M. (2007). Temporal and spatial variation in age-specific survival rates of a long-lived mammal, the Hawaiian monk seal. *Proceedings of the Royal Society B*, **274**, 407–415.

Balance, L.T., Pitman, R.L., and Fiedler, P.C. (2006). Oceanographic influences on seabirds and cetaceans of the eastern tropical Pacific: A review. *Progress in Oceanography*, **69**, 360–390.

Ballou, J.D., Gilpin, M., and Foose, J.T. (eds), (1995). *Population Management for Survival and Recovery*. Columbia University Press, New York.

Barker, R.J. (1997). Joint modelling of live-recapture, tag-resight and tag-recovery data. *Biometrics*, **53**, 666–677.

Barlow, J. and Clapham, P.J. (1997). A new birth-interval approach to estimating demographic parameters of humpback whales. *Ecology*, **78**, 535–546.

Barlow, J. and Taylor, B.L. (2005). Estimates of sperm whale abundance in the northeastern temperate Pacific from a combined acoustic and visual survey. *Marine Mammal Science*, **21**, 429–445.

Barlow, J., Oliver, C.W., Jackson, T.D., and Taylor, B.L. (1988). Harbor porpoise, *Phocoena phocoena*, abundance estimation for California, Oregon and Washington: aerial surveys. *Fishery Bulletin*, **86**, 433–444.

Barlow, J., Gerrodette, T., and Forcada, J. (2001). Factors affecting perpendicular sighting distances on shipboard line transect surveys for cetaceans. *Journal of Cetacean Research and Management*, **3**, 201–212.

Barrett-Lennard, L.G. (2000). Population structure and mating patterns of killer whales (*Orcinus orca*) as revealed by DNA analysis. PhD thesis, University of British Columbia, Vancouver, BC.

Bartumeus, F., Da Luz, M.G.E., Viswanathan, G.M., and Catalan, J. (2005). Animal search strategies: A quantitative. random-walk analysis. *Ecology*, **86**, 3078–3087.

Bas, W., Beekmans, P.M., Whitehead, H., Huele, R., Steiner, L., and Steenbeek, A.G. (2005). Comparison of two computer-assisted photo-identification methods applied to sperm whales (*Physeter macrocephalus*). *Aquatic Mammals*, **31**, 243–247.

Baumgartner, M.F. and Mate, B.R. (2003). Summertime foraging ecology of North Atlantic right whales. *Marine Ecology—Progress Series*, **264**, 123–135.

Baumgartner, M.F. and Mate, B.R. (2005). Summer and fall habitat of North Atlantic right whales (*Eubenaena glacialis*) inferred from satellite telemetry. *Canadian Journal of Fisheries and Aquatic Science*, **62**, 527–543.

Baur, D.C., Bean, M.J., and Gosliner, M.L. (1999). The laws governing marine mammal conservation in the United States. In J.R. Twiss and R.R. Reeves (eds), *Conservation and Management of Marine Mammals*, pp. 48–86. Smithsonian Institution Press, Washington, DC.

Bayly, K.L., Evans, C.S., and Taylor, A. (2006). Measuring social structure: a comparison of eight dominance indices. *Behavioural Processes*, **73**, 1–12.

Bearzi, M. (2005a). Habitat partitioning by three species of dolphins in Santa Monica bay, California. *Bulletin of the Southern California Academy of Science*, **104**, 113–124.

Bearzi, M. (2005b). Dolphin sympatric ecology. *Marine Biology Research*, **1**, 165–175.

Beasley, I., Robertson, K., and Arnold, P.W. (2005). Description of a new dolphin, the Australian snubfin dolphin *Orcaella heinsohni* sp. n. (Cetacea, Delphinidae). *Marine Mammal Science*, **21**, 365–400.

Beaumont, M.A. and Nichols, R.A. (1996). Evaluating loci for use in the genetic analysis of population structure. *Proceedings of the Royal Society of London Series B*, **263**, 1619–1626.

Beauplet, G., Barbraud, C., Dabin, W., Kussener, C., and Guinet, C. (2006). Age-specific survival and reproductive performances in fur seals: evidence of senescence and individual quality. *Oikos*, **112**, 430–441.

Beck, C.A., Bowen, W.D., and Iverson, S.J. (2000). Seasonal changes in buoyancy and diving behaviour of adult grey seals. *Journal of Experimental Biology*, **203**, 2323–2330.

Beck, C.A., McMillan, J.I., and Bowen, W.D. (2002). An algorithm to improve geolocation positions using sea surface temperature and diving depth. *Marine Mammal Science*, **18**, 940–951.

Beck, C.A., Bowen, W.D., McMillan, J.I., and Iverson, S.J. (2003a). Sex differences in diving at multiple temporal scales in a size-dimorphic capital breeder. *Journal of Animal Ecology*, **72**, 979–993.

Beck, C.A., Bowen, W.D., and Iverson, S.J. (2003b). Seasonal energy storage and expenditure in a phocid seal: evidence of sex-specific trade-offs. *Journal of Animal Ecology*, **72**, 280–291.

Beck, C.A., Iverson, S.J., and Bowen, W.D. (2005). Blubber fatty acids of grey seals reveal sex differences in diet of a size-dimorphic marine carnivore *Canadian Journal of Zoology*, **83**, 377–388.

Beck, C.A., Iverson, S.J., Bowen, W.D., and Blanchard, W. (2007). Sex differences in grey seal diet reflect seasonal variation in foraging behaviour and reproductive expenditure: evidence from quantitative fatty acid signature analysis. *Journal of Animal Ecology*, **76**, 490–502.

Beddington, J. and Taylor, D. (1973). Optimum age specific harvesting of a population. *Biometrics*, **29**, 801–809.

Beerli, P. and Felsenstein, J. (1999). Maximum-likelihood estimation of migration rates and effective population numbers in two populations using a coalescent approach. *Genetics*, **152**, 763–773.

Beerli, P. and Felsenstein, J. (2001). Maximum likelihood estimation of a migration matrix and effective population sizes in n subpopulations by using a coalescent approach. *Proceedings of the National Academy of Sciences of the United States of America*, **98**, 4563–4568.

Begon, M., Harper, J., and Townsend, C. (1996). *Ecology: Communities, Populations and Individuals*. Blackwell Scientific, Oxford.

Beissinger, S.R. and McCullough, D.R. (eds). (2002). *Population Viability Analysis*. University of Chicago Press, Chicago, IL

Bejder, L. and Dawson, S. (2001). Abundance, residency, and habitat utilisation of Hector's dolphins (*Cephalorhynchus hectori*) in Porpoise Bay, New Zealand. *New Zealand Journal of Marine and Freshwater Research*, **35**, 277–287.

Bejder, L., Fletcher, D., and Bräger, S. (1998). A method for testing association patterns of social animals. *Animal Behaviour*, **56**, 719–725.

Bejder, L., Samuels, A., Whitehead, H., Gales, N., Mann, J., Connor, R., Heithaus, M., Watson-Capps, J., Flaherty, C., and Krutzen, M. (2006). Decline in relative abundance of bottlenose dolphins exposed to long-term disturbance. *Conservation Biology*, **20**, 1791–1798.

Bekkby, T. and Bjorge, A. (1998). Variation in stomach temperature as indicator of meal size in harbor seals, *Phoca vitulina*. *Marine Mammal Science*, **14**, 627–637.

Bekoff, M. (2002). Ethics and marine mammals. In W.F. Perrin, B. Würsig, and J.G.M. Thewissen (eds), *Encyclopedia of Marine Mammals*, pp. 398–404. Academic Press, San Diego, CA.

Bell, C.M., Hindell, M.A., and Burton, H.R. (1997). Estimation of body mass in the southern elephant seal, Mirounga leonina, by photogrammetry and morphometrics. *Marine Mammal Science*, **13**, 669–682.

Benhamou, S. (2004). How to reliably estimate the tortuosity of an animal's path: straightness, sinuosity, or fractal dimension? *Journal of Theoretical Biology*, **229**, 209–220.

Benoit-Bird, K.J. and Au, W.W.L. (2003). Prey dynamics affect foraging by a pelagic predator (*Stenella longirostris*) over a range of spatial and temporal scales. *Behavioral Ecology and Sociobiology*, **53**, 364–373.

Bensasson, D., Zhang, D.X., Hartl, D.L., and Hewitt, G.M. (2001). Mitochondrial pseudo-genes: evolution's misplaced witnesses. *Trends in Ecology and Evolution*, **16**, 314–321.

Bensch, S. and Akesson, M. (2005). Ten years of AFLP in ecology and evolution: why so few animals? *Molecular Ecology*, **14**, 2899–2914.

Benson, A. and Trites, A. (2002). Ecological effects of regime shifts in the Bering Sea and eastern North Pacific Ocean. *Fish and Fisheries*, **3**, 95–113.

Berg, J. (1979). Discussion of the methods of investigating the food of fishes with reference to a preliminary study of the food of *Gobiusculus flavescent* (Gobidae). *Marine Biology*, **50**, 263–273.

Bernard, H.J. and Hohn, A.A. (1989). Differences in feeding habits between pregnant and lactating spotted dolphins (*Stenella attenuata*). *Journal of Mammalogy*, **70**, 211–215.

Bernt, K.E., Hammill, M.O., and Kovacs, K.M. (1996). Age determination of grey seals (*Halichoerus grypus*) using incisors. *Marine Mammal Science*, **12**, 476–482.

Best, P.B. (1979). Social organization in sperm whales, Physeter macrocephalus. In H.E. Winn and B.L. Olla (eds), *Behavior of Marine Animals*, pp. 227–289. Plenum, New York.

Best, P.B. and Schell, D.M. (1996). Stable isotopes in southern right whale (*Eubalaena australis*) baleen as indicators of seasonal movements, feeding and growth. *Marine Biology*, **124**, 483–494.

Bigg, M. (1982). An assessment of killer whale (*Orcinus orca*) stocks off Vancouver Island, British Columbia. *Report of the International Whaling Commission*, **32**, 655–666.

Bigg, M.A. and Fawcett, I. (1985). Two biases in diet determination of northern fur seals (*Callorhinus ursinus*). In J.R. Beddington, R.J.H. Beverton, and D.M. Lavigne (eds), *Marine Mammals and Fisheries*, pp. 284–291. George Allen and Unwin Ltd, London.

Bigg, M.A. and Olesiuk, P.F. (1990). An enclosed elutriator for processing marine mammal scats. *Marine Mammal Science*, **6**, 350–355.

Bigg, M.A. and Perez, M.A. (1985). Modified volume: A frequency-volume method to assess marine mammal food habits. In J.R. Bedding-ton, R.J.H. Beverton, and D.M. Lavigne (eds), *Marine Mammals and Fisheries*, pp. 277–283. George Allen and Unwin Ltd, London.

Bigg, M.A., Olesiuk, P.F., Ellis, G.M., Ford, J.K. B., and Balcomb, K.C. (1990). Social organizations and genealogy of resident killer whales (*Orcinus orca*) in the coastal waters of British Columbia and Washington State. *Report of the International Whaling Commission*, Special Issue **12**, 383–405.

Bininda-emonds, O.R.P. (2004). The evolution of supertrees. *Trends in Ecology and Evolution*, **19**, 315–322.

Biuw, M., McConnell, B., Bradshaw, C.J.A., Burton, H., and Fedak, M. (2003). Blubber and buoyancy: monitoring the body condition of free-ranging seals using simple dive characteristics. *Journal of Experimental Biology*, **206**, 3405–3423.

Biuw, E., Boehme, L., Guinet, C., Hindell, M., Costa, D., Charrassin, J-B., Roquet, F., Bailleul, F., Meredith, M., Thorpe, S., Tremblay, Y., McConnell, B., Park, Y-H., Rintoul, S., Bindoff, N., Goebel, M., Crocker, D., Lovell, P., Nicholson, J., Monks, F., and Fedak, M.A. (2007). Variations in behavior and condition of a Southern Ocean top predator in relation to in situ oceanographic conditions. *Proceedings of the National Academy of Sciences of the United States of America*, **104**, 13705–13710.

Bjorge, A. and Donovan, G.P. (1995). *Biology of the Phocoenids, Special Issue 16* International Whaling Commission, Cambridge, UK.

Blackwell, S.B., Haverl, C.A., Le Boeuf, B.J., and Costa. D.P. (1999). A method for calibrating swim-speed recorders. *Marine Mammal Science*, **15**, 894–905.

Blaxter, K. (1989). *Energy Metabolism in Animals and Man*. Cambridge University Press, Cambridge, UK.

Bligh, E.G. and Dyer, W.J. (1959). A rapid method of total lipid extraction and purification. *Canadian Journal of Biochemistry and Physiology*, **37**, 911–917.

Block, B.A., Costa, D.P., Boehlert, G.W., and Kochevar, R.E. (2002). Revealing pelagic habitat use: the tagging of Pacific pelagics program. *Oceanologica Acta*, **25**, 255–266.

Blundell, G.M. and Pendleton, G.W. (2008). Estimating age of harbor seals (*Phoca vitulina*)

with incisor teeth and morphometrics. *Marine Mammal Science*, **24**, 577–590.

Boehme, L., Thorpe, S.E., Biuw, M., Fedak, M., and Meredith, M.P. (2008). Monitoring Drake Passage with elephant seals: Frontal structures and snapshots of transport. *Limnology and Oceanography*, **53**, 2350–2360.

Boily, F., Beaudoin, S., and Measures L.N. (2006). Hematology and serum chemistry of harp (*Phoca groenlandica*) and hooded seals (*Cystophora cristata*) during the breeding season, in the Gulf of St. Lawrence, Canada. *Journal of Wildlife Diseases*, 42, 115–132.

Boily, P. and Lavigne, D.M. (1995). Resting metabolic rates and respiratory quotients of gray seals (*Halichoerus grypus*) in relation to time of day and duration of food deprivation. *Physiological Zoology*, **68**, 1181–1193.

Boltnev, A.I. and York, A.E. (1998). Northern fur seals young: interrelationships among birth size, growth, and survival. *Canadian Journal of Zoology*, **76**, 843–854.

Boness, D.J. (1991). Determinants of mating systems in the Otariidae (Pinnipedia). In D. Renouf (ed.), *Behaviour of Pinnipeds*, pp. 1–44. Chapman and Hall, London.

Boness, D.J., Bowen, W.D., and Oftedal, O.T. (1994). Evidence of a maternal foraging cycle resembling that of otariid seals in a small phocid the harbor seal. *Sociobiology*, **34**, 95–104.

Boness, D.J., Bowen, W.D., and Iverson, S.J. (1995). Does male harassment of females contribute to reproductive synchrony in the grey seal by affecting maternal performance? *Behavioral Ecology and Sociobiology*, **36**, 1–10.

Bonner, W. (1989). *The Natural History of Seals*. Christopher Helm (Publishers) Ltd, Bromley, Kent.

Borchers, D.L. (1999). Composite mark–recapture line-transect surveys. In G.W. Garner, S.C. Armstrup, J.L. Laake, B.F.J. Manly, L.L. McDonald, and D.G. Robertson (eds), *Marine Mammal Survey and Assessment Methods*, pp. 115–126. Balkema, Rotterdam.

Borchers, D.L., Buckland, S.T., and Zucchini, W. (2002). *Estimating Animal Abundance: Closed Populations*. Springer-Verlag, London.

Bordino, P., Wells, R.S., and Stamper, M.A. (2007). Site fidelity of Franciscana dolphins *Pontoporia blainvillei* off Argentina. *17th Biennial Conference on the Biology of Marine Mammals*, 29 November–3 December, Cape Town, South Africa.

Born, E.W., Outridge, P., Riget, F.F., Hobson, K.A., Dietz, R., Øien, N., and Haug, T. (2003). Population substructure of North Atlantic minke whales (*Balaenoptera acutostrata*) inferred from regional variation of elemental and stable isotopic signatures in tissues. *Journal of Marine Systems*, **43**, 1–17.

Boschi, E., Fishbach, C., and Iorio, M. (1992). Catálogo ilustrado de los crustáceos estomatópodos y decápodos marinos de Argentina. *Frente Marítimo*, **10**, 56–57.

Boveng, P.L., Bengston, J.L., Withrow, D.E., Cesarone, J.C., Simpkins, M.L., Frost, K.J., and Burns, J.J. (2003). The abundance of harbor seals in the Gulf of Alaska. *Marine Mammal Science*, **19**, 111–127.

Bovet, P. and Benhamou, S. (1988). Spatial-analysis of animals movements using a correlated random-walk model. *Journal of Theoretical Biology*, **131**, 419–433.

Bowen, W.D. (1990). Population biology of sealworm (*Pseudoterranova decipiens*) in relation to its intermediate and seal hosts *Canadian Bulletin of Fisheries and Aquatic Sciences*, **222**, 243–254.

Bowen, W.D. (1991). Behavioral ecology of pinniped neonates. In D. Renouf (ed.), *Behaviour of Pinnipeds*, pp. 66–117. Chapman and Hall, London.

Bowen, W.D. (1997). Role of marine mammals in aquatic ecosystems. *Marine Ecology Progress Series*, **158**, 267–274.

Bowen, W.D. (2000). Reconstruction of pinniped diets: accounting for complete digestion of otoliths and cephalopod beaks. *Canadian Journal of Fisheries and Aquatic Science*, **57**, 898–905.

Bowen, W.D. and Harrison, G.D. (1994). Offshore diet of grey seals *Halichoerus grypus* near Sable Island, Canada. *Marine Ecology Progress Series*, **112**, 1–11.

Bowen, W.D. and Iverson, S.J. (1998). Estimation of total body water in pinnipeds using hydrogen-isotope dilution. *Physiological Zoology*, **71**, 329–332.

Bowen, W.D. and Sergeant, D.E. (1983). Mark–recapture estimates of harp seal pup (*Phoca groenlandica*) production in the northwest Atlantic. *Canadian Journal of Fisheries and Aquatic Sciences*, **40**, 728–742.

Bowen, W.D., Capstick, C.K. and Sargent, D.E. (1981). Temporal changes in the reproductive potential of female harp seals (*Pagophilus*

groenlandicus). *Canadian Journal of Fisheries and Aquatic Sciences*, **38**, 495–503.

Bowen, W.D., Sergeant, D.E. and Øritsland, T. (1983). Validation of age determination in the harp seal, *Phoca groenlandica*, using dentinal annuli. *Canadian Journal of Fisheries and Aquatic Science*, **40**, 1430–1441.

Bowen, W.D., Oftedal, O.T., and Boness, D.J. (1992). Mass and energy transfer during lactation in a small phocid, the harbor seal (*Phoca vitulina*). *Physiological Zoology*, **65**, 844–866.

Bowen, W.D., Boness, D.J., and Iverson, S.J. (1998). Estimation of total body water in harbor seals: how useful is bioelectrical impedance analysis? *Marine Mammal Science*, **14**, 765–777.

Bowen, W.D., Beck, C.A., and Iverson, S.J. (1999). Bioelectrical impedance analysis as a means of estimating total body water in grey seals. *Canadian Journal of Zoology*, **77**, 418–422.

Bowen, W.D., Iverson, S.J., Boness, D.J., and Oftedal, O.T. (2001). Foraging effort, food intake and lactation performance depend on maternal mass in a small phocid seal. *Functional Ecology*, **15**, 325–334.

Bowen, W.D., Tully, D., Boness, D.J., Bulheier, B.M., and Marshall, G.J. (2002). Prey-dependent foraging tactics and prey profitability in a marine mammal. *Marine Ecology Progress Series*, **244**, 235–245.

Bowen, W.D., McMillan, J.I., and Mohn, R. (2003). Sustained exponential population growth of the grey seal on Sable Island. *ICES Journal of Marine Science*, **60**, 1265–1274.

Bowen, W.D., Iverson, S.J., McMillan, J.I., and Boness, D.J. (2006). Reproductive performance in grey seals: age-related improvement and senescence in a capital breeder. *Journal of Animal Ecology*, **75**, 1340–1351.

Bowen, W.D., McMillan, J.I., and Blanchard, W. (2007). Reduced population growth of gray seals at Sable Island: evidence from pup production and age of primiparity. *Marine Mammal Science*, **23**, 48–64.

Boyce, M.S. and McDonald, L.L. (1999). Relating populations to habitats using resource selection functions. *Trends in Ecology and Evolution*, **14**, 268–272.

Boyce, M.S., Vernier, P.R., Nielsen, S.E., and Schmiegelow, F.K.A. (2002). Evaluating resource selection functions. *Ecological Modelling*, **157**, 281–300.

Boyd, I.L. (1984). Development and regression of the corpus luteum in grey seal (*Halichoerus grypus*) ovaries and its use in determining fertility rates. *Canadian Journal of Zoology*, **62**, 1095–1100.

Boyd, I.L. (1993). Selecting sampling frequency for measuring dive behavior. *Marine Mammal Science*, **9**, 424–430.

Boyd, I.L. (1996). Temporal scales of foraging in a marine predator. *Ecology*, **77**, 426–434.

Boyd, I.L. (1999). Foraging and provisioning in Antarctic fur seals: Interannual variability in time-energy budgets. *Behavioral Ecology*, **10**, 198–208.

Boyd, I.L. (2002a). Energetics: consequences for fitness. In A.R. Hoelzel (ed.), *Marine Mammal Biology: An Evolutionary Approach*, pp 247–277. Blackwell Publishing, Oxford.

Boyd, I.L. (2002b). Estimating food consumption of marine predators: Antarctic fur seals and macaroni penguins. *Journal of Applied Ecology*, **39**, 103–119.

Boyd, I.L. (2002c). The cost of information: should black rhinos be immobilized? *Journal of Zoology, London*, **258**, 277.

Boyd, I.L. (2002d). Integrated environment–prey–predator interactions off South Georgia: implications for management of fisheries. *Aquatic Conservation*, **12**, 119–126.

Boyd, I.L. and McCann, T.S. (1989). Pre-natal investment in reproduction by female Antarctic fur seals. *Behavioural Ecology and Sociobiology*, **24**, 377–385.

Boyd, I.L. and Murray, A.W.A. (2001). Monitoring a marine ecosystem using responses of upper trophic level predators. *Journal of Animal Ecology*, **70**, 747–760.

Boyd. I.L. and Roberts, J.P. (1993). Tooth growth in male Antarctic fur seals from South Georgia: an indicator of long-term growth history. *Journal of Zoology, London*, **229**, 177–190.

Boyd, I.L., Arnould, J.P.Y., Barton, T., and Croxall, J.P. (1994). Foraging behaviour of Antarctic fur seals during periods of contrasting prey abundance. *Journal of Animal Ecology*, **63**, 703–713.

Boyd, I.L., Woakes, A.J., Butler, P.J., Davis, R.W., and Williams, T.M. (1995a). Validation of heart rate and doubly labelled water as measures of metabolic rate during swimming in California sea lions. *Functional Ecology*, **9**, 151–160.

Boyd, I.L., Reid, K., and Bevan, R.M. (1995b). Swimming speed and allocation of time during the dive cycle in Antarctic fur seals. *Animal Behaviour*, **50**, 679–784.

Boyd, I.L., Croxall, J.P., Lunn, N.J., and Reid, K. (1995c). Population demography of Antarctic fur seals: the costs of reproduction and implications for life histories. *Journal of Animal Ecology*, **64**, 505–518.

Boyd, I.L., McCafferty, D.J., and Walker, T.R. (1997). Variation in foraging effort by lactating Antarctic fur seals: Response to simulated increased foraging costs. *Behavioral Ecology and Sociobiology*, **40**, 135–144.

Boyd, I.L., Lockyer, C., and Marsh, H.D. (1999a). Reproduction in marine mammals. In J.E. Reynolds and S.A. Rommel (eds), *Biology of Marine Mammals*. Smithsonian Institution, Washington DC.

Boyd, I.L., Bevan, R.M., Woakes, A.J., and Butler, P.J. (1999b). Heart rate and behavior of fur seals: implications for measurement of field energetics. *American Journal of Physiology—Heart and Circulatory Physiology*, **276**, H844–H857.

Boyd, I.L., Hawker, E.J., Brandon, M.A., and Staniland, I.J. (2001). Measurement of ocean temperatures using instruments carried by Antarctic fur seals. *Journal of Marine Systems*, **27**, 277–288.

Bradford, A.L., Wade, P.R., Weller, D.W., Burdin, A.M., Ivashchenko, Y.V., Tsidulko, G.A., VanBlaricom, G.R., and Brownell Jr, R.L. (2006). Survival estimates of western gray whales *Eschrichtius robustus* incorporating individual heterogeneity and temporary emigration. *Marine Ecology Progress Series*, **315**, 293–307.

Bradshaw, C.J.A., Barker, R.J., and Davis, L.S. (2000). Modeling tag loss in New Zealand fur seal pups. *Journal of Agricultural Biological and Environmental Statistics*, **5**, 475–485.

Bradshaw, C.J.A., Higgins, J., Michael, K.J., Wotherspoon, S.J., and Hindel, M.A. (2004). At-sea distribution of female southern elephant seals relative to variation in ocean surface properties. *ICES Journal of Marine Science*, **61**, 1014–1027.

Bradshaw, C.J.A., Sims, D.W., and Hays, G.C. (2007). Measurement error causes scale-dependent threshold erosion of biological signals in animal movement data. *Ecological Applications*, **17**, 628–638.

Branch, T.A. (2007). Abundance of Antarctic blue whales south of 60°S from three complete circumpolar sets of surveys. *Journal of Cetacean Research and Management*, **9**, 253–262.

Brault, S. and Caswell, H. (1993). Pod-specific demography of killer whales (*Orcinus orca*). *Ecology*, **74**, 1444–1454.

Breed, G.A., Bowen, W.D., McMillan, J.I., and Leonard, M.L. (2006). Sexual segregation of seasonal foraging habitats in a non-migratory marine mammal. *Proceedings of the Royal Society B—Biological Sciences*, **273**, 2319–2326.

Breed, G.A., Jonsen, I.D., Myers, R.A., Bowen, W.D., and Leonard, M.L. (2009). Sex-specific, seasonal foraging tactics of adult grey seals (*Halichoerus grypus*) revealed by state–space analysis. *Ecology*, **90**, 3209–3221.

Brook, F., Bonn, W.V., and Jensen, E.D. (2001). Ultrasonography. In L.A. Dierauf and F.M. D. Gulland (eds), *CRC Handbook of Marine Mammal Medicine*, 2nd edn, pp. 593–620. CRC Press, Boca Raton, FL.

Broussard, D.R., Michener, G.R., Risch, T.S., and Dobson, F.S. (2005). Somatic senescence: evidence from female Richarson's ground squirrels. *Oikos*, **108**, 591–601.

Brown, E.G. and Pierce, G.J. (1997). Diet of common seals at Mousa Shetland during the third quarter of 1994. *Journal of the Marine Biological Association (UK)*, **77**, 539–555.

Brown, M.R., Corkeron, P.J., Hale, P.T., Schultz, K.W., and Bryden, M.M. (1995). Evidence for a sex-segregated migration in the humpback whale (*Megaptera novaeangliae*). *Proceedings of the Royal Society London, B*, **259**(1355), 229–234.

Brown, M.W., Kraus, S.D., Slay, K.C., and Garrison, L.P. (2007). Surveying for discovery, science, and management. In S.D. Kraus and R.M. Rolland (eds), *The Urban Whale: North Atlantic Right Whales at the Crossroads*, pp. 105–137. Harvard University Press, Cambridge, MA.

Brown, R.F., Wright, B.E., Riemer, S.D., and Laake, J. (2005). Trends in abundance and current status of harbor seals in Oregon: 1977–2003. *Marine Mammal Science*, **21**, 657–670.

Brownell, R.L. and Ralls, K. (1986). Potential for sperm competition in baleen whales. *Reports of the International Whaling Commission* (Special Issue), **8**, 97–112.

Brownie, C., Anderson, D.R., Burnham, K.P., and Robson, D.S. (1985). *Statistical Inference from Band Recovery Data—A Handbook*, 2nd edn. US Fish and Wildlife Service, Resource Publication 156. Washington DC.

Brownie, C., Hines, J.E., Nichols, J.D., Pollock, K.H., and Hestbeck, J.B. (1993). Capture–recapture studies for multiple strata including non-Markovian transitions. *Biometrics*, **49**, 1173–1187.

Brumfield, R.T., Beerli, P., Nickerson, D.A., and Edwards, S.V. (2003). Single nucleotide polymorphisms (snps) as markers in phylogeography. *Trends in Ecology and Evolution*, **18**, 249–256.

Bryden, M.M. (1972). Body size and composition of elephant seals (*Mirounga leonina*): absolute measurements and estimates from bone dimensions. *Journal of Zoology, London*, **167**, 265–276.

Buck, J.D., Wells, R.S, Rhinehart, H.L., and Hansen, L.J. (2006). Aerobic microorganisms associated with free-ranging bottlenose dolphins in coastal Gulf of Mexico and Atlantic Ocean waters. *Journal of Wildlife Diseases*, **42**, 536–544.

Buckland, S.T. (1992). Fitting density functions with polynomials. *Applied Statistics*, **41**, 63–76.

Buckland, S.T., Burnham, K.P., and Augustin, N.H. (1997). Model selection: An integral part of inference. *Biometrics*, **53**, 603–618.

Buckland, S.T., Gouldie, I.B.J., and Borchers, D.L. (2000). Wildlife population assessment: past developments and future directions. *Biometrics*, **56**, 1–12.

Buckland, S.T., Anderson, D.R., Burnham, K.P., Laake, J.L., Borchers, D.L., and Thomas, L. (2001). *Introduction to Distance Sampling: Estimating Abundance of Biological Populations*. Oxford University Press, Oxford.

Buckland, S.T., Anderson, D.R., Burnham, K.P., Laake, J.L., Borchers, D.L., and Thomas, L. (2002). *Introduction to Distance Sampling: Estimating Abundance of Biological Populations*. Elsevier, Amsterdam.

Buckland, S.T., Anderson, D.R., Burnham, K.P., Laake, J.L., Borchers, D.L., and Thomas, L. (eds), (2004). *Advanced Distance Sampling: Estimating Abundance of Biological Populations*. Oxford University Press.

Buckland, S.T., Newman, K.B., Fernández, C., Thomas, L., and Harwood, J. (2007). Embedding population dynamics models in inference. *Statistical Science*, **22**, 44–58.

Budge, S.M., Iverson, S.J., Bowen, W.D., and Ackman, R.G. (2002). Among- and within-species variability in fatty acid signatures of marine fish and invertebrates on the Scotian Shelf Georges Bank and southern Gulf of St Lawrence. *Canadian Journal of Fisheries and Aquatic Science*, **59**, 886–898.

Budge, S.M., Iverson, S.J., and Koopman, H.N. (2006). Studying trophic ecology in marine ecosystems using fatty acids: a primer on analysis and interpretation. *Marine Mammal Science*, **22**, 759–801.

Bundy, A., Heymans, J.J., Morissette, L., and Savenkoff, C. (2009). Seals, cod and forage fish: A comparative exploration of variations in the theme of stock collapse and ecosystem change in four Northwest Atlantic ecosystems. *Progress in Oceanography*, **81**, 188–206.

Burgman, M.A., Maslin, B.R., Andrewartha, D., Keatley, M.R., Boek, C., and McCarthy M. (2000). Inferring threat from scientific collections: Power tests and an application to Western Australian *Acacia* species. In S. Ferson and M.A. Burgman (eds), *Quantitiative Methods for Conservation Biology*, pp. 7–26. Springer, New York.

Burgman, M.A., Breininger, D.R., Duncan, B.W., and Ferson, S. (2001). Setting reliability bounds on habitat suitability indices. *Ecological Applications*, **11**, 70–78.

Burnell, S.R. and Shanahan, D. (2001). A note on a prototype system for simple computer-assisted matching of individually identified southern right whales, *Eubalaena australis*. *Journal of Cetacean Research and Management*, Special Issue **2**, 297–300.

Burnham, K.P. (1993). A theory for combined analysis of ring recovery and recapture data. In J-D. Lebreton and P.M. North (eds), *Marked Individuals in the Study of Bird Populations* pp. 199–213. Birkhauser Verlag, Basel, Switzerland.

Burnham, K. and Anderson, D. (2002). *Model Selection and Multi-Model Inference*, 2nd edn. Spring-Verlag, New York.

Burns, J.J. (1981). The bearded seal (*Erignathus barbatus*, Erxleben, 1777). In S.H. Ridgway and R.J. Harrison (eds), *Handbook of Marine Mammals*, Vol. 2, pp. 145–170. Academic Press, London.

Burns, J.M. (1999). The development of diving behavior in juvenile Weddell seals: Pushing physiological limits in order to survive. *Canadian Journal of Zoology*, **77**, 737–747.

Burns, J.M. and Castellini, M.A. (1998). Dive data from satellite tags and time-depth recorders: A comparison in Weddell seal pups. *Marine Mammal Science*, **14**, 750–764.

Burns, J.M., Schreer, J.F., and Castellini, M.A. (1997). Physiological effects on dive patterns and foraging strategies in yearling Weddell seals (Leptonychotes weddellii). *Canadian Journal of Zoology*, **75**, 1796–1811.

Burns, J.M., Costa, D.P, Fedak, M.A., Hindell, M.A., Bradshaw, C.J.A., Gales, N.J., McDonald, B., Trumble, S.J., and Crocker, D.E. (2004). Winter habitat use and foraging behavior of crabeater seals along the Western Antarctic Peninsula. *Deep-Sea Research Part II—Topical Studies in Oceanography*, **51**, 2279–2303.

Busch, B. (1985). *The War Against The Seals*. McGill–Queen's University Press, Toronto.

Bustamante, P., Curant, F., Fowler, S.W., and Miramand, P. (1998). Cephalopods as a vector for the transfer of cadmium to top marine predators in the north-east Atlantic Ocean. *Science of the Total Environment*, **220**, 71–80.

Butler, P.J., Woakes, A.J., Boyd, I.L., and Kanatous, S. (1992). Relationship between heart rate and oxygen consumption during steady-state swimming in California sea lions. *Journal of Experimental Biology*, **170**, 35–42.

Butler, P.J., Green, J.A., Boyd, I.L., and Speakman, J.R. (2004). Measuring metabolic rate in the field: the pros and cons of the doubly labelled water and heart rate methods. *Functional Ecology*, **18**, 168–183.

Butti, C., Corain, L., Cozzi, B., Podesta, M., Pirone, A., Affonte, M., and Zotti, A. (2007). Bone density of the arm and forearm as an age indicator in specimens of stranded striped dolphins (*Stenella coeruleoalba*). *Journal of Anatomy*, **211**, 639–646.

Caiafa, C.F., Proto, A.N., Vergani, D.F., and Stanganelli, Z. (2005). Development of individual recognition of female southern elephant seals, *Mirounga leonina*, from Punta Norte Península Valdés, applying principal components analysis. *Journal of Biogeography*, **32**, 1257–1266.

Cairns, S.J. and Schwager, S.J. (1987). A comparison of association indices. *Animal Behaviour*, **35**, 1454–1469.

Calambokidis, J. and Barlow, J. (2004). Abundance of blue and humpback whales in the eastern North Pacific estimated by capture–recapture and line-transect methods. *Marine Mammal Science*, **20**, 63–85.

Calambokidis, J., Steiger, G.H., Cubbage, J.C., Balcomb, K.C., Ewald, C., Kruse, S., Wells, R., and Sears, R. (1990). Sightings and movements of blue whales off central California 1986–88 from photo-identification of individuals. *Report of the Whaling Commission*, **12**, 343–348.

Calenge, C., Dufour, A.B., and Maillard, D. (2005). K-select analysis: a new method to analyse habitat selection in radio-tracking studies. *Ecological Modelling*, **186**, 143–153.

Calkins, D.G. and Pitcher, K.W. (1982). Population assessment, ecology and trophic relationships of Steller sea lions in the Gulf of Alaska. *Environmental Assessment of the Alaskan Continental Shelf, Final Report*, **19**, 455–546.

Calkins, D.G., Becker, E.F., and Pitcher, K.W. (1998). Reduced body size of female Steller seal lions from a declining population in the Gulf of Alaska. *Marine Mammal Science*, **14**, 232–244.

Cameron, E.Z. (1998). Is suckling behaviour a useful predictor of milk intake? *Animal Behaviour*, **56**, 521–532.

Cameron, M.F. and Siniff, D.B. (2004). Age-specific survival, abundance, and immigration rates of a Weddell seal (*Leptonychotes weddellii*) population in McMurdo Sound, Antarctica. *Canadian Journal of Zoology*, **82**, 601–615.

Campagna, C., Fedak, M.A., and McConnell, B.J. (1999). Post-breeding distribution and diving behavior of adult male southern elephant seals from Patagonia. *Journal of Mammalogy*, **80**, 1341–1352.

Campana, S.E. (2001). Accuracy, precision amd quality control in age determination, including a review of the use and abuse of age validation methods. *Journal of Fish Biology*, **59**, 197–242.

Campana, S.E. (2004). *Photographic Atlas of Fish Otoliths of the Northwest Atlantic Ocean.* NRC Research Press, Ottawa, Ontario.

Campana, S.E., Annand, M.C., and McMillan, J.I. (1995). Graphical and statistical methods for determining the consistency of age determination. *Transactions of the American Fisheries Society*, **124**, 131–138.

Campbell, D. (2007). Need to distinguish science (good and bad) from ethics. *Nature*, **446**, 24.

Campbell, G.S., Gisiner, R.C., Helweg, D.A., and Milette, L.L. (2002). Acoustic identification of female Steller sea lions (*Eumetopias jubatus*). *Journal of the Acoustical Society of America*, **111**, 2920–2928.

Cañadas, A. and Hammond, P.S. (2006). Model-based abundance estimates for bottlenose dolphins (*Tursiops truncatus*) off southern Spain: implications for management. *Journal of Cetacean Research and Management*, **8**, 13–27.

Cañadas, A. and Hammond, P.S. (2008). Abundance and habitat preferences of the short-beaked common dolphin (*Delphinus delphis*) in the South-western Mediterranean: implications for conservation. *Endangered Species Research*, **4**, 309–331.

Cañadas, A., Sagarminaga, R., and García-Tiscar, S. (2002). Cetacean distribution related with depth and slope in the Mediterranean waters off southern Spain. *Deep-sea Research I*, **49**, 2053–2073.

Cañadas, A., Desportes, G., and Borchers D.L. (2004). The estimation of the detection function and g(0) for shortbeaked common dolphins (*Delphinus delphis*), using double platform data collected during the NASS-95 Faroese survey. *Journal of Cetacean Research and Management*, **6**, 191–198.

Cañadas, A., Sagarminaga, R., De Stephanis, R., Urquiola, E., and Hammond, P.S. (2005). Habitat preference modelling as a conservation tool: proposals for marine protected areas for cetaceans in southern Spanish waters. *Aquatic Conservation: Marine and Freshwater Ecosystems*, **15**, 495–521.

Cannon, D.Y. (1987). *Marine Fish Oestology: A Manual for Archaeologists.* Department of Anthropology, Simon Fraser University, BC, Canada. Publication No 18.

Carbone, C. and Houston, A.I. (1996). The optimal allocation of time over the dive cycle: An approach based on aerobic and anaerobic respiration. *Animal Behaviour*, **51**, 1247–1255.

Casper, R.M., Gales, N.J., Hindell, M.A., and Robinson, S.M. (2006). Diet estimation based on an integrated mixed prey feeding experiment using *Arctocephalus* seals. *Journal of Experimental Marine Biology and Ecology*, **328**, 228–239.

Casper, R.M., Jarman, S.N., Gales, N.J., and Hindell, M.A. (2007a). Combining DNA and morphological analyses of faecal samples improves insight into trophic interactions: a case study using a generalist predator. *Marine Biology*, **152**, 815–825.

Casper, R.M., Jarman, S.N., Deagle, B.W., Gales, N.J., and Hindell, M.A. (2007b). Detecting prey from DNA in predator scats: a comparison with morphological analysis using *Arctocephalus* seals fed a known diet. *Journal of Experimental Marine Biology and Ecology*, **347**(1–2), 144–154.

Casteel, R.W. (1976). *Fish Remains in Archaeology and Paleo-Environmental Studies*. Academic Press, New York.

Castellini, M.A. and Calkins, D.G. (1993). Mass estimates using body morphology in Steller sea lions. *Marine Mammal Science*, **9**, 48–54.

Castellini, M.A., Kooyman, G.L., and Ponganis, P.J. (1992). Metabolic rates of freely diving Weddell seals: correlations with oxygen stores, swim velocity and diving duration. *Journal of Experimental Biology*, **165**, 181–194.

Caswell, H. (1978). A general formula for the sensitivity of population growth rate to changes in life history parameters. *Theoretical Population Biology*, **14**, 215–230.

Caswell, H. (2001). *Matrix Population Models. Construction, Analysis, and Interpretation*, 2nd edn. Sinauer Associates, Sunderland, MA.

Caswell, H., Brault, S., Read, A.J., and Smith, T.D. (1998). Harbor porpoise and fisheries: an uncertainty analysis of incidental mortality. *Ecological Applications*, **8**, 1226–1238.

Catry, P., Phillips, R.A., Phalan, B., Silk, J.R.D., and Croxall, J.P. (2004). Foraging strategies of grey-headed albatrosses *Thalassarche chrysostoma*: integration of movements, activity and feeding events. *Marine Ecology Progress Series*, **280**, 261–273.

Cetap (1982). *A Characterization of Marine Mammals and Turtles in the Mid- and North Atlantic Areas of the US Outer Continental Shelf*. Cetacean and Turtle Assessment Program, University of Rhode Island US Bureau of Land Management, Contract No. AA551-CTB-48., Washington, DC.

Chakraborty, R., Stivers, D.N., Su, B., Zhong, Y., and Budowle, B. (1999). The utility of short tandem repeat loci beyond human identification: implications for development of new DNA typing systems. *Electrophoresis*, **20**, 1682–1696.

Chaloupka, M., Osmond, M., and Kaufman, G. (1999). Estimating seasonal abundance trends and survival probabilities of humpback whales in Hervey Bay (east coast Australia). *Marine Ecology Progress Series*, **184**, 291–301.

Chambellant, M. and Ferguson, J.H. (2009). Ageing live ringed seals (*Phoca hispida*): Which tooth to pull? *Marine Mammal Science*, **25**, 478–486.

Chapman, D.G. and Johnson, A.M. (1968). Estimation of fur seal pup populations by randomized sampling. *Transactions of the American Fisheries Society*, **97**, 264–270.

Charrassin, J.B., Hindell, M., Rintoul, S.R., Roquet, F., Sokolov, S., Biuw, M., Costa, D., Boehme, L., Lovell, P., Coleman, R., Timmermann, R., Meijers, A., Meredith, M., Park, Y.H., Bailleul, F., Goebel, M., Tremblay, Y., Bost, C.A., McMahon, C.R., Field, I.C., Fedak, M.A., and Guinet, C. (2008). Southern Ocean frontal structure and sea-ice formation rates revealed by elephant seals. *Proceedings of the National Academy of Sciences of the United States of America*, **105**, 11634–11639.

Chen, C., Durand, E., Forbes, F., and Francois, O. (2007). Bayesian clustering algorithms ascertaining spatial population structure: a new computer program and a comparison study. *Molecular Ecology Notes*, **7**, 747–756.

Cherel, Y., Hobson, K.A., and Weimerskirch, H. (2005). Using stable isotopes to study resource acquisition and allocation in procellariiform seabirds. *Oecologia*, **145**, 533–540.

Chittleborough, R.G. (1959). *Balaenoptera brydei* Olsen on the west coast of Australia. *Norsk Hvalfangsttid*, **48**, 62–66.

Chivers, S., Dizon, A., Gearin, P., and Robertson, K. (2002). Small-scale population structure of eastern North Pacific harbor porpoise, Phocoena phocoena, indicated by molecular genetic analyses. *Journal of Cetacean Research and Management*, **4**, 111–122.

Christie, W.W. (1982). *Lipid Analysis*. Pergamon, Oxford.

Churchill, G.A. (2002). Fundamentals of experimental design for cDNA microarrays. *Nature Genetics*, **32**(Suppl.), 490–495.

Clark, C.W. (1976). *Mathematical Bioeconomics: The Optimal Management of Renewable Resources*. John Wiley and Sons, New York.

Clark, C.W. and Gagnon, C.G. (2004). Low-frequency vocal behaviors of baleen whales in the North Atlantic: Insights from IUSS detections, locations and tracking from 1992 to 1996. *Journal of Underwater Acoustics*, **52**, 651–668.

Clark, J.S. (2007). *Models for Ecological Data*. Princeton University Press, Princeton, NJ.

Clark, W.G. (2004). Nonparametric estimates of age misclassification from paired readings. *Canadian Journal of Fisheries and Aquatic Sciences*, **61**, 1881–1889.

Clarke, M.R. (1986). *A Handbook for the Identification of Cephalopod Beaks*. Clarendon Press, Oxford.

Clementz, M.T. and Koch, P.L. (2004). Differentiating aquatic mammal habitat and foraging ecology with stable isotopes in tooth enamel. *Oecologia*, **129**, 461–72.

Clutton-Brock, T.H. and Pemberton, J.M. (2004). *Soay Sheep: Dynamics and Selection in an Island Population*. Cambridge University Press, Cambridge, U.K.

Clutton-Brock, T.H., Guinness, F.E., and Albon, S.D. (1982). *Red Deer: Behaviour and Ecology of Two Sexes*. Chicago University Press, Chicago.

Cochran, W. (1997). *Sampling Techniques*, 3rd edn. Wiley, New York.

Cody, M.L. and Smallwood, J. (1996). *Long-Term Studies in Vertebrate Communities*. Academic Press, San Diego, CA.

Connor, R.C. (2001). Social relationships in a big-brained aquatic mammal. In L.A. Dugatkin (ed.), *Model Systems in Behavioral Ecology*. Princeton University Press, Princeton, NJ.

Connor, R.C., Mann, J., Tyack, P.L., and Whitehead, H. (1998). Social evolution in toothed whales. *Trends in Ecology and Evolution*, **13**, 228–232.

Conradt, L. and Roper, T.J. (2005). Consensus decision making in animals. *Trends in Ecology and Evolution*, **20**, 449–456.

Cooch, E. and White G. (2008). *Program MARK: A Gentle Introduction* (version 17/6/08). http://www.phidot.org/software/mark/docs/book/

Cook, H.W. (1996). Fatty acid desaturation and chain elongation in eukaryotes. In D.E. Vance and J.E. Vance (eds), *Biochemistry of Lipids and Membranes*, pp. 129–152. Elsevier, Amsterdam.

Cooke, F., Rockwell, R.F., and Lank, D.B. (1995). *The Snow Geese of La Perouse Bay*. Oxford University Press, Oxford.

Cooper, M.H. (2004). Fatty acid metabolism in marine carnivores: implications for quantitative estimation of predator diets. PhD thesis, Dalhousie University Halifax, NS.

Cooper, M.H., Iverson, S.J., and Heras, H. (2005). Dynamics of blood chylomicron fatty acids in a marine carnivore: implications for lipid metabolism and quantitative estimation of predator diets. *Journal of Comparative Physiology B*, **175**, 133–145.

Corander, J., Waldmann, P., and Sillanpaa, M.J. (2003). Bayesian analysis of genetic differentiation between populations. *Genetics*, **163**, 367–374.

Corander, J., Waldmann, P., Marttinen, P., and Sillanpaa, M.J. (2004). Baps2: enhanced possibilities for the analysis of genetic population structure. *Bioinformatics*, **20**, 2363–2369.

Corkeron, P.J. (2004). Fishery management and culling. *Science*, **306**, 1891–1892.

Corkeron, P.J. (2006). Opposing views of the 'ecosystem approach' to fisheries management. *Conservation Biology*, **20**, 617–619.

Cormack, R.M. (1964). Estimates of survival from the sighting of marked animals. *Biometrika*, **51**, 429–438.

Costa, D.P. and Gentry, R.L. (1986). Free ranging energetics of northern fur seals. In R.L. Gentry and G.L. Koymann (eds), *Fur Seals: Maternal Strategies on Land and at Sea*, pp. 79–101. Princeton University Press, Princeton, NJ.

Costa, D.P. and Kooyman, G.L. (1984). Contribution of specific dynamic action to heat balance and thermoregulation in the sea otter, *Enhydra lutris*. *Physiological Zoology*, **57**, 199–203.

Costa, D.P. and Williams, T.M. (1999). Marine mammal energetics. In J.E. Reynolds and S.A. Rommel (eds), *Biology of Marine Mammals*. Smithsonian Institution Press, Washington DC.

Costa, D.P., Croxall, J.P., and Duck, C.D. (1989). Foraging energetics of antarctic fur seals in relation to changes in prey availability. *Ecology*, **70**, 596–606.

Cottrell, P.E. and Trites, A.W. (2002). Classifying prey hard part structures recovered from fecal remains of captive Steller sea lions (*Eumetopias jubatus*). *Marine Mammal Science*, **18**, 525–539.

Cottrell, P.E., Trites, A.W., and Miller, E.H. (1996). Assessing the use of hard parts in faeces to identify harbour seal prey: results of captive-feeding trials. *Canadian Journal of Zoology*, **74**, 875–880.

Cox, T.M., Lewison, R.L., Zydelis, R., Crowder, L.B., Safina, C., and Read, A.J. (2007). Comparing effectiveness of experimental and implemented bycatch reduction measures: the ideal and the real. *Conservation Biology*, **21**, 1155–1164.

Coyne, M.S. and Godley, B.J. (2005). Satellite Tracking and Analysis Tool (STAT): an integrated system for archiving, analyzing and mapping animal tracking data. *Marine Ecology Progress Series*, **301**, 1–7.

Craig, B.A. and Reynolds, J.E. III. (2004). Determination of manatee population trends along the Atlantic coast of Florida using a Bayesian approach with temperature-adjusted aerial survey data. *Marine Mammal Science*, **20**, 386–400.

Cramer, E.M. (1985). Multicollinearity. In S. Kotz, N.L. Johnson, and C.B. Read (eds), *Encyclopedia of Statistics*, pp. 639–643. John Wiley and Sons, New York.

Crida, A. and Manel, S. (2007). Wombsoft: an R package that implements the Wombling method to identify genetic boundary. *Molecular Ecology Notes*, **7**, 588–591.

Crocker, D.E., Le Boeuf, B.J., and Costa, D.P. (1996). Drift diving in female northern elephant seals: Implications for food processing. *Canadian Journal of Zoology*, **75**, 27–39.

Crocker, D.E., LeBoeuf, B.J., and Costa, D.P. (1997). Drift diving in female northern elephant seals: Implications for food processing. *Canadian Journal of Zoology*, **75**, 27–39.

Crocker, D.E., Williams, J.D., Costa, D.P., and Le Boeuf, B.J. (2001). Maternal traits and reproductive effort in northern elephant seals. *Ecology*, **82**, 3451–3555.

Crocker, D.E., Costa, D.P., Le Boeuf, B.J., Webb, P.M., and Houser, D.S. (2006). Impact of El Nino on the foraging behavior of female northern elephant seals. *Marine Ecology Progress Series*, **309**, 1–10.

Croft, D.P., James, R., and Krause, J. (2008). *Exploring Animal Social Networks*. Princeton University Press, Princeton, NJ.

Croll, D.A., Tershy, B.R., Hewitt, R.P., Demer, D.A., Fiedler, P.C., Smith, S.E., Armstrong, W., Popp, J.M., Kiekhefer, T., Lopez, V.R., Urban, J., and Gendron, D. (1998). An integrated approach to the foraging ecology of marine birds and mammals. *Deep-Sea Research Part II—Topical Studies in Oceanography*, **45**, 1353–1371.

Croll, D.A., Clark, C.W., Calambokidis, J., Ellison, W.T., and Tershy, B.R. (2001). Effect of anthropogenic low-frequency noise on the foraging ecology of Balaenoptera whales. *Animal Conservation*, **4**, 13–27.

Croll, D.A., Marinovic, B., Benson, S., Chavez, F.P., Black, N., Ternullo, R., and Tershy, B.R. (2005). From wind to whales: trophic links in a coastal upwelling system. *Marine Ecology Progress Series*, **289**, 117–130.

Croxall, J.P. (1993). Diet. In R.M. Laws (ed.), *Antarctic Seals: Research Methods and Techniques*, pp. 268–290. Cambridge University Press, Cambridge.

Croxall, J.P. and Hiby, L. (1983). Fecundity, survival and site fidelity in Weddell seals *Leptonychotes weddellii*. *Journal of Applied Ecology*, **20**, 19–32.

Croxall, J.P., McCann, T.S., Prince, P.A., and Rothery, P. (1988). Reproductive performance of seabirds and seals at South Georgia and Signy Island, South Orkneys Islands, 1976–1987: Implications for Southern Ocean monitoring studies. In D. Sahrhage (ed.), *Antarctic Ocean Resource Variability*, pp. 261–285. Springer-Verlag, Berlin, Heidelberg.

Cunningham, L. (2007). Investigating monitoring options for harbour seals in Special Areas of Conservation in Scotland. PhD dissertation, University of St Andrews, Scotland.

Cunningham, L. (2009). Using computer-assisted photo-identification and capture–recapture techniques to monitor the conservation status of harbour seals (*Phoca vitulina*). *Aquatic Mammals*, **35**, 319–329.

Curgus, C.S., McAllister, D., Raum-Suryan, K., Pitcher, K., and Cunningham, W. (2001).

Live-capture method for Steller sea lions using SCUBA. *14th Biennial Conference on the Biology of Marine Mammals,* November 28–December 3, 2001, Vancouver, BC, Canada, pp. 51. [Abstract]

D'Agrosa, C., Lennert-Cody, C.E., and Vidal, O. (2000). Vaquita bycatch in Mexico's artisanal gillnet fisheries: driving a small population to extinction. *Conservation Biology,* 14, 1110–1119.

Dahl, T.M., Lydersen, C., Kovacs, K., Falk-Petersen, S., Sargent, J., Gjertz, I., and Gulliksen, B. (2000). Fatty acid composition of the blubber of white whales (*Delphinapterus leucas*). *Polar Biology,* 23, 401–409.

Dalebout, M.L., Mead, J., Baker, C.S., Baker, A., and Van Helden, A. (2002). A new species of beaked whale Mesoplodon perrini sp. n. (cetacea, Ziphiidae) discovered through phylogenetic analyses of mitochondrial DNA sequences. *Marine Mammal Science,* 18, 1–32.

Dalsgaard, J., St John, M., Kattner, G., Müller-Navarra, D.C., and Hagen, W. (2003). Fatty acid trophic markers in the pelagic marine environment. *Advances in Marine Biology,* 46, 225–340.

Dalton, R. (2005). Is this any way to save a species? *Nature,* 436, 14–16.

Danil, K. and Chivers, S.J. (2007). Growth and reproduction of female short-beaked common dolphins, *Delphinus delphis,* in the eastern tropical Pacific. *Canadian Journal of Zoology,* 85, 108–121.

Das, K., Lepoint, G., Loizeau, V., Debacker, V., Dauby, P., and Bouquegneau, J.M. (2000). Tuna and dolphin associations in the northeast Atlantic: evidence of different ecological niches from stable isotope and heavy metal measurements. *Marine Pollution Bulletin,* 40, 102–109.

da Silva, V.M.F. and Martin, A.R. (2000). A study of the Boto, or Amazon River dolphin (*Inia geoffrensis*), in the Mamiraua Reserve, Brazil: Operation and techniques. In R.R. Reeves, B.D. Smith, and T. Kasuya (eds), *Biology and Conservation of Freshwater Cetaceans in Asia,* pp. 121–131, Occasional Paper of the IUCN Species Survival Commission No. 23. Gland, Switzerland and Cambridge, UK.

Davies, C.E., Kovacs, K.M., Lydersen, C., and Van Parijs, S.M. (2006). Development of display behavior in young captive bearded seals. *Marine Mammal Science,* 22, 952–965.

Davis, R.W., William, T.M., and Kooyman, G.L. (1985). Swimming metabolism of yearling and adult harbor selas, *Phoca vitulina. Physiological Zoology,* 58, 590–596.

Davis, R.W., Fuiman, L.A., Williams, T.M., Collier, S.O., Hagey, W.P., Kanatous, S.B., Kohin, S., and Horning, M. (1999). Hunting behavior of a marine mammal beneath the Antarctic fast ice. *Science,* 283, 993–996.

Davis, R.W., Fuiman, L.A., Williams, T.M., and Le Boeuf, B.J. (2001). Three-dimensional movements and swimming activity of a northern elephant seal. *Comparative Biochemistry and Physiology A—Molecular and Integrative Physiology,* 129, 759–770.

Davis, R.W., Fuiman, L.A., Williams, T.M., Horning, M., and Hagey, W. (2003). Classification of Weddell seal dives based on 3-dimensional movements and video-recorded observations. *Marine Ecology Progress Series,* 264, 109–122.

Davison, A. and Chiba, S. (2003). Laboratory temperature variation is a previously unrecognized source of genotyping error during capillary electrophoresis. *Molecular Ecology Notes,* 3, 321–323.

Dawson, K.J. and Belkhir, K. (2001). A Bayesian approach to the identification of panmictic populations and the assignment of individuals. *Genetical Research,* 78, 59–77.

Dawson, S., Wade, P., Slooten, E., and Barlow, J. (2008). Design and field methods for sighting surveys of cetaceans in coastal and riverine habitats. *Mammal Review,* 38, 19–49.

Dayton, P.K. (1998). Reversal of the burden of proof in fisheries management. *Science,* 279, 821–822.

Deagle, B.E. and Tollit, D.J. (2007). Quantitative analysis of prey DNA in pinniped faeces: potential to estimate diet composition? *Conservation Genetics,* 8, 743–747.

Deagle, B.E., Tollit, D.J., Jarman, S.N., Hindell, M. A., Trites, A.W., and Gales, N.J. (2005). Molecular scatology as a tool to study diet: analysis of prey DNA in scats from captive Steller sea lions. *Molecular Ecology,* 14, 1831–1842.

Deagle, B.E., Kirkwood, R., and Jarman, S.N. (2009). Analysis of Australian fur seal diet by pyrosequencing prey DNA in faeces. *Molecular Ecology,* 18, 2022–2038.

De'Ath, G. (2002). Multivariate regression trees: A new technique for modelling species–environment relationships. *Ecology*, **83**, 1105–1117.

De'Ath, G. and Fabricius, K.E. (2000). Classification and regression trees: A powerful yet simple technique for ecological data analysis. *Ecology*, **81**, 3178–3192.

Deeke, V.J., Ford, J., and Slater, P. (2005). The vocal behaviour of mammal-eating killer whales: communicating with costly calls. *Animal Behaviour*, **69**, 385–405.

de Kroon, H., Plaisier, A., van Groenendael, J., and Caswell, H. (1986). Elasticity: the relative contribution of demographic parameters to population growth rate. *Ecology*, **67**, 1427–1431.

Dellinger, T. and Trillmich, F. (1988). Estimating diet composition from scat analysis in otariid seals (*Otariidae*): is it reliable? *Canadian Journal of Zoology*, **66**, 1865–1870.

DeMaster, D.P. (1978). Calculation of the average age of sexual maturity in marine mammals. *The Journal of Fisheries Resident Board of Canada*, **35**, 912–915.

DeMaster, D.P. and Drevenak, J.K. (1988). Survivorship patterns in three species of captive cetaceans. *Marine Mammal Science*, **4**, 297–311.

DeMaster, D., Angliss, R., Cochrane, J., Mace, P., Merrick, R., Miller, M., Rumsey, S., Taylor, B., Thompson, G., and Waples, R. (2004). *Recommendations to NOAA Fisheries: ESA Listing Criteria by the Quantitative Working Group*. US Department of Commerce NOAA Technical Memorandum NMFS-F/SPO-67.

Dendrinos, P., Karamanlidis, A.A., Kotomatas, S., Legakis, A., Tounta, E., and Matthiopoulos, J. (2007). Pupping habitat use in the Mediterranean monk seal: A long-term study. *Marine Mammal Science*, **23**, 615–628.

Dennis, B. (2002). Allee effects in stochastic populations. *Oikos*, **96**, 389–401.

Dennis, B., Munholland, P.L., and Scott, J.M. (1991). Estimation of growth and extinction parameters for endangered species. *Ecological Monographs*, **61**, 115–143.

de Queiroz, A. and Gatesy, J. (2007). The supermatrix approach to systematics. *Trends in Ecology and Evolution*, **22**, 34–41.

Descartes, R. (1997). Discourses: Part V. In Enrique Chavez-Arvizo, (ed.), *Key Philosophical Writings* (transl. E.S. Haldane and G.R.T. Ross), pp 97–108. Wordsworth Editions Ltd, Ware, Herts.

Deutsch, C.J., Bonde, R.K., and Reid, J.P. (1998). Radio-tracking manatees from land and space: Tag design, implementation, and lessons learned from long-term study. *Marine Technology Society Journal*, **32**, 18–29.

de Vries, H. (1998). Finding a dominance order most consistent with a linear hierarchy: a new procedure and review. *Animal Behaviour*, **55**, 827–843.

de Vries, H., Stevens, J.M.G., and Vervaecke, H. (2006). Measuring and testing the steepness of dominance hierarchies. *Animal Behaviour*, **71**, 585–592.

Dierauf, L.A. and Gulland, F. M.D. (2001). *CRC Handbook of Marine Mammal Medicine*, 2nd edn. CRC Press, Boca Raton, FL.

Diggle, P.J., Twan, J.A., and Moyeed, R.A. (1998). Model-based geostatistics (with discussion). *Applied Statistics*, **47**, 299–350.

Dixon, K.R. and Chapman, J.A. (1980). Harmonic mean measure of animal activity areas. *Ecology*, **61**, 1040–1044.

Doniol-Valcroze, T., Berteaux, D., Larouche, P., and Sears, R. (2007). Influence of thermal fronts on habitat selection by four rorqual whale species in the Gulf of St. Lawrence. *Marine Ecology Progress Series*, **335**, 207–216.

Doubleday, W.G. and Bowen, W.D. (1980). *Inconsistencies in Reading the Age of Harp Seal (Pagophilus groenlandicus) Teeth, their Consequences, and Means of Reducing Resulting Biases*, Rep. No. NAFO SCR Doc. 80/XI/160 Ser. No. N247. Northwest Atlantic Fisheries Organization, Ottawa.

Draper, D. (1995). Assessment and propagation of model uncertainty. *Journal of the Royal Statistical Society*, Series B., **57**, 45–97.

Drews, C. (1993). The concept and definition of dominance in animal behaviour. *Behaviour*, **125**, 283–311.

Duffield, D.A. and Wells, R.S. (2002). The molecular profile of a resident community of bottlenose dolphins, *Tursiops truncatus*. In C.J. Pfeiffer (ed.), *Molecular and Cell Biology of Marine Mammals*, pp. 3–11. Krieger Publishing Company, Melbourne, FL.

Dungan, J.L., Perry, J.N., Dale, M.R.T., Legendre, P., Citron-Pousty, S., Fortin, M.-J., Jakomulska, A., Miriti, M., and Rosenberg, M.S. (2002). A balanced view of scale in spatial statistical analysis. *Ecography*, **25**, 626–640.

Dunshea, G. (2009). DNA-based diet analysis for any predator. *PLoS ONE*, **4**, e5252 doi:101371/journalpone0005252.

Eberhardt, L.L. and Siniff, D.B. (1977). Population dynamics and marine mammal management policies. *Journal of the Fisheries Research Board of Canada*, **34**, 183–190.

Edwards, A.M. (2008). Using likelihood to test for Levy flight search patterns and for general power-law distributions in nature. *Journal of Animal Ecology*, **77**, 1212–1222.

Eguchi, T. (2008). An introduction to Bayesian statistics without using equations. *Marine Turtle Newsletter*, **122**, 1–5.

Ellner, S.P., Fieberg, J., Ludwig, D., and Wilcox, C. (2002). Precision of population viability analysis. *Conservation Biology*, **16**, 258–261.

Engler, R., Guisan, A., and Rechsteiner, L. (2004). An improved approach for predicting the distribution of rare and endangered species from occurrence and pseudo-absence data. *Journal of Applied Ecology*, **41**, 263–274.

Erickson, A.W. and Bester, M.N. (1993). Immobilization and capture. In R.M. Laws (ed.), *Antarctic Seals: Research Methods and Techniques*, pp. 46–88. Cambridge University Press, Cambridge, UK.

Ersts, P.J. and Rosenbaum, H.C. (2003). Habitat preference reflects social organization of humpback whales (*Megaptera novaeangliae*) on a wintering ground. *Journal of Zoology, London*, **260**, 337–345.

Estes, J.A. and Duggins, D.O. (1995). Sea otters and kelp forests in Alaska: generality and variation in a community ecological paradigm. *Ecological Monographs*, **65**, 75–100.

Estes, J.A. and Palmisano, J.F. (1974). Sea otters: their role in structuring nearshore communities. *Science*, **185**, 1058–1060.

Estes, J.A., Tinker, M.T., Williams, T.M., and Doak, D.F. (1998). Killer whale predation on sea otters linking oceanic and nearshore ecosystems. *Science*, **282**, 473–476.

Estes, J.A., Danner, E.M., Doak, D.F., Konar, B., Springer, A.M., Steinberg, P.D., Tinker, M.T., and Williams, T.M. (2004). Complex trophic interactions in kelp forest ecosystems. *Bulletin of Marine Science*, **74**, 621–638.

Estes, J.A., Lindberg, D.R., and Wray, C. (2005). Evolution of large body size in abalones (Haliotis): patterns and implications. *Paleobiology*, **31**, 591–606.

Evans, W.E., Hall, J.D., Irvine, A.B., and Leatherwood, S.J. (1972). Methods for tagging small cetaceans. *Fishery Bulletin*, **70**, 61–65.

Fadely B.S., Worthy, G.A.J., and Costa, D.P. (1990). Assimilation efficiency of northern fur seals determined using dietary manganese. *Journal of Wildlife Management*, **54**, 246–251.

Fahlman, A., Wilson, R., Svard, C., Rosen, D.A.S., and Trites, A.W. (2008). Activity and diving metabolism correlate in Steller sea lion Eumetopias jubatus. *Aquatic Biology*, **2**, 75–84.

Falush, D., Stephens, M., and Pritchard, J.K. (2003). Inference of population structure using multilocus genotype data: linked loci and correlated allele frequencies. *Genetics*, **164**, 1567–1587.

Farley, S.D. and Robbins, C.T. (1994). Development of two methods to estimate body composition of bears. *Canadian Journal of Zoology*, **72**, 220–226.

Farnsworth, E.J. and Rosovsky, J. (1993). The ethics of ecological field experimentation. *Conservation Biology*, **7**, 463–472.

Fauchald, P. and Tveraa, T. (2003). Using first-passage time in the analysis of area-restricted search and habitat selection. *Ecology*, **84**, 282–288.

Fauchald, P. and Tveraa, T. (2006). Hierarchical patch dynamics and animal movement pattern. *Oecologia*, **149**, 383–395.

Fauchald, P., Erikstad, K.E., and Skarsfjord, H. (2000). Scale-dependent predator–prey interactions: The hierarchical spatial distribution of seabirds and prey. *Ecology*, **81**, 773–783.

Fea, N. and Harcourt, R. (1997). Assessing the use of faecal and regurgitate analysis as a means of determining the diet of New Zealand fur seals. In M. Hindell and C. Kemper (eds), *Marine Mammal Research in the Southern Hemisphere: Status Ecology and Medicine*, pp. 143–150. Surrey Beatty and Sons Ltd, Chipping Norton, NSW, Australia.

Fedak, M.A. (1986). Diving and exercise in seals: a benthic perspective. In A.D. Brubakk, J.W. Kanwisher, and G. Sundnes (eds), *Diving in Animals and Man*, pp. 11–32. Tapir Publishers, Trondheim.

Fedak, M.A., Rome, L., and Seeherman, H.J. (1981). One-step N2 dilution technique for calibrating open-circuit VO_2 measurement systems. *Journal of Applied Physiology*, **51**, 772–776.

Fedak, M.A., Pullen, M.R., and Kanwisher, J. (1988). Circulatory responses of seals to periodic breathing—heart-rate and breathing during exercise and diving in the laboratory and open sea. *Canadian Journal of Zoology—Revue Canadienne De Zoologie*, **66**, 53–60.

Fedak, M.A., Lovell, P., and McConnell, B.J. (1996). MAMVIS: A marine mammal behaviour visualization system. *Journal of Visualization and Computer Animation*, **7**, 141–147.

Fedak, M., Lovell, P., McConnell, B., and Hunter, C. (2002). Overcoming the constraints of long range radio telemetry from animals: Getting more useful data from smaller packages. *Integrative and Comparative Biology*, **42**, 3–10.

Federal Register. (1996). Policy regarding the recognition of distinct vertebrate population segments under the Endangered Species Act. *Federal Register*, **61**, 4721–4725.

Federal Register. (2007). Endangered and threatened species: Proposed endangered status for the Cook Inlet beluga whale. *Federal Register*, **72**, 19 854–19 862.

Felsenstein, J. (1985). Phylogenies and the comparative method. *American Naturalist*, **125**, 1–15.

Ferguson, S.H., Stirling I., and McLoughlin, P. (2005). Climate change and ringed seal (*Phoca hispida*) recruitment in western Hudson Bay. *Marine Mammal Science*, **21**, 121–135.

Fernandez, A., Edwards, J.F., Rodriguez, F., Espinosa de los Monteros, A., Herraez, P., Castro, P., Jaber, J.R., Martin, V., and Arbelo, M. (2005). 'Gas and fat embolic syndrome' involving a mass stranding of beaked whales (family Ziphiidae) exposed to anthropogenic sonar signals. *Veterinary Pathology*, **42**, 446–457.

Ferrero, R.C., Moore, S.E., and Hobbs, R.C. (2000). Development of beluga, *Delphinapterus leucas*, capture and satellite tagging protocol in Cook Inlet, Alaska. *Marine Fisheries Review*, **62**, 112–123.

Ferrero, R.C., Hobbs, R.C., and VanBlaricom, G.R. (2002). Indications of habitat use patterns among small cetaceans in the central North Pacific based on fisheries observer data. *Journal of Cetacean Research and Management*, **4**, 311–321.

Fick, A. (1870). Uber die Messung des Blutquantums in den Herzventrikeln. *Verhandlungen der Physikalisch-Medizinschen Gesellschaft zu Wurzberg*, **2**, 16.

Fieberg, J. and Kochanny C.O. (2005). Quantifying home-range overlap: The importance of the utilisation distribution. *Journal of Wildlife Management*, **69**, 1346–1359.

Fiedler, P.L. and Kareiva, P.M. (1998). *Conservation Biology for the Coming Decade*, 2nd edn. Chapman and Hall, New York.

Fiedler, P.C., Reilly, S.B., Hewitt, R.P., Demer, D., Philbrick, V.A., Smith, S., Armstrong, W., Croll, D.A., Tershy, B.R., and Mate, B.R. (1998). Blue whale habitat and prey in the California channel Islands. *Deep Sea Research Part II: Topical Studies in Oceanography*, **45**, 1781–1801.

Fisher, M., Rannala, B., Chaturvedi, V., and Taylor, J. (2002). Disease surveillance in recombining pathogens: Multilocus genotypes identify sources of human Coccidiodes infections. *Proceedings of the National Academy of Sciences USA*, **99**, 9067–9071.

Fisk, A.T., Hobson, K.A., and Norstrom, R.J. (2001). Influence of chemical and biological factors on trophic transfer of persistent organic pollutants in the Northwater Polynya marine food web. *Environmental Science and Technology*, **35**, 732–738.

Flack, J.C., Girvan, M., de Waal, F.B.M., and Krakauer, D.C. (2006). Policing stabilizes construction of social niches in primates. *Nature*, **439**, 426–429.

Flemming, J.E.M., Field, C.A., James, M.C., Jonsen, I.D., and Myers, R.A. (2006). How well can animals navigate? Estimating the circle of confusion from tracking data. *Environmetrics*, **17**, 351–362.

Flewelling, L., Naar, J., Abbott, J., Baden, D., Barros, N., Bossart, G.D., Bottein, M., Hammond, D., Haubold, E., Heil, C., Henry, M., Jacocks, H., Leighfield, T. Pierce, R., Pitchford, T., Rommel, S.A. Scott, P., Steidinger, K., Truby, E., Van Dolah, F., and Landsberg, J. (2005). Brevetoxicosis: red tides and marine mammal mortalities. *Nature*, **435**, 755–756.

Folch, J., Lees, M., and Sloane-Stanley, G.H. (1957). A simple method for the isolation and purification of total lipids from animal tissues. *Journal of Biological Chemistry*, **226**, 497–509.

Forcada, J. and Aguilar, A. (2000). Use of photographic identification in capture–recapture studies of Mediterranean monk seals. *Marine Mammal Science*, **16**, 767–793.

Forcada, J. and Robinson, S.L. (2006). Population abundance, structure and turnover estimates for leopard seals during winter dispersal combining tagging and photo-identification data. *Polar Biology*, **29**, 1052–1062.

Forcada, J., Gazo, M., Aguilar, A., Gonzalvo, J., and Fernandez-Contrera, M. (2004). Bottlenose dolphin abundance in the NW Mediterranean: addressing heterogeneity in distribution. *Marine Ecology Progress Series*, **275**, 275–287.

Forcada, J., Trathan, P.N., Reid, K., and Murphy, E.J. (2005). The effects of global climate variability in pup production of Antarctic fur seals. *Ecology*, **86**, 2408–2417.

Ford, E. (1937). Vertebral variation in teleostean fishes. *Journal of the Marine Biological Association, UK*, **22**, 1–60.

Ford, J.K.B. (1991). Vocal traditions among resident killer whales (Orcinus orca) in coastal waters of British Columbia. *Canadian Journal of Zoology*, **69**, 1454–1483.

Ford, J.K.B. and Ellis, G.M. (2006). Selective foraging by fish-eating killer whales *Orcinus orca* in British Columbia. *Marine Ecology Progress Series*, **316**, 185–199.

Ford, J.K.B., Ellis, G.M., Barrett-Lennard, L.G., Morton, A.B., Palm, R.S., and Balcomb III, K.C. (1998). Dietary specialization in two sympatric populations of killer whales (Orcinus orca) in coastal British Columbia and adjacent waters. *Canadian Journal of Zoology*, **76**, 1456–1471.

Ford, J.K.B., Ellis, G.M., and Balcomb, K.C. (2000). *Killer Whales: The Natural History and Genealogy of Orcinus orca in British Columbia and Washington*, 2nd edn. UBC Press, Vancouver.

Forney, K.A. (1999). Trends in harbour porpoise abundance off central California, 1986–95: evidence for interannual changes in distribution? *Journal of Cetacean Research and Management*, **1**, 73–80.

Forney, K.A. (2000). Environmental models of cetacean abundance: reducing uncertainty in population trends. *Conservation Biology*, **14**, 1271–1286.

Forrow, L., Wartman, S.A., and Brock, D.W. (1988). Science, ethics, and the making of clinical decisions. *Journal of the American Medical Association*, **259**, 3161–3167.

Fortin, M.-J. and Dale, M. (2005). *Spatial Analysis: A Guide for Ecologists*. Cambridge University Press. Cambridge.

Fossi, M.C., Marsili, L., Junin, M., Castello, H., Lorenzani, J.A. Casini, S., Savelli, C., and Leonzio, C. (1997). Use of nondestructive biomarkers and residue analysis to assess the health of endangered species of pinnipeds in the south-west Atlantic. *Marine Pollution Bulletin*, **34**, 157–162.

Fowler, C.W. (1981). Density dependence as related to life history strategy. *Ecology*, **62**, 602–610.

Fox, J. (1997). *Applied Regression Analysis, Linear Models, and Related Methods*. Sage Publications Ltd, London.

France, R. (1995). Differentiation between littoral and pelagic food webs in lakes using stable carbon isotopes. *Limnology and Oceanography*, **40**, 1310–1313.

Francis, R. and Shooton (1997). 'Risk' in fisheries management. *Canadian Journal of Fisheries and Aquatic Sciences*, **54**, 1699–1715.

Francois, O., Ancelet, S., and Guillot, G. (2006). Bayesian clustering using hidden Markov random fields in spatial population genetics. *Genetics*, **174**, 805–816.

Frasier, T.R., McLeod, B.A., Bower, R.M., Brown, M.W., and White, B.N. (2007). Right whales past and present as revealed by their genes. In S.D. Kraus and R.M. Rolland (eds), *The Urban Whale: North Atlantic Right Whales at the Crossroads*, pp. 200–232. Harvard University Press, Cambridge, MA.

Freitas, C., Kovacs, K.M., Lydersen, C., and Ims, R.A. (2008). A novel method for quantifying habitat selection and predicting habitat use. *Journal of Applied Ecology*, **45**, 1213–1220.

Friday, N.A., Smith, T.D., Stevick, P.T., Allen, J., and Fernauld, T. (2008). Balancing bias and precision in capture–recapture estimates of abundance. *Marine Mammal Science*, **24**, 253–275.

Fujiwara, M. and Caswell, H. (2001). Demography of the endangered North Atlantic right whale. *Nature*, **414**, 537–541.

Fulman, L.A., Madden, K.M., Williams, T.M., and Davis, R.W. (2007). Structure of foraging dives by Weddell seals at an offshore isolated hole in the Antarctic fast-ice environment. *Deep-Sea Research Part II—Topical Studies in Oceanography*, **54**, 270–289.

Fulton, E.A., Smith, A.D.M., and Johnson, C.R. (2003). Effect of complexity on marine eco-system models. *Marine Ecology Progress Series*, **253**, 1–16.

Fung, W.K. (2003). User-friendly programs for easy calculations in paternity testing and kinship determinations. *Forensic Science International*, **136**, 22–34.

Gales, N.J. (1989). Chemical restraint and anaesthesia of pinnipeds: A review. *Marine Mammal Science*, **5**, 228–256.

Gales, N.J. (2000). *A Field Review of the Macquarie Island Elephant Seal Hot Iron Branding Program: December 2000*. A report prepared for the Antarctic Animal Ethics Committee. Australian Antarctic Division, Tasmania, Australia.

Gales, N.J. and Burton, H.R. (1987). Ultrasonic measurement of blubber thickness of the southern elephant seal, *Mirounga leonina* (Linn). *Australian Journal of Zoology*, **35**, 207–217.

Gales, N.J. and Cheal, A.J. (1992). Estimates diet composition of the Australian sea-lion (*Neophoca cinerea*) from scat analysis—an unreliable technique. *Wildlife Research*, **19**, 447–451.

Gales, N.J. and Jarman, S.N. (2002). A non-lethal genetic method for identifying whale prey. *International Whaling Commission Background Paper*, pp. 1–12.

Gales, N.J., Brennan, A., and Barker, R. (2003a). Ethics and marine mammal research. In N.J. Gales, M. Hindell, R. Kirkwood (eds), *Marine Mammals: Fisheries, Tourism and Management Issues*, pp. 321–329. CSIRO Publishing, Victoria, Australia.

Gales, N., Hindell, M., and Kirkwood, R. (eds), (2003b). *Marine Mammals: Fisheries, Tourism and Management Issues*. CSIRO Publishing, Collingwood, Australia.

Gales, N.J., Barnes, J., Chittick, B., Gray, M., Robinson, S., Burns, J., and Costa, D. (2005a). Effective, field-based inhalational anaesthesia for ice seals. *Marine Mammal Science*, **21**, 717–727.

Gales, N.J., Kasuya, T., Clapham, P.J., and Brownell, Jr, R.L. (2005b). Japan's whaling plan under scrutiny. *Nature*, **435**, 883–884.

Gales, N.J., Bowen, W.D., Johnston, D.W., Kovacs, K.M., Littnan, C.L., Perrin, W.F. Reynolds, J.E. III, and Thompson, P.M.

(2009). Guidelines for the treatment of marine mammals in field research. *Marine Mammal Science*, **25**, 725–736.

Gales, R. and Renouf, D. (1993). Detecting and measuring food and water intake in captive seals using temperature telemetry. *Journal of Wildlife Management*, **57**, 514–519.

Gales, R. and Renouf, D. (1994). Assessment of body condition in harp seals. *Polar Biology*, **14**, 381–387.

Gales, R., Renouf, D., and Noseworthy, E. (1994a). Body condition of harp seals. *Canadian Journal of Zoology*, **72**, 545–551.

Gales, R., Renouf, D., and Worthy, G.A.J. (1994b). Use of bioelectrical impedance analysis to assess body composition of seals. *Marine Mammal Science*, **10**, 1–12.

Galtier, N., Enard, D., Radondy, Y., Bazin, E., and Belkhir, K. (2006). Mutation hot spots in mammalian mitochondrial DNA. *Genome Research*, **16**, 215–222.

Garcia, S.M. and Cochrane, K.L. (2005). Ecosystem approach to fisheries: a review of implementation guidelines. *ICES Journal of Marine Science*, **62**, 311–318.

Garcia-Aguilar, M.D.L. and Morales-Bojorquez, E. (2005). Estimating the haul-out population size of a colony of northern elephant seals *Mirounga angustirostris* in Mexico, based on mark–recapture data. *Marine Ecology Progress Series*, **297**, 297–302.

Gärdenfors, U., Hilton-Taylor, C., Mace, G., and Rodriguez, J. (2001). The application of IUCN red list criteria at regional levels. *Conservation Biology*, **15**, 1206–1212.

Gardiner, K.J., Boyd, I.L., Racey, P.A., Reijnders, P.J., and Thompson, P.M. (1996). Plasma progesterone concentrations measured using an enzyme-linked immunsorbent assay useful for diagnosing pregnancy in harbor seals *Phoca vitulina*. *Marine Mammal Science*, **12**, 265–273.

Garner, G.W., Armstrup, S.C., Laake, J.L., Manly, B.F.J., McDonald, L.L., and Robertson, D.G. (eds), (1999). *Marine Mammal Survey and Assessment Methods*. Balkema, Rotterdam.

Garvey, J.E., Marschall, E.A., and Wright, R.A. (1998). From star charts to stoneflies: Detecting relationships in continuous bivariate data. *Ecology*, **79**, 442–447.

Gatesy, J., Baker, R.H., and Hayashi, C. (2004). Inconsistencies in arguments for the supertree approach: supermatrices versus supertrees of Crocodylia. *Systems Biology*, **53**, 342–355.

Gatesy, J., Desalle, R., and Wahlberg, N. (2007). How many genes should a systematist sample? Conflicting insights from a phylogenomic matrix characterized by replicated incongruence. *Systems Biology*, **56**, 355–363.

Gaydos, J.K., Balcomb, K.C., Osborne, R.W., and Dierauf, L. (2004). Evaluating potential infectious disease threats for southern resident killer whales, *Orcinus orca*: a model for endangered species. *Biological Conservation*, **117**, 253–262.

Gearin, P.J., Jeffries, S.J., Gosho, M.E., Thomason, J.R., DeLong, R.L., Wilson, M., Lambourn, D., Hanson, B., Osmek, S., and Melin, S.R. (1996). *Capture and Marking California Sea Lions in Puget Sound, Washington During 1994–95: Distribution, Abundance and Movement Patterns*, 34 pp. Unpublished report. Available National Marine Mammal Laboratory, 7600 Sand Point Way, NE, Seattle, WA 98115.

Geertsen, B.M., Teilmann, J., Kastelein, R.A., Vlemmix, H.N.J., and Miller, L.A. (2004). Behaviour and physiological effects of transmitter attachments on a captive harbour porpoise (*Phocoena phocoena*). *Journal of Cetacean Research and Management*, **6**, 139–146.

Gelatt, T., Davis, C., Cameron, M., Siniff, D., and Strobeck, C. (2000). The old and the new: integrating population ecology and population genetics of Weddell seals. In W. Davison, C. Howard-Williams, and P. Broady (eds), *Antarctic Ecosystems: Models For Wider Ecological Understanding*, pp. 63–70. New Zealand Natural Sciences, University of Canterbury, NZ.

Gentry, R.L. and Holt, J.R. (1982). *Equipment and Techniques for Handling Northern Fur Seals.* US Department of Commerce, NOAA Technical Report NMFS SSRF 758, 16 pp.

George, J.C., Bada, J., Zeh, J., Scott, L., Brown, S.E., O'Hara, T., and Sydeman, W.J. (1999). Age and growth estimates of bowhead whales (*Balaena mysticetus*) via aspartic acid racemization. *Canadian Journal of Zoology*, **77**, 571–580.

George, J.C., Zeh, J., Suydam, R., and Clark, C. (2004). Abundance and population trend (1978–2001) of western Arctic bowhead whales surveyed near Barrow, Alaska. *Marine Mammal Science*, **20**, 755–773.

Georges, J.-Y., Bonadonna, F., and Guinet, C. (2000). Foraging habitat and diving activity of lactating Subantarctic fur seals in relation to sea-surface temperatures at Amsterdam Island. *Marine Ecology—Progress Series*, **196**, 291–304.

Georges, J.Y., Groscolas, R., Guinet, C., and Robin, J.P. (2001). Milking strategy in subantarctic fur seals *Arctocephalus tropicalis* breeding on Amsterdam Island: evidence from changes in milk composition. *Physiological and Biochemical Zoology*, **74**, 548–559.

Geraci, J.R. and Lounsbury, V.J. (1993). *Marine Mammals Ashore; a Field Guide for Strandings.* Texas A&M University Sea Grant College Program, Publication TAMU-SG-923-601.

Geraci, J.R. and Lounsbury, V.J. (2005). *Marine Mammals Ashore. A Field Guide for Strandings.* National Aquarium, Baltimore, MD.

Geraci, J.R. and Ridgway, S.H. (1991). On disease transmission between cetaceans and humans. *Marine Mammal Science*, **7**, 191–194.

Gerber, L.R., Tinker, M.T., Doak, D.F., Estes, J.A., and Jessup, D.A. (2004). Mortality sensitivity in life-stage simulation analysis: a case study of southern sea otters. *Ecological Applications*, **14**, 1554–1565.

Gerber, L.R., Keller, A.C., and DeMaster, D.P. (2007). Ten thousand and increasing: Is the western Arctic population of bowhead whale endangered? *Biological Conservation*, **137**, 577–583.

Gero, S. and Whitehead, H. (2007). Suckling behavior in sperm whale calves: observations and hypotheses. *Marine Mammal Science*, **23**, 398–413.

Gero, S., Bejder, L., Whitehead, H., Mann, J., and Connor, R.C. (2005). Behaviourally specific preferred associations in bottlenose dolphins, *Tursiops* sp. *Canadian Journal of Zoology*, **83**, 1566–1573.

Gerondeau, M., Barbraud, C., Ridoux, V., and Vincent, C. (2007). Abundance estimate and seasonal patterns of grey seal (*Halichoerus grypus*) occurrence in Brittany, France, as assessed by photo-identification and capture–mark–recapture. *Journal of the Marine Biological Association of the United Kingdom*, **87**, 365–372.

Gerrodette, T. and Forcada, J. (2005). Non-recovery of two spotted and spinner dolphin populations in the eastern tropical Pacific

Ocean. *Marine Ecology Progress Series*, **291**, 1–21.

Giannasi, N., Thorpe, R.S., and Malhotra, A. (2001). The use of amplified fragment length polymorphism in determining species trees at fine taxonomic levels: analysis of a medically important snake, *Trimeresurus albolabris*. *Molecular Ecology*, **10**, 419–426.

Gibson, Q.A. and Mann, J. (2008). The size, composition and function of wild bottlenose dolphin (*Tursiops* sp.) mother–calf groups in Shark Bay, Australia. *Animal Behaviour*, **76**, 389–405.

Gilks, W.R., Richardson, S., and Spiegelhalter, D. (1995). *Markov Chain Monte Carlo in Practice*. Chapman and Hall, New York.

Ginsberg, J.R. and Young, T.P. (1992). Measuring association between individuals or groups in behavioural studies. *Animal Behaviour*, **44**, 377–379.

Glaubitz, J.C., Rhodes, O.E., and Dewoody, J.A. (2003). Prospects for inferring pairwise relationships with single nucleotide polymorphisms. *Molecular Ecology*, **12**, 1039–1047.

Goetz, K., Rugh, D.J., Read, A.J., and Hobbs, R.C (2007). Habitat use in a marine ecosystem: beluga whales *Delphinapterus leucas* in Cook inlet, Alaska. *Marine Ecology Progress Series*, **330**, 247–256.

Goldstein, J.D., Reese, E., Reif, J.S., Varela, R.A., McCulloch, S.D., Defran, R.H., Fair, P. A., and Bossart, G.D. (2006). Hematologic, biochemical, and cytologic findings from apparently healthy Atlantic bottlenose dolphins (*Tursiops truncatus*) inhabiting the Indian River Lagoon, Florida, USA. *Journal of Wildlife Diseases*, **42**, 447–454.

Goldstein, T., Mazet, J.A.K., Zabka, T.S., Langlois, G., Colegrove, K.M., Silver, M., Bargu, S., Van Dolah, F., Leighfield, T., Conrad, P.A., Barakos, J., Williams, D.C., Dennison, S., Haulena, M., and Gulland F.M.D. (2008). Novel symptomatology and changing epidemiology of domoic acid toxicosis in California sea lions (*Zalophus californianus*): an increasing risk to marine mammal health. *Proceedings of the Royal Society B—Biological Sciences*, **275**, 267–276.

Gómez de Segura, A., Hammond, P.S., Cañadas, A., and Raga, J.A. (2007). Comparing cetacean abundance estimates derived from spatial models and design-based line transect

methods. *Marine Ecology Progress Series*, **329**, 289–299.

Goodman, D. (2004). Methods for joint inference from multiple data sources for improved estimates of population size and survival rates. *Marine Mammal Science*, **20**, 401–423.

Gordon, J.C.D. (1987). Sperm whale groups and social behaviour observed off Sri Lanka. *Reports of the International Whaling Commission*, **37**, 205–217.

Gordon, J. (2001). Measuring the range to animals at sea from boats using photographic and video images. *Journal of Applied Ecology*, **38**, 879–887.

Gotelli, N.J. (2001). *A Primer of Ecology*, 3rd edn. Sinauer Associates, Inc., Sunderland, MA.

Goudet, J., Raymond, M., De Meeus, T., and Rousset, F. (1996). Testing differentiation in diploid populations. *Genetics*, **144**, 1933–1940.

Goulet, A-M., Hammill, M.O., and Barrette, C. (2001). Movements and diving of grey seal females (*Halichoerus grypus*) in the Gulf of St. Lawrence, Canada. *Polar Biology*, **21**, 432–439.

Graham, B.S., Koch, P.L., Newsome, S.D., McMahon, K.W., and Aurioles, D. (2009). Using isoscapes to trace the movements and foraging behavior of top predators in oceanic ecosystems. In G.J. Bowen, J. West, K. Tu, and T. Dawson (eds), *Isoscapes: Isotope Mapping and its Applications*. Springer-Verlag, New York.

Grant, P.R. and Grant, B.R. (2000). Non-random fitness variation in two populations of Darwin's finches. *Proceedings of Royal Society of London, B*, **267**, 131–138.

Greaves, D.K., Hughson, R.L., Topor, Z., Schreer, J.F., Burns, J.M., and Hammill, M. O. (2004). Changes in heart rate variability during diving in young harbor seals, Phoca vitulina. *Marine Mammal Science*, **20**, 861–871.

Green, J.A., White, C.R., and Butler, P.J. (2005). Allometric estimation of metabolic rate from heart rate in penguins. *Comparative Biochemistry and Physiology A*, **142**, 478–484.

Green, J.A., Frappell, P.B., Clark, T.D., and Butler, P.J. (2006). Physiological response to feeding in little penguins. *Physiological and Biochemical Zoology*, **79**, 1088–1097.

Green J.A., Haulena, M., Boyd, I.L., Calkins, D., Gulland, F., Woakes, A.J., and Butler, P.J.

(2009). Trial implantation of heart rate data loggers in two species of pinnped. *Journal of Wildlife Management*, **73**, 115–121.

Green, K. (1997). Diving behaviour of Antarctic fur seals *Arctocephalus gazella* Peters around Heard Island. In M. Hindell and C. Kemper (eds), *Marine Mammal Research in the Southern Hemisphere*, Vol. 1—*Status, Ecology and Medicine*, pp. 97–104. Surrey Beatty and Sons, Chipping Norton, NSW, Australia.

Gregr, E.J. and Trites, A.W. (2001). Predictions of critical habitat for five whale species in the waters of coastal British Columbia. *Canadian Journal of Fisheries and Aquatic Science*, **58**, 1265–1285.

Grellier, K. and Hammond, P.S. (2005). Feeding method affects otolith digestion in captive grey seals: implications for diet composition estimation. *Marine Mammal Science*, **21**, 296–306.

Grellier, K. and Hammond, P.S. (2006). Robust digestion and passage rate estimates for hard parts of grey seal (*Halichoerus grypus*) prey. *Canadian Journal of Fisheries and Aquatic Science*, **63**, 1982–1998.

Grenfell, B. and Dobson, A. (1995). *Ecology of Infectious Diseases in Natural Populations*. Cambridge University Press, Cambridge.

Griffin, R.B. and Griffin, N. (2003). Distribution, habitat partitioning and abundance of atlantic spotted dolphins, bottlenose dolphins and loggerhead sea turtles on the eastern gulf of Mexico continental shelf. *Gulf of Mexico Science*, **1**, 23–34.

Grimm, V. and Railsback, S.F. (2005). *Individual-based Modeling and Ecology*. Princeton University Press. Princeton, NJ.

Grue Nielsen, H. (1972). Age determination of the harbour porpoise *Phocoena phocoena* (L.) (Cetacea). *Videnskabelige Meddelelser Dansk Naturhistorisk Forening*, **135**, 61–84.

Gudmundson, C.J., Zeppelin, T.K., and Ream, R.R. (2006). Application of two methods for determining diet in northern fur seals (*Callorhinus ursinus*). *Fishery Bulletin*, **104**, 445–455.

Guillot, G., Mortier, F., and Estoup, A. (2005). Geneland: a computer package for landscape genetics. *Molecular Ecology Notes*, **5**, 712–715.

Guinet, C., Dubroca, L., Lea, M.A., Goldsworthy, S., Cherel, Y., Duhamel, G., Bonadonna, F., and Donnay, J.P. (2001). Spatial distribution of foraging in female Antarctic fur seals Arctocephalus gazella in relation to oceanographic variables: a scale-dependent approach using geographic information systems. *Marine Ecology Progress Series*, **219**, 251–264.

Guisan, A. and Theurillat, J.-P. (2000). Equilibrium modelling of alpine plant distribution: how far can we go. *Phytocoenologia*, **30**, 353–384.

Guisan, A. and Zimmermann, N.E. (2000). Predictive habitat distribution models in ecology. *Ecological Modelling*, **135**, 147–186.

Guisan, A., Edwards, T.C., and Hastie, T. (2002). Generalized linear and generalized additive models in studies of species distributions: setting the scene. *Ecological Modelling*, **157**, 89–100.

Gulland, F.M. (1999). Stranded seals: important sentinels. *Journal of the American Veterinary Medical Association*, **214**, 1191–1192.

Gulland, F.M.D. and Hall, A.J. (2007). Is marine mammal health deteriorating? Trends in the global reporting of marine mammal disease. *EcoHealth*, **4**, 135–150.

Gulland, F.M.D., Dierauf, L.A., and Rowles T.K. (2001a). Marine mammal stranding networks. In L.A. Dierauf and F.M.D. Gulland (eds), *Marine Mammal Medicine*, pp. 45–69. CRC Press, Boca Raton, FL.

Gulland, F.M.D., Haulena, M., and Dierauf, L.A. (2001b). Seals and sea lions. In L.A. Dierauf and F.M.D. Gulland (eds), *CRC Handbook of Marine Mammal Medicine*, pp. 907–922. CRC Press, New York.

Hadley, G.L., Rotella, J.J., Garrott, R.A., and Nichols, J.D. (2006). Variation in probability of first reproduction of Weddell seals. *Journal of Animal Ecology*, **75**, 1058–1070.

Hain, J.H.W. (1991). Airships for marine mammal research: evaluation and recommendations. *NTIS publication*, Pb92–128271.

Hairston, N.G., Smith, F.E., and Slobodkin, L.B. (1960). Community structure, population control, and competition. *The American Naturalist*, **94**, 421.

Haley, M.P., Deutsch, C.J., and Le Boeuf, B.J. (1991). A method for estimating mass of large pinnipeds. *Marine Mammal Science*, **7**, 157–164.

Hall, A.J. and McConnell, B.J. (2007). Measuring changes in juvenile grey seal body condition. *Marine Mammal Science*, **23**, 650–665.

Hall, A.J., Moss, S., and McConnell, B.J. (2000). A new tag for identifying seals. *Marine Mammal Science*, **16**, 254–257.

Hall, A.J., McConnell, B.J., and Barker, R.J. (2001). Factors affecting first-year survival in grey seals and their implications for life history strategy. *Journal of Animal Ecology*, **70**, 138–149.

Hall A.J., McConnell B.J., and Barker R.J. (2002). The effect of total immunoglobulin levels, mass and condition on the first-year survival of Grey Seal pups. *Functional Ecology*, **16**, 462–474.

Hall, A.J., McConnell, B.J., Rowles, T.K., Aguilar, A., Borrell, A., Schwacke, L., Reijnders, P.J.H., and Wells, R.S. (2006a). Individual-based model framework to assess population consequences of polychlorinated biphenyl exposure in bottlenose dolphins. *Environmental Health Perspectives*, **114**, 60–64.

Hall, A.J., Hugunin, K., Deaville, R., Law, R.J., Allchin, C.R., and Jepson P.D. (2006b). The risk of infection from polychlorinated biphenyl exposure in harbor porpoise (*Phocoena phocoena*)—A case-control approach. *Environmental Health Perspectives*, **114**, 704–711.

Hall, A.J., Wells, R.S., Sweeney, J.C., Townsend, F.I., Balmer, B.C., Hohn, A.A., and Rhinehart, H.L. (2007). Annual, seasonal and individual variation in hematology and clinical blood chemistry profiles in bottlenose dolphins (*Tursiops truncatus*) from Sarasota Bay, Florida. *Comparative Biochemistry and Physiology, Part A*, **148**, 266–277.

Hall, L.S., Krausman, P.R., and Morrison, M. (1997). The habitat concept and a plea for standard terminology. *Wildlife Society Bulletin*, **25**, 173–182.

Hall-Aspland, S.A., Rogers, T.L., and Canfield, R.B. (2005). Stable carbon and nitrogen isotope analysis reveals seasonal variation in the diet of leopard seals. *Marine Ecology Progress Series*, **305**, 249–259.

Halpin, P.N., Read, A.J., Best, B.D., Hyrenbach, K.D., Fujioka, E., Coyne, M.S., Crowder, L.B., Freeman, S.A., and Spoerri, C. (2006). OBIS-SEAMAP: Developing a biogeographic research data commons for the ecological studies of marine mammals, seabirds and sea turtles. *Marine Ecology Progress Series*, **316**, 239–246.

Hamilton, P.K. and Marx M.K. (2005). Skin lesions on North Atlantic right whales: categories, prevalence and change in occurrence in the 1990s. *Diseases of Aquatic Organisms*, **68**, 71–82.

Hamilton, P.K., Knowlton, A.R., and Marx, M.K. (2007). Right whales tell their own stories: The photo-identification catalog. In S.D. Kraus and R.M. Rolland (eds), *The Urban Whale: North Atlantic Right Whales at the Crossroads*, pp. 75–104. Harvard University Press, Harvard, MA.

Hamilton, W.D. (1964). The genetical evolution of social behaviour. *Journal of Theoretical Biology*, **7**, 1–52.

Hammill, M.O. and Stenson, G.B. (2000). Estimated prey consumption by harp seals (*Phoca groenlandica*), hooded seals (*Cystophora cristata*), grey seals (*Halichoerus grypus*) and harbour seals (*Phoca vitulina*) in Atlantic Canada. *Journal of Northwest Atlantic Fishery Science*, **26**, 1–23.

Hammill, M. and Stenson, G. (2007). Application of the precautionary approach and conservation reference points to management of Atlantic seals. *ICES Journal of Marine Science*, **64**, 702–706.

Hammill, M.O., Lesage, V., and Carter, P. (2005). What do harp seals eat? Comparing diet composition from different compartments of the digestive tract with diets estimated from stable isotope ratios. *Canadian Journal of Zoology*, **83**, 1365–1372.

Hammond, P.S. (1986). Estimating the size of naturally marked whale populations using capture–recapture techniques. *Reports of the International Whaling Commission*, Special Issue **8**, 253–282.

Hammond, P.S. (1990a). Capturing whales on film—estimating cetacean population parameters from individual recognition data. *Mammal Review*, **20**, 17–22.

Hammond, P.S. (1990b). Heterogeneity in the Gulf of Maine? Estimating humpback whale population size when capture probabilities are not equal. *Reports of the International Whaling Commission*, Special Issue **12**, 135–139.

Hammond, P.S. (1995). Estimating the abundance of marine mammals: a North Atlantic perspective. In A.S. Blix, L. Walløe, and Ø. Ulltang (eds), *Whales, Seals, Fish and Man*, Vol. 4: *Developments in Marine Biology*, pp. 3–12. Elsevier Science BV, Amsterdam.

Hammond, P.S. (2009). Mark–recapture. In W.F. Perrin, B. Würsig, and J.G.M. Thewissen, (eds), *Encyclopedia of Marine Mammals*, 2nd edn. Elsevier, San Diego, CA.

Hammond, P.S. and Fedak, M.A. (1994). *Grey Seals in the North Sea and their Interactions with Fisheries*. Sea Mammal Research Unit, Cambridge, UK, Final Report to The Ministry of Agriculture, Fisheries, and Food, Contract MF 0503.

Hammond, P.S. and Grellier, K. (2005). *Grey Seal Diet Composition and Fish Consumption in the North Sea*. Department for Environment, Food and Rural Affairs under project MF0319, pp. 1–54. London.

Hammond, P.S. and Rothery, P. (1996). Application of computer sampling in the estimation of seal diet. *Journal of Applied Statistics*, **23**, 525–533.

Hammond, P.S., Mizroch, S.A., and Donovan, G.P. (eds) (1990). Individual recognition of cetaceans: use of photo-identification and other techniques to estimate population parameters. *Reports of the International Whaling Commission*, Special Issue **12**. IWC, Cambridge, UK.

Hammond, P.S., Berggren, P., Benke, H., Borchers, D.L., Collet, A., Heide-Jørgensen, M.P., Heimlich, S., Hiby, A.R., Leopold, M. F., and Øien, N (2002). Abundance of harbour porpoise and other cetaceans in the North Sea and adjacent waters. *Journal of Applied Ecology*, **39**, 361–376.

Hansel, H.C., Duke, S.D., Lofy, P.T., and Gray, G.A. (1988). Use of diagnostic bones to identify and estimate original lengths of ingested prey fishes. *Transactions of the American Fisheries Society*, **117**, 55–62.

Hanson, M.B., Baird, R.W, and DeLong, R.L. (1999). Movements of a tagged harbor porpoise in inland Washington waters from June 1998 to January 1999. In A.L. Lopez and D. P. DeMaster (eds), *Marine Mammal Protection Act and Endangered Species Act Implementation Program 1998*, pp. 85–95. Unpublished report. Available National Marine Mammal Laboratory, 7600 Sand Point Way, NE, Seattle, WA 98115, 305 pp.

Hanson, N., Wurster, C., Bird, M., Reid, K., and Boyd, I.L. (2009). Intrinsic and extrinsic forcing in life histories: patterns of growth and stable isotopes in male Antarctic fur seal

teeth. *Marine Ecology Progress Series*, **388**, 263–272.

Harcourt, R. and Davis, I. (1997). The use of satellite telemetry to determine fur seal foraging areas. In M. Hindell and C. Kemper (eds), *Marine Mammal Research in the Southern Hemisphere*, Volume 1: *Status, Ecology and Medicine*, pp. 137–142. Surrey Beatty & Sons, Chipping Norton, NSW, Australia 2170.

Harcourt, R.G., Hindell, M.A., Bell, D.G., and Waas, J.R. (2000). Three-dimensional dive profiles of free-ranging Weddell seals. *Polar Biology*, **23**, 479–487.

Harcourt, R.G., Bradshaw, C.J.A., Dickson, K., and Davis, L.S. (2002). Foraging ecology of a generalist predator, the female New Zealand fur seal. *Marine Ecology Progress Series*, **227**, 11–24.

Härkönen, T. (1986). *Guide to the Otoliths of the Bony Fishes of the North-east Atlantic*, pp. 256. Danbiu ApS Biological Consultants, Hellerup, Denmark.

Harrison, R.J. (1969). Reproduction and reproductive organs. In S.D Kraus and R.M. Rolland (eds), *The Biology of Marine Mammals*, pp. 253–348. Academic Press, New York.

Harrison, R.J., Brownell, R.L., and Boice, R.C. (1972). Reproductive and gonadal appearances in some odontocetes. In R.J. Harrison (ed.), *Functional Anatomy of Marine Mammals*, Vol. 1, pp. 362–429. Academic Press, London.

Hart, J.L. (1973). Pacific fishes of Canada. *Fisheries Research Board of Canada Bulletin*, **180**, 1–740.

Harting, A.L. (2002). Stochastic simulation model for the Hawaiian monk seal. PhD thesis, Montana State University.

Harting, A.L., Baker, J.D., and Johanos, T.C. (2007). Reproductive patterns of the Hawaiian monk seal. *Marine Mammal Science*, **23**, 553–573.

Harvey, J.T. (1989). Assessment of errors associated with harbor seal (*Phoca vitulina*) fecal sampling. *Journal of Zoology (London)*, **219**, 101–111.

Harvey, J.T. and Antonelis, G.A. (1994). Biases associated with non-lethal methods of determining the diet of northern elephant seals. *Marine Mammal Science*, **10**, 178–187.

Harvey, J.T., Loughlin, T.R., Perez, M.A., and Oxman, D.S. (2000). *Relationship between Fish Size and Otolith Length for 63 Species of Fishes from the Eastern North Pacific Ocean*. NOAA Technical Report NMFS 150, US Department of Commerce, Seattle, WA. pp. 38.

Harwood, J. (2007). Is there a role for ecologists in an ecosystem approach to the management of marine resources? *Aquatic Conservation: Marine and Freshwater Ecosystems*, **17**, 1–4.

Harwood, J. and Stokes, K. (2003). Coping with uncertainty in ecological advice: lessons from fisheries. *Trends in Ecology and Evolution*, **18**, 617–622.

Harwood, J. and Wilson, B. (2001). The implications of developments on the Atlantic frontier for marine mammals. *Continental Shelf Research*, **21**, 1073–1093.

Harwood, J., Anderson, S.S, Fedak, M.A., Hammond, P.S., Hiby, A.R., McConnell, B.J., Martin, A.R., and Thompson, D. (1989). New approaches for field studies of mammals: experiences with marine mammals. *Biological Journal of the Linnean Society*, **38**, 103–111.

Hassrick, J.L., Crocker, D.E., Zeno, R.L., Blackwell, S.B., Costa, D.P., and Le Boeuf, B.J. (2007). Swimming speed and foraging strategies of northern elephant seals. *Deep-Sea Research Part II—Topical Studies in Oceanography*, **54**, 369–383.

Hastie, G.D., Barton, T.R., Grellier, K., Hammond, P.S., Swift, R.J., Thompson, P.M., and Wilson, B. (2003). Distribution of small cetaceans within a candidate Special Area of Conservation; implications for management. *Journal of Cetacean Research and Management*, **5**, 261–266.

Hastie, G.D., Wilson, B., Wilson, L.J., Parsons, K.M., and Thompson, P.M. (2004). Functional mechanisms underlying cetacean distribution patterns: hotspots for bottlenose dolphins are linked to foraging. *Marine Biology*, **144**, 397–403.

Hastie, G.D., Swift, R., Slesser, G., Thompson, P.M., and Turrell, W.R. (2005). Environmental models for predicting oceanic dolphin habitat in the Northeast Atlantic. *ICES Journal of Marine Science*, **62**, 760–770.

Hastie, G.D., Rosen, D.A.S., and Trites, A.W. (2007). Reductions in oxygen consumption during dives and estimated submergence limitations of Steller sea lions (*Eumetopias jubatus*). *Marine Mammal Science*, **23**, 272–286.

Hastie, T.J. and Tibshirani R.J. (1990). *Generalized Additive Models*. Chapman and Hall, London.

Haulena, M. and Heath, R.B. (2001). Marine mammal anaesthesia. In L.A. Dierauf and F.M.D. Gulland (eds), *CRC Handbook of Marine Mammal Medicine*, pp. 655–688. CRC Press, Boca Raton, FL.

Hayes, J.P. and Steidl, R.J. (1997). Statistical power analysis and amphibian population trends. *Conservation Biology*, **11**, 273–275.

Hays, G.C., Akesson, S., Godley, B.J., Luschi, P., and Santidrian, P. (2001). The implications of location accuracy for the interpretation of satellite-tracking data. *Animal Behaviour*, **61**, 1035–1040.

Hays, G.C., Hobson, V.J., Metcalfe, J.D., Righton, D., and Sims, D.W. (2006). Flexible foraging movements of leatherback turtles across the north Atlantic Ocean. *Ecology*, **87**, 2647–2656.

Hearn, W.S., Leigh, G.M., and Beverton, R.J.H. (1991). An examination of a tag-shedding assumption, with application to southern bluefin tuna. *ICES Journal of Marine Science.* **48**, 41–51.

Heath, R.B., Calkins, D., McAllister, D., Taylor, W., and Spraker, T. (1996). Telazol and isoflurane field anesthesia in free-ranging Steller sea lions (*Eumetopias jubatus*). *Journal of Zoo and Wildlife Medicine*, **27**, 35–43.

Hedd, A., Gales, R., and Renouf, D. (1996). Can stomach temperature telemetry be used to quantify prey consumption by seals? A re-examination. *Polar Biology*, **16**, 261–270.

Hedley, S.T., Buckland, S.T., and Borchers, D.L. (1999). Spatial modelling from line transect data. *Journal of Cetacean Research and Management*, **1**, 255–264.

Hedley, S.L., Buckland, S.T., and Borchers, D.L. (2004). Spatial distance sampling models. In S.T. Buckland, D.R. Anderson, K.P. Burnham, J.L. Laake, D.L. Borchers, and L. Thomas (eds), *Advanced Distance Sampling: Estimating Abundance of Biological Population*, pp 48–70. Oxford University Press, Oxford.

Heegaard, E. (2002). The outer border and central border for species–environmental relationships estimated by non-parametric generalised additive models. *Ecological Modelling*, **157**, 131–139.

Heide-Jørgensen, M.P. (2004). Aerial digital photographic surveys of narwhals, *Monodon monoceros*, in Northwest Greenland. *Marine Mammal Science*, **20**, 246–261.

Heide-Jorgensen, M.P., Kleivane, L., Oien, N., Laidre, K.L., and Jensen, M.V. (2001). A new technique for deploying satellite transmitters on baleen whales: Tracking a blue whale (*Balaenoptera musculus*) in the North Atlantic. *Marine Mammal Science*, **17**, 949–954.

Heide-Jorgensen, M.P., Laidre, K., Borchers, D., Samarra, F., and Stern, H. (2007). Increasing abundance of bowhead whales in West Greenland. *Biology Letters*, **3**, 577–580.

Heide-Jorgensen, M.P., Laidre, K.L., Hansen, R.G., Rasmussen, M., Burt, M.L. and Borchers, D.L. Rate of increase and current abundance of humpback whales in West Greenland. *Journal of Cetacean Research and Management*. [In press.]

Heinsohn, G.E. (1981). Methods of taking measurements, other data and specimen material from dugong carcasses. In H. Marsh (ed.), *The Dugong: Proceedings of a Seminar/Workshop held at James Cook University* 8–13 May, 1979, pp. 228–238. James Cook University, Townsville, Australia.

Heisey, D.M. and Patterson, B.R. (2006). A review of methods to estimate cause-specific mortality in presence of competing risks. *Journal of Wildlife Management*, **70**, 1544–1555.

Heithaus, M.R. and Frid, A. (2003). Optimal diving under the risk of predation. *Journal of Theoretical Biology*, **223**, 79–92.

Heithaus, M., Frid, A., Wirsig, A., and Worm, B. (2008). Predicting ecological consequences of marine top predator declines. *Trends in Ecology and Evolution*, **23**, 202–209.

Hemson, G., Johnson, P., South, A., Kenward, R., Ripley, R., and MacDonald, D. (2005). Are kernels the mustard? Data from global positioning system (GPS) collars suggests problems for kernel home-range analyses with least-squares cross-validation. *Journal of Animal Ecology*, **64**, 455–463.

Henderson, N. and Sutherland, W.J. (1996). Two truths about discounting and their environmental consequences. *Trends in Ecology and Evolution*, **11**, 527–528.

Herman, D.P., Burrows, D.G., Wade, P.R., Durban, J.W., Matkin, C.O., LeDuc, R.G., Barrett-Lennard, L.G., and Krahn, M.M. (2005). Feeding ecology of eastern North Pacific killer whales *Orcinus orca* from fatty acid stable isotope and organochlorine analyses of blubber biopsies. *Marine Ecology Progress Series*, **302**, 275–291.

Herman, D.P., Matkin, C.O., Ylitalo, G.M., Durban, J.W., Hanson, M.B., Dahlheim, M. E., Straley, J.M., Wade, P.R., Tilbury, K.L., Boyer, R.H., Pearce, R.W., and Krahn, M.M. (2008). Assessing age distributions of killer whale *Orcinus orca* populations from the composition of endogenous fatty acids in their outer blubber layers. *Marine Ecology Progress Series*, **372**, 289–302.

Herraeza, D.L., Schafer, H., Mosner, J., Fries, H.R., and Wink, M. (2005). Comparison of microsatellite and single nucleotide polymorphism markers for the genetic analysis of a Galloway cattle population. *Verlag der Zeitschrift für Naturforschung* [c], **60**, 637–643.

Hey, J., Waples, R.S., Arnold, M.L., Butlin, R.K., and Harrison, R.G. (2003). Understanding and confronting species uncertainty in biology and conservation. *Trends in Ecology and Evolution*, **18**, 597–603.

Heyning, J.E. and Perrin, W.F. (1994). Evidence for two species of common dolphins (genus *Delphinus*) fron the eastern North Pacific. *Natural History Museum of Los Angeles County Contributions in Science*, **442**, 1–35.

Hiby, A.R. and Hammond, P.S. (1989). Survey techniques for estimating abundance of cetaceans. *Reports of the International Whaling Commission*, Special Issue, **11**, 47–80.

Hiby, L. (1999). The objective identification of duplicate sightings in aerial survey for porpoise. In G.W. Garner, S.C. Armstrup, J.L. Laake, B.F.J. Manly, L.L. McDonald, and D. G. Robertson (eds), *Marine Mammal Survey and Assessment Methods*, pp. 179–189. Balkema, Rotterdam.

Hiby, A.R. and Lovell, P. (1998). Using aircraft in tandem formation to estimate abundance of harbour porpoises. *Biometrics*, **54**, 1280–1289.

Hiby, L. and Lovell, P. (1990). Computer-aided matching of natural markings: a prototype system for grey seals. *Reports of the International Whaling Commission*, Special Issue, **12**, 57–61.

Hicks, B.D., St. Aubin, D.J., Geraci, J.R., and Brown, W.R. (1985). Epidermal growth in the bottlenose dolphin, *Tursiops truncatus*. *Journal of Investigative Dermatology*, **85**, 60–63.

Hilborn, R., Orsensanz, J.M., and Parma, A.M. (2005). Institutions, incentives and the future of fisheries. *Philosophical Transactions of the Royal Society B: Biological Sciences*, **360**, 47–57.

Hildebrand, J. (2005). Impacts of anthropogenic sound. In J.E. Reynolds, W.F. Perrin, R.R. Reeves, S. Montgomery, and T.J. Ragen (eds), *Marine Mammal Research: Conservation Beyond Crisis*, pp. 101–123. The Johns Hopkins University Press, Baltimore, MD.

Hill, K.E. and Binford, M.W. (2002). The role of category definition in habitat models: Practical and logical limitations of using Boolean, indexed, probabilistic and fuzzy categories. In J.M. Scott, P.J. Heglund, and M.L. Morrison (eds), *Predicting Species Occurrences: Issues of Accuracy and Scale*, pp. 97–106. Island Press, Washington DC.

Hill, M.O. (1991). Patterns of species distribution in Britain elucidated by canonical correspondence analysis. *Journal of Biogeography*, **18**, 247–255.

Hill, R.D. (1994). Theory of geolocation by light levels. In B.J. Le Boeuf (ed.), *Elephant Seals: Population Ecology, Behavior, and Physiology*, pp. 227–236. University of California Press, Berkeley, CA.

Hill, R.D. and Braun, M.J. (2001). Geolocation by light-level. The next step: lattitude. In J. Sibert and J. Nielsen (eds), *Electronic Tagging and Tracking in Marine Fisheries Reviews: Methods and Technologies in Fish Biology and Fisheries*, pp. 443–456. Kluwer Academic Press, Dordrecht.

Hillman, G.R. (2005). Photo-identification of Steller sea lions. In T.R. Loughlin, S. Atkinson, and D.G. Calkins (eds), *Synopsis of Research on Steller Sea Lions: 2001–2005*, pp. 275–283.Alaska SeaLife Center's Steller Sea Lion Program. Sea Script Company, Seattle, WA.

Hillman, G.R., Kehtarnavaz, N., Würsig, B., Araabi, B., Gailey, G., Weller, D., Mandava, S., and Tagare, H. (2002). 'Finscan', a computer system for photographic identification of marine animals. *Second Joint EMBS–BMES Conference Proceedings*, Vol. **2**, 1065–1066.

Hillman, G.R., Würsig, B., Gailey, G.A., Kehtarnavaz, N., Drobyshevsky, A., Araabi, B.N., Tagare, H.D., and Weller, D.W. (2003). Computer-assisted photo-identification of individual marine vertebrates: A multi-species system. *Aquatic Mammals*, **29**, 117–123.

Hinde, R.A. (1976). Interactions, relationships and social structure. *Man*, **11**, 1–17.

Hindell, M. (2008). To breathe or not to breathe: optimal strategies for finding prey in a dark, three-dimensional environment. *Journal of Animal Ecology*, **77**, 847–849.

Hindell, M.A., Slip, D.J., and Burton, H.R. (1991). The diving behavior of adult male and female southern elephant seals, *Mirounga leonina* (Pinnipedia, *Phocidae*). *Australian Journal of Zoology*, **39**, 595–619.

Hindell, M.A., Lea, M.A., Morrice, M.G., and MacMahon, C.R. (2000). Metabolic limits on dive duration and swimming speed in the southern elephant seal Mirounga leonina. *Physiological and Biochemical Zoology*, **73**, 790–798.

Hindell, M.A., Harcourt, R., Waas, J.R., and Thompson, D. (2002). Fine-scale three-dimensional spatial use by diving, lactating female Weddell seals *Leptonychotes weddellii*. *Marine Ecology Progress Series*, **242**, 275–284.

Hindell, M.A., Bradshaw, C.J.A., Harcourt, R., and Guinet, C. (2003). Ecosystem monitoring: Are seals a potential tool for monitoring change in marine systems? In N.J. Gales, M.A. Hindell, and R. Kirkwood (eds), *Marine Mammals. Fisheries, Tourism and Management Issues*, pp 330–343. CSIRO Publishing, Melbourne, Australia.

Hirzel, A.H., Hausser, J., Chessel, D., and Perrin, N. (2002). Ecological-niche factor analysis: how to compute habitat suitability maps without absence data? *Ecology*, **83**, 2027–2036.

Hjermann, D.O. (2000). Analyzing habitat selection in animals without well-defined home ranges. *Ecology*, **81**, 1462–1468.

Hoberecht, L.K. (2006). Investigating the use of blubber fatty acids to detect steller sea lion (Eumetopias jubatus) foraging on ephemeral high-quality prey. PhD thesis, University of Washington, School of Aquatic and Fishery Sciences, Seattle, WA.

Hobs, N.T. and Hilborn, R. (2006). Alternatives to statistical hypothesis testing in ecology: A guide to self teaching. *Ecological Applications*, **16**, 5–19.

Hobson, B.M. and Boyd, I.L. (1984). Concentrations of placental gonadotrophin, placental progesterone and maternal plasma progesterone during pregnancy in the grey seal (*Halichoerus grypus*). *Journal of Reproduction and Fertility*, **72**, 521–528.

Hobson, K.A. and Sease, J. (1998). Stable isotope analysis of tooth annuli reveal temporal dietary records: an example using Steller sea lions. *Marine Mammal Science*, **14**, 116–129.

Hobson, K.A. and Welch, H.E. (1992). Determination of trophic relationships within a high Arctic marine food web using $\delta^{13}C$ and $\delta^{15}N$ analysis. *Marine Ecology Progress Series*, **84**, 9–18.

Hobson, K.A., Schell, D.M., Renouf, D., and Noseworthy, E. (1996). Stable-carbon and nitrogen isotopic fractionation between diet and tissues of captive seals: implications for dietary reconstructions involving marine mammals. *Canadian Journal of Fisheries and Aquatic Sciences*, **53**, 528–533.

Hobson, K.A., Sease, J.L., Merrick, R.L., and Piatt, J.F. (1997a). Investigating trophic relationships of pinnipeds in Alaska and Washington using stable isotope ratios of nitrogen and carbon. *Marine Mammal Science*, **13**, 114–132.

Hobson, K.A., Gibbs, H.L., and Gloutney, M.L. (1997b). Preservation of blood and tissue samples for stable-carbon and stable-nitrogen isotope analysis. *Canadian Journal of Zoology*, **75**, 1720–1723.

Hobson, K.A., Fisk, A.T., Karnovsky, N., Holst, M., Gagnon, J.M., and Fortier, M. (2002). A stable isotope (^{13}C, ^{15}N) model for the North Water Polynya foodweb: implications for evaluating trophodynamics and the flow of energy and contaminants. *Deep Sea Research Part II: Topical Studies in Oceanography*, **49**, 5131–5150.

Hobson, K.A., Riget, F.F., Outridge, P.M., Dietz, R., and Born, E. (2004a). Baleen as a biomonitor of mercury content and dietary history of North Atlantic Minke Whales (*Balaenopetra acutorostrata*): Combining elemental and stable isotope approaches. *Science of the Total Environment*, **331**, 69–82.

Hobson, K.A., Sinclair, E.H., York, A.E., Thomason, J., and Merrick, R.E. (2004b). Retrospective isotopic analyses of Steller's sea lion tooth annuli and seabird feathers: a cross-taxa approach to investigating regime and dietary shifts in the Gulf of Alaska. *Marine Mammal Science*, **20**, 621–638.

Hobson, K.A., Barnett-Johnson, R., and Cerling, T. (2010). Using isoscapes to track animal migration. In G.J. Bowen, J. West, K. Tu, and T. Dawson (eds), *Isoscapes: Isotope Mapping and its Applications*, pp. 273–298. Springer-Verlag, New York.

Hodgson, A.J. and Marsh, H. (2007). Response of dugongs to boat traffic: The risk of disturbance and displacement. *Journal of Experimental Marine Biology and Ecology*, **340**, 50–61.

Hodgson, A.J., Marsh, H., Delean, S., and Marcus, L. (2007). Is attempting to change marine mammal behaviour a generic solution to the bycatch problem? A dugong case study. *Animal Conservation*, **10**, 263–273.

Hoelzel, A.R., Goldsworthy, S.D., and Fleischer, R.C. (2002). Population genetic structure. In A.R. Hoelzel (ed.), *Marine Mammal Biology: An Evolutionary Approach*, pp. 325–352. Blackwell Science, Oxford.

Hoffman, J.I. and Amos, W. (2005). Microsatellite genotyping errors: detection approaches, common sources and consequences for paternal exclusion. *Molecular Ecology*, **14**, 599–612.

Hoffman, J.I., Forcada, J., Trathan, P.N., and Amos, W. (2007). Female fur seals show active choice for males that are heterozygous and unrelated. *Nature*, **445**, 912–914.

Hohn, A.A. (1989). Variation in life-history traits: the influence of introduced variation. PhD dissertation, University of California, Los Angeles, CA.

Hohn, A.A. (2002). Age estimation. In W.F. Perrin, B. Wursig, and H.G.M. Thewissen (eds), *Encyclopedia of Marine Mammals*, pp. 6–13. Academic Press, San Diego, CA.

Hohn, A.A. and Fernandez, F. (1999). Biases in dolphin age structure due to age estimation technique. *Marine Mammal Science*, **15**, 1124–1132.

Hohn, A.A., Scott, M.D., Wells, R.S., Sweeney, J.C., and Irvine, A.B. (1989). Growth layers in the teeth from known-age, free-ranging bottlenose dolphins. *Marine Mammal Science*, **5**, 315–342.

Holmes, E.E. (2004). Beyond theory to application and evaluation: diffusion approximations for population viability analysis. *Ecological Applications*, **14**, 1272–1293.

Holmes E.E., Fritz L.W., York A.E., and Sweeney K. (2007). Age-structured modeling reveals long-term declines in the natality of Western Steller sea lions. *Ecological Applications*, **17**, 2214–2232.

Hooker, S.K. and Baird, R.W. (2001). Diving and ranging behaviour of odontocetes: a methodological review and critique. *Mammal Review*, **31**, 81–105.

Hooker, S.K. and Boyd, I.L. (2003). Salinity sensors on seals: using marine predators to carry CTD data loggers. *Deep-Sea Research (Part 1)*, **50**, 927–939.

Hooker, S.K., Iverson, S.J., Ostrom, P., and Smith, S.C. (2001a). Diet of northern bottlenose whales inferred from fatty-acid and stable-isotope analyses of biopsy samples. *Canadian Journal of Zoology*, **79**, 1442–1454.

Hooker, S.K., Baird, R.W., Al-Omari, S., Gowans, S., and Whitehead, H. (2001b). Behavioral reactions of northern bottlenose whales (*Hyperoodon ampullatus*) to biopsy darting and tag attachment procedures. *Fishery Bulletin*, **99**, 303–308.

Hooker, S.K., Boyd, I.L., Jessopp, M., Cox, O., Blackwell, J., Boveng, P.L., and Bengtson, J.L. (2002a). Monitoring the prey-field of marine predators: Combining digital imaging with datalogging tags. *Marine Mammal Science*, **18**, 680–697.

Hooker, S.K., Whitehead, H., Gowans, S., and Baird, R.W. (2002b). Fluctuations in distribution and patterns of individual range use of northern bottlenose whales. *Marine Ecology Progress Series*, **225**, 287–297.

Hooker, S.K., Whitehead, H., and Gowans, S. (2002c). Ecosystem consideration in conservation planning: energy demand of foraging bottlenose whales (*Hyperoodon ampullatus*) in a marine protected area. *Biological Conservation*, **104**, 51–58.

Horne, J.S., Garton, E.O., Krone, S.M., and Lewis, J.S. (2007). Analyzing animal movements using Brownian bridges. *Ecology*, **88**, 2354–2363.

Horning, M. and Hill, R.D. (2005). Designing an archival satellite transmitter for life-long deployments on oceanic vertebrates: The Life History Transmitter. *IEEE Journal of Oceanic Engineering*, **30**, 807–817.

Horning, M. and Trillmich, F. (1997). Ontogeny of diving behaviour in the Galapagos fur seal. *Behaviour*, **134**, 1211–1257.

Horsburgh, J.M., Morrice, M., Lea, M.A., and Hindell, M.A. (2008). Determining feeding events and prey encounter rates in a southern elephant seal: a method using swim speed and stomach temperature. *Marine Mammal Science*, **24**, 207–217.

Houston, A.I. and Carbone, C. (1992). The optimal allocation of time during the diving cycle. *Behavioral Ecology*, **3**, 255–265.

Hoyt, E. (2001). *Whale Watching 2001*. International Fund for Animal Welfare, London.

Huber, H.R., Rovetta, A.C., Fry, L.A., and Johnston, S. (1991). Age-specific natality of northern elephant seals at the south Farallon Islands, California. *Journal of Mammalogy*, **72**, 525–534.

Huber, H.R., Jeffries, S.J., Brown, R.F., DeLong, R.L., and VanBlaricom, G. (2001). Correcting aerial survey counts of harbor seals (*Phoca vitulina richardsi*) in Washington and Oregon. *Marine Mammal Science*, **17**, 276–293.

Hudson, P.J., Rizzoli, A.P., Grenfell, B.T., Heesterbeek, H., and Dobson, A.P. (eds), (2002). *The Ecology of Wildlife Diseases*. Oxford Univerity Press, Oxford.

Hudson, R.R., Boos, D.D., and Kaplan, N.L. (1992). A statistical test for detecting geographic subdivision. *Molecular Biology and Evolution*, **9**, 138–151.

Hughes-Hanks, J., Rickard, L., Panuska, C., Saucier, J., O'Hara, T., Dehn, L.A., and Rolland, R. (2005). Prevalence of *Cryptosporidium* spp. and *Giardia* spp. in five marine mammal species. *Journal of Parasitology*, **91**, 1225–1228.

Hunter, C.M. and Caswell, H. (2005). The use of the vec-permutation matrix in spatial matrix population models. *Ecological Modelling*, **188**, 15–21.

Hurley, J.A. and Costa, D.P. (2001). Standard metabolic rate at the surface and during trained submersions in adult California sea lions (*Zalophus californianus*). *Journal of Experimental Biology*, **204**, 3273–3281.

Hyslop, E.J. (1980). Stomach contents analysis: a review of methods and their application. *Journal of Fish Biology*, **17**, 411–429.

ICES (2004). *Report of the ICES/NAFO Working Group on Harp and Hooded Seals*. ICES CM 2004/ACFM:06. International Council for the Exploration of the Sea, Copenhagen.

ICES (2006). *Report of the ICES/NAFO Working Group on Harp and Hooded Seals.* ICES CM 2006/ACFM:32. International Council for the Exploration of the Sea, Copenhagen.

Insley, S.J., Phillips, A.V., and Charrier, I. (2003). A review of social recognition in pinnipeds. *Aquatic Mammals*, **29**, 181–201.

Insley, S.J., Robson, B.W., Yack, T., Ream, R.R., and Burgess, W.C. (2007). Acoustic determination of activity and flipper stroke rate in foraging northern fur seal females. *Endangered Species Research*, **4**, 147–155.

Ireland, D., Garrott, R.A., Rotella, J., and Banfield, J. (2006). Development and application of a mass-estimation method for Weddell seals. *Marine Mammal Science*, **22**, 361–378.

Irish, K.E. and Norse, E.A. (1996). Scant emphasis on marine biodiversity. *Conservation Biology*, **10**, 680.

Irvine, A.B. and Wells, R.S. (1972). Results of attempts to tag Atlantic bottlenose dolphins (*Tursiops truncatus*). *Cetology*, **13**, 1–5.

Irvine, A.B., Scott, D.M., Wells, R.S., and Kaufmann, J.H. (1981). Movements and activities of the Atlantic bottlenose dolphin, *Tursiops truncatus*, near Sarasota, Florida. *Fisheries Bulletin*, **79**, 671–688.

Irvine, A.B., Wells, R.S., and Scott, M.D. (1982). An evaluation of techniques for tagging small odontocete cetaceans. *Fisheries Bulletin US*, **80**, 135–143.

Irvine, L.G., Hindell, M.A., van den Hoff, J., and Burton, H.R. (2000). The influence of body size on dive duration of underyearling southern elephant seals (*Mirounga leonina*). *Journal of Zoology, London*, **251**, 463–471.

Isaaks, E.H. and Srivastava, R.M. (1990). *Applied Geostatistics.* Oxford University Press, Oxford.

IUCN (International Union for the Conservation of Nature) and Natural Resources (2008). *IUCN Red List of Threatened Species.* http://www.iucnredlist.org. [Downloaded on 17 May 2008.]

Iverson, L.R. and Prasad, A.M. (1998). Predicting abundance of 80 tree species following climate change in the eastern United States. *Ecological Monographs*, **68**, 465–485.

Iverson, S.J. (1988). Composition intake and gastric digestion of milk lipids in pinnipeds. PhD thesis, University of Maryland, College Park, MD.

Iverson, S.J. (1993). Milk secretion in marine mammals in relation to foraging: can milk fatty acids predict diet? *Symposium of the Zoological Society of London*, **66**, 263–291.

Iverson, S.J. (2009a). Blubber. In W.F. Perrin, B. Wursig, and H.G.M Thewissen (eds), *Encyclopedia of Marine Mammals*, 2nd edn, pp. 115–120. Academic Press, Elsevier, San Diego, CA.

Iverson, S.J. (2009b). Tracing aquatic food webs using fatty acids: from qualitative indicators to quantitative determination. In M.T. Arts, M.T. Brett, and M.J. Kainz (eds), *Lipids in Aquatic Ecosystems*, pp. 281–307. Springer, New York.

Iverson, S.J., Bowen, W.D., Boness, D.J., and Oftedal, O.T. (1993). The effect of maternal size and milk energy output on pup growth in gray seals (*Halichoerus grypus*). *Physiological Zoology*, **66**, 61–88.

Iverson, S.J., Arnould, J.P.Y., and Boyd, I.L. (1997a). Milk fatty acid signatures indicate both major and minor shifts in diet of lactating Antarctic fur seals. *Canadian Journal of Zoology*, **75**, 188–197.

Iverson, S.J., Frost, K.J., and Lowry, L.L. (1997b). Fatty acid signatures reveal fine scale structure of foraging distribution of harbor seals and their prey in Prince William Sound, Alaska. *Marine Ecology Progress Series*, **151**, 255–271.

Iverson, S.J., Lang, S., and Cooper, M. (2001). Comparison of the Bligh and Dyer and Folch methods for total lipid determination in a broad range of marine tissue. *Lipids*, **36**, 1283–1287.

Iverson, S.J., Frost, K.J., and Lang, S. (2002). Fat content and fatty acid composition of forage fish and invertebrates in Prince William Sound Alaska: factors contributing to among and within species variability. *Marine Ecology Progress Series*, **241**, 161–181.

Iverson, S.J., Field, C., Bowen, W.D., and Blanchard, W. (2004). Quantitative fatty acid signature analysis: a new method of estimating predator diets. *Ecological Monographs*, **74**, 211–235.

Iverson, S.J., Stirling, I., and Lang, S. (2006). Spatial and temporal variation in the diets of polar bears across the Canadian arctic: indicators of changes in prey populations and environment. In I.L. Boyd, S. Wanless, and C.J. Camphuysen (eds), *Top Predators in Marine Ecosystems*, pp. 98–117. Cambridge University Press, Cambridge.

Iverson, S.J., Springer, A.M., and Kitaysky, A.S. (2007). Seabirds as indicators of food web structure and ecosystem variability: qualitative and quantitative diet analyses using fatty acids. *Marine Ecology Progress Series*, **352**, 235–244.

Iwasa, Y., Higashi, M., and Yamamura, N. (1981). Prey distribution as a factor determining the choice of optimal foraging strategy. *American Naturalist*, **117**, 710–723.

IWC (2004). Report of the Workshop to design simulation-based performance tests for evaluation of methods used to infer population structure from genetic data. *Journal of Cetacean Research and Management*, **6**, 469–485.

IWC (2005). Requirements and guidelines for conducting surveys and analysing data within the Revised Management Scheme. *Journal of Cetacean Research and Management*, **7** (Suppl.), 92–101.

IWC (2009). Catch limits and catches taken: Information on recent catches taken by commercial, aboriginal and scientific permit whaling. http://www.iwcoffice.org/conservation/catches.htm

Jameson, R.J., Kenyon, K.W., Johnson, A.M., and Wight, H.M. (1982). History and status of translocated sea otter populations in North America. *Wildlife Society Bulletin*, **10**, 100–107.

Janik, V., Sayigh, L.S., and Wells, R.S. (2006). Signature whistle shape conveys identity information to bottlenose dolphins. *Proceedings of The National Academy of Sciences USA*, **103**, 8293–8297.

Jaramillo-Legorreta, A.M., Rojas-Bracho, L., and Gerrodette, T. (1999). A new abundance estimate for vaquitas: first step for recovery. *Marine Mammal Science*, **15**, 957–973.

Jaramillo-Legoretta, A., Rojas-Bracho, L., Brownell, R.L. Jr, Read, A.J., Reeves, R.R., Ralls, K., and Taylor, B.L. (2007). Saving the vaquita: immediate action, not more data. *Conservation Biology*, **21**, 1653–1655.

Jardine, T.D. and Cunjak, R.A. (2005). Analytical error in stable isotope ecology. *Oecologia*, **144**, 528–533.

Jarman, S.N., Gales, N.J., Tierney, M., Gill, P.C., and Elliott, N.G. (2002). A DNA-based method for identification of krill species and its application to analysing the diet of marine vertebrate predators. *Molecular Ecology*, **11**, 2679–2690.

Jarman, S.N., Deagle, B.E., and Gales, N.J. (2004). Group-specific polymerase chain reaction for DNA-based analysis of species diversity and identity in dietary samples. *Molecular Ecology*, **13**, 1313–1322.

Jarne, P. and Lagoda, P.J.L. (1996). Microsatellites, from molecules to populations and back. *Trends in Ecology and Evolution*, **11**, 424–429.

Jay, C.V. and Garner, G.W. (2002). Performance of a satellite-linked GPS on Pacific walruses (*Odobenus rosmarus divergens*). *Polar Biology*, **25**, 235–237.

Jay, C.V., Heide-Jorgensen, M.P., Fischbach, A. S., Jensen, M.V., Tessler, D.F., and Jensen, A. V. (2006). Comparison of remotely deployed satellite radio transmitters on walruses. *Marine Mammal Science*, **22**, 226–236.

Jefferson, T., Karczmarski, L., Laidre, K., O'Corry-Crowe, G., Reeves, R., Rojas-Bracho, L., Secchi, E., Slooten, E., Smith, B., Wang, J., and Kaiya, Z. (2008). Delphinapterus leucas *(Pallas, 1776). Beluga Whale*. Cetacean Specialist Group, IUCN, Gland, Switzerland.

Jeffries, S.J., Brown, R.F., and Harvey, J.T. (1993). Techniques for capturing, handling and marking harbour seals. *Aquatic Mammals*, **19**, 21–25.

Jenkins, S.G., Partridge, S.T., Stephenson, T.R., Farley, S.D., and Robbins, C.T. (2001). Nitrogen and carbon isotope fractionation between mothers, neonates, and nursing offspring. *Oecologia*, **129**, 336–341.

Jennings, J.G., Coe, J.M., and Gandy, W.F. (1981). A corral system for examining pelagic dolphin schools. *Marine Fisheries Review*, **43**, 16–20.

Jennings, S., Warr, K.J., and Mackinson, S. (2002). Use of size-based production and stable isotope analyses to predict trophic transfer efficiencies and predator–prey body mass ratios in food webs. *Marine Ecology Progress Series*, **240**, 11–20.

Jepson, P.D. (2003). *Pathology and Toxicology of Stranded Harbour Porpoises (Phocoena phocoena) in UK Waters*. University of London, London.

Jiménez, I. (2005). Development of predictive models to explain the distribution of the West Indian manatee *Trichechus manatus* in tropical watercourses. *Biological Conservation*, **125**, 491–503.

Johanos, T.C., Becker, B.L., and Ragen, T.J. (1994). Annual reproductive cycle of the female Hawaiian monk seal (*Monachus schauinslandi*). *Marine Mammal Science*, **10**, 13–30.

Johnson, A.J., Kraus, S.D., Kenney, J.F., and Mayo, C.A. (2007). The entangled lives of right whales and fishermen: Can they coexist? Right whale mortality: a message from the dead to the living. In S.D. Kraus and R.M. Rolland (eds), *The Urban Whale: North Atlantic Right Whales at the Crossroads*, pp. 380–408. Harvard University Press, Cambridge, MA.

Johnson, D.H. (1999). The insignificance of statistical significance testing. *Journal of Wildlife Management*, **63**, 763–772.

Johnson, D.S., London, J.M., Lea, M.A., and Durban, J.W. (2008). Continuous-time correlated random walk model for animal telemetry data. *Ecology*, **89**, 1208–1215.

Johnson, M.P. and Tyack, P.L. (2003). A digital acoustic recording tag for measuring the response of wild marine mammals to sound. *IEEE Journal of Oceanic Engineering*, **28**, 3–12.

Johnson, M., Madsen, P.T., Zimmer, W.M.X., de Soto, N.A., and Tyack, P.L. (2004). Beaked whales echolocate on prey. *Proceedings of the Royal Society of London, Series B*, **271**, S383–S386.

Johnson, M., Madsen, P.T., Zimmer, W.M.X., de Soto, N.A., and Tyack, P.L. (2006). Foraging Blainville's beaked whales (Mesoplodon densirostris) produce distinct click types matched to different phases of echolocation. *Journal of Experimental Biology*, **209**, 5038–5050.

Johnston, D., Meisenheimer, P., and Lavigne, D. (2000). An evaluation of management objectives for Canada's commercial harp seal hunt, 1996–1998. *Conservation Biology*, **14**, 729–737.

Johnston, D.W., Thorne, L.H., and Read, A.J. (2005a). Fin whales *Balaenoptera physalus* and minke whales *Balaenoptera acutorostrata* exploit a tidally driven island wake ecosystem in the Bay of Fundy. *Marine Ecology Progress Series*, **305**, 287–295.

Johnston, D.W., Westgate, A.J., and Read, A.J. (2005b). Effects of fine-scale oceanographic features on the distribution and movements of harbour porpoises *Phocoena phocoena* in the Bay of Fundy. *Marine Ecology Progress Series*, **295**, 279–293.

Jolliffe, I.T. (2002). *Principal Components Analysis*. Springer-Verlag, New York.

Jolly G.M. (1965). Explicit estimates from capture–recapture data with both death and immigration—Stochastic model. *Biometrika*, **52**, 225–247.

Jonsen, I.D., Myers, R.A., and Flemming, J.M. (2003). Meta-analysis of animal movement using state–space models. *Ecology*, **84**, 3055–3063.

Jonsen, I.D., Flemming, J.M., and Myers, R.A. (2005). Robust state–space modeling of animal movement data. *Ecology*, **86**, 2874–2880.

Jonsen, I.D., Myers, R.A., and James, M.C. (2006). Robust hierarchical state–space models reveal diel variation in travel rates of migrating leatherback turtles. *Journal of Animal Ecology*, **75**, 1046–1057.

Jonsen, I.D., Myers, R.A., and James, M.C. (2007). Identifying leatherback turtle foraging behaviour from satellite telemetry using a switching state–space model. *Marine Ecology Progress Series*, **337**, 255–264.

Joy, R., Tollit, D.J., Laake, J.L., and Trites, A.W. (2006). Using simulations to evaluate reconstructions of sea lion diet from scat. In A.W. Trites, S.K. Atkinson, D.P. DeMaster, L.W. Fritz, T.S. Gelatt, L.D. Rea, and K.M. Wynne (eds), *Sea Lions of the World*, pp. 205–222. Alaska Sea Grant College Program, University of Alaska, Fairbanks, AK.

Kafadar, K. and Horn, P.S. (2002). Smoothing. In A.H. El-Shaarawi and W.W. Piegorsch (eds), *Encyclopaedia of Environmetrics*, pp. 2014–2020. John Wiley and Sons, Chichester.

Kalinowski, S.T. (2002). How many alleles per locus should be used to estimate genetic distances? *Heredity*, **88**, 62–65.

Kaneko, H. and Lawler, I.R. (2006). Can near infrared spectroscopy be used to improve assessment of marine mammal diets via fecal analysis? *Marine Mammal Science*, **22**, 261–275.

Karczmarski, L., Cockcroft, V.G., and McLachlan, A. (2000). Habitat use and preferences of Indo-Pacific humpback dolphins *Sousa chinensis* in Algoa bay south Africa. *Marine Mammal Science*, **16**, 65–79.

Karczmarski, L., Würsig, B., Gailey, G., Larson, K.W., and Vanderlip, C. (2005). Spinner dolphins in a remote Hawaiian atoll: social grouping and population structure. *Behavioral Ecology*, **16**, 675–685.

Karlsson, O., Hiby, L., Lundberg, T., Jüssi, M., Jüssi, I., and Helander, B. (2005). Photo-identification, site fidelity, and movement of female gray seals (*Halichoerus grypus*) between haul-outs in the Baltic Sea. *Ambio*, **34**, 628–634.

Kaschner, K., Watson, R., Trites, A.W., and Pauly, D. (2006). Mapping world-wide distributions of marine mammal species using relative environmental suitability (RES) model. *Marine Ecology Progress Series*, **316**, 285–310.

Kastelle, C.R., Sheldon, K.E.W., and Kimura, D.K. (2003). Age determination of mysticete whales using ^{210}Pb/^{226}Ra disequilibria. *Canadian Journal of Zoology*, **81**, 21–32.

Kasuya, T., Brownell, R.L., and Balcomb, K.C. (1988). Preliminary analysis of life history of Baird's beaked whales off the Pacific coast of central Japan. *Reports of the International Whaling Commission*, **39**, 465.

Katona, S., Baxter, B., Brazer, O., Kraus, S., Perkins, J., and Whitehead, H. (1979). Identification of humpback whales from fluke photographs. In H.E. Winn and B. Olla (eds), *Behavior of Marine Mammals—Current Perspectives in Research*, pp. 33–44. Plenum Press, New York.

Keating, K.A. and Cherry, S. (2004). Use and interpretation of logistic regression in habitat selection studies. *Journal of Wildlife Management*, **68**, 774–789.

Keith, D.A., McCarthy, M.A., Regan, H., Regan, T., Bowles, C., Drill, C., Craig, C., Pellow, B., Burgman, M.A., Master, L.L., Ruckelshaus, M., Mackenzie, B., Andelman, S.L., and Wade, P.R. (2004). Protocols for listing species can forecast extinction. *Ecology Letters*, **7**, 1101–1108.

Keitt, T.H., Bjørnstad, O.N., Dixon, P.M., and Citron-Pousty, S. (2002). Accounting for spatial pattern when modeling organism–environment interactions. *Ecography*, **25**, 616–625.

Keiver, K.M., Ronald, K., and Beamish, F.W.H. (1984). Metabolizable energy requirements for maintenance and faecal and urinary losses of juvenile harp seals (*Phoca groenlandica*). *Canadian Journal of Zoology*, **62**, 769–776.

Kellar, N.M., Trego, M.L., Marks, C.I., and Dizon, A.E. (2006). Determining pregnancy from the blubber in three species of delphinids. *Marine Mammal Science*, **22**, 1–16.

Kendall W.L. and Bjorkland R. (2001). Using open robust design models to estimate temporary emigration from capture–recapture data. *Biometrics*, **57**, 1113–1122.

Kendall W.L., Pollock K.H., and Brownie C. (1995). A likelihood-based approach to capture–recapture estimation of demographic parameters under the robust design. *Biometrics*, **51**, 293–308.

Kendall W.L., Nichols J.D., and Hines J.E. (1997). Estimating temporary emigration using capture–recapture data with Pollock's robust design. *Ecology*, **78**, 563–578.

Kendall, W.L., Langtimm, C.A., Beck, C.A., and Runge, M.C. (2004). Capture–recapture analysis for estimating manatee reproductive rates. *Marine Mammal Science*, **20**, 424–437.

Kenward, R.E. (2000). *Wildlife Radio Tagging: Equipment, Field Techniques and Data Analysis*. Academic Press, London.

Kenward, R.E., Clarke, R.T., Hodder, K.H., and Walls, S.S. (2001). Density and linkage estimators of home range: Nearest neighbor clustering defines multinuclear cores. *Ecology*, **82**, 1905–1920.

Kernohan, B.J., Millspaugh, J.J., Jenks, J.A., and Naugle, D.E. (1998). Use of an adaptive kernel home-range estimator in a GIS environment to calculate habitat use. *Journal of Environmental Management*, **53**, 83–89.

Kerr, M.K. and Churchill, G.A. (2001). Experimental design for gene expression microarrays. *Biostatistics*, **2**, 183–201.

Keyes, M.C. and Farrell, R.K. (1979). Freeze marking of northern fur seal. In L. Hobbs and J. Russel (eds), *Report on Pinniped and Sea Otter Tagging Workshop*, 18–19 January, 1979. National Marine Mammal Laboratory, 7600 Sand Point Way, NE, Seattle WA. American Institute of Biological Sciences. Arlington VA.

Khaemba, W.M. and Stein, A. (2000). Use of GIS for a spatial and temporal analysis of

Kenyan wildlife with generalised linear modelling. *International Journal of Geographical Information Science*, **14**, 833–853.

King, R.A., Read, D.S., Traugott, M., and Symondson, W.O.C. (2008). Molecular analysis of predation: a review of best practice for DNA-based approaches. *Molecular Ecology*, **17**, 947–963.

Kingston, S.E. and Rosel, P.E. (2004). Genetic differentiation among recently diverged delphinid taxa determined using Aflp markers. *Journal of Heredity*, **95**, 1–10.

Kirsch, P.E., Iverson S.J., and Bowen, W.D. (2000). Effect of a low-fat diet on body composition and blubber fatty acids in captive harp seals (*Phoca groenlandica*). *Physiological and Biochemical Zoology*, **73**, 45–59.

Klasson Wehler, E., Mörck, A., and Hakk, H. (2001). Metabolism of polybrominated diphenyl ethers in the rat. In A. Bergman (ed.), *Second International Workshop on Brominated Flame Retardants*, 14–16 May, 2001, pp. 145–149. Swedish Chemical Society, Stockholm, Sweden.

Kleiber, M. (1975). *The Fire of Life: An Introduction to Animal Energetics*. Robert E. Krieger Publ. Co., New York.

Klem, A. (1935). Studies in the biochemistry of whale oils. *Hvalradets Skr Norske*, **11**, 49–108.

Klevezal, G.A. (1996). *Recording Structures of Mammals. Determination of Age and Reconstruction of Life History* (transl. from Russian by M.V. Mina and A.V. Oreshkin). A.A. Balkema, Rotterdam.

Klevezal, G.A. and Kleinenberg, S.E. (1969). Age determination of mammals from annual layers in teeth and bones. *Israel Program for Scientific Translation*, TT 69–55033, **1**, 1–128. [Transl. from Russian.]

Knowlton, A.R. and Brown, M.W. (2007). Running the gauntlet: Right whales and vessel strikes. In S.D. Kraus and R.M. Rolland (eds), *The Urban Whale: North Atlantic Right Whales at the Crossroads*, pp. 409–435. Harvard University Press, Cambridge, MA.

Kocher, T.D., Thomas, W.K., Meyer, A., Edwards, S.V., Paabo, S., Villablanca, F.X., and Wilson, A.C. (1989). Dynamics of mitochondrial DNA evolution in animals: amplification and sequencing with conserved primers. *Proceedings National Academy Science USA*, **86**, 6196–6200.

Koopman, H.N. (2007). Phylogenetic ecological and ontogenetic factors influencing the biochemical structure of the blubber of odontocetes. *Marine Biology*, **151**, 277–291.

Koopman, H.N., Iverson, S.J., and Gaskin, D.E. (1996). Stratification and age-related differences in blubber fatty acids of the male harbour porpoise (*Phocoena phocoena*). *Journal of Comparative Physiology B*, **165**, 628–639.

Koopman, H.N., Pabst, D.A., McLellan, W.A., Dillaman, R.M., and Read, A.J. (2002). Changes in blubber distribution and morphology associated with starvation in the harbour porpoise (*Phocoena phocoena*): evidence for regional differences in blubber structure and function. *Physiological and Biochemical Zoology*, **75**, 498–512.

Kooyman, G. (2007). Animal-borne instrumentation systems and the animals that bear them: Then (1939) and now (2007). *Marine Technology Society Journal*, **41**, 6–8.

Kooyman, G.L., Kerem, D.H., Campbell, W.B., and Wright, J.J. (1973). Pulmonary gas exchange on freely diving Weddell seals (*Leptonychotes weddellii*). *Respiratory Physiology*, **17**, 283–290.

Kooyman, G.L., Gentry, R.L., and Urquhart, D.L. (1976). Northern fur seal diving behaviour; a new approach to its study. *Science*, **193**, 411–412.

Kooyman, G.L., Wahrenbrock, E.A., Castellini, M.A., Davis, R.W., and Sinnett, E.E. (1980). Aerobic and anaerobic metabolism during voluntary diving in Weddell seals: evidence of preferred pathways from blood chemistry and behavior. *Journal of Comparative Physiology*, **138**, 335–346.

Kovacs, K.M. (1989). Maternal behaviour and early behavioural ontogeny of harp seals, Phoca groenlandica. *Animal Behaviour*, **35**, 844–855.

Krafft, B.A. (1999). Diving behaviour of lactating bearded seals (*Erignathus barbatus*) in the Svalbard area. MSc. University of Tromso, Tromso.

Kramer, D.L. (1988). The behavioural ecology of air breathing by aquatic animals. *Canadian Journal of Zoology*, **66**, 89–94.

Kraus, S.D. and Rolland, R.M. (2007). *The Urban Whale: North Atlantic Right Whales at the Crossroads*. Harvard University Press, Cambridge, MA.

Kraus, S.D., Moore, K.E., Price, C.A., Crone, M.J., Watkins, W.A., Winn, H.E., and Prescott, J.H. (1986). The use of photographs to identify individual North Atlantic right whales (*Eubalaena glacialis*). In R.L. Brownell, P.B. Best, and J.H. Prescott (eds), Right Whales: Past and Present Status. *Proceedings of the Workshop on the Status of Right Whales*, pp. 145–151. Report of the International Whaling Commission, Special Issue 10.

Kraus, S.D., Read, A.J., Anderson, E., Baldwin, K., Solow, A., Spradlin, T., and Williamson, J. (1997). Acoustic alarms reduce porpoise mortality. *Nature*, **388**, 525.

Kraus, S.D., Pace, I., Rolland, R.M., and Frasier, T.R. (2007). High investment, low return: The strange case of reproduction in Eubalaena glacialis. In S.D. Kraus and R.M. Rolland, (eds), *The Urban Whale: North Atlantic Right Whales at the Crossroads*, pp. 172–199. Harvard University Press, Cambridge, MA.

Krawczak, M. (1999). Informativity assessment for biallelic single nucleotide polymorphisms. *Electrophoresis*, **20**, 1676–1681.

Krebs, C.J. (1972). *Ecology: The Experimental Analysis of Distribution and Abundance*. Harper Row, New York.

Kreho, A., Kehtarnavaz, N., Araabi, B., Hillman, G., Würsig, B., and Weller, D. (1999). Assisting manual dolphin identification by computer extraction of dorsal ratio. *Annals of Biomedical Engineering*, **27**, 830–838.

Kriete, B. (1995). Bioenergetics of the killer whale, *Orcinus orca*. PhD thesis. University of British Columbia, Vancouver, BC.

Kruggel, F., Pélégrini-Isaac, M., and Benali, H. (2002). Estimating the effective degrees of freedom in univariate multiple regression analysis. *Medical Image Analysis*, **6**, 63–75.

Krützen, M., Barre, L.M., Moller, L.M., Heithaus, M.R., Simms, C., and Sherwin, W.B. (2002). A biopsy system for small cetaceans: Darting success and wound healing in *Tursiops* spp. *Marine Mammal Science*, **18**, 863–878.

Krützen, M., Mann, J., Heithaus, M.R., Connor, R.C., Bejder, L., and Sherwin, W.B. (2005). Cultural transmission of tool use in bottlenose dolphins. *Proceedings of the National Academy of Sciences USA*, **102**, 8939–8943.

Krzanowski, W.J. (1998). *An Introduction to Statistical Modelling*. Arnold Texts in Statistics, London.

Kucey, L. and Trites, A.W. (2006). A review of the potential effects of disturbance on sea lions: assessing response and recovery. In A.W. Trites, S.K.Atkinson, D.P. DeMaster, L.W. Fritz, T.S. Gelatt, L.D. Rea, and K.M. Wynne (eds), *Sea Lions of the World*, pp. 581–589. Alaska Sea Grant College Program, University of Alaska, Fairbanks, AK.

Kucklick, J.R., Christopher, S.J., Becker, P.R., Pugh, R.S., Porter, B.J., Schantz, M.M., Mackey, E.A., Wise, S.A., and Rowles, T.K. (2002). *Description and Results of the 2000 NIST/NOAA Inter-laboratory Comparison Exercise Program for Organic Contaminants and Trace Elements in Marine Mammal Tissues*, NIS-TIR 6849. National Institute for Standards and Technology, Charleston, South Carolina.

Kuhn, C.E. and Costa, D.P. (2006). Identifying and quantifying prey consumption using stomach temperature change in pinnipeds. *Journal of Experimental Biology*, **209**, 4524–4532.

Kuhn, C.E., Crocker, D.E., Tremblay, Y., and Costa, D.P. (2009). Time to eat: measurements of feeding behaviour in a large marine predator, the northern elephant seal (*Mirounga angustirostris*). *Journal of Animal Ecology*, **78**, 513–523.

Kuiken, T. and Garcia Hartmann, M.G. (1991). Cetacean pathology: dissection techniques and tissue sampling. *The First European Cetacean Society Workshop*, 13–14 September, 1991. European Cetacean Society, Leiden, The Netherlands.

Kurle, C.M. (2002). Stable-isotope ratios of blood components from captive northern fur seals (*Callorhinus ursinus*) and their diet: applications for studying the foraging ecology of wild otariids. *Canadian Journal of Zoology*, **80**, 902–909.

Kurle, C.M. and Gudmundson, C.J. (2007). Regional differences in foraging of young-of-the eyar Steller sea lions *Eumatpias jubatus* in Alaska: stable carbon and nitrogen isotope ratios in blood. *Marine Ecology Progress Series*, **342**, 303–310.

Laake, J.L. (1999). Distance sampling with independent observers: reducing bias from heterogeneity by weakening the conditional independence assumption. In G.W. Garner, S.C. Armstrup, J.L. Laake, B.F.J. Manly, L.L. McDonald, and D.G. Robertson (eds), *Marine*

Mammal Survey and Assessment Methods, pp. 137–148. Balkema, Rotterdam.

Laake, J.L. and Borchers, D.L. (2004). Methods for incomplete detection at distance zero. In S.T. Buckland, D.R. Anderson, K.P. Burnham, J.L. Laake, D.L. Borcher, and L. Thomas (2004). *Advanced Distance Sampling: Estimating Abundance of Biological Populations*, pp 108–189. Oxford University Press, Oxford.

Laake, J. and Rexstad, E. (2008). RMark—an alternative approach to building linear models in MARK. In E. Cooch and G. White, *Program MARK: A Gentle Introduction* [version 17/6/08]. Appendix C, 115 pp. (http://www.phidot.org/software/mark/docs/book/)

Laake, J.L., Calambokidis, J., Osmek, S.D., and Rugh D.J. (1997). Probability of detecting harbor porpoise from aerial surveys: estimating g(0). *Journal of Wildlife Management*, **61**, 63–75.

Laake, J.L., Browne, P., DeLong, R.L., and Huber, H.R. (2002). Pinniped diet composition: a comparison of estimation models. *Fishery Bulletin*, **100**, 434–477.

Lahood, E.S., Moran, P., Olsen, J., Grant, W.S., and Park, L.K. (2002). Microsatellite allele ladders in two species of Pacific salmon: preparation and field-test results. *Molecular Ecology Notes*, **2**, 187–190.

Laidre, K.L., Heide-Jorgensen, M.P., Dietz, R., Hobbs, R.C., and Jorgensen, O.A. (2003). Deep-diving by narwhals *Monodon monoceros*: differences in foraging behaviour between wintering areas. *Marine Ecology Progress Series*, **261**, 269–281.

Laidre, K.L., Heide-Jørgensen, M.P., Logdson, M.L., Hobbs, R.C., Heagerty, P., Dietz, R., Jørgensen, O.A., and Treble, M.A. (2004a). Seasonal narwhal habitat associations in the High Arctic. *Marine Biology*, **145**, 821–831.

Laidre, K.L., Heide-Jorgensen, M.P., Logsdon, M.L., Hobbs, R.C., Dietz, R., and VanBlaricom, G.R. (2004b). Fractal analysis of narwhal space use patterns. *Zoology*, **107**, 3–11.

Laidre, K.L., Stirling, I., Lowry, L.F., Wiig, Ø., Heide-Jorgensen, M.P., and Ferguson, S.H. (2008). Quantifying the sensitivity of Arctic marine mammals to climate-induced habitat change. *Ecological Applications*, **18**, S97–S125.

Laland, K.N. and Janik, V.M. (2006). The animal cultures debate. *Trends in Ecology and Evolution*, **21**, 542–547.

Lance, M.M., Orr, A.J., Riemer, S.D., Weise, M.J., and Laake, J.L. (2001). Pinniped food habits and prey identification techniques protocol. *AFSC Processed Report 2001–04*. pp. 1–36. Alaska Fisheries Science Center, part of National Marine Fisheries Service, National Oceanic and Atmospheric Administration, Department of Commerce, Seattle, WA 98115.

Landau, H.G. (1951). On dominance relations and the structure of animal societies: I. Effect of inherent characteristics. *Bulletin of Mathematical Biophysics*, **13**, 1–19.

Lande, R. and Orzack, S.H. (1988). Extinction dynamics of age-structured populations in a fluctuating environment. *Proceedings of National Academy of Sciences USA*, **85**, 7418–7421.

Lander, M.E., Westgate, A.J., Bonde, R.K., and Murray, M.J. (2001). Tagging and tracking. In L.A. Dierauf and F.M.D. Gulland (eds), *CRC Handbook of Marine Mammal Medicine*, 2nd edn, pp. 851–880. CRC Press, London.

Lander, M.E., Harvey, J.T., and Gulland, F.M.D. (2003). Hematology and serum chemistry comparisons between free-ranging and re-habilitated harbor seal (*Phoca vitulina richardsi*) pups. *Journal of Wildlife Diseases*, **39**, 600–609.

Lander, M.E., Haulena, M., Gulland, F.M.D., and Harvey, J.T. (2005). Implantation of sub-cutaneous radio transmitters in the harbor seal (*Phoca vitulina*). *Marine Mammal Science*, **21**, 154–161.

Landsberg, J.H. (2002). The effects of harmful algal blooms on aquatic organisms. *Review of Fisheries Science*, **10**, 113–390.

Lang, S.L.C., Iverson, S.J., and Bowen, W.D. (2009). Repeatability in lactation performance and the consequences for maternal reproductive success in gray seals. *Ecology*, **90**, 2513–2523.

Lanyon, J.M., Slade, R.W., Sneath, H.L., Broderick, D., Kirkwood, J.M., Limpus, D., Limpus, C., and Jessop, T. (2006). A method for capturing dugongs (*Dugong dugon*) in open water. *Aquatic Mammals*, **32**, 196–201.

Larkin, I.L.V., Fowler, V.F., and Reep, R.L. (2007). Digesta passage rates in the Florida manatee (*Trichechus manatus latirostris*). *Zoo Biology*, **26**, 503–515.

Larsen, F. and Hammond, P.S. (2004). Distribution and abundance of West Greenland humpback whales (*Megaptera novaeangliae*). *Journal of Zoology, London*, **263**, 343–358.

Last, J.M., Spasoff, R.A., and Harris, S.S (eds), (2000). *A Dictionary of Epidemiology*, 4th edn. Oxford University Press, New York.

Latch, E.K., Dharmarajan, G., Glaubitz, J.C., and Rhodes Jr, O.E. (2006). Relative performance of Bayesian clustering software for inferring population substructure and individual assignment at low levels of population differentiation. *Conservation Genetics*, **7**, 295–302.

Lavigne, D.M., Innes, S., Worthy, G.A.J., Kovacs, K.M., Schmitz, O.J., and Hickie, J.P. (1986). Metabolic rates of seals and whales. *Canadian Journal of Zoology*, **64**, 279–284.

Lavigne, D.M., Scheffer, V.B., and Kellert, S.R. (1999). The evolution of North American attitudes toward marine mammals. In J.R. Twiss and R.R. Reeves (eds), *Conservation and Management of Marine Mammals*, pp. 10–48. Smithsonian Institution Press, Washington DC.

Laws, E.A., Popp, B.N., Bidigarem, R.R, Kennikutat, C., and Macko, S. (1995). Dependence of phytoplankton carbon isotopic composition on growth rate and (CO_2)aq: Theoretical considerations and experimental results. *Geochimica et Cosmochimica Acta*, **59**, 1131–1138.

Laws, R.M. (1952). A new method of age determination for mammals. *Nature*, **169**, 972–973.

Laws, R.M. (1977). The significance of vertebrates in the Antarctic marine ecosystem. In G.A. Llano (ed.), Adaptation within Antarctic ecosystems. *Proceedings of 3rd SCAR Symposium on Antarctic Biology*, pp. 411–438. Smithsonian Institution, Washington DC.

Laws, R.M. (1993). Development of technology and research needs. In R.M Laws (ed.), *Antarctic Seals: Research Methods and Techniques*. Cambridge University Press, Cambridge.

Laws, R.M. and Purves, P.E. (1956). The ear plug of the Mysticeti as an indication of age with special reference to the North Atlantic fin whale (*Balaenoptera physalus* Linn.). *Norsk Kvanlfangst-Tidende*, **45**, 413–425.

Laws, R.M. and Sinha, A.A. (1993). Reproduction. In R.M Laws (ed.), *Antarctic Seals: Research Methods and Techniques*. Cambridge University Press, Cambridge.

Laws, R.M., Baird, A., and Bryden, M.M. (2002). Age estimation in crabeater seals (*Lobodon carcinophagus*). *Journal of Zoology, London*, **258**, 197–203.

Lawson, J.W., Harrison, G.D., and Bowen, W.D. (1992). Factors affecting accuracy of age determination in the harp seal *Phoca groenlandica*. *Marine Mammal Science*, **8**, 169–171.

Lawson, J.W., Miller, E.H., and Noseworthy, E. (1997). Variation in assimilation efficiency and digestive efficiency of captive harp seals (*Phoca groenlandica*) on different diets. *Canadian Journal of Zoology*, **75**, 1285–1291.

Layton, H., Rouvinen, K., and Iverson, S.J. (2000). Body composition in mink (*Mustela vison*) kits during 21–42 days postpartum using estimates of hydrogen isotope dilution and direct carcass analysis. *Comparative Biochemistry and Physiology A*, **126**, 295–303.

Lea, M.A. and Dubroca, L. (2003). Fine-scale linkages between the diving behaviour of Antarctic fur seals and oceanographic features in the southern Indian Ocean. *ICES Journal of Marine Science*, **60**, 990–1002.

Lea, M.A., Cherel, Y., Guinet, C., and Nichols, P.D. (2002). Antarctic fur seals foraging in the Polar Frontal Zone: inter-annual shifts in diet as shown from fecal and fatty acid analyses. *Marine Ecology Progress Series*, **245**, 281–297.

Lea, M.A., Guinet, C., Cherel, Y., Duhamel, G., Dubroca, L., Pruvost, P., and Hindell, M. (2006). Impacts of climatic anomalies on provisioning strategies of a Southern Ocean predator. *Marine Ecology Progress Series*, **310**, 77–94.

Leatherwood, S., Reeves, R.R., Perrin, W.F., and Evans, W.E. (1988). *Whales, Dolphins, and Porpoises of the Eastern North Pacific and Adjacent Arctic Waters: A Guide to their Identification*. Dover Publications, Mineola, NY.

Le Boeuf, B.J. (1974). Male–male competition and reproductive success in elephant seals. *American Zoologist*, **14**, 163–176.

Le Boeuf, B.J. (1991). Pinniped mating systems on land, ice and in the water: emphasis on the Phocidae. In D. Renouf (ed.), *Behaviour of Pinnipeds*, pp. 45–65. Chapman and Hall, London.

Le Boeuf, B.J. and Laws, R.M. (1994). Elephant seals: An introduction to the genus. In B.J. Le Boeuf and R.M. Laws (eds), *Elephant Seals: Population Ecology, Behavior and Physiology*. University of California Press. Los Angeles, CA.

Le Boeuf, B.J. and Reiter, J. (1988). Lifetime reproductive success in northern elephant seals. In T.H. Clutton-Brock (ed.), *Reproductive*

Success, pp. 344–362. The University of Chicago Press, Chicago, IL.

Le Boeuf, B.J., Naito, Y., Asaga, T., Crocker, D.E., and Costa, D.P. (1992). Swim speed in a female northern elephant seal: metabolic and foraging implications. *Canadian Journal of Zoology*, **70**, 786–795.

Le Boeuf, B.J., Morris, P.A., Blackwell, S.B., Crocker, D.E., and Costa, D.P. (1996). Diving behaviour of juvenile northern elephant seals. *Canadian Journal of Zoology*, **74**, 1632–1644.

Le Boeuf, B.J., Crocker, D.E., Costa, D.P., Blackwell, S.B., Webb, P.M., and Houser, D.S. (2000). Foraging ecology of northern elephant seals. *Ecological Monographs*, **70**, 353–382.

Lebreton, J-D. and Pradel, R. (2002). Multistate recapture models: modeling incomplete individual histories. *Journal of Applied Statistics*, **29**, 353–369.

Lebreton, J-D., Burnham, K.P., Clobert, J., and Anderson, D.R. (1992). Modeling survival and testing biological hypotheses using marked animals. A unified approach with case studies. *Ecological Monographs*, **62**, 67–118.

Lederberg, J. (2000). The dawning of molecular genetics. *Trends in Microbiology*, **8**, 194–195.

LeDuc, R.G., Perrin, W.F., and Dizon, A.E. (1999). Phylogenetic relationships among the Delphinid Cetaceans based on full cytochrome B sequences. *Marine Mammal Science*, **15**, 619–648.

Lefkovitch, L.P. (1965). The study of population growth in organisms grouped by stages. *Biometrics*, **21**, 1–18.

Legendre, P. and Anderson, M.J. (1999). Distance-based redundancy analysis: Testing multispecies responses in multifactorial ecological experiments. *Ecological Monographs*, **69**, 1–24.

Lehner, P.N. (1998). *Handbook of Ethological Methods*. Cambridge University Press, Cambridge.

Leopold, M.F., van Damme, C.J.D., Philippart, C.J.M., and Winter, C.J.N. (2001). *Otoliths of North Sea Fish—Fish Identification Key by Means of Otoliths and Other Hard Parts*, CDROM Version 10. Expert Centre for Taxonomic Identification, University of Amsterdam, Amsterdam, The Netherlands.

Lesage, V., Hammill, M.O., and Kovacs, K.M. (1999). Functional classification of harbor seal (Phoca vitulina) dives using depth profiles, swimming velocity, and an index of foraging success. *Canadian Journal of Zoology, London*, **77**, 74–87.

Leslie, P.H. (1945). On the use of matrices in certain population mathematics. *Biometrika*, **33**, 183–212.

Levin, S.A. (1989). The problem of pattern and scale in ecology. *Ecology*, **73**, 1943–1967.

Lewis, T., Gillespie, D., Lacey, C., Matthews, J., Danbolt, M., Leaper, R., McLanaghan, R., and Moscrop, A. (2007). Sperm whale abundance estimates from acoustic surveys of the Ionian Sea and Straits of Sicily in 2003. *Journal of the Marine Biological Association of the United Kingdom*, **87**, 353–357.

Lichstein, J.W., Simons, T.R., Shriner, S.A., and Franzreb, K.E. (2002). Spatial autocorrelation and autoregressive models in ecology. *Ecological Monographs*, **72**, 445–463.

Liebsch, N., Wilson, R.P., Bornemann, H., Adelung, D., and Plotz, J. (2007). Mouthing off about fish capture: Jaw movement in pinnipeds reveals the real secrets of ingestion. *Deep-Sea Research Part II—Topical Studies in Oceanography*, **54**, 256–269.

Lifson, N., Gordon, G.B., and McClintock, R. (1955). Measurement of total carbon dioxide production by means of D_2 ^{18}O. *Journal of Applied Physiology*, **7**, 704–710.

Lindberg, M.S. and Walker, J. (2007). Satellite telemetry in avian research and management: Sample size considerations. *Journal of Wildlife Management*, **71**, 1002–1009.

Lindgren, A., Mascolo, C., Lonergan, M., and McConnell, B. (2008). Seal–2–Seal: A delay-tolerant protocol for contact logging in wildlife monitoring sensor networks. *Proceedings of the Fifth IEEE International Conference on Mobile Ad-hoc and Sensor Systems* (MASS 2008), Sept 29–Oct 2, 2008, Atlanta, GA, pp. 321–327.

Littaye, A., Gannier, A., Laran, S., and Wilson, J.P.F. (2004). The relationship between summer aggregation of fin whales and satellite-derived environmental conditions in the northwestern Mediterranean sea. *Remote Sensing of Environment*, **90**, 44–52.

Littnan, C.L., Baker, J.D., Parrish, F.A., and Marshall, G.J. (2004). Effects of video camera attachment on the foraging behavior of immature Hawaiian monk seals. *Marine Mammal Science*, **20**, 345–352.

Littnan, C.L., Stewart, B.S., Yochem, P.K., and Braun, R. (2006). Survey for selected pathogens and evaluation of disease risk factors for endangered Hawaiian monk seals in the main Hawaiian Islands. *EcoHealth*, **3**, 232–244.

Liu, J.S. (2008). *Monte Carlo Strategies in Scientific Computing*, Springer-Verlag, New York.

Lockyer, C. (1972). The age at sexual maturity of the southern fin whale (*Balaenoptera physalus*) using annual layer counts in the ear plug. *Journal Du Conseil*, **34**, 276–294.

Lockyer, C. (1984). Age determination by means of the earplug in baleen whales. *Scientific Report of the International Whaling Commission*, **34**, 692–698.

Lockyer, C. and Waters, T. (1986). Weights and anatomical measurements of northeast Atlantic fin (*Balaenoptera physalis*, Linnaeus) and sie (*B. borealis*, Lesson) whales. *Marine Mammal Science*, **2**, 169–185.

Lockyer, C.H., McConnell, L.C., and Waters, T.D. (1985). Body condition in terms of anatomical assessment of body fat in North Atlantic fin and sei whales. *Canadian Journal of Zoology*, **63**, 2328–2338.

Lonergan, M., Fedak, M., and McConnell, B. (2009). The effects of interpolation error and location quality on animal track reconstruction. *Marine Mammal Science*, **25**, 275–282.

Lowry, L., O'Corry-Crowe, G., and Goodman, D. (2006). *Delphinapterus leucas* (Cook Inlet subpopulation). In *IUCN 2007: IUCN Red List of Threatened Species*. www.iucnredlist.org. [Downloaded 17 May 2008.]

Lowry, L., Laist, D.W., and Taylor, E. (2007). *Endangered, Threatened and Depleted Marine Mammals in U.S. Waters. A Review of Classification Systems and Listed Species*. US Marine Mammal Commission. Bethesda, MD.

Lubetkin, S.C., Zeh, J.E., Rosa, C., and George, J.C. (2008). Age estimation for young bowhead whales (*Balaena mysticetus*) using annual baleen growth increments. *Canadian Journal of Zoology*, **86**, 525–538.

Ludwig, D. (1999). Is it meaningful to estimate a probability of extinction? *Ecology*, **80**, 298–310.

Luikart, G. and England, P.R. (1999). Statistical analysis of microsatellite DNA data. *Trends in Ecology and Evolution*, **14**, 253–256.

Luikart, G., England, P.R., Tallmon, D., Jordan, S., and Taberlet, P. (2003). The power and promise of population genomics: from genotyping to genome typing. *Nature Reviews Genetics*, **4**, 981–994.

Lukaski, H.C. (1987). Methods for the assessment of human body composition: traditional and new. *American Journal of Clinical Nutrition*, **46**, 537–556.

Lunn, N.J. and Boyd, I.L. (1991). Pupping-site fidelity of Antarctic fur seals at Bird Island, South Georgia. *Journal of Mammalogy*, **72**, 202–206.

Lunn, N.J., Boyd I.L., and Croxall, J.P. (1994). Reproductive performance of female Antarctic fur seals: the influence of age, breeding experience, environmental variation and individual quality. *Journal of Animal Ecology*, **63**, 827–840.

Luque, P.L., Learmonth, J.A., Santos, M.B., Ieno, E., and Pierce, G.J. (2009). Comparison of two histological techniques for age determination in small cetaceans. *Marine Mammal Science*, **25**, 902–919.

Lurton, X. (2002). *Underwater Acoustics: An Introduction*. Springer-Verlag, Berlin and Heidelberg GmbH and Co.K.

Lusk, J.J., Guthrey, F.S., and DeMaso, S.J. (2002). A neural network model for predicting northern bobwhite abundance in the rolling red plains of Oklahoma. In J.M. Scott, P.J. Heglund, and M.L. Morrison (eds), *Predicting Species Occurrences: Issues of Accuracy and Scale*, pp. 345–355. Island Press, Washington DC.

Lusseau, D. (2003). The emergent properties of a dolphin social network. *Proceedings of the Royal Society of London B*, **270**, S186–S188.

Lusseau, D. and Bejder, L. (2007). The long-term consequences of short-term responses to disturbance experiences from whale-watching impact assessment. *International Journal of Comparative Psychology*, **20**, 228–236.

Lusseau, D. and Higham, J.E.S. (2004). Managing the impacts of dolphin-based tourism through the definition of critical habitats: the case of bottlenose dolphins (*Tursiops* spp.) in Doubtful Sound, New Zealand. *Tourism Management*, **25**, 657–667.

Lusseau, D. and Newman, M.E.J. (2004). Identifying the role that animals play in social networks. *Proceedings of the Royal Society of London B*, **271**, S477–481.

Lusseau, D., Whitehead, H., and Gero, S. (2008). Incorporating uncertainty into the study of animal social networks. *Animal Behaviour*, **75**, 1093–1099.

Lütolf, M., Kienast, F., and Guisan, A. (2006). The ghost of past species occurrence: improving species distribution models for presence-only data. *Journal of Applied Ecology*, **43**, 802–815.

Lydersen, C., Nost, O.A., Lovell, P., McConnell, B.J., Gammelsrod, T., Hunter, C., Fedak, M.A., and Kovacs, K.M. (2002). Salinity and temperature structure of a freezing Arctic fjord—monitored by white whales (*Delphinapterus leucas*). *Geophysical Research Letters*, **29**, 2793–2796.

Lyons, L.A., Laughlin, T.F., Copeland, N.G., Jenkins, N.A., Womack, J.E., and O'Brien, S.J. (1997). Comparative anchor tagged sequences (cats for integrative mapping of mammalian genomes. *Nature Genetics*, **15**, 47–56.

McCafferty, D.J., Boyd, I.L., and Taylor, R.I. (1998). Diving behaviour of Antarctic fur seal (*Arctocephalus gazella*) pups. *Canadian Journal of Zoology*, **76**, 513–520.

McCafferty, D.J., Walker, T.R., and Boyd, I.L. (2004). Using time–depth–light recorders to measure light levels experienced by a diving marine mammal. *Marine Biology*, **146**, 191–199.

McCafferty, D.J., Currie, J., and Sparling, C.E. (2007). The effect of instrument attachment on the surface temperature of juvenile grey seals (*Halichoerus grypus*) as measured by infrared thermography. *Deep-Sea Research Part II—Topical Studies in Oceanography*, **54**, 424–436.

McConkey, S.D. (1999). Photographic identification of the New Zealand sea lion: a new technique. *New Zealand Journal of Marine and Freshwater Research*, **33**, 63–66.

McConnell, B.J., Chambers, C., and Fedak, M.A. (1992). Foraging ecology of southern elephant seals in relation to the bathymetry and productivity of the Southern-Ocean. *Antarctic Science*, **4**, 393–398.

McConnell, B., Bryant, E., Hunter, C., Lovell, P., and Hall, A. (2004). Phoning home—a new GSM mobile phone telemetry system to collect mark–recapture data. *Marine Mammal Science*, **20**, 274–283.

McCullagh, P. and Nelder, J.A. (1989). *Generalized Linear Models*, 2nd edn. Chapman and Hall, London.

McCune, B. (2006). Non-parametric habitat models with automatic interactions. *Journal of Vegetation Science*, **17**, 819–830.

McDonald, B.I., Crocker, D.E., Burns, J.M., and Costa, D.P. (2008). Body condition as an index of winter foraging success in crabeater seals (*Lobodon carcinophaga*). *Deep-Sea Research II*, **55**, 515–522.

Macintosh, N. (1965). *The Stocks of Whales*. Fishing News (Books) Ltd, London.

McIntosh, R. (2007). Life history and population demographics of the Australian sea lion. PhD dissertation, LaTrobe University, Victoria, Australia.

McIntosh, R.R., Shaughnessy, P.D., and Goldsworthy, S.D. (2006). Mark–recapture estimates of pup production for the Australian sea lion (*Neophoca cinera*) at Seal Bay Conservation Park, South Australia. In A.W. Trites, S.K. Atkinson, D.P. DeMaster, L.W. Fritz, T.S. Gelatt, L.D. Rea, and K.M Wynne (eds), *Sea Lions of the World*, pp. 353–367. Alaska Sea Grant College, University of Alaska, Fairbanks. AK-SG-06-01.

MacKay, D.J.C. (2003). *Information Theory, Inference and Learning Algorithms*. Cambridge University Press, Cambridge.

MacKenzie, D.I., Nichols, J.D., Royle, J.A., Pollock, K.H., Bailey, L.L., and Hines, J.E. (2006). *Occupancy Estimation and Modeling. Inferring Patterns and Dynamics of Species Occurrence*. Academic Press, Amsterdam.

Mackey, B.L., Durban, J.W., Middlemas, J.J., and Thompson, P.M. (2008). A Bayesian estimate of harbour seal survival using sparse photo-identification data. *Journal of Zoology, London*, **274**, 18–27.

Mackinson, S., Blanchard, J.L., Pinnegar, J.K., and Scott, R. (2003). Consequences of alternative functional response formulations in models exploring whale-fishery interactions. *Marine Mammal Science*, **19**, 661–681.

McLaren, I.A. (1993). Growth in pinnipeds. *Biology Review*, **68**, 1–79.

McLaren, I.A. and Smith, T.G. (1985). Population ecology of seals: retrospective and prospective views. *Marine Mammal Science*, **1**, 54–83.

MacLean, G. (2009). Weak GPS signal detection in animal tracking. *The Journal of Navigation*, **62**, 1–21.

McLean, J.A. and Tobin, G. (1987). *Animal and Human Caloritmetry*. Cambridge University Press, New York.

MacLeod, C.D. (1998). Intraspecific scarring in odontocete cetaceans: an indicator of male 'quality' in aggressive social interactions? *Journal of Zoology, London*, **244**, 71–77.

MacLeod, C.D., Hauser, N., and Peckham, H. (2004). Diversity, relative density and structure of the cetacean community in summer months east of Great Abaco, Bahamas. *Journal of the Marine Biological Association of the U.K.*, **84**, 469–474.

MacLeod, K., Fairbairns, R., Gill, A., Fairbairns, B., Gordon, J., Blair-Myers, C., and Parsons, E.C. (2004). Seasonal distribution of Minke whales, *Balaenoptera acutorostrata* in relation to physiography and prey off the Isle of Mull, Scotland. *Marine Ecology Progress Series*, **277**, 263–274.

McMahon, C.R., Burton, H., McLean, S., Slip, D., and Bester, M. (2000). Field immobilization of southern elephant seals with intravenous tiletamine and zolazepam. *The Veterinary Record*, **146**, 251–254.

McMahon, C.R., Hindell, M.A., Burton, H.R., and Bester, M.N. (2005). Comparison of southern elephant seal populations, and observations of a population on a demographic knife-edge. *Marine Ecology Progress Series*, **288**, 273–283.

McMahon, C.R., Burton, H.R., Van den Hoff, J., Woods, R., and Bradshaw, C.J.A. (2006). Assessing hot-iron and cryo-branding for permanently marking southern elephant seals. *Journal of Wildlife Management*, **70**, 1484–1489.

McMahon, C.R., Bradshaw, C.J.A., and Hays, G.C. (2007). Applying the heat to research techniques for species conservation. *Conservation Biology*, **21**, 271–273.

McNamara, J.M., Green, R.F., and Olsson, O. (2006). Bayes' theorem and its applications in animal behaviour. *Oikos*, **112**, 243–251.

McPhee, J.M., Rosen, D.A.S., Andrews, R.D., and Trites, A.W. (2003). Predicting metabolic rate from heart rate in juvenile Steller sea lions *Eumetopius jubatus*. *Journal of Experimental Biology*, **206**, 1941–1951.

Mačuhová, J., Tančin, V., and Bruckmaier, R.M. (2004). Effects of oxytocin administration on oxytocin release and milk ejection. *Journal of Dairy Science*, **87**, 1236–1244.

Madsen, P.T., Payne, R., Kristiansen, N.U., Wahlberg, M., Kerr, I., and Mohl, B. (2002). Sperm whale sound production studied with ultrasound time/depth-recording tags. *Journal of Experimental Biology*, **205**, 1899–1906.

Madsen, P.T., Johnson, M., Miller, P.J.O., Soto, N.A., Lynch, J., and Tyack, P. (2006). Quantitative measures of air-gun pulses recorded on sperm whales (Physeter macrocephalus) using acoustic tags during controlled exposure experiments. *Journal of the Acoustical Society of America*, **120**, 2366–2379.

Maguire, L.A. (1986). Using decision analysis to manage endangered species populations. *Journal of Environmental Management*, **22**, 345–360.

Maguire, L.A. (1987). Decision analysis: A tool for tiger conservation and management. In R.L. Tilson and U.S. Seal (eds), *Tigers of the World—The Biology, Biopolitics, Management, and Conservation of an Endangered Species*, pp. 475–486. Noyes/William Andrew Publishing, Norwich, NY.

Mancia, A., Lundqvist, M.L., Romano, T.A., Peden-Adams, M.M., Fair, P.A., Kindy, M.S., Ellis, B.C., Gattoni-Celli, S., McKillen, D.J., Trent, H.F., Chen, Y.A., Almeida, J.S., Gross, P.S., Chapman, R.W., and Warr, G.W. (2007). A dolphin peripheral blood leukocyte cDNA microarray for studies of immune function and stress reactions. *Developmental and Comparative Immunology*, **31**, 520–529.

Manel, S., Dias, J.-M., and Ormerod, S.J. (1999). Comparing discriminant analysis, neural networks and logistic regression for predicting species distributions: a case study with a Himalayan river bird. *Ecological Modelling*, **120**, 337–347.

Manel, S., Schwartz, M.K., Luikart, G., and Taberlet, P. (2003). Landscape genetics: combining landscape ecology and population genetics. *Trends in Ecology and Evolution*, **18**, 189–197.

Mangel, M. and Clark, C.W. (1986). Towards a unified foraging theory. *Ecology*, **67**, 1127–1138.

Manly, B.F.J., McDonald, L., Thomas, D.L., McDonald, T.L., and Erickson, W.P. (2002). *Resource Selection by Animals: Statistical Design and Analysis for Field Studies*, 2nd edn. Chapman and Hall, London.

Mann, J. (1999). Behavioral sampling methods for cetaceans: A review and critique. *Marine Mammal Science*, **15**, 102–122.

Mann, J. (2000). Unraveling the dynamics of social life: long-term studies and observational methods. In J. Mann, R.C. Connor, P.L. Tyack, and H. Whitehead (eds), *Cetacean Society Field Studies of Dolphins and Whales*. The University of Chicago Press, Chicago, IL.

Mann, J. and Smuts, B.B. (1998). Natal attraction: allomaternal care and mother–infant separations in wild bottlenose dolphins. *Animal Behaviour*, **55**, 1097–1113.

Mansfield, A.W. (1991). Accuracy of age determination in the grey seal *Halichoerus grypus* of eastern Canada. *Marine Mammal Science*, **7**, 44–49.

Mansour, A.A.H., McKay, D.W., Lien, J., Orr, J.C., Banoub, J.H., Oien, N., and Stenson, G. (2002). Determination of pregnancy status from blubber samples in minke whales (*Balaenoptera acutorostrata*). *Marine Mammal Science*, **18**, 112–120.

Marcovecchio, J.E., Mreno, V.J., Bastida, R.O., Gerpe, M.S., and Rodriguez, D.H. (1990). Tissue distribution of heavy metals in small cetaceans from the southwestern Atlantic Ocean. *Marine Pollution Bulletin*, **21**, 299–303.

Marcus, J., Bowen, W.D., and Eddington, J.D. (1998). Effects of meal size on otolith recovery from fecal samples of gray and harbor seal pups. *Marine Mammal Science*, **14**, 789–802.

Margulies, M., Egholm, M., Altman, W.E., Attiya, S., Bader, J.S., Bemben, L.A., Berka, J., Braverman, M.S., Chen, Y.J., Chen, Z., Dewell, S.B., Du, L., Fierro, J.M., Gomes, X.V., Godwin, B.C., He, W., Helgesen, S., Ho, C.H., Irzyk, G.P., Jando, S.C., Alenquer, M.L., Jarvie, T.P., Jirage, K.B., Kim, J.B., Knight, J.R., Lanza, J.R., Leamon, J.H., Lefkowitz, S.M., Lei, M., Li, J., Lohman, K.L., Lu, H., Makhijani, V.B., Mcdade, K.E., Mckenna, M.P., Myers, E.W., Nickerson, E., Nobile, J.R., Plant, R., Puc, B.P., Ronan, M.T., Roth, G.T., Sarkis, G.J., Simons, J.F., Simpson, J.W., Srinivasan, M., Tartaro, K.R., Tomasz, A., Vogt, K.A., Volkmer, G.A., Wang, S.H., Wang, Y., Weiner, M.P., Yu, P., Begley, R.F., and Rothberg, J.M. (2005). Genome sequencing in microfabricated high-density picolitre reactors. *Nature*, **437**, 376–380.

Marine Mammal Commission. (2007). *Report of the Workshop on Assessing the Population Viability of Endangered Marine Mammals in U.S. Waters*. US Marine Mammal Commission, Bethesda, MD.

Marine Mammal Commission. (2008). *The Biological Viability of the Most Endangered Marine Mammals and the Cost-Effectiveness of Protection Programs*. US Marine Mammal Commission. Bethesda, MD.

Marino, L. (1996). What can dolphins tell us about primate evolution? *Evolutionary Anthropology*, **5**, 81–85.

Markussen, N.H., Bjorge, A., and Oritsland, N.A. (1989). Growth in harbour seals (Phoca vitulina) on the Norwegian coast. *Journal of Zoology, London*, **219**, 433–440.

Marmontel, M.M., O'Shea, T.J., Kochman, H.I., and Humphrey, S.R. (1996). Age determination in manatees using growth-layer counts in bone. *Marine Mammal Science*, **12**, 54–88.

Marques, F.F.C. and Buckland, S.T. (2004). Covariate models for the detection function. In S.T. Buckland, D.R. Anderson, K.P. Burnham, J.L. Laake, D.L. Borchers, and L. Thomas (eds), *Advanced Distance Sampling: Estimating Abundance of Biological Populations*, pp. 31–47. Oxford University Press, Oxford.

Marsh, H. (1980). Age determination of the dugong (*Dugong dugon* Müller, 1776) Northern Australia and its biological implication. *International Whaling Commission*, Special Issue, **3**, 181–200.

Marsh, H. and Kasuya, T. (1984). Changes in the ovaries of the short-finned pilot whale, *Globicephala macrorhynchus*, with age and reproductive activity. In W.F. Perrin, R.L. Brownell, Jr, and D.P. DeMaster (eds), *Reproduction in Whales, Dolphins, and Porpoises*. International Whaling commission, Special Issue 6, Cambridge University Press, Cambridge.

Marsh, H. and Sinclair, D.F. (1989). Correcting for visibility bias in strip transect aerial surveys of aquatic fauna. *Journal of Wildlife Management*, **53**, 1017–1024.

Marshall, G.J. (1998). CRITTERCAM: An animal-borne imaging and data logging system. *Marine Technology Society Journal*, **32**, 11–17.

Marshall, G., Bakhtiari, M., Shepard, M., Tweedy, J., Rasch, D., Abernathy, K., Joliff, B., Carrier, J.C., and Heithaus, M.R. (2007). An advanced solid-state animal-borne video and environmental data-logging Device('CRITTERCAM') for marine research. *Marine Technology Society Journal*, **41**, 31–38.

Mårtensson P.E., Nordøy, E.S., and Blix, A.S. (1994). Digestibility of crustaceans and capelin in harp seals (*Phoca groenlandica*). *Marine Mammal Science*, **10**, 325–331.

Martien, K. and Taylor, B.L. (2003). Limitations of hypothesis-testing in defining management units for continuously distributed species. *Journal of Cetacean Research and Management*, **5**, 213–218.

Martien, K., Gregovich, D., Bravington, M., Punt, A., Strand, A., Tallmon, D., and Taylor, B.L. (2009). TOSSM: an R package for assessing performance of genetic analytical methods in a management context. *Molecular Ecology Resources*, **9**, 456–459.

Martin, A.R. and da Silva, V.M.F. (2004). River dolphins and flooded forest: seasonal habitat use and sexual segregation of botos (*Inia geoffrensis*) in an extreme cetacean environment. *Journal of Zoology, London*, **263**, 295–305.

Martin, A.R., da Silva, V.M.F., and Rothery, P.R. (2006). Does radio tagging affect the survival or reproduction of small cetaceans? A test. *Marine Mammal Science*, **22**, 17–24.

Mate, B., Mesecar, R., and Lagerquist, B. (2007). The evolution of satellite-monitored radio tags for large whales: One laboratory's experience. *Deep-Sea Research Part II—Topical Studies in Oceanography*, **54**, 224–247.

Matejusová, I., Doig, F., Middlemas, S.J., Mackay, S., Douglas, A., Armstrong, J.D., Cunningham, C.O., and Snow, M. (2008). Using quantitative real-time PCR to detect salmonid prey in scats of grey *Halichoerus grypus* and harbour *Phoca vitulina* seals in Scotland—an experimental and field study. *Journal of Applied Ecology*, **45**, 630–640.

Matthiopoulos, J. (2003a). Model-supervised kernel smoothing for the estimation of spatial usage. *Oikos*, **102**, 367–377.

Matthiopoulos, J. (2003b). The use of space by animals as a function of accessibility and preference. *Ecological Modelling*, **159**, 239–268.

Matthiopoulos, J., McConnell, B., Duck, C., and Fedak, M. (2004). Using satellite telemetry and aerial counts to estimate space use by grey seals around the British Isles. *Journal of Applied Ecology*, **41**, 476–491.

Matthiopoulos, J., Harwood, J., and Thomas, L. (2005). Metapopulation consequences of site fidelity for colonially breeding mammals and birds. *Journal of Animal Ecology*, **74**, 716–727.

Matthiopoulos, J., Smout, S., Winship, A.J., Thompson, D., Boyd, I.L., and Harwood, J. (2008). Getting beneath the surface of marine mammal—fisheries competition. *Mammal Review*, **38**, 167–188.

Mattlin, R.H., Gales, N.J., and Costa, D.P. (1998). Seasonal dive behaviour of lactating New Zealand fur seals (Arctocephalus forsteri). *Canadian Journal of Zoology*, **76**, 350–360.

Mayden, R. (1997). A hierarchy of species concepts: the denouement in the saga of the species problem. In M.F. Claridge, H.A. Dawah, and M.R. Wilson (eds), *Species: The Units of Biodiversity*. Chapman and Hall, London.

Mead, J.I., Spiess, A.E., and Sobolik, K.D. (2000). Skeleton of extinct North American sea mink (*Mustela macrodon*). *Quaternary Research*, **53**, 247–262.

Melin, S.R., Ream, R.R., and Zeppelin, T.K. (2006). *Report of the Alaska Region and Alaska Fisheries Science Center's Northern Fur Seal Tagging and Census Workshop*, 6th–9th September, 2005, Seattle, Washington. AFSC Processed Rep. 2006–15, 59 pp. Alaska Fisheries Science Center, NOAA, National Marine Fisheries Service, 7600 Sand Point Way NE, Seattle WA 98115.

Mellish, J.A.E., Iverson, S.J., and Bowen, W.D. (1999). Variation in milk production and lactation performance in grey seals and consequences for pup growth and weaning characteristics. *Physiological and Biochemical Zoology*, **72**, 677–690.

Mellish, J.A.E., Tuomi, P.A., and Horning, M. (2004). Assessment of ultrasound imaging as a non-invasive measure of blubber thickness in pinnipeds. *Journal of Zoo and Wildlife Medicine*, **35**, 116–118.

Mellish, J.A.E., Hennen, D., Thomton, J., Petrauskas, L., Atkinson, S., and Calkins, D. (2007). Permanent marking in an endangered species: physiological response to hot branding in Steller sea lions (*Eumetopias jubatus*). *Wildlife Research*, **34**, 43–47.

Mendes, S., Newton, J., Reid, R.J., Frantzis, A., and Pierce, G.J. (2007). Stable isotope profiles in sperm whale teeth: variations between areas and sexes. *Journal of the Marine Biological Association (UK)*, **87**, 621–627.

Merrick, R.L., Loughlin, T.R., and Calkins, D.G. (1996). *Hot Branding: A Technique for Long-term*

Marking of Pinnipeds. US Department of Commerce, NOAA Technical Memorandum NMFS-AFSC-68.

Merton, R. (1936). The unanticipated consequences of purposive social action. *American Sociological Review,* **1**, 894–904.

Meyer, M., Stenzel, U., and Hofreiter, M. (2008). Parallel tagged sequencing on the 454 platform. *Nature Protocols,* **3**, 267–278.

Meynier, L. (2004). Food and feeding ecology of the common dolphin, *Delphinus delphis,* in the Bay of Biscay: intraspecific dietary variation and food transfer modelling. MSc thesis, University of Aberdeen, Scotland.

Meynier, L. (2009). Feeding ecology of the New Zealand sea lion *Phocarctos hookeri.* PhD dissertation, Massey University, Palmerston North, New Zealand.

Miller, M. (2005). Allele In Space (AIS): computer software for the joint analysis of inter-individual sspatial and genetic information. *Journal of Heredity,* **96**, 722–724.

Miller, P.J.O., Johnson, M.P., and Tyack, P.L. (2004a). Sperm whale behaviour indicates the use of echolocation click buzzes 'creaks' in prey capture. *Proceedings of the Royal Society of London Series B—Biological Sciences,* **271**, 2239–2247.

Miller, P.J.O., Johnson, M.P., Tyack, P.L., and Terray, E.A. (2004b). Swimming gaits, passive drag and buoyancy of diving sperm whales Physeter macrocephalus. *Journal of Experimental Biology,* **207**, 1953–1967.

Millspaugh, J.J. and Marzluff, J.M. (2001). *Radio Tracking and Animal Populations.* Academic Press, San Diego, CA.

Mitani, Y., Watanabe, Y., Sato, K., Cameron, M.F., and Naito, Y. (2004). 3D diving behavior of Weddell seals with respect to prey accessibility and abundance. *Marine Ecology Progress Series,* **281**, 275–281.

Mizroch, S.A. and Harness, S.A.D. (2003). A test of computer-assisted matching using the North Pacific humpback whale, *Megaptera novaeangliae,* tail flukes photograph collection. *Marine Fisheries Review,* **65**, 25–37.

Mizroch, S.A., Beard, J., and Lynde, M. (1990). Computer-assisted photo-identification of humpback whales. In P.S. Hammond, S.A. Mizroch, and G.P. Donovan (eds), *Individual Recognition of Cetaceans: Use of Photo-Identification and Other Techniques to Estimate*

Population Parameters, pp. 63–70. Report of the International Whaling Commission, Special Issue 12, Cambridge, UK.

Mohl, B., Wahlberg, M., Madsen, P.T., Miller, L.A., and Surlykke, A. (2000). Sperm whale clicks: Directionality and source level revisited. *Journal of the Acoustical Society of America,* **107**, 638–648.

Mohl, B., Wahlberg, M., Madsen, P.T., Heerfordt, A., and Lund, A. (2003). The monopulsed nature of sperm whale clicks. *Journal of the Acoustical Society of America,* **114**, 1143–1154.

Mohn, R.K. (2009). The uncertain future of assessment uncertainty. In R.J. Beamish and B.J. Rothschild (eds), *The Future of Fisheries Science in North America,* pp. 495–504. Springer, New York.

Mohn, R. and Bowen, W.D. (1996). Grey seal predation on the eastern Scotian Shelf: modelling the impact on Atlantic cod. *Canadian Journal of Fisheries and Aquatic Science,* **53**, 2722–2738.

Moll, R.J., Millspaugh, J.J., Beringer, J., Sartwell, J., and He, Z. (2007). A new 'view' of ecology and conservation through animal-borne video systems. *Trends in Ecology and Evolution,* **22**, 660–668.

Möller, L.M., Beheregaray, L.B., Allen, S.J., and Harcourt, R.G. (2006). Association patterns and kinship in female Indo-Pacific bottlenose dolphins (Tursiops aduncus) of southeastern Australia. *Behavioural Ecology and Sociobiology,* **69**, 109–117.

Monamy, V. (2007). Hot iron branding of seals and sea lions: why the ban will remain. *Australian Veterinary Journal,* **85**, 485–486.

Monestiez, P., Dubroca, L., Bonnin, E., Dubrec, J.-P., and Guinet., C. (2006). Geostatistical modelling of spatial distribution of Balaenoptera physalus in the Northwestern Mediterranean Sea from sparse count data and heterogeneous observation efforts. *Ecological Modelling,* **193**, 615–628.

Monmonier, M. (1973). Maximum-difference barriers: an alternative numerical regionalization method. *Geographical Analysis,* **3**, 245–261.

Moore, M.G., Early, G., Touhey, K., Barco, S., and Gulland, F. (2007). Wells rehabilitation and release of marine mammals in the United States: Risks and benefits. *Marine Mammal Science,* **23**, 731–750.

Moore, M.J., Miller, C.A., Morss, M.S., Arthur, R., Lange, W., Prada, K.G., Marx, M.K., and Frey, E.A. (2001). Ultrasonic measurement of blubber thickness in right whales. *Journal of Cetacean Research and Management*, Special Issue, **2**, 301–309.

Moore, M.J., McLellan, W.A., Daoust, P.-Y., Bonde, R.K., and Knowlton, A.R. (2007). Right whale mortality: a message from the dead to the living. In S.D. Kraus and R.M. Rolland (eds), *The Urban Whale: North Atlantic Right Whales at the Crossroads*, pp. 358–379. Harvard University Press, Cambridge, MA.

Moore, S.E. (2005). Long-term environmental change. In J.E. Reynolds, W.F. Perrin, R.R. Reeves, S. Montgomery, and T.J. Ragen (eds), *Marine Mammal Research: Conservation Beyond Crisis*, pp. 137–147. The Johns Hopkins University Press, Baltimore, MD.

Moore, S.S., Sargeant, L.L., King, T.J., Mattick, J.S., Georges, M., and Hetzel, D.J.S. (1991). The conservation of dinucleotide microsatellites among mammalian genomes allows the use of heterologous PCR primer pairs in closely related species. *Genomics*, **10**, 654–660.

Morales, J.M., Haydon, D.T., Frair, J., Holsiner, K.E., and Fryxell, J.M. (2004). Extracting more out of relocation data: Building movement models as mixtures of random walks. *Ecology*, **85**, 2436–2445.

Moran, M.D. (2003). Arguments for rejecting the sequential Bonferroni in ecological studies. *Oikos*, **100**, 403–406.

Morell, V. (2008). Puzzling over a Steller whodunit. *Science*, **320**, 44–45.

Mori, M., Wurtz, M., Bonaccorsi, R., and Lauriano, G. (1992). Crustacean remains from the stomachs and faeces of some Mediterranean cetaceans: an illustrated sheet. *European Research on Cetaceans*, **6**, 192–193.

Mori, Y. (1998a). Optimal choice of foraging depth in divers. *Journal of Zoology*, **245**, 279–283.

Mori, Y. (1998b). The optimal patch use in divers: Optimal time budget and the number of dive cycles during bout. *Journal of Theoretical Biology*, **190**, 187–199.

Mori, Y. (1999). The optimal allocation of time and respiratory metabolism over the dive cycle. *Behavioral Ecology*, **10**, 155–160.

Mori, Y. (2002). Optimal diving behaviour for foraging in relation to body size. *Journal of Evolutionary Biology*, **15**, 269–276.

Mori, Y. and Boyd, I.L. (2004). The behavioral basis for nonlinear functional responses and optimal foraging in antarctic fur seals. *Ecology*, **85**, 398–410.

Mori, Y., Yoda, K., and Sato, K. (2001). Defining dive bouts using a sequential differences analysis. *Behaviour*, **138**, 1451–1466.

Mori, Y., Takahashi, A., Mehlum, F., and Watanuki, Y. (2002). An application of optimal diving models to diving behaviour of Brunnich's guillemots. *Animal Behaviour*, **64**, 739–745.

Mori, Y., Watanabe, Y., Mitani, Y., Sato, K., Cameron, M.F., and Naito, Y. (2005). A comparison of prey richness estimates for Weddell seals using diving profiles and image data. *Marine Ecology Progress Series*, **295**, 257–263.

Morin, P.A. and McCarthy, M. (2007). Highly accurate SNP genotyping from historical and low-quality samples. *Molecular Ecology Notes*, **7**, 937–946.

Morin, P.A., Moore, J.J., Wallis, J., and Woodruff, D.S. (1994). Paternity exclusion in a community of wild chimpanzees using hypervariable simple sequence repeats. *Molecular Ecology*, **3**, 469–478.

Morin, P.A., Luikart, G., Wayne, R.K., and SNP working group (2004). SNPs in ecology, evolution and conservation. *Trends in Ecology and Evolution*, **19**, 208–216.

Morin, P.A., Aitken, N.C., Rubio-Cisneros, N., Dizon, A.E., and Mesnick, S.L. (2007). Characterization of 18 SNP markers for sperm whale (*Physeter macrocephalus*). *Molecular Ecology Notes*, **7**, 626–630.

Morin, P.A., Leduc, R.G., Archer, F.I., Martien, K.K., Huebinger, R., Bickham, J.W., and Taylor, B.L. (2009a). Significant deviations from Hardy–Weinberg equilibrium caused by low levels of microsatellite genotyping errors. *Molecular Ecology Resources*, **9**, 498–504.

Morin, P.A., Martien, K.K., and Taylor, B.L. (2009b). Assessing statistical power of SNPs for population structure and conservation studies. *Molecular Ecology Resources*, **9**, 66–73.

Moritz, C. (1994). Defining 'evolutionary significant units' for conservation. *Trends in Ecology and Evolution*, **9**, 373–375.

Morris, W.F. and Doak, D.F. (2002). *Quantitative Conservation Biology*. Sinauer Associates, Inc. Sunderland, MA.

Morris, W.F. and Doak, D.F. (2006). Population viability analysis and conservation decision making. In M.J. Groom, G.K. Meffe, and C.R. Carroll (eds), *Principles of Conservation Biology*, pp. 433–435. Sinauer Associates Inc., Sunderland, MA.

Mostman-Liwanag, H.E., Williams, T.M., Costa, D.P. Kanatous, S., Davis, R.W., and Boyd, I.L. (2009). The effects of water temperature on the energetic costs of juvenile and adult California sea lions (*Zalophus californianus*): The importance of recycling endogenous heat during activity. *Journal of Experimental Biology*, 212, 3977–3984.

Mrosovsky, N. (1997). IUCN's credibility critically endangered. *Nature*, **389**, 436.

Muelbert, M.M.C., Bowen, W.D., and Iverson, S.J. (2003). Weaning mass affects changes in body composition and food intake in harbour seal pups during the first month of independence. *Physiological and Biochemical Zoology*, **76**, 418–427.

Mujib, K.A. (1967). The cranial osteology of the Gadidae. *Journal of the Fisheries Research Board of Canada*, **24**, 1315–1375.

Murie, D.J. (1987). Experimental approaches to stomach content analysis of piscivorous marine mammals. In A.C. Huntley, D.P. Costa, G.A.J. Worthy, and M.A. Castellini (eds), *Approaches to Marine Mammal Energetics*, pp. 147–163. Society for Marine Mammals Special Publication **1**, 253 pp., Lawrence, Kansas.

Murie, D.J. and Lavigne, D.M. (1985). A technique for the recovery of otoliths from stomach contents of piscivorous pinnipeds. *Journal of Wildlife Management*, **49**, 910–912.

Murphy, W.J., Eizirik, E., Johnson, W.E., Zhang, Y.P., Ryder, O.A., and O'Brien, S.J. (2001). Molecular phylogenetics and the origins of placental mammals. *Nature*, **409**, 614–618.

Murray, D.L. (2006). On improving telemetry-based survival estimation. *Journal of Wildlife Management*, **70**, 1530–1543.

Murray, D.L. and Fuller, M.R. (2000). A critical review of the effects of marking on the biology of vertebrates. In L. Boitani and T.K. Fuller (eds), *Research Techniques in Animal Ecology: Controversies and Consequences*, pp. 15–65. Columbia University Press, New York.

Musyl, M.K., Brill, R.W., Curran, D.S., Gunn, J.S., Hartog, J.R., Hill, R.D., Welch, D.W.,

Eveson, J.P., Boggs, C.H., and Brainard, R.E. (2001). Ability of archival tags to provide estimates of geographical position based on light intensity. In J. Sibert and J. Nielsen (eds), *Electronic Tagging and Tracking in Marine Fisheries Reviews: Methods and Technologies in Fish Biology and Fisheries*, pp. 343–368. Kluwer Academic Press, Dordrecht.

Myrick, A.C. Jr, Shallenberger, E.W., Kang, I., and MacKay, D.B. (1984). Calibration of dental layers in seven captive Hawaiian spinner dolphins, *Stenella longirostris*, based on tetracycline labeling. *Fishery Bulletin*, **82**, 207–225.

Mysterud, A. and Ims, R.A. (1999). Relating populations to habitats. *Trends in Ecology and Evolution*, **14**, 490–491.

Nagy, K.A. and Costa, D.P. (1980). Water flux in animals: analysis of potential errors in the tritiated water method. *American Journal of Physiology*, **238**, R454–R465.

Nakagawa, S. (2004). A farewell to Bonferroni: the problems of low statistical power and publication bias. *Behavioral Ecology*, **15**, 1044–1045.

Narum, S.R., Banks, M., Beacham, T.D., Bellinger, M.R., Campbell, M.R., Dekoning, J., Elz, A., Guthrie, C.M. III, Kozfkay, C., Miller, K.M., Moran, P., Phillips, R., Seeb, L.W., Smith, C.T., Warheit, K., Young, S. F., and Garza, J.C. (2008). Differentiating salmon populations at broad and fine geographic scales with microsatellites and SNPs. *Molecular Ecology*, **17**, 3464–3477.

National Marine Fisheries Service. (2008). *Marine Mammal Stock Assessment Reports*. http://www.nmfs.noaa.gov/pr/sars/ [Downloaded 17 May, 2008.]

Needham, D.J. (1997). The role of stones in the sea lion stomach: investigations using contrast radiography and fluoroscopy. In M.A. Hindell and C.M. Kemper (eds), *Marine Mammal Research in the Southern Hemisphere*, pp.164–169. Surrey Beatty and Sons, Chipping Norton, New South Wales, Australia.

Neimanis, A.S., Koopman, H.N., Westgate, A.J., Murison, L.D., and Read, A.J. (2004). Entrapment of harbour porpoises (*Phocoena phocoena*) in herring weirs in the Bay of Fundy, Canada. *Journal of Cetacean Research and Management*, **6**, 7–17.

Newman, M.E.J. (2003). The structure and function of complex networks. *Society for*

Industrial and Applied Mathematics Review, **45,** 167–256.

Newman, M.E.J. (2004). Analysis of weighted networks. *Physical Review E,* **70,** 056131.

Newman, M.E.J. (2006). Modularity and community structure in networks. *Proceedings of the National Academy of Sciences USA,* **103,** 8577–8582.

Newsome, G.E. (1977). Use of opercular bones to identify and estimate lengths of prey consumed by piscivores. *Canadian Journal of Zoology,* **55,** 733–736.

Newsome, S.D., Koch, P.L., Etnier, M.A., and Aurioles-Gamboa, D. (2006). Using carbon and nitrogen isotope values to investigate maternal strategies in northeast Pacific otariids. *Marine Mammal Science,* **22,** 556–572.

Newsome, S.D., Etnier, M.A., Kurle, C.M., Waldbauer, J.R., Chambers, C.P., and Koch, P.L. (2007). Historic decline in primary productivity in western Gulf of Alaska and eastern Bering Sea: Isotopic analysis of northern fur seal teeth. *Marine Ecology Progress Series,* **332,** 211–224.

Nichol, R.K. and Wild, C.J. (1984). Numbers of individuals in faunal analysis: The decay of fish bone in archaeological sites. *Journal of Archaeological Science,* **11,** 35–51.

Nicholopoulos, J. (1999). The endangered species listing program. *Endangered Species Bulletin,* **24,** 6–9.

Nichols J.D. and Kendall, W.L. (1995). The use of multi-state capture–recapture models to address questions in evolutionary ecology. *Journal of Applied Statistics,* **22,** 835–846.

Nielsen, A., Bigelow, K.A., Musyl, M.K., and Sibert, J.R. (2006). Improving light-based geolocation by including sea surface temperature. *Fisheries Oceanography,* **15,** 314–325.

Nielsen, S.E., Johnson, C.J., Heard, D.C., and Boyce, M.S. (2005). Modeling species occurrence and abundance; does probability of occurrence reflect population density? *Ecography,* **28,** 197–208.

Nilssen, K.T., Pedersen, O., Folkow, L.P., and Haug, T. (2000). Food consumption estimates of Barents Sea harp seals. *NAMMCO (North Atlantic Marine Mammal Commission) Scientific Publication,* **2,** 9–27.

Nishiwaki, M. and Yagi, T. (1953). On the age and growth of teeth in a dolphin—*Prodelphinus caeruleo-albus. Scientific Report of the Whales Research Institute, Tokyo,* **8,** 133–146.

NMFS (National Marine Fisheries Service) (2007). *Final Programmatic Environmental Impact Statement: Steller Sea Lion And Northern Fur Seal Research.* US Department of Commerce, NOAA Fisheries, Washington DC. [Available at: http://www.nmfs.noaa.gov/pr/permits/eis/steller.htm]

Norden, C.R. (1961). Comparative osteology of representative salmonid fishes with particular reference to the grayling (*Thymallus arcticus*) and its phylogeny. *Journal of the Fisheries Research Board of Canada,* **18,** 679–791.

Nordstrom, C.A., Wilson, L.J., Iverson, S.J., and Tollit, D.J. (2008). Evaluating quantitative fatty acid signature analysis (QFASA) using harbour seals (*Phoca vitulina richardsi*) in captive feeding studies. *Marine Ecology Progress Series,* **360,** 245–263.

Noren, S.R. and Wells, R.S. (2009). Blubber deposition during ontogeny in free-ranging bottlenose dolphins (*Tursiops truncatus*): balancing disparate roles of insulation and locomotion. *Journal of Mammalogy,* **90,** 629–637.

Norman, S.A., Hobbs, R.C., Foster, J., Schroeder, J.P., and Townsend, F.I. (2004). A review of animal and human health concerns during capture–release, handling and tagging of odontocetes. *Journal of Cetacean Research and Management,* **6,** 53–62.

Norris, K.S. and Dohl, T.P. (1980). The behavior of the Hawaiian spinner porpoise, *Stenella longirostris. Fishery Bulletin,* **77,** 821–849.

Norris, K.S. and Mammalogists, T.C.o.M.M.A.S.o. (1961). Standardized methods for measuring and recording data on the smaller cetaceans. *Journal of Mammalogy,* **42,** 471–476.

Norris, K.S. and Pryor, K.W. (1970). A tagging method for small cetaceans. *Journal of Mammalogy,* **51,** 609–610.

North, A.W., Croxall, J.P., and Doidge, D.W. (1983). Fish prey of the Antarctic fur seal *Arctocephalus gazella* at South Georgia. *British Antarctic Survey Bulletin,* **61,** 27–37.

Noss, R. (1986). Dangerous simplifications in conservation biology. *Bulletin of the Ecological Society of America,* **67,** 278–279.

Nowacek, D.P. (2002). Sequential foraging behaviour of bottlenose dolphins, Tursiops truncatus, in Sarasota Bay, FL. *Behaviour,* **139,** 1125–1145.

Nowacek, D.P., Tyack, P.L., and Wells, R.S. (2001a). A platform for continuous behavioral and acoustic observation of free-ranging marine mammals: Overhead video combined with underwater audio. *Marine Mammal Science*, **17**, 191–199.

Nowacek, D.P., Johnson, M.P., Tyack, P.L., Shorter, K.A., McLellan, W.A., and Pabst, D.A. (2001b). Buoyant balaenids: the ups and downs of buoyancy in right whales. *Proceedings of the Royal Society of London, Series B*, **268**, 1811–1816.

Nowacek, D.P., Johnson, M.P., and Tyack, P.L. (2004). North Atlantic right whales (*Eubalaena glacialis*) ignore ships but respond to alerting stimuli. *Proceedings of the Royal Society of London, Series B*, **271**, 227–231.

Nowacek, D.P., Thorne, L.H., Johnston, D.W., and Tyack, P.L. (2007). Responses of cetaceans to anthropogenic noise. *Mammal Review*, **37**, 81–115.

O'Corry-Crowe, G., Taylor, B.L., Gelatt, T., Loughlin, T.R., Bickham, J., Basterretche, M., Pitcher, K.W., and Demaster, D.P. (2006). Demographic independent along ecosystem boundaries in Steller sea lions revealed by mtDNA analysis: implications for management of an endangered species. *Canadian Journal of Zoology*, **84**, 1796–1809.

Odum, H.T. and Allee, W.C. (1954). A note on the stable point of populations showing both intraspecific cooperation and disoperation. *Ecology*, **35**, 95–97.

Oftedal, O.T. (1984). Milk composition, milk yield and energy output at peak lactation: a comparative review. *Symposia of the Zoological Society of London*, **51**, 33–85.

Oftedal, O.T. (1997). Lactation in whales and dolphins: evidence of divergence between baleen- and toothed-species. *Journal of Mammary Gland Biology and Neoplasia*, **2**, 205–230.

Oftedal, O.T. and Iverson, S.J. (1987). Hydrogen isotope methodology for measurement of milk intake and energetics of growth in suckling young. In A.D. Huntley, D.P. Costa, G.A.J. Worthy, and M.A. Castellini (eds), *Approaches to Marine Mammal Energetics*, pp. 67–96. Allen Press, Lawrence, Kansas.

Oftedal, O.T. and Iverson, S.J. (1995). Phylogenetic variation in the gross composition of mammalian milks. In R.G. Jensen and M. Thompson (eds), *The Handbook of Milk Composition*, pp. 790–827. Academic Press, San Diego, CA.

Oftedal, O.T., Boness, D.J., and Tedman, R.A. (1987a). The behavior, physiology, and anatomy of lactation in the pinnipedia. In H.H. Genoways (ed.), *Current Mammology*, Vol. 1, pp. 175–245. Plenum Publishing Corporation, New York.

Oftedal, O.T., Iverson, S.J., and Boness, D.J. (1987b). Milk and energy intakes of suckling California sea lion *Zalophus californianus* pups in relation to sex, growth, and predicted maintenance requirements. *Physiological Zoology*, **60**, 560–575.

Oftedal, O.T., Bowen, W.D., and Boness, D.J. (1988). The composition of hooded seal (*Cystophora cristata*) milk: an adaptation for postnatal fattening. *Canadian Journal of Zoology*, **66**, 318–322.

Oftedal, O.T., Bowen, W.D., Widdowson, E.M., and Boness, D.J. (1989). Effects of suckling and the postweaning fast on weights of the body and internal organs of harp and hooded seal pups. *Biology of the Neonate*, **56**, 283–300.

O'Hagan, A. (1995). Fractional Bayes factors for model comparison. *Journal of the Royal Statistical Society, Series B*, **57**, 99–138.

Olesiuk, P.F. (1993). Annual prey consumption by harbour seals (*Phoca vitulina*) in the Strait of Georgia, British Columbia. *Fisheries Bulletin*, **91**, 491–515.

Olesiuk, P.F., Bigg, M.A., and Ellis, G.M. (1990a). Life history and population dynamics of resident killer whales (*Orcinus orca*) in the coastal waters of British Columbia and Washington State. *Report of the International Whaling Commission*, Special Issue **12**, 209–243.

Olesiuk, P.F., Bigg, M.A, Ellis, G.M., Crockford, S.J., and Wigen, R.J. (1990b). An assessment of the feeding habits of harbour seals (*Phoca vitulina*) in the Strait of Georgia British Columbia based on scat analysis. *Canadian Technical Reports of Fisheries and Aquatic Sciences*, **1730**, 135.

Oliver, G.W., Morris, P.A., Thorson, P.H., and Le Boeuf, B.J. (1998). Homing behavior of juvenile northern elephant seals. *Marine Mammal Science*, **14**, 245–256.

Olsen, E. (2002). Errors in age estimates of North Atlantic minke whales when counting growth zones in bulla tympanica. *Journal of Cetacean Research and Management* **4**, 185–192.

Olsen, E. and Sunde, J. (2002). Age determination of minke whales (*Balaenoptera acutorostrata*) using the aspartic acid racemization technique. *Sarsia*, **87**, 1–8.

Oosthuizen, W.H. (1997). Evaluation of an effective method to estimate age of Cape fur seals using ground tooth sections. *Marine Mammal Science*, **13**, 683–693.

Orr, A.J. and Harvey, J.T. (2001). Quantifying errors associated with using fecal samples to determine the diet of the California sea lion (*Zalophus californianus*). *Canadian Journal of Zoology*, **79**, 1080–1087.

Orr, A.J., Laake, J.L., Druv, M.I., Banks, A.S., DeLong, R.L., and Huber, H.R. (2003). Comparison of processing pinniped scat samples using a washing machine and nested sieves. *Wildlife Society Bulletin*, **31**, 253–257.

Ortiz C.L., Costa, D.P., and LeBoeuf, B.J. (1978). Water and energy flux in elephant seal pups fasting under natural conditions. *Physiological Zoology*, **51**, 166–178.

Osborne, P.E. and Suárez-Seoane, S. (2002). Should data be partitioned spatially before building large-scale distribution models? *Ecological Modelling*, **157**, 249–259.

O'Shea, T.J. (1999). Environmental contaminants and marine mammals. In J.E.I. Reynolds and S.A. Rommel (eds), *Biology of Marine Mammals*, pp. 485–564. Smithsonian Institution Press, Washington DC.

Otis, D.L., Burnham, K.P., White, G.C., and Anderson, D.R. (1978). Statistical inference from capture data on closed animal populations. *Wildlife Monographs*, **62**, 1–135.

Ottensmeyer, C.A. and Whitehead, H. (2003). Behavioural evidence for social units in long-finned pilot whales. *Canadian Journal of Zoology*, **81**, 1327–1338.

Outridge, P.M. and Stewart, R.E.A. (1999). Stock discrimination of Atlantic Walrus (*Odobenus rosmarus rosmarus*) in the eastern Canadian Arctic using lead isotope and elemental signatures in teeth. *Canadian Journal of Fisheries and Aquatic Science*, **56**, 105–112.

Oxnard, C.E. (1978). One biologist's view of morphometrics. *Annual Review of Ecology and Systematics*, **9**, 219–241.

Pace, N. and Rathbun, E.N. (1945). Studies on body composition. 3. The body water and chemically combined nitrogen content in relation to fat content. *Journal of Biological Chemistry*, **158**, 685–691.

Packer, C., Tatar, M., and Collins, A. (1998). Reproductive cessation in female mammals. *Nature*, **392**, 807–811.

Page, B., McKenzie, J., and Goldsworthy, S.D. (2005a). Inter-sexual differences in New Zealand fur seal diving behaviour. *Marine Ecology Progress Series*, **304**, 249–264.

Page, B., McKenzie, J., Hindell, M.A., and Goldsworthy, S.D. (2005b). Drift dives by male New Zealand fur seals (Arctocephalus forsteri). *Canadian Journal of Zoology*, **83**, 293–300.

Paine, R.T. (1976). Size-limited predation—observational and experimental approach with Mytilus–Pisaster Interaction. *Ecology*, **57**, 858–873.

Palka, D. (1995a). Abundance estimate of the Gulf of Maine harbor porpoise. *Reports of the International Whaling Commission*, Special Issue **15**, 27–50.

Palka, D. (1995b). Influences on spatial patterns of Gulf of Maine harbor porpoises. In A.S. Blix, L. Walløe, and Ø. Ulltang (eds), *Whales, Seals, Fish and Man*, pp. 69–75. Elsevier, Amsterdam.

Palka, D.L. and Hammond, P.S. (2001). Accounting for responsive movement in line transect estimates of abundance. *Canadian Journal of Fisheries and Aquatic Sciences*, **58**, 777–787.

Pallen, M.J. and Wren, B.W. (2007). Bacterial pathogenomics. *Nature*, **449**, 835–842.

Palmer, M.W. (1993). Putting things in even better order: The advantages of canonical correspondence analysis. *Ecology*, **74**, 2215–2230.

Palsbøll, P., Allen, J., Bérubé, M., Clapham, P.J., Feddersen, T.P., Hammond, P.S., Hudson, R.R., Jørgensen, H., Katona, S., Larsen, A. H., Larsen, F., Lien, J., Matttila, D.K., Sigurjónsson, J., Sears, R., Smith, T., Sponer, R., Stevick, P., and Øien, N. (1997). Genetic tagging of humpback whales. *Nature*, **388**, 767–769.

Palumbi, S.R. and Baker, C.S. (1994). Contrasting population structure from nuclear intron sequences and mtDNA of humpback whales. *Molecular Biology Evolution*, **11**, 426–435.

Panigada, S., Di Sciara, G.N., Zanardelli Panigada, M., Airoldi, S., Borsani, J.F., and

Jahoda, M. (2005). Fin whales (Balaenoptera physalus) summering in the Ligurian Sea: distribution, encounter rate, mean group size and relation to physiographic variables. *Journal of Cetacean Research and Management*, **7**, 137–145.

Paradiso, J.A. and Starner, T. (2005). Energy scavenging for mobile and wireless electronics. *IEEE Pervasive Computing*, **4**, 18–27.

Parra, G.J. (2006). Resource partitioning in sympatric delphinids: space use and habitat preferences of Australian snubfin and Indo-Pacific humpback dolphins. *Journal of Applied Ecology*, **75**, 862–874.

Parra, G.J., Schick, R., and Corkeron, P.J. (2006). Spatial distribution and environmental correlates of Australian snubfin and Indo-Pacific humpback dolphins. *Ecography*, **29**, 396–406.

Parrish, C.C. (1999). Determination of total lipid classes and fatty acids in aquatic samples. In M.T. Arts and B.C. Wainman (eds), *Lipids in Freshwater Ecosystems*, pp. 4–20. Springer, New York.

Parrish, F.A., Craig, M.P., Ragen, T.J., Marshall, G.J., and Buhleier, B.M. (2000). Identifying diurnal foraging habitat of endangered Hawaiian monk seals using a seal-mounted video camera. *Marine Mammal Science*, **16**, 392–412.

Parrish, F.A., Marshall, G.J., Littnan, C.L., Heithaus, M., Canja, S., Becker, B., Braun, R., and Antonelis, G.A. (2005). Foraging of juvenile monk seals at French Frigate Shoals, Hawaii. *Marine Mammal Science*, **21**, 93–107.

Parsons, K.M., Dallas, J.F., Claridge, D.E., Durban, J.W., Balcomb, K.C., Thompson, P.M., and Noble, L.R. (1999). Amplifying dolphin mitochondrial DNA from faecal plumes. *Molecular Ecology*, **8**, 1766–1768.

Parsons, K.M., Piertney, S.B., Middlemas, S.J., Hammond, P.S., and Armstrong, J.D. (2005). DNA-based identification of salmonid prey species in seal faeces. *Journal of Zoology, London*, **266**, 275–281.

Pascal, J.C. and Ackman, R.G. (1976). Long chain monoethylenic alcohol and acid isomers in lipids of copepods and capelin. *Chemistry and Physics of Lipids*, **16**, 219–223.

Patterson, T.A., Thomas, L., Wilcox, C., and Ovaskainen, O. (2008). State–space models of individual animal movement. *Trends in Ecology and Evolution*, **23**, 87–94.

Patterson, T.A., McConnell, B.J., Fedak, M.A., Bravington, M.V., and Hindell, M.A. (2010). Using GPS data to evaluate the accuracy of state–space methods for correction of Argos satellite telemetry error. *Ecology*, **91**, 273–285.

Pavlov, V.V., Wilson, R.P., and Lucke, K. (2007). A new approach to tag design in dolphin telemetry: Computer simulations to minimise deleterious effects. *Deep-Sea Research Part II—Topical Studies in Oceanography*, **54**, 404–414.

Payne, R., Brazier, O., Dorsey, E.M., Perkins, J.S., Rowntree, V.J., and Titus, A. (1983). External features in southern right whales (*Eubalaena australis*) and their use in identifying individuals. In R. Payne (ed.), *Communication and Behavior of Whales*, pp. 371–445. Westview Press, Boulder, CO.

Peaker, M. and Goode, J.A. (1978). The milk of the fur seal, *Arctocephalus tropicalis gazella*: in particular the composition of the aqueous phase [Correct species: *Mirounga leonina*]. *Journal of Zoology, London*, **185**, 469–476.

Pendleton, G.W., Pitcher, K.W., Fritz, L.W., York, A.E., Raum-Suryan, K.L., Loughlin, T.R., Calkins, D.G., Hastings, K.K., and Gelatt, T.S. (2006). Survival of Steller sea lions in Alaska: A comparison of increasing and decreasing populations. *Canadian Journal of Zoology*, **84**, 1163–1172.

Pepin, D., Adrados, C., Mann, C., and Janeau, G. (2004). Assessing real daily distance traveled by ungulates using differential GPS locations. *Journal of Mammalogy*, **85**, 774–780.

Perneger, T.V. (1998). What's wrong with Bonferroni adjustments. *British Medical Journal*, **316**, 1236–1238.

Perrin, D.R. (1958). The calorific value of milk of different species. *Journal of Dairy Research*, **25**, 215–220.

Perrin, W.F. and Donovan, G.P. (1984). Report of the Workshop. In W.F. Perrin, R.L. Brownell, Jr, and D.P. DeMaster (eds), *Reproduction in Whales, Dolphins, and Porpoises*. Report of the International Whaling Commission, Special Issue **6**. Cambridge University Press, Cambridge.

Perrin, W.F. and Myrick, A.C., Jr. (1980). *Age Determination of Toothed Whales and Sirenians*. Report of the International Whaling Commission, Special Issue **3**, Cambridge, UK.

Perrin, W.F. and Reilly, S.B. (1984). Reproductive parameters of dolphins and small whales of the family Delphinidae. In W.F. Perrin, R. L. Brownell, Jr, and D.P. DeMaster (eds), *Reproduction in Whales, Dolphins, and Porpoises*. Report of the International Whaling Commission, Special Issue **6**, Cambridge University Press, Cambridge.

Perrin, W.F., Coe, J.M., and Zweifel, J.R. (1976). Growth and reproduction of the spotted porpoise, *Stenella attenuata*, in the offshore eastern tropical Pacific. *Fishery Bulletin*, **74**, 229–269.

Perrin, W.F., Holts, D.B., and Miller, R.B. (1977). Growth and reproduction of the eastern spinner dolphin, a geographical form of *Stenella longirostris* in the eastern tropical Pacific. *Fisheries Bulletin*, **75**, 725–750.

Perrin, W.F., Dolar, M.L.L., Chan, C.M., and Chivers, S.J. (2005). Length–weight relationships in the spinner dolphin (*Stenella longirostris*). *Marine Mammal Science*, **21**, 765–778.

Perry, J.N., Liebhold, A.M., Rosenberg, M.S., Dungan, J., Miriti, M., Jakomulska, A., and Citron-Pousty, S. (2002). Illustrations and guidelines for selecting statistical methods for quantifying spatial pattern in ecological data. *Ecography*, **25**, 578–600.

Perryman, W.L. and Lynn, M.S. (1993). Identification of geographic forms of common dolphin (*Delphinus delphis*) from aerial photogrammetry. *Marine Mammal Science*, **9**, 119–137.

Perryman, W.L., Donahue, M.A., Perkins, P.C., and Reilly, S.B. (2002). Gray whale calf production 1994–2000: are observed fluctuations related to changes in seasonal ice cover? *Marine Mammal Science*, **18**, 121–144.

Peters, C.N. and Marmorek, D.R. (2001). Application of decision analysis to evaluate recovery actions for threatened Snake River spring and summer chinook salmon (*Oncorhynchus tshawytscha*). *Canadian Journal of Fisheries and Aquatic Sciences*, **58**, 2431–2446.

Peters, C.N., Marmorek, D.R., and Deriso, R.B. (2001). Application of decision analysis to evaluate recovery actions for threatened Snake River fall chinook salmon (*Oncorhynchus tshawytscha*). *Canadian Journal of Fisheries and Aquatic Sciences*, **58**, 2447–2458.

Pettis, H.M., Rolland, R.M., Hamilton, P.K., Brault, S., Knowlton, A.R., and Kraus, S.D. (2004). Visual health assessment of North Atlantic right whales (*Eubalaena glacialis*) using photographs. *Canadian Journal of Zoology*, **82**, 8–19.

Phillips, D.L. (2001). Mixing models in analyses of diet using multiple stable isotopes: A critique. *Oecologia*, **127**, 166–170.

Phillips, D.L. and Gregg, J.W. (2003). Source partitioning using stable isotopes: coping with too many sources. *Oecologia*, **136**, 261–269.

Phillips, E.M. and Harvey, J.R. (2009). A captive feeding study with the Pacific harbor seal (*Phoca vitulina richardii*): Implications for scat analysis. *Marine Mammal Science*, **25**, 373–391.

Pierce, G.J. and Boyle, P.R. (1991). A review of methods for diet analysis in piscivorous marine mammals. *Oceanography and Marine Biology: An Annual Review*, **29**, 409–486.

Pierce, G.J., Diack, J.S.W., and Boyle, P.R. (1990). Application of serological methods to identification of fish prey in diets of seals and dolphins. *Journal of Experimental Marine Biology and Ecology*, **137**, 123–140.

Pierce, G.J., Boyle, P.R., Watt, J., and Grisley, M. (1993). Recent advances in diet analysis of marine mammals. In Marine mammals: advances in behavioural and population ecology. *Symposia of the Zoological Society of London*, **66**, 241–261.

Pierce, G.J., Santos, M.B., Learmonth, J.A., Mente, E., and Stowasser, G. (2004). Methods for dietary studies on marine mammals. In Investigating the roles of cetaceans in marine ecosystems, *CIESM Workshop Monographs*, **25**, pp. 29–36. Monaco.

Pierce, G.J., Santos, M.B., and Cerviño, S. (2007). Assessing sources of variation underlying estimates of cetacean diet composition: a simulation study on analysis of harbour porpoise diet in Scottish (UK) waters. *Journal of the Marine Biological Association of the United Kingdom*, **87**, 213–221.

Pikitch, E., Santora, C., Babcock, E., Bakun, A., Bonfil, R., Conover, D., Dayton, P., Doukakis, P., Fluharty, D., Heneman, B., Houde, E., Link, J., Livingston, P., Mangel, M., McAllister, M., Pope, J., and Sainsbury, K. (2004a). Ecosystem-based fishery management. *Science*, **305**, 346–347.

Pikitch, E., Santora, C., Babcock, E., Bakun, A., Bonfil, R., Conover, D., Dayton, P., Doukakis, P., Fluharty, D., Heneman, B., Houde, E., Livingston, P., Mangel, M., McAllister, M., Pope, J., and Sainsbury, K. (2004b). Fishery management and culling: response [to Corkeron 2004]. *Science*, **306**, 1891–1892.

Pinaud, D. and Weimerskirch, H. (2002). Ultimate and proximate factors affecting the breeding performance of a marine top-predator. *Oikos*, **99**, 141–150.

Pinaud, D. and Weimerskirch, H. (2005). Scale-dependent habitat use in a long-ranging central place predator. *Journal of Animal Ecology*, **74**, 852–863.

Pinaud, D. and Weimerskirch, H. (2007). At-sea distribution and scale-dependent foraging behaviour of petrels and albatrosses: a comparative study. *Journal of Animal Ecology*, **76**, 9–19.

Pinheiro, J.C. and Bates, D.M. (2000). *Mixed-effects Models in S and S-Plus*. Springer-Verlag, New York.

Pinkas, L.M., Oliphant, S., and Iverson, I.L.K. (1971). Food habits of albacore bluefin tuna and bonito in Californian waters. *California Department of Fish and Game, Fishery Bulletin*, **152**, 1–105.

Pistorius P.A., Bester M.N., Kirkman S.P., and Boveng P.L. (2000). Evaluation of age- and sex-dependent rates of tag loss in southern elephant seals. *Journal of Wildlife Management*, **64**, 373–380.

Pitcher, K.W., Rehberg, M.J., Pendleton, G.W., Raum-Suryan, K.L., Gelatt, T.S., Swain, U. G., and Sigler, M.F. (2005). Ontogeny of dive performance in pup and juvenile Steller sea lions in Alaska. *Canadian Journal of Zoology*, **83**, 1214–1231.

Pitt, J.A., Larivière, S., and Messier, F. (2006). Condition indices and bioelectrical impedance analysis to predict body condition of small carnivores. *Journal of Mammalogy*, **87**, 717–722.

Plowright, R.K., Sokolow, S.H., Gorman, M.E., Daszak P., and Foley, J.E. (2008). Causal inference in disease ecology: investigating ecological drivers of disease emergence. *Frontiers in Ecology and the Environment*, **6**, 420–429.

Polischuk, S.C., Hobson, K.A., and Ramsay, M. A. (2001). Use of stable-carbon and nitrogen isotopes to assess weaning and fasting in female polar bears and their cubs. *Canadian Journal of Zoology*, **79**, 499–511.

Pollock, K.H. (1982). A capture–recapture design robust to unequal probability of capture. *Journal of Wildlife Management*, **46**, 757–760.

Pollock, K., Marsh, H.D., Lawler, I.R., and Alldredge, M.W. (2006). Estimating animal abundance in heterogeneous environments: an application to aerial surveys for dugongs. *Journal of Wildlife Management*, **70**, 255–262.

Pomeroy, P.P., Anderson, S.S., Twiss, S.D., and McConnell, B.J. (1994). Dispersion and site fidelity of breeding female grey seals (*Halichoerus grypus*) on North Rona, Scotland. *Journal of Zoology, London*, **233**, 429–448.

Pomeroy, P.P., Fedak, M.A., Rothery, P., and Anderson, S.S. (1999). Consequences of maternal size for reproduction expenditure and pupping success of grey seals at North Rona, Scotland. *Journal of Animal Ecology*, **68**, 235–253.

Pompe, S., Simon, J., Wiedemann, P.M., and Tannert, C. (2005). Future trends and challenges in pathogenomics. A Foresight study. *EMBO Reports*, **6**, 600–605.

Ponce de Leon, A. (1984). Lactation and quantitative composition of the milk of the South American fur seals, *Arctocephalus australias* (Zimmerman, 1783). Industria Lobera y Pesquera del Estado, pp. 43–58, Montevideo, Uruguay, Anales.

Pond, C.M., Mattacks, C.A., Colby, R.H., and Ramsay, M.A. (1992). The anatomy, chemical composition, and metabolism of adipose tissue in wild polar bears (*Ursus maritimus*). *Canadian Journal of Zoology*, **70**, 326–341.

Ponganis, P.J. (2007). Bio-logging of physiological parameters in higher marine vertebrates. *Deep-Sea Research Part II—Topical Studies in Oceanography*, **54**, 183–192.

Ponganis, P.J., Ponganis, E.P., Ponganis, K.V., Kooyman, G.L., Gentry, G.L., and Trillmich, F. (1990). Swimming velocities in otariids. *Canadian Journal of Zoology*, **68**, 2105–2112.

Popp, B.N., Graham, B.S., Olson, R.J., Hannides, C.C.S., Lott, M.J., Lopez-Ibarra, G.A., Galvan-Magana, F., and Fry, B. (2007). Insight into the trophic ecology of Yellowfin Tuna, *Thunnus albacares*, from compound-specific nitrogen isotope analysis of proteinaceous amino acids. In T.E. Dawson and

R.T.W. Siegwolf (eds), *Stable Isotopes as Indicators of Ecological Change*, pp. 173–190. Academic Press, London.

Possingham, H.P., Andelman, S.J., Noon, B.R., Trombulak, S., and Pulliam, H.R. (2001). Making smart conservation decisions. In M.E. Soulé and G.H. Orians (eds), *Conservation Biology: Research Priorities for the Next Decade*, pp. 225–244. Island Press, Washington DC.

Post, D.M. (2002). Using stable isotopes to estimate trophic position: Models, methods and assumptions. *Ecology*, **83**, 703–718.

Preen, A. (2004). Distribution, abundance and conservation status of dugongs and dolphins in the southern and western Arabian Gulf. *Biological Conservation*, **118**, 205–218.

Prenda, J., Freitas, D., Santos-Reis, M., and Collares-Pereira, M.J. (1997). Guía para la identificación de restos óseos pertenecientes a algunos peces comunes en las aguas continentales de la Península ibérica para el estudio de la dieta de depredadores ictiófagos. Doñana. *Acta Vertebrata*, **24**, 155–180.

Prime, J.H. (1979). *Observation on the Digestion of some Gadoid Fish Otoliths by a Young Common Seal*, ICES CM 1979/N:14. International Council for the Exploration of the Sea, Copenhagen, Denmark.

Prime, J.H. and Hammond, P.S. (1987). Quantitative assessment of gray seal from faecal analysis. In A.C. Huntley, D.P. Costa, G.A.J. Worthy, and M.A. Castellini (eds), *Approaches to Marine Mammal Energetics*, pp. 165–182. Society for Marine Mammals Special Publication 1, Lawrence, Kansas.

Pritchard, J.K., Stephens, M., and Donnelly, P. (2000). Inference of population structure using multilocus genotype data. *Genetics*, **155**, 945–959.

Proffitt, K.M., Garrott, R.A., Rotella, J.J., and Banfield, J. (2007). The importance of considering prediction variance in analyses using photogrammetric mass estimates. *Marine Mammal Science*, **23**, 65–76.

Promislow, D.E.L. (1991). Senescence in natural populations of mammals: A comparative study. *Evolution*, **45**, 1869–1887.

Pulster, E.L., Smalling, K.L., and Maruya K.A. (2005). Polychlorinated biphenyls and toxaphene in preferred prey fish of coastal southeastern US bottlenose dolphins (*Tursiops truncatus*). *Environmental Toxicology and Chemistry*, **24**, 3128–3136.

Punt, A. and Donovan, G. (2007). Developing management procedures that are robust to uncertainty: lessons from the International Whaling Commission. *ICES Journal of Marine Science*, **64**, 603–612.

Punt, A. and Hilborn, R. (1997). Fisheries stock assessment and decision analysis: the Bayesian approach. *Reviews in Fish Biology and Fisheries*, **7**, 35–63.

Purcell, M., Mackey, G., LaHood, E., Huber, H.R., and Park, L. (2004). Molecular methods for genetic identification of salmonid prey from Pacific harbour seal (*Phoca vitulina richardsii*) scat. *Fishery Bulletin*, **102**, 213–220.

Pusineri, C., Magnin, V., Meynier, L., Spitz, J., Hassani, S., and Ridoux, V. (2007). Food and feeding ecology of the common dolphin (*Delphinus delphis*) in the oceanic northeast Atlantic and comparison with its diet in neritic areas. *Marine Mammal Science*, **23**, 30–47.

Quinn, G.P. and Keough, M.J. (2002). *Experimental Design and Data Analysis for Biologists*. Cambridge University Press, Cambridge.

R Development Core Team (2009). *R: A Language and Environment for Statistical Computing*. R Foundation for Statistical Computing, Vienna, Austria.

Ragen, T.J., Huntington, H.P., and Hovelsrud, G.K. (2008). Conservation of arctic marine mammals faced with climate change. *Ecological Applications*, **18**, S166–S174.

Ramp, C., Bérubé, M., Hagen, W., and Sears, R. (2006). Survival of adult blue whales *Balaenoptera musculus* in the Gulf of St. Lawrence, Canada. *Marine Ecology Progress Series*, **319**, 287–295.

Rau, G.H., Sweeney, R.E., and Kaplan, I.R. (1982). Plankton $^{13}C/^{12}C$ ratio changes with latitude: differences between northern and southern oceans. *Deep Sea Research*, **29**, 1035–1039.

Read, A.J. (1990). Age at sexual maturity and pregnancy rates of harbour porpoises *Phocoena phocoena* from the Bay of Fundy. *Canadian Journal of Fisheries and Aquatic Science*, **47**, 561–565.

Read, A.J. (2005). By-catch and depredation. In J.E. Reynolds, W.F. Perrin, R.R. Reeves, S. Montgomery, and T.J. Ragen (eds), *Marine Mammal Research: Conservation beyond Crisis*, pp. 5–17. The Johns Hopkins University Press, Baltimore, MD.

Read, A.J. (2008). The looming crisis: interactions between marine mammals and fisheries. *Journal of Mammalogy*, **89**, 541–548.

Read, A.J. and Gaskin, D.E. (1990). Changes in growth and reproduction of harbour porpoises, *Phocena phocoena*, from the Bay of Fundy. *Canadian Journal of Fisheries and Aquatic Science*, **47**, 2158–2163.

Read, A.J. and Hohn, A.A. (1995). Life in the fast lane—the life-history of harbor porpoises from the Gulf of Maine. *Marine Mammal Science*, **11**, 423–440.

Read, A.J., Wells, R.S., Hohn, A.A., and Scott, M.D. (1993). Patterns of growth in wild bottlenose dolphins, *Tursiops truncatus. Journal of Zoology, London*, **231**, 107–123.

Read, A.J., Urian, K.W., Wilson, B., and Waples, D.M. (2003). Abundance of bottlenose dolphins in the bays, sounds, and estuaries of North Carolina. *Marine Mammal Science*, **19**, 59–73.

Read, A.J., Drinker, P.B., and Northridge, S.P. (2006). By-catches of marine mammals in U. S. fisheries and a first estimate of global marine mammal by-catch. *Conservation Biology*, **20**, 163–169.

Reddy, M.L., Reif, J.S., Bachand, A., and Ridgway, S.H. (2001). Opportunities for using Navy marine mammals to explore associations between organochlorine contaminants and unfavorable effects on reproduction. *Science of the Total Environment*, **274**, 171–182.

Redfern, J.V., Ferguson, M.C., Becker, E.A., Hyrenbach, K.D., Good, C., Barlow, J., Kaschner, K., Baumgartner, M.F., Forney, K.A., Balance, L.T., Fauchald, P., Halpin, P., Hamazaki, T., Pershing, A.J., Qian, S.S., Read, A., Reilly, S.B., Torres, L., and Werner, F. (2006). Techniques for cetacean-habitat modeling. *Marine Ecology Progress Series*, **310**, 271–295.

Reed, J.M. and Blaustein, A.R. (1997). Biologically significant population declines and statistical power. *Conservation Biology*, **11**, 281–282.

Reed, J.Z., Chambers, C., Fedak, M.A., and Butler, P.B. (1994). Gas exchange of freely diving grey seals (*Halichoerus grypus*). *Journal of Experimental Biology*, **191**, 1–18.

Reed, J.Z., Tollit, D.J., Thompson, P.M., and Amos, W. (1997). Molecular scatology: the use of molecular genetic analysis to assign species sex and individual identity to seal faeces. *Molecular Ecology*, **6**, 225–234.

Reeves, R., Perrin, W.F., Taylor, B.L., Baker, C., and Mesnick, S. (2004). *Report of the Workshop on Shortcomings of Cetacean Taxonomy in Relation to Needs of Conservation and Management*, April 30–May 2, 2004. NOAA Technical Memorandum NMFS. La Jolla, CA.

Reeves, R.R., Read, A.J., Lowry, L., Katona, S., and Boness, D.J. (2007). *Report of the North Atlantic Right Whale Program Review.* Marine Mammal Commission, Bethesda, MD.

Regehr, E.V., Lunn, N.J., Amstrup, S.C., and Stirling, I. (2007). Effects of earlier sea ice breakup on survival and population size of polar bears in western Hudson Bay. *Journal of Wildlife Management*, **71**, 2673–2683.

Rehtanz, M., Ghim, S.J., Rector, A., Van Ranst, M., Fair, P.A., Bossart, G.D., and Jenson, A. B. (2006). Isolation and characterization of the first American bottlenose dolphin papillomavirus: *Tursiops truncatus* papillomavirus type 2. *Journal of General Virology*, **87**, 3559–3565.

Reid, J.P., Bonde, R.K., and O'Shea, T.J. (1995). Reproduction and mortality of radio-tagged and recognizable manatees on the Atlantic coast of Florida. In T.J. O'Shea, B.B. Ackerman, and H.F. Percival (eds), *Population Biology of the Florida Manatee*. Information and technology report 1. US Department of the Interior, National Biological Service, Reston, VA.

Reid, K. (1995). The diet of Antarctic fur seals (*Arctocephalus gazella*, Peters 1875) during winter at South Georgia. *Antarctic Science*, **7**, 241–249.

Reid, K., Barlow, K.E., Croxall, J.P., and Taylor, R.I. (1999). Predicting changes in the Antarctic krill, *Euphausia superba*, population at South Georgia. *Marine Biology*, **135**, 647–652.

Reid, K., Jessopp, M.J., Barrett, M.S., Kawaguchi, S., Seigel, V., and Goebel, M.E. (2004). Widening the net: spatio-temporal variability in the krill population structure across the Scotia Sea. *Deep Sea Research Part II—Topical Studies in Oceanography*, **51**, 1275–1287.

Reijnders, P.J.H., Aguilar, A., and Donovan, G.P. (eds). (1999). *Chemical Pollutants and*

Cetaceans. International Whaling Commission, Cambridge, UK.

Reilly, J.J., and Fedak, M.A. (1990). Measurement of living grey seals (*Halichoerus grypus*) by hydrogen isotope dilution. *Journal of Applied Physiology*, **69**, 885–891.

Reilly, S.B. and Barlow, J.P. (1986). Rates of increase in dolphin population size. *Fisheries Bulletin* **84**, 527–533.

Reilly, S.B. and Fiedler, P.C. (1994). Inter annual variability of dolphin habitats in the eastern tropical Pacific. 1. Research vessel surveys, 1986–1990. *Fisheries Bulletin*, **92**, 434–450.

Rendell, L. and Whitehead, H. (2001). Culture in whales and dolphins. *Behavioral and Brain Sciences*, **24**, 309–324.

Renouf, D. and Gales, R. (1994). Seasonal variation in the metabolic rate of harp seals: unexpected energetic economy in the cold ocean. *Canadian Journal of Zoology*, **72**, 1625–1632.

Rettie, W.J. and McLoughlin, P.D. (1999). Overcoming radiotelemetry bias in habitat-selection studies. *Canadian Journal of Zoology*, **77**, 1175–1184.

Reynolds, J.C. and Aebischer, N.J. (1991). Comparison and quantification of carnivore diet by fecal analysis—a critique with recommendations based on a study of the fox, *Vulpes vulpes. Mammal Review*, **21**, 97–122.

Reynolds, J.E. III, Perrin, W.F., Reeves, R.R., Montgomery, S., and Ragen, T.J. (eds), (2005). *Marine Mammal Research: Conservation Beyond Crisis.* The Johns Hopkins University Press, Baltimore, MD.

Rice, D.W. (1999). *Marine Mammals of the World: Systematics and Distribution.* Society for Marine Mammalogy, Special Publication Number 4. Allen Press, Lawrence, KS.

Rice, J.C. and Rochet, M.-J. (2005). A framework for selecting a suite of indicators for fisheries management. *ICES Journal of Marine Science*, **62**, 516–527.

Ridoux, V. (1994). The diets and dietary segregation of seabirds at the subantarctic Crozet Islands. *Marine Ornithology*, **22**, 1–192.

Ries, E.H., Hiby, L.R., and Reijnders, P.J.H. (1998). Maximum likelihood population size estimation of harbour seals in the Dutch Wadden Sea based on a mark–recapture experiment. *Journal of Applied Ecology*, **35**, 332–339.

Ripley, B.D. (2004). *Spatial Statistics.* John Wiley and Sons, New York.

Robertson, M.P., Caithness, N., and Villet, M.H. (2001). A PCA-based modelling technique for predicting environmental suitability for organisms from presence records. *Diversity and Distributions*, **7**, 15–27.

Robertson, M.P., Peter, C.I., Villet, M.H., and Ripley, B.S. (2003). Comparing models for predictive species' potential distributions: a case study using correlative and mechanistic predictive modelling techniques. *Ecological Modelling*, **164**, 153–167.

Robinson, P.W., Tremblay, Y., Antolos, M., Crocker, D.E., Kuhn, C.E., Shaffer, S.A., Simmons, S., and Costa, D.P. (2007). A comparison of indirect measures of feeding behavior based on ARGOS tracking data. *Deep Sea Research Part II—Topical Studies in Oceanography*, **54**, 356–368.

Rodhouse, P.G., Arnbom, T.R., Fedak, M.A., Yeatman, J., and Murray, A.W.A. (1992). Cephalopod prey of the southern elephant seal *Mirounga leonina* L. *Canadian Journal of Zoology*, **70**, 1007–1015.

Rodrigues, A.S.L., Pilgrim, J.D., Lamoreux, J.F., Hoffman, M., and Brooks, T.M. (2006). The value of the IUCN Red List for conservation. *Trends in Ecology and Evolution*, **21**, 71–76.

Roff, D.A. (1992). *The Evolution of Life Histories.* Routledge, Chapman and Hall. New York.

Roff, D.A. and Bowen, W.D. (1982). Population dynamics and management of the northwest atlantic harp seal (*Phoca groenlandica*). *Canadian Journal of Fisheries and Aquatic Science*, **40**, 919–932.

Roff, D. and Bowen, W.D. (1986). Analysis of population trends in northwest Atlantic harp seals (*Phoca groenlandica*) from 1967–1983. *Canadian Journal of Fisheries and Aquatic Science*, **43**, 553–564.

Rohrer, G.A., Freking, B.A., and Nonneman, D. (2007). Single nucleotide polymorphisms for pig identification and parentage exclusion. *Animal Genetics*, **38**, 253–258.

Rolland, R.M., Hunt, K.E., Kraus, S.D., and Wasser, S.K. (2005). Assessing reproductive status of right whales (*Eubalaena glacialis*) using fecal hormone metabolites. *General and Comparative Endocrinology*, **142**, 308–317.

Rolland, R.M., Hamilton, P.K., Kraus, S.D., Davenport, B., Gillett, R.M., and Wasser, S.

K. (2006). Faecal sampling using detection dogs to study reproduction and health in North Atlantic right whales (*Eubalaena glacialis*). *Journal of Cetacean Research and Management*, **8**, 121–125.

Rolland, R.M., Hunt, K.E., Doucette, G.J., Rickard, L.G., and Wasser, S.K. (2007). The inner whale: hormones, biotoxins, and parasites. In S.D. Kraus and R.M. Rolland (eds), *The Urban Whale: North Atlantic Right Whales at the Crossroads*, pp. 232–272. Harvard University Press, Cambridge, MA.

Ronald, K., Keiver, K.M., Beamish, F.W.H., and Frank, R. (1984). Energy requirements for maintenance and faecal and urinary losses of the grey seal (Halichoerus grypus). *Canadian Journal of Zoology*, **62**, 1101–1105.

Ronquist, F. (2004). Bayesian inference of character evolution. *Trends in Ecology and Evolution*, **19**, 475–481.

Rooney, S.M., Wolfe, A., and Haydenm, T.J. (1998). Autocorrelated data in telemetry studies: time to independence and the problem of behavioural effects. *Mammal Review*, **28**, 89–98.

Ropert-Coudert, Y. and Wilson, R.P. (2005). Trends and perspectives in animal-attached remote sensing. *Frontiers in Ecology and the Environment*, **3**, 437–444.

Rorres, C. and Fair, W. (1975). Optimal harvesting policy for an age-specific population. *Mathematical Biosciences*, **24**, 31–47.

Rosel, P.E., France, S.C., Wang, J.Y., and Kocher, T.D. (1999). Genetic structure of harbour porpoise *Phocoena phocoena* populations in the northwest Atlantic based on mitochondrial and nuclear markers. *Molecular Ecology*, **8**, S41–S54.

Rosello Izquierdo, E. (1986). *Contribución al Atlas Osteológico de los Teleósteos Ibéricos I Dentario y Articular*. Universidad Autónoma de Madrid, Ed Siglo XXI, Spain.

Rosen, D.A.S. and Trites, A.W. (2000). Digestive efficiency and dry-matter digestibility in Steller sea lions fed herring, pollock, squid and salmon. *Canadian Journal of Zoology*, **78**, 234–239.

Rosenbaum, H.C., Brownell, R.L., Brown, M. W., Schaeff, C., Portway, V., White, B.N., Malik, S., Pastene, L.A., Patenaude, N.J., Baker, C.S., Goto, M., Best, P.B., Clapham, P.J., Hamilton, P., Moore, M., Payne, R.,

Rowntree, V., Tynan, C.T., Bannister, J. L., and Desalle, R. (2000). World-wide genetic differentiation of Eubalaena: questioning the number of right whale species. *Molecular Ecology*, **9**, 1793–1802.

Rosenblum, E.B. and Novembre, J. (2007). Ascertainment bias in spatially structured populations: a case study in the eastern fence lizard. *Journal of Heredity*, **98**, 331–336.

Ross, H.A., Lento, G.M., Dalebout, M.L., Goode, M., Ewing, G., Mclaren, P., Rodrigo, A.G., Lavery, S., and Baker, C.S. (2003). DNA surveillance: web-based molecular identification of whales, dolphins, and porpoises. *Journal of Heredity*, **94**, 111–114.

Ross, P.S., Ellis, G.M., Ikonumou, M.G., Barrett-Lennard, L.G., and Addison, R.F. (2000). High PCB concentrations in free-ranging Pacific killer whales, Orcinus orca: effects of age, sex and dietary preference. *Marine Pollution Bulletin*, **40**, 504–515.

Rothman, K.J. and Greenland, S. (2005). *Modern Epidemiology*, 2nd edn. Lippincott–Raven, Philadelphia, PA.

Royer, F. and Lutcavage, M. (2008). Filtering and interpreting location errors in satellite telemetry of marine animals. *Journal of Experimental Marine Biology and Ecology*, **359**, 1–10.

Royer, F., Fromentin, J.M., and Gaspar, P. (2005). A state–space model to derive blue-fin tuna movement and habitat from archival tags. *Oikos*, **109**, 473–484.

Rugh, D.J., Hobbs, R.C., Lerczak, J.A., and Breiwick, J.M. (2005). Estimates of abundance of the eastern North Pacific stock of gray whales (*Eschrichtius robustus*) 1997–2002. *Journal of Cetacean Research and Management*, **7**, 1–12.

Ruiz-Cooley, R.I., Gendron, D., Aguiiga, S., Mesnick, S., and Carriquiy, J.D. (2004). Trophic relationships between sperm whales and jumbo squid using stable isotopes of C and N. *Marine Ecology Progress Series*, **277**, 275–283.

Runge, M.C., Langtimm, C.A., and Kendall, W. L. (2004). A stage-based model of manatee population dynamics. *Marine Mammal Science*, **20**, 361–385.

Rushton, S.P., Lurz, P.W.W., Fuller, R., and Garson, P.J. (1997). Modelling the distribution of the red and grey squirrel at the landscape scale: a combined GIS and population dynamics approach. *Journal of Applied Ecology*, b, 1137–1154.

Russel, W.M.S. and Burch, R.L. (1959). *The Principle of Humane Experimental Technique*. Methuen, London.

Ruud, J.T. (1940). The surface structure of the baleen plates as a possible clue to age in whales. *Hvalrådets Skrifter*, **23**, 1–23.

Ryg, M.L., Lydersen, C., Markussen, N.H., Smith, T.G., and Øritsland, N.A. (1990). Estimating the blubber content of phocid seals. *Canadian Journal of Fisheries and Aquatic Science*, **47**, 1223–1227.

Ryman, N. and Jorde, P.E. (2001). Statistical power when testing for genetic differentiation. *Molecular Ecology*, **10**, 2361–2373.

Ryman, N. and Palm, S. (2006). POWSIM: a computer program for assessing statistical power when testing for genetic differentiation. *Molecular Ecology Notes*, **6**, 600–602.

Ryman, N., Palm, S., André, C., Carvalho, G.R., Dahlgren, T.G., Jorde, P.E., Laikre, L., Larsson, L.C., Palmé, A., and Ruzzante, D. E. (2006). Power for detecting genetic divergence: differences between statistical methods and marker loci. *Molecular Ecology*, **15**, 2031–2045.

St. Aubin, D.J. (2001). Endocrinology. In L.A. Dierauf and F.M.D. Gulland (eds), *CRC Handbook of Marine Mammal Medicine*, 2nd edn. CRC Press, Boca Raton, FL.

St. Aubin, D.J. and Dierauf, L.A. (2001). Stress and marine mammals. In L.A. Dierauf and F. M.D. Gulland, *CRC Handbook of Marine Mammal Medicine*, 2nd edn, pp. 253–269. CRC Press, London.

St. Aubin, D.J., Smith, T.G., and Geraci, J.R. (1990). Seasonal epidermal molt in beluga, *Delphinapterus leucas*. *Canadian Journal of Zoology*, **68**, 359–367.

Samuels, A. and Gifford, T. (1997). A quantitative assessment of dominance relations among bottlenose dolphins. *Marine Mammal Science*, **13**, 70–99.

Sanderson, M.J., Purvis, A., and Henze, C. (1998). Phylogenetic supertrees: Assembling the trees of life. *Trends in Ecology and Evolution*, **13**, 105–109.

Santos, M.B., Clarke, M.R., and Pierce, G.J. (2001). Assessing the importance of cephalopods in the diets of marine mammals and other top predators: problems and solutions. *Fisheries Research*, **52**, 121–139.

Sato, K., Mitani, Y., Cameron, M.F., Siniff, D.B., Watanabe, Y., and Naito, Y. (2002). Deep foraging dives in relation to the energy depletion of Weddell seal (*Leptonychotes weddellii*) mothers during lactation. *Polar Biology*, **25**, 696–702.

Sato, K., Mitani, Y., Cameron, M.F., Siniff, D.B., and Naito, Y. (2003). Factors affecting stroking patterns and body angle in diving Weddell seals under natural conditions. *Journal of Experimental Biology*, **206**, 1461–1470.

SCANS-II (2008). *Small Cetaceans in the European Atlantic and North Sea*. Final Report to the European Commission under project LIFE04NAT/GB/000245. Sea Mammal Research Unit, Gatty Marine Laboratory, University of St Andrews, St Andrews, Fife, Scotland.

Scheffer, V.B. (1950). Growth layers on the teeth of pinnipeds as an indication of age. *Science*, **112**, 309–311.

Scheffer, V.B., Fiscus, C.H., and Todd, E.I. (1984). *History of Scientific Study and Management of the Alaskan Fur Seal, Callorhinus ursinus, 1786–1964*. US Department of Commerce, NOAA Technical Report NMFS SSRF-780.

Schell, D.M., Saupe, S.M., and Haubenstock, N. (1989). Bowhead whale (*Balaena mysticetus*) growth and feeding as estimated by del13C techniques. *Marine Biology*, **103**, 433–443.

Schell, D.M., Barnett, B.A., and Vinnette, K. (1998). Carbon and nitrogen isotope ratios in zooplankton of the Bering, Chukchi and Beaufort seas. *Marine Ecology Progress Series*, **162**, 11–23.

Schick, R.S. and Urban, D.L. (2000). Spatial components of bowhead whale (*Balaena mysticetus*) distribution in the Alaskan Beaufort Sea. *Canadian Journal of Fisheries and Aquatic Science*, **57**, 2193–2200.

Schipper, J., Chanson, J.S., Chiozza, F., *et al.* (2008). The status of the world's land and marine mammals: diversity, threat, and knowledge. *Science*, **322**, 225–230.

Schlötterer, C. and Pemberton, J. (1998). The use of microsatellites for genetic analysis of natural populations—a critical review. In R. Desalle and B. Schierwater (eds), *Molecular Approaches to Ecology and Evolution*. Birkhäuser Verlag, Berlin.

Schmidt-Nielsen, K. (1980). *Animal Physiology*, 2nd edn. Cambridge University Press, Cambridge.

Schneider, K., Baird, R.W., Dawson, S., Visser, I., and Childerhouse, S. (1998). Reactions of bottlenose dolphins to tagging attempts using a remotely-deployed suction-cup tag. *Marine Mammal Science*, **14**, 316–324.

Schreer, J.F. and Testa, J.W. (1995). Statistical classification of diving behavior. *Marine Mammal Science*, **11**, 85–93.

Schusterman, R.J., Reichmuth, C.J., and Kastak, D. (2000). How animals classify friends and foes. *Current Directions in Psychological Science*, **9**, 1–6.

Schwacke, L.H., Voit, E.O., Hansen, L.J., Wells, R.S., Mitchum, G.B., Hohn, A.A., and Fair, P.A. (2002). Probabilistic risk assessment of reproductive effects of polychlorinated biphenyls on bottlenose dolphins (*Tursiops truncatus*) from the southeast United States coast. *Environmental Toxicology and Chemistry*, **21**, 2752–2764.

Schwacke, L.H., Hall, A.J., Townsend, F.I., Wells, R.S., Hansen, L.J., Hohn, A.A., Bossart, G.D., Fair, P.A., and Rowles, T.K. (2009). Hematology and clinical blood chemistry reference intervals for free-ranging common bottlenose dolphins (*Tursiops truncatus*) and variation related to geographic sampling site. *American Journal of Veterinary Research*, **70**, 973–985.

Schwarz, C.J. and Stobo, W.T. (1997). Estimating temporary migration using the robust design. *Biometrics*, **53**, 178–194.

Schwarz, C.J., Schweigert, J.F., and Arnason, A.N. (1993). Estimating migration rates using tag-recovery data. *Biometrics*, **49**, 177–193.

Schweder, T. (1999). Line transecting with difficulties: lessons from surveying minke whales. In G.W. Garner, S.C. Armstrup, J.L. Laake, B.F.J. Manly, L.L. McDonald, and D. G. Robertson (eds), *Marine Mammal Survey and Assessment Methods*, pp. 149–166. Balkema, Rotterdam.

Schweder, T., Hagen, G., Helgeland, J., and Koppervik, I. (1996). Abundance estimation of northeastern Atlantic minke whales. *Reports of the International Whaling Commission*, **46**, 391–408.

SCOS (2008). *Scientific Advice on Matters Related to the Management of Seal Populations: 2008*. Sea Mammal Research Unit, Gatty Marine Laboratory, University of St Andrews, St Andrews, Fife, Scotland.

Scott, M.D. and Chivers, S.J. (1999). Distribution and herd structure of bottlenose dolphins in the eastern tropical Pacific Ocean. In S. Leatherwood and R.R. Reeves (eds), *The Bottlenose Dolphin*, pp. 387–402. Academic Press, San Diego, CA.

Scott, M.D. and Chivers, S.J. (2009). Movements and diving behavior of pelagic spotted dolphins. *Marine Mammal Science*, **25**, 137–160.

Scott, M.D., Wells, R.S., Irvine, A.B., and Mate, B.R. (1990a). Tagging and marking studies on small cetaceans. In S. Leatherwood and R.R. Reeves (eds), *The Bottlenose Dolphin*, pp. 489–514. Academic Press, San Diego, CA.

Scott, M.D., Wells, R.S., and Irvine, A.B. (1990b). A long-term study of bottlenose dolphins on the west coast of Florida. In S. Leatherwood and R.R. Reeves (eds), *The Bottlenose Dolphin*, pp. 235–244. Academic Press, San Diego, CA.

Seaman, D.E. and Powell, R.A. (1996). An evaluation of the accuracy of kernel density estimators for home range analysis. *Ecology*, **77**, 2075–2085.

Seber, G.A.F. (1965). A note on the multiple recapture census. *Biometrika*, **52**, 249–259.

Seber, G.A.F. (1970). Estimating time-specific survival and reporting rates for adult birds from band returns. *Biometrika*, **57**, 313–318.

Seber, G.A.F. (1973). *The Estimation of Animal Abundance and Related Parameters*. Griffin, London.

Seddon, J.M., Parker, H.G., Ostrander, E.A., and Ellegren, H. (2005). SNPs in ecological and conservation studies: a test in the Scandinavian wolf population. *Molecular Ecology*, **14**, 503–511.

Selkoe, K.A. and Toonen, R.J. (2006). Microsatellites for ecologists: a practical guide to using and evaluating microsatellite markers. *Ecology Letters*, **9**, 615–629.

Sellas, A.B., Wells, R.S., and Rosel, P.E. (2005). Mitochondrial and nuclear DNA analyses reveal fine scale geographic structure in bottlenose dolphins (*Tursiops truncatus*) in the Gulf of Mexico. *Conservation Genetics*, **6**, 715–728.

Sergeant, D. (1991). *Harp Seals, Man and Ice.* Canadian Special Publication of Fisheries and Aquatic Sciences, Ottawa.

Shaffer, H.B. and Thomson, R.C. (2007). Delimiting species in recent radiations. *Systems Biology*, **56**, 896–906.

Shao, Q., Wilson, M.D., Romanek, C.S., and Hobson, K.A. (2004). Time series analyses of elemental and isotopic data from biomineralized whale tissue. *Environmental and Ecological Statistics*, **11**, 323–327.

Sharples, R.J. MacKenzie, M.L., and Hammond, P.S. (2009). Estimating the seasonal abundance of seal populations from counts and telemetry data. *Marine Ecology Progress Series*, **378**, 289–298.

Shaughnessy, P.D., Troy, S.K., Kirkwood, R., and Nicholls, A.O. (2000). Australian fur seals at Seal Rocks, Victoria: pup abundance by mark recapture estimation shows continued increase. *Wildlife Research*, **27**, 629–633.

Shelton, P.A., Warren, W.B., Stenson, G.B., and Lawson, J.W. (1997). Quantifying some of the major sources of uncertainty associated with estimates of harp seal prey consumption. Part II: Uncertainty in consumption estimates associated with population size, residency, energy requirement and diet. *Journal of Northwest Atlantic Fishery Science*, **22**, 303–315.

Sheppard, J.K., Preen, A.R., Marsh, H., Lawler, I.R., Whiting, S.D., and Jones, R.E. (2006). Movement heterogeneity of dugongs, *Dugong dugon* (Muller), over large spatial scales. *Journal of Experimental Marine Biology and Ecology*, **334**, 64–83.

Shrader-Frechette, K. (ed.). (1994). *Ethics of Scientific Research*. Rowman and Littlefield Publishers, Inc., Lanham, MD.

Sibert, J.R., Musyl, M.K., and Brill, R.W. (2003). Horizontal movements of bigeye tuna (*Thunnus obesus*) near Hawaii determined by Kalman filter analysis of archival tagging data. *Fisheries Oceanography*, **12**, 141–151.

Sibert, J.R., Lutcavage, M.E., Nielsen, A., Brill, R.W., and Wilson, S.G. (2006). Interannual variation in large-scale movement of Atlantic bluefin tuna (*Thunnus thynnus*) determined from pop-up satellite archival tags. *Canadian Journal of Fisheries and Aquatic Sciences*, **63**, 2154–2166.

Sibly, R.M., Nott, H.M.R., and Fletcher, D.J. (1990). Splitting behavior into bouts. *Animal Behaviour*, **39**, 63–69.

Sigler, M.F., Tollit, D.J., Vollenweider, J.J., Thedinga, J.F., Womble, J.N., Wong, M.A., Csepp, D.J., and Trites, A.W. (2009). Foraging response of a marine predator to prey availability. *Marine Ecology Progress Series*, **38**, 243–261.

Silverman, B.W. (1986). *Density Estimation for Statistics and Data Analysis*. Chapman and Hall, London.

Simpkins, M.A., Kelly, B.P., and Wartzok, D. (2001). Three-dimensional diving behaviors of ringed seals (Phoca hispida). *Marine Mammal Science*, **17**, 909–925.

Simpkins, M.A., Withrow, D.E., Cesarone, J.C., and Boveng, P.L. (2003). Stability in the proportion of harbor seals hauled out under locally ideal conditions. *Marine Mammal Science*, **19**, 791–805.

Sims, D.W., Righton, D., and Pitchford, J.W. (2007). Minimizing errors in identifying Levy flight behaviour of organisms. *Journal of Animal Ecology*, **76**, 222–229.

Siniff, D.B., DeMaster, D.P., Hofman, R.J., and Eberhardt, L.L. (1977). An analysis of the dynamics of a Weddell seal population. *Ecological Monographs*, **47**, 319–335.

Siniff, D.B., Garrott, R.A., Rotella, J.J., Fraser, W.R., and Ainley, D.G. (2008). Projecting the effects of environmental change on Antarctic seals. *Antarctic Science*, **20**, 425–435.

Sites, J., Jr and Marshall, J. (2003). Deliming species: a Renaissance issue in systematic biology. *Trends in Ecology and Evolution*, **18**, 462–470.

Sjöberg, M. and Ball, J.P. (2000). Grey seal, *Halichoerus grypus*, habitat selection around haulout sites in the Baltic Sea: bathymetry or central place foraging? *Canadian Journal of Zoology*, **78**, 1661–1667.

Skalski, J.R., Ryding, K.E., and Millspaugh, J.J. (2005). *Wildlife Demography: Analysis of Sex, Age and Count Data*. Elsevier Academic Press, Amsterdam.

Slater, P.J.B. and Lester, N.P. (1982). Minimizing errors in splitting behavior into bouts. *Behaviour*, **79**, 153–161.

Slip, D.J., Burton, H.R., and Gales, N.J. (1992). Determining blubber mass in the southern elephant seal *Mirounga leonina* by ultrasonic and isotopic techniques. *Australian Journal of Zoology*, **40**, 143–152.

Smale, M.J., Watson, G., and Hecht, T. (1995). Otolith atlas of Southern African marine fishes. *Ichthyological Monographs of the J.L.B. Smith Institute of Ichthyology 1*, Grahamstown, South Africa.

Small, R.J., Lowry, L.F., Hoef, J.M.V., Frost, K. J., DeLong, R.A., and Rehberg, M.J. (2005). Differential movements by harbor seal pups in contrasting Alaska environments. *Marine Mammal Science*, **21**, 671–694.

Smirnov, N. (1924). On the eastern harp seal *Phoca (pagophoca) groenlandica* var. *oceanica* Lepechin. *Tromsø Museum Archives*, **47**, 1–11.

Smith, C.T., Templin, W.D., Seeb, J. E., and Seeb, L.W. (2005). Assessing statistical power of SNPs for population structure and conservation studies. *North American Journal of Fisheries Management*, **25**, 944–953.

Smith, C.T., Antonovich, A., Templin, W.D., Elfstrom, C.M., Narum, S.R., and Seeb, L. W. (2007). Impacts of marker class bias relative to locus-specific variability on population inferences in chinook salmon: A comparison of single-nucleotide polymorphisms with short tandem repeats and allozymes. *Transactions of the American Fisheries Society*, **136**, 1674–1687.

Smith, D.R., Burnham, K.P., Kahn, D.M., He, X., Goshorn, C.J., Hattala, K.A., and Kahnle, A.W. (2000). Bias in survival estimates from tag-recovery models where catch-and-release is common, with an example from Atlantic striped bass (*Morone saxatilis*). *Canadian Journal of Fisheries and Aquatic Science*, **57**, 886–897.

Smith, S., Iverson, S.J., and Bowen, W.D. (1997). Fatty acid signatures and classification trees: new tools for investigating the foraging ecology of seals. *Canadian Journal of Fisheries and Aquatic Sciences*, **54**, 1377–1386.

Smith, T. and Polachek, T. (1981). Reexamination of the life table for Northern Fur Seals with implications about population regulatory mechanisms. In C.W. Fowler and T.D. Smith (eds), *Dynamics of Large Mammal Populations*, pp. 99–120. John Wiley and Sons, New York.

Smith, T.D., Allen, J., Clapham, P.J., Hammond, P.S., Katona, S., Larsen, F., Lien, J., Mattila, D., Palsboll, P.J., Sigurjonsson, J., Stevick, P.T.,

and Øien, N. (1999). An ocean-basin-wide mark–recapture study of the North Atlantic humpback whale (*Megaptera novaeangliae*). *Marine Mammal Science*, **15**, 1–32.

Smith, T.G. (1973). *Population Dynamics of the Ringed Seal in the Canadian Eastern Arctic.* Fisheries Research Board of Canada, Ottawa.

Smith, T.G., Beck, B., and Sleno, G.A. (1973). Capture, handling, and branding of ringed seals. *Journal of Wildlife Management*, **37**, 579–583.

Smout, S.C. (2006). Modelling the multispecies functional response of generalist marine predators. PhD thesis, University of St Andrews, Scotland.

Solberg, H. (1983). The theory of reference values. Part 5. Statistical treatment of collected reference values. Determination of reference limits. *Journal of Clinical Chemistry and Clinical Biochemistry*, **21**, 749–760.

Søreide, J.E., Tamelander, T., Hop, H., Hobson, K.A., and Johansen, I. (2007). Sample preparation effects on stable C and N isotope values: a comparison of methods in Arctic marine food web studies. *Marine Ecology Progress Series*, **86**, 33–44.

Sorenson, M.D. and Fleischer, R.C. (1996). Multiple independent transpositions of mitochondrial DNA control region sequences to the nucleus. *Proceedings of the National Academy of Science USA*, **93**, 15 239–15 243.

Soulé, M. (ed.). (1987). *Viable Populations for Conservation.* Cambridge University Press, New York.

Southwell, C., Paxton, C.G.M., Borchers, D., Boveng, P., and De la Mare, W. (2008). Taking account of dependent species in management of the Southern Ocean krill fishery: estimating crabeater seal abundance off east Antarctica. *Journal of Applied Ecology*, **45**, 622–631.

Soutullo, A., Cadahia, L., Urios, V., Ferrer, M., and Negro, J.J. (2007). Accuracy of lightweight satellite telemetry: a case study in the Iberian Peninsula. *Journal of Wildlife Management*, **71**, 1010–1015.

Sparling, C.E. and Fedak, M.A. (2004). Metabolic rates of captive grey seals during voluntary diving. *Journal of Experimental Biology*, **207**, 1615–1624.

Sparling, C.E., Speakman, J.R., and Fedak, M.A. (2006). Seasonal variation in the metabolic rate and body composition of female grey seals: Fat conservation prior to high-cost reproduction in a capital breeder? *Journal of Comparative Physiology B*, **176**, 505–512.

Sparling, C.E., Thompson, D., and Fedak, M. (2007). Estimating field metabolic rates of pinnipeds: DLW gets the seal of approval. *Comparative Biochemistry and Physiology*, **146**, S173.

Sparling, C.E., Thompson, D., Fedak, M.A., Gallon, S., and Speakman, J. (2008). Estimating field metabolic rates of pinnipeds: Doubly-labelled water gets the seal of approval. *Functional Ecology*, **22**, 245–254.

Speakman, J.R. (1995). Estimation of precision in the DLW studies using the two-point methodology. *Obesity Research*, **3**(Suppl. 1), 31–39.

Speakman, J.R. (1997). *Doubly Labelled Water: Theory and Practice*. Chapman and Hall, London.

Speakman, J.R. (2001). *Body Composition Analysis in Animals: A Handbook of Non-destructive Methods*. Cambridge University Press, Cambridge.

Spiegelhalter, D.J., Thomas, A., and Best, N.G. (1999). *WinBUGS Version 1.2 User Manual*. MRC Biostatistics Unit, Cambridge.

Spiegelhalter, D.J., Best, N.G., Carlin, B.P., and van der Linde, A. (2002). Bayesian measures of model complexity and fit. *Journal of the Royal Statistical Society, Series B*, **64**, 583–639.

Springer, A.M., Estes, J.A., van Vliet, G.B., Williams, T.M., Doak, D.F., Danner, E.M., Forney, K.A., and Pfister, B. (2003). Sequential megafaunal collapse in the North Pacific Ocean: An ongoing legacy of industrial whaling? *Proceedings of the National Academy of Sciences USA*, **100**, 12 223–12 228.

Springer, A.M., Estes, J.A., Van Vliet, G.B., Williams, T.M., Doak, D.F., Danner, E.M., and Pfister, B. (2008). Mammal-eating killer whales, industrial whaling, and the sequential megafaunal collapse in the North Pacific Ocean: a reply to the critics of Springer *et al.* 2003. *Marine Mammal Science*, **24**, 414–442.

Staniland, I.J. (2002). Investigating the biases in the use of hard prey remains to identify diet composition using Antarctic fur seals (*Arctocephalus Gazella*) in captive feeding trials. *Marine Mammal Science*, **18**, 223–243.

Staniland, I.J., Taylor, R.I., and Boyd, I.L. (2003). An enema method for obtaining fecal material from known individual seals on land. *Marine Mammal Science*, **19**, 363–370.

Stanton, T.K., Lavery, A.C., Johnson, M.P., Madsen, P.T., and Tyack, P.L. (2008). Classification of broadband echoes from prey of a foraging Blainville's beaked whale. *Journal of the Acoustical Society of America*, **123**, 1753–1762.

Staples, D.F., Taper, M.L., and Dennis, B. (2004). Estimating population trend and process variation for PVA in the presence of sampling error. *Ecology*, **85**, 923–929.

Starfield, A.M., Roth, J.D., and Ralls, K. (1995). 'Mobbing' in Hawaiian monk seals (*Monachus schauinslandi*): the value of simulation modeling in the absence of apparently crucial data. *Conservation Biology*, **9**, 166–174.

Stearns, S.C. (1992). *The Evolution of Life Histories*. Oxford University Press, Oxford.

Steinberg, P.D., Estes, J.A., and Winter, F.C. (1995). Evolutionary consequences of food chain length in kelp forest communities. *Proceeding of the National Academy of Science USA*, **92**, 8145–8148.

Stensland, E., Carlen, I., Sarnblad, A., Bignert, A., and Berggren, P. (2006). Population size, distribution, and behavior of indo-pacific bottlenose (*Tursiops aduncus*) and humpback (*Sousa chinensis*) dolphins off the south coast of Zanzibar. *Marine Mammal Science*, **22**, 667–682.

Stenson, G.B. and Hammill, M.O. (2004). Quantifying uncertainty in estimates of Atlantic cod (*Gadus morhua*) consumption by harp seals (*Phoca groenlandica*). DFO Canadian Science Advisory Research Document 2004/089.

Stenson, G.B., Hammill, M.O., and Lawson, J.W. (1997). Predation by harp seals in Atlantic Canada: preliminary consumption estimates for Arctic cod, capelin and Atlantic cod. *Journal of Northwest Atlantic Fisheries Science*, **22**, 137–154.

Stephens, D.W. and Krebbs, J.R. (1986). *Foraging Theory*. Princeton University Press, Princeton, NJ.

Stephenson, C.M., Matthiopoulos, J., and Harwood, J. (2007). Evaluating the influence of aggression and topography on the distribution of breeding grey seals (*Halichoerus grypus*). *Ecological Informatics*, **2**, 308–317.

Stevens, C.E. and Hume, I.D. (1995). *Comparative Physiology of the Vertebrate Digestive System*. Cambridge University Press, Cambridge.

Stevick, P.T., Palsbøll, P., Smith, T.D, Bravington, M.V., and Hammond, P.S. (2001). Errors in identification using natural markings: rates, sources and effects on capture–recapture estimates of abundance. *Canadian Journal of Fisheries and Aquatic Sciences*, **58**, 1861–1870.

Stevick, P.T., Allen, J., Clapham, P.J., Friday, N., Katona, S.K., Larsen, F., Lien, J., Mattila, D. K., Palsbøll, P.J., Sigurjónsson, J., Smith, T. D., Øien, N., and Hammond, P.S. (2003). North Atlantic humpback whale abundance and rate of increase four decades after protection from whaling. *Marine Ecology Progress Series*, **258**, 263–273.

Stewart, C. (2005). Inference on the diet of predators using fatty acid signatures. PhD thesis, Dalhousie University, Halifax, Nova Scotia.

Stewart, R.E.A., Stewart, B., Stirling, I., and Street, E. (1996). Counts of growth layer groups in cementum and dentine in ringed seals (*Phoca hispida*). *Marine Mammal Science*, **12**, 383–401.

Stewart, R.E.A., Campana, S.E., Jones, C.M., and Stewart, B.E. (2006). Bomb radiocarbon dating calibrates beluga (*Delphinapterus leucas*) age estimates. *Canadian Journal of Zoology*, **84**, 1840–1852.

Stirling, I. (1966). A technique for handling live seals. *Journal of Mammalogy*, **47**, 543–544

Stirling, I. (1983). The evolution of mating systems in pinnipeds. In J.F. Eisenberg and D.G. Kleiman (eds), *Advances in the Study of Mammalian Behavior*, pp. 489–527. American Society of Mammalogists, Shippensburg, PA.

Stirling, I. (2002). Polar bears and seals in the Eastern Beaufort Sea and Amundsen Gulf: A synthesis of population trends and ecological relationships over three decades. *Arctic*, **55**, 59–76.

Stirling, I. (2005). Reproductive rates of ringed seals and survival of pups in Northwestern Hudson Bay, Canada, 1991–2000. *Polar Biology*, **28**, 381–387.

Stirling, I. and Øritsland, N.A. (1995). Relationships between estimates of ringed seal (*Phoca*

hispida) and polar bear (*Ursus maritimus*) populations in the Canadian Arctic. *Canadian Journal of Fisheries and Aquatic Sciences*, **52**, 2594–2612.

Stirling, I. and Parkinson, C.L. (2006). Possible effects of climate warming on selected populations of polar bears (*Ursus maritimus*) in the Canadian Arctic. *Arctic*, **59**, 261–275.

Stirling, I., Lunn, N.J., and Iacozza, J. (1999). Long-term trends in the population ecology of polar bears in western Hudson Bay in relation to climatic change. *Arctic*, **52**, 294–306.

Stobo, W. and Horne, J.K. (1994). Tag loss in grey seals (*Halichoerus grypus*) and potential effects on population estimates. *Canadian Journal of Zoology*, **72**, 555–561.

Stone, B.J. (1984). An asymptotically optimal window selection rule for kernel density estimates. *The Annals of Statistics*, **4**, 1285–1297.

Stone, L.R. (1990). Diagnostic ultrasound in marine mammals. In L.A. Dierauf (ed.), *CRC Handbook of Marine Mammal Medicine: Health, Disease, and Rehabilitation*, 1st edn, pp. 235–264. CRC Press, Boca Raton, FL.

Strandberg, U., Kakela, A., Lyderson, C., Kovacs, K.M., Grahl-Nielson, O., Hyvarinien, H., and Kaela, R. (2008). Stratification composition and function of marine mammal blubber: the ecology of fatty acids in marine mammals. *Physiological and Biochemical Zoology*, **81**, 473–485.

Strindberg, S., Buckland, S.T., and Thomas, L. (2004). Design of distance sampling surveys and geographic information systems. In S.T. Buckland, D.R. Anderson, K.P. Burnham, J. L. Laake, D.L. Borchers, and L. Thomas (eds), *Advanced Distance Sampling: Estimating Abundance of Biological Populations*, pp. 190–228. Oxford University Press, Oxford.

Sumich, J.L. (1983). Swimming velocities, breathing patterns, and estimated costs of locomotion in migrating gray whales, *Eschrichtius robustus*. *Canadian Journal of Zoology*, **61**, 647–652.

Sumich, J.L. (2001). Direct and indirect measures of oxygen extraction, tidal lung volumes and respiratory rates in a rehabilitating gray whale calf. *Aquatic Mammals*, **27**, 279–283.

Sunderland, K.D. (1988). Quantitative methods for detecting invertebrate predation occurring in the field. *Annals of Applied Biology*, **112**, 201–224.

Sunnucks, P. (2000). Efficient genetic markers for population biology. *Trends in Ecology and Evolution*, **15**, 199–206.

Sutherland, W.J. (1998). The importance of behavioural studies in conservation biology. *Animal Behaviour*, **56**, 801–809.

Sydeman, W.J., Huber, H.R., Emslie, S.D., Ribic, C. A., and Nur, N. (1991). Age-specific weaning success of northern elephant seals in relation to previous breeding experience. *Ecology*, **72**, 2204–2217.

Syrjala, S.E. (1996). A statistical test for a difference between the spatial distributions of two populations. *Ecology*, **77**, 75–80.

Taylor, B.L. (2005). Identifying units to conserve. In J.E. Reynolds, W.F. Perrin, R.R. Reeves, S. Montgomery, and T.J. Ragen (eds), *Marine Mammal Research: Conservation Beyond Crisis*, pp. 149–162. The Johns Hopkins University Press, Baltimore, MD.

Taylor, B. and DeMaster, D. (1993). Implications of non-linear density dependence. *Marine Mammal Science*, **9**, 360–371.

Taylor, B.L. and Dizon, A.E. (1996). The need to estimate power to link genetics and demography for conservation. *Conservation Biology*, **10**, 661–664.

Taylor, B.L. and Dizon, A.E. (1999). First policy then science: Why a management unit based solely on genetic criteria can't work. *Molecular Ecology*, **8**, S11–S16.

Taylor, B.L. and Gerrodette, T. (1993). The uses of statistical power in conservation biology: the vaquita and northern spotted owl. *Conservation Biology*, **7**, 489–500.

Taylor, B.L., Chivers, S., and Dizon, A.E. (1997). Using statistical power to interpret genetic data to define management units for marine mammals. In A.E. Dizon, S. Chivers, and W.F. Perrin (eds), *Molecular Genetics of Marine Mammals*. Society for Marine Mammalogy, Lawrence, Kansas.

Taylor, B.L., Wade, P.R., DeMaster, D.P., and Barlow, J. (2000). Incorporating uncertainty into management models for marine mammals. *Conservation Biology*, **14**, 1243–1252.

Taylor, L.R., Woiwod, I.P., and Perry, J.N. (1978). The density-dependence of spatial behaviour and the rarity of randomness. *Journal of Animal Ecology*, **47**, 383–406.

Teo, S.L.H., Boustany, A., Blackwell, S., Walli, A., Weng, K.C., and Block, B.A. (2004). Validation of geolocation estimates based on light level and sea surface temperature from electronic tags. *Marine Ecology Progress Series*, **283**, 81–98.

TerBraak, C.J.F. (1986). Canonical correspondence analysis: a new eigenvector technique for multivariate direct gradient analysis. *Ecology*, **67**, 1167–1179.

Testa, J.W. (1994). Over-winter reproductive movements and diving behavior of female Weddell seals (Leptonychotes weddellii) in the south-western Ross Sea, Antarctica. *Canadian Journal of Zoology*, **72**, 1700–1710.

Testa, J.W. and Rothery, P. (1992). Effectiveness of various cattle ear tags as markers for Weddell seals. *Marine Mammal Science*, **8**, 344–353.

Thiemann, G.W., Budge, S.M., and Iverson, S.J. (2004). Determining blubber fatty acid composition: a comparison of *in situ* direct and traditional methods. *Marine Mammal Science*, **20**, 284–295.

Thiemann, G.W., Iverson, S.J., and Stirling, I. (2006). Seasonal sexual and anatomical variability in the adipose tissue composition of polar bears (*Ursus maritimus*). *Journal of Zoology, London*, **269**, 65–76.

Thiemann, G.W., Budge, S.M., Iverson, S.J., and Stirling, I. (2007). Unusual fatty acid biomarkers reveal age- and sex-specific foraging in polar bears (*Ursus maritimus*). *Canadian Journal of Zoology*, **85**, 505–517.

Thiemann, G.W., Iverson, S.J., and Stirling, I. (2008). Variation in blubber fatty acid composition among marine mammals in the Canadian Arctic. *Marine Mammal Science*, **24**, 91–111.

Thiemann, G.W., Iverson, S.J., and Stirling, I. (2009). Using fatty acids to study marine mammal foraging: the evidence from an extensive and growing literature. *Marine Mammal Science*, **25**, 243–249.

Thomas, L. and Harwood, J. (2008). *Estimating the Size of the UK Grey Seal Population between 1984 and 2007*. SCOS Briefing Paper 08/02. Sea Mammal Research Unit, Gatty Marine Laboratory, University of St Andrews, St Andrews, Fife, Scotland.

Thomas, L., Buckland, S.T., Newman, K.B., and Harwood, J. (2005). A unified framework for modeling wildlife population dynamics. *Australian and New Zealand Journal of Statistics*, **47**, 19–34.

Thomas, L., Laake, J.L., Rexstad, E., Strindberg, S., Marques, F.F.C., Buckland, S.T., Borchers, D.L., Anderson, D.R., Burnham, K.P., Burt, M.L., Hedley, S.L., Pollard, J.H., Bishop, J.R.B., and Marques, T.A. (2006). *Distance 6.0. Release 1.* Research Unit for Wildlife Population Assessment, University of St Andrews, Scotland. http://www.ruwpa.st-and.ac.uk/distance/

Thompson, D. and Fedak, M.A. (2001). How long should a dive last? A simple model of foraging decisions by breath-hold divers in a patchy environment. *Animal Behaviour*, **61**, 287–296.

Thompson, D., Moss, S.E.W., and Lovell, P. (2003). Foraging behaviour of South American fur seals Arctocephalus australis: extracting fine scale foraging behaviour from satellite tracks. *Marine Ecology Progress Series*, **260**, 285–296.

Thompson, P.M. and Wheeler, H. (2008). Photo-ID-based estimates of reproductive patterns in female harbor seals. *Marine Mammal Science*, **24**, 138–146.

Thompson, P.M., Tollit, D.J., Wood, D., Corpe, H.M., Hammond, P.S., and MacKay, A. (1997). Estimating harbour seal abundance and status in an estuarine habitat in N.E. Scotland. *Journal of Applied Ecology*, **34**, 43–52.

Thomson, J. (1986). *Rights, Restitution and Risk.* Harvard University Press, Cambridge, MA.

Thums, M., Bradshaw, C.J.A., and Hindell, M.A. (2008a). Tracking changes in relative body composition of southern elephant seals using swim speed data. *Marine Ecology Progress Series*, **370**, 249–261.

Thums, M., Bradshaw, C.J.A., and Hindell, M.A. (2008b). A validated approach for supervised dive classification in diving vertebrates. *Journal of Experimental Marine Biology and Ecology*, **363**, 75–83.

Tierney, M., Hindell, M., Lea, M.A., and Tollit, D. (2001). A comparison of techniques used to estimate body condition of southern elephant seals (Mirounga leonina). *Wildlife Research*, **28**, 581–588.

Tirasin, E.M. and Jorgensen, T. (1999). An evaluation of the precision of diet description. *Marine Ecology Progress Series*, **182**, 243–252.

Todd, S.K., Ostrom, P., Lein, J., and Abrajano, J. (1997). Use of biopsy samples of humpback whale (*Megaptera novaenliae*) skin for stable isotope (δ^{13}C) determination. *Journal of Northwest Atlantic Fisheries Science*, **22**, 71–76.

Toft, C.A. and Shea, P.J. (1983). Detecting community-wide patterns: estimating power strengthens statistical inference. *The American Naturalist*, **122**, 618–625.

Tolkamp, B.J. and Kyriazakis, I. (1999). To split behaviour into bouts, log-transform the intervals. *Animal Behaviour*, **57**, 807–817.

Tollit, D.J., Steward, M., Thompson, P.M., Pierce, G.J, Santos, M.B., and Hughes, S. (1997). Species and size differences in the digestion of otoliths and beaks; implications for estimates of pinniped diet composition. *Canadian Journal of Fisheries and Aquatic Science*, **54**, 105–119.

Tollit, D.J., Black, A.D., Thompson, P.M., Mackay, A., Corpe, H.M., Wilson, B., VanParijs, S.M., Grellier, K., and Parlane, S. (1998). Variations in harbour seal *Phoca vitulina* diet and dive-depths in relation to foraging habitat. *Journal of Zoology, London*, **244**, 209–222.

Tollit, D.J., Wong, M, Winship, A.J., Rosen, D.A.S., and Trites, A.W. (2003). Quantifying errors associated with using prey skeletal structures from fecal samples to determine the diet of the Steller sea lion (*Eumetopias jubatus*). *Marine Mammal Science*, **19**, 724–744.

Tollit, D.J., Heaslip, S.G., Zeppelin, T.K., Joy, R., Call, K.A., and Trites, A.W. (2004). Improving size estimates of walleye pollock and Atka mackerel consumed by Steller sea lions using digestion correction factors and bones recovered in scats. *Fishery Bulletin*, **102**, 498–508.

Tollit, D.J., Heaslip, S.G., Deagle, B., Iverson, S.J., Joy, R., Rosen, D.A.S., and Trites, A.W. (2006). Estimating diet composition in sea lions: which technique to choose? In A.W. Trites, S.K. Atkinson, D.P. DeMaster, L.W. Fritz, T. Gelatt, L.D. Rea, and K.M. Wynne, *Sea Lions of the World*, pp. 293–308. Alaska Sea Grant College Program, University of Alaska Fairbanks, AK.

Tollit, D.J., Heaslip, S.G., Barrick, R.L., and Trites, A.W. (2007). Impact of diet index selection and the digestion of prey hard parts on determining the diet of the Steller sea lion (*Eumetopias jubatus*). *Canadian Journal of Zoology*, **85**, 1–15.

Tollit, D.J., Schulze, A.D., Trites, A.W., Olesiuk, P., Crockford, S.J., Gelatt, T., Ream, R.R., and Miller, K.M. (2009). Development and application of DNA techniques for validating and improving pinniped diet estimates. *Ecological Applications*, **19**, 889–905.

Tomilin, A.G. (1945). The age of whales as determined from their baleen apparatus. *CR Academy of Science, USSR*, **49**, 460–463.

Tønnessen, J. and Johnsen, A. (1982). *The History of Modern Whaling*. C. Hurst and Co, London.

Tremblay, Y., Shaffer, S.A., Fowler, S.L., Kuhn, C.E., McDonald, B.I., Weise, M.J., Bost, C. A., Weimerskirch, H., Crocker, D.E., Goebel, M.E., and Costa, D.P. (2006). Interpolation of animal tracking data in a fluid environment. *Journal of Experimental Biology*, **209**, 128–140.

Tremblay, Y., Roberts, A.J., and Costa, D.P. (2007). Fractal landscape method: an alternative approach to measuring area-restricted searching behavior. *Journal of Experimental Biology*, **210**, 935–945.

Trites, A.W. and Bigg, M.A. (1992). Changes in body growth of northern fur seals from 1958 to 1974: density effects or changes in the ecosystem? *Fisheries Oceanography*, **1**, 127–136.

Trites, A.W. and Joy, R. (2005). Dietary analysis from fecal samples: how many scats are enough? *Journal of Mammology*, **86**, 704–712.

Trites, A.W. and Pauly, D. (1998). Estimating mean body masses of marine mammals from maximum body lengths. *Canadian Journal of Zoology*, **76**, 886–896.

Troy, S., Middleton, D., and Phelan, J. (1997). On capture, anaesthesia and branding of adult male New Zealand fur seals *Arctocephalus forsteri*. In M. Hindell and C. Kemper (eds), *Marine Mammal Research in the Southern Hemisphere*, Volume 1—*Status, Ecology and Medicine*, pp. 179–183. Surrey Beatty and Sons, Chipping Norton, NSW, Australia.

Trumble, S.J. and Castellini, M.A. (2005). Diet mixing in an aquatic carnivore, the harbour seal. *Canadian Journal of Zoology*, **83**, 851–859.

Trzcinski, M.K., Mohn, R., and Bowen, W.D. (2006). Continued decline of an Atlantic cod population: How important is gray seal predation? *Ecological Applications*, **16**, 2276–2292.

Tucker, S., Bowen, W.D., and Iverson, S.J. (2008). Convergence of diet estimates derived from fatty acids and stable isotopes within individual grey seals. *Marine Ecology Progress Series*, **354**, 267–276.

Tucker, S., Bowen, W.D., Iverson, S.J., Blanchard, W., and Stenson, G.B. (2009). Sources of variation in diets of harp and hooded seals estimated from quantitative fatty acid signature analysis (QFASA). *Marine Ecology Progress Series*, **384**, 287–302.

Turakulov, R. and Easteal, S. (2003). Number of SNPs loci needed to detect population structure. *Human Heredity*, **55**, 37–45.

Turchin, P. (1996). Fractal analyses of animal movement: A critique. *Ecology*, **77**, 2086–2090.

Turchin, P. (1998). *Quantitative Analysis of Movement*. Sinauer Associates Inc, Sunderland, MA.

Turvey, S.T., Pitman, R.L., Taylor, B.L., Barlow, J., Akamatsu, T., Barrett, L.A., Zhao, X., Reeves, R.R., Stewart, B.S., Wang, K., Wei, Z., Zhang, X., Pusser, L.T., Richlen, M., Brandon, J.R., and Wang, D. (2007). First human-caused extinction of a cetacean species? *Biology Letters*, **3**, 537–540.

Tuset, V.M., Lombarte, A., and Assis, C.A. (2008). Otolith atlas for the western Mediterranean, north and central eastern Atlantic. *Scientia Marina*, **72**, 1–203.

Twiss, J.R. Jr and Reeves, R.R. (eds), (1999). *Conservation and Management of Marine Mammals*. Smithsonian Institution Press, Washington DC.

Tyack, P. (1999). Communication and cognition. In J.E. Reynolds and S.A. Rommel (eds), *Biology of Marine Mammals*, pp. 287–323. Smithsonian Institution Press, Washington DC.

Tyack, P.L. and Recchia, C.A. (1991). A datalogger to identify vocalizing dolphins. *Journal of the Acoustical Society of America*, **90**, 1668–1671.

Tyack, P.L., Johnson, M.P., Zimmer, W.M.X., de Soto, N.A., and Madsen, P.T. (2006). Acoustic behavior of beaked whales, with implications for acoustic monitoring. *Oceans 2006*, **1–4**, 509–514.

USFWS (US Fish and Wildlife Service) (1996). Policy regarding the recognition of distinct vertebrate population segments under the

Endangered Species Act. *The Federal Register*, **61** (February 7), 4722.

Valone, T.J. (2006). Are animals capable of Bayesian updating? An empirical review. *Oikos*, **112**, 252–259.

Van den Hoff, J., Burton, H., and Davies, R. (2003). Diet of male southern elephant seals (*Mirounga leonina* L) hauled out at Vincennes Bay East Antarctica. *Polar Biology*, **26**, 27–31.

Van den Hoff, J., Sumner, M.D., Field, I.C., Bradshaw, J.A., Burton, H.R., and McMahon, C.R. (2004). Temporal changes in the quality of hot-iron brands on elephant seal (*Mirounga leonina*) pups. *Wildlife Research*, **31**, 619–629.

Vanderklift, M.A. and Ponsard, S. (2003). Sources of variation in consumer-diet $\delta^{15}N$ enrichment: a meta-analysis. *Oecologia*, **136**, 169–182.

Vanderzwaag, D.L. and Hutchings, J.A. (2005). Canada's marine species at risk: Science and law at the helm, but a sea of uncertainties. *Ocean Development and International Law*, **36**, 219–259.

van Hooff, J.A.R.A.M. and Wensing, J.A.B. (1987). Dominance and its behavioral measures in a captive wolf pack. In H. Frank (ed.), *Man and Wolf*, pp. 219–252. Junk, Dordrecht.

van Oosterhout, C., Hutchinson, W.F., Wills, D.P.M., and Shipley, P. (2004). Micro-checker: software for identifying and correcting genotyping errors in microsatellite data. *Molecular Ecology Notes*, **4**, 535–538.

Van Parijs, S.M. and Clark, C.W. (2006). Long-term mating tactics in an aquatic-mating pinniped, the bearded seal, *Erignathus barbatus*. *Animal Behaviour*, **72**, 1269–1277.

Van Parijs, S.M., Smith, J., and Corkeron, P.J. (2002). Using calls to estimate the abundance of inshore dolphins: a case study with Pacific humpback dolphins *Sousa chinensis*. *Journal of Applied Ecology*, **39**, 853–864.

van Utrecht, W.L. (1965). On the growth of the baleen plate of the fine whale and the blue whale. *Bijdragen tot de Dierkunde*, **35**, 3–38.

Vayssiéres, M.P., Plant, R.E., and Allen-Diaz, B.H. (2000). Classification trees: An alternative non-parametric approach for predicting species distributions. *Journal of Vegetation Science*, **11**, 679–694.

Venables, W.N. and Ripley, B.D. (2003). *Modern Applied Statistics with S*. Springer, Berlin.

Vincent, C., McConnell, B.J., Ridoux, V., and Fedak, M.A. (2002). Assessment of Argos location accuracy from satellite tags deployed on captive gray seals. *Marine Mammal Science*, **18**, 156–166.

Viswanathan, G.M., Buldyrev, S.V., Havlin, S., da Luz, H.G.E., Raposo, E.P., and Stanley, H.E. (1999). Optimizing the success of random searches. *Nature*, **401**, 911–914.

Vos, J.G., Bossart, G.D., Fournier, M., and O'Shea, T.J. (eds), (2003). *Toxicology of Marine Mammals (New Perspectives: Toxicology and Environment)*. Taylor Francis, London.

Wade, P.R. (1998). Calculating limits to the allowable human-caused mortality of cetaceans and pinnipeds. *Marine Mammal Science*, **14**, 1–37.

Wade, P.R. (2001). The conservation of exploited species in an uncertain world: novel methods and the failure of traditional techniques. In J.D. Reynolds, G.M. Mace, K.H. Redford, and J.G. Robinson, *Conservation of Exploited Species*, pp. 110–144. Conservation Biology Series No. 6. Cambridge University Press. Cambridge.

Wahlberg, M. (2002). The acoustic behaviour of diving sperm whales observed with a hydrophone array. *Journal of Experimental Marine Biology and Ecology*, **281**, 53–62.

Walker, P.A. and Cocks, K.D. (1991). HABITAT: a procedure for modelling a disjoint environmental envelope for a plant or animal species. *Global Ecology and Biogeography Letters*, **1**, 108–118.

Walker, T.R, Boyd, I.L., McCafferty, D.J., Huin, N., Taylor, R.I., and Reid, K. (1998). Seasonal occurrence of leopard seals (*Hydrurga leptonyx*) at Bird Island, South Georgia. *Antarctic Science*, **10**, 75–82.

Walsberg, G.E. and Hoffman, T.C.M. (2005). Direct calorimetry reveals large errors in respirometric estimates of energy expenditure. *Journal of Experimental Biology*, **208**, 1035–1043.

Walton, M.J., Henderson, R.J., and Pomeroy, P.P. (2000). Use of blubber fatty acid profiles to distinguish dietary differences between grey seals *Halichoerus grypus* from two UK breeding colonies. *Marine Ecology Progress Series*, **193**, 201–208.

Wand, M.P. and Jones, M.C. (1995). *Kernel Smoothing. Monographs on Statistics and Applied Probability*. Chapman and Hall, London.

Wang, D., Zhang, X., Wang, K., Wei, Z., Würsig, B., Braulik, G.T., and Ellis, S. (2006). Conservation of the baiji: no simple solution. *Conservation Biology*, **20**, 623–625.

Wang, J. (2004). Estimating pairwise relatedness from dominant genetic markers. *Molecular Ecology*, **13**, 3169–3178.

Wang, J.Y., Ghou, L.S., and White, B.N. (2000). Differences in the external morphology of two sympatric species of bottlenose dolphins (Genus *Tursiops*) in the waters of China. *Journal of Mammalogy*, **81**, 1157–1165.

Wang Ding, L., Renjun, Z., Xianfeng, Y., Jian, W., Zhou, K., Qingzhong, Z., and Xiaoqiang, W. (2000). Status and conservation of the finless porpoise. In R.R. Reeves, B.D. Smith, and T. Kasuya (eds), *Biology and Conservation of Freshwater Cetaceans in Asia*. Occasional Paper of the IUCN Species Survival Commission, Number 23, pp. 81–85. IUCN, Gland, Switzerland.

Waples, R.S. and Gaggiotti, O. (2006). What is a population? An empirical evaluation of some genetic methods for identifying the number of gene pools and their degree of connectivity. *Molecular Ecology*, **15**, 1419–1439.

Ward, J., Morrissey, R., Moretti, D., DiMarzio, N., Jarvis, S., Johnson, M., Tyack, P., and White, C. (2008). Passive acoustic detection and localization of *Mesoplodon densirostris* (Blainville's beaked whale) vocalizations using distributed, bottom-mounted hydrophones in conjunction with a Digital Tag (DTag) recording. *Canadian Acoustics*, **36**, 60–66.

Ware, C., Arsenault, R., and Plumlee, M. (2006). Visualizing the underwater behavior of humpback whales. *IEEE Computer Graphics and Applications*, **26**, 14–18.

Wartzok, D., Sayegh, S., Stone, H., Barchak, J., and Barnes, W. (1992). Acoustic tracking system for monitoring under-ice movements of polar seals. *Journal of the Acoustical Society of America*, **92**, 682–687.

Watanabe, Y., Baranov, E.A., Sato, K., Naito, Y., and Miyazaki, N. (2006a). Body density affects stroke patterns in Baikal seals. *Journal of Experimental Biology*, **209**, 3269–3280.

Watanabe, Y., Bornemann, H., Liebsch, N., Plotz, J., Sato, K., Naito, Y., and Miyazaki, N. (2006b). Seal-mounted cameras detect invertebrate fauna on the underside of an Antarctic ice shelf. *Marine Ecology Progress Series*, **309**, 297–300.

Watkins, W. (1980). Acoustics and the behavior of sperm whales. In R. Busnel and J.F. Fish (eds), *Animal Sonar Systems*, pp. 291–297, Plenum, New York.

Watkins, W.A. and Daher, M.A. (1984). Variable spectra and nondirectional characteristics of clicks from near-surface sperm whales *Physeter catodon*. In J.A. Thomas, C.F. Moss, and M. Vater (eds), *Echolocation in Bats and Dolphins*, pp. 410–413. The University of Chicago Press, Chicago, IL.

Watt, J., Pierce, G.J., and Boyle, P.R, (1997). *A Guide to the Identification of North Sea Fish using Premaxillae and Vertebrae*. Co-operative Research Report No 220. International Council for the Exploration of the Sea, Copenhagen, Denmark.

Watwood, S.L., Miller, P.J.O., Johnson, M., Madsen, P.T., and Tyack, P.L. (2006). Deep-diving foraging behaviour of sperm whales (*Physeter macrocephalus*). *Journal of Animal Ecology*, **75**, 814–825.

Waugh, J. (2007). DNA barcoding in animal species: progress, potential and pitfalls. *Bioessays*, **29**, 188–197.

Webb, P.M., Crocker, D.E., Blackwell, S.B., Costa, D.P., and Le Boeuf, B.J. (1998). Effects of buoyancy on the diving behavior of northern elephant seals. *Journal of Experimental Biology*, **201**, 2349–2358.

Weimerskirch, H. (1992). Reproductive effort in long-lived birds: age-specific patterns of condition, reproduction and survival in the wandering albatross. *Oikos*, **64**, 464–473.

Weir, C.R., Stockin, K.A., and Pierce, G.J. (2007). Spatial and temporal trends in the distribution of harbour porpoises, white-beaked dolphins and minke whales off Aberdeenshire (UK), north-western North Sea. *Journal of the Marine Biology Association of the UK*, **87**, 327–338.

Welch, D.W., Boehlert, G.W., and Ward, B.R. (2002). POST—the Pacific Ocean salmon tracking project. *Oceanologica Acta*, **25**, 243–253.

Wells, D.E. and De Boer, J. (1994). The 1993 QUASIMEME laboratory performance study—chlorobiphenyls in fish oil and standard solutions. *Marine Pollution Bulletin*, **29**, 174–184.

Wells, R.S. (1991). The role of long-term study in understanding the social structure of a bottlenose dolphin community. In K. Pryor and K.S. Norris (eds), *Dolphin Societies: Discoveries and Puzzles*, pp. 199–225. University of California Press, Berkeley, CA.

Wells, R.S. (2002). Identification methods. In W. F. Perrin, B. Würsig, and J.G.M Thewissen (eds), *Encyclopedia of Marine Mammals*, pp. 601–608. Academic Press, San Diego, CA.

Wells, R.S. (2003). Dolphin social complexity: Lessons from long-term study and life history. In F.B.M. de Waal and P.L. Tyack (eds), *Animal Social Complexity: Intelligence, Culture, and Individualized Societies*, pp. 32–56. Harvard University Press, Cambridge, MA.

Wells, R.S. and Scott, M.D. (1990). Estimating bottlenose dolphin population parameters from individual identification and capture–release techniques. *Report of the International Whaling Commission*, Special issue **12**, 407–415.

Wells, R.S., Scott, M.D., and Irvine, A.B. (1987). The social structure of free-ranging bottlenose dolphins. *Current Mammalogy*, **1**, 247–305.

Wells, R.S., Rhinehart, H.L., Hansen, L.J., Sweeney, J.C., Townsend, F.I., Stone, R., Casper, D., Scott, M.D., Hohn, A.A., and Rowles, T.K. (2004). Bottlenose dolphins as marine ecosystem sentinels: Developing a health monitoring system. *EcoHealth*, **1**, 246–254.

Wells, R.S., Tornero, V., Borrell, A., Aguilar, A., Rowles, T.K., Rhinehart, H.L., Hofmann, S., Jarman, W.M., Hohn, A.A., and Sweeney, J. C. (2005). Integrating potential life-history and reproductive success data to examine relationships with organochlorine compounds for bottlenose dolphins (*Tursiops truncatus*) in Sarasota Bay, Florida. *Science of the Total Environment*, **349**, 106–119.

Wells, R.S., Allen, J.B., Hofmann, S., Bassos-Hull, K., Fauquier, D.A., Barros, N. B., DeLynn, R.E., Hurst, G., Socha, V., and Scott, M.D. (2008). Consequences of injuries on survival and reproduction of common bottlenose dolphins (*Tursiops truncatus*) along the west coast of Florida. *Marine Mammal Science*, **24**, 774–794.

Welsh, A.H., Cunningham, R.B., Donnelly, C.F., and Lindermayer, D.B. (1996). Modelling the abundance of rare species: Statistical models for counts with extra zeros. *Ecological Modelling*, **88**, 297–308.

Westgate, A.J. and Read, A.J. (2006). Life history of short-beaked common dolphins (*Delphinus delphis*) from the western North Atlantic. *Marine Biology*, **150**, 1011–1024.

Westgate, A.J., McLellan, W.A., Wells, R.S., Scott, M.D., Meagher, E.M., and Pabst, D. A. (2007). A new device to remotely measure heat flux and skin temperature from free-swimming dolphins. *Journal of Experimental Marine Biology and Ecology*, **346**, 45–59.

Wey, T., Blumstein, D.T., Shen, W., and Jordán, F. (2008). Social network analysis of animal behaviour: a promising tool for the study of sociality. *Animal Behaviour*, **75**, 333–344.

Whelan, A. and Regan, F. (2006). Antifouling strategies for marine and riverine sensors. *Journal of Environmental Monitoring*, **8**, 880–886.

White, G.C. and Garrott, R.A. (1990). *Analysis of Wildlife Radio-tracking Data*. Academic Press, London.

White, M.J. Jr, Jennings, J.G., Gandy, W.F., and Cornell, L.H. (1981). An evaluation of tagging, marking, and tattooing techniques for small delphinids. *US Department of Commerce, NOAA Technical Memorandum* NMFS-SWFC-16.

White, T.E. (1953). Observations of the butchering techniques of some aboriginal peoples. *American Antiquity*, **19**, 160–164.

Whitehead, H. (1990a). Computer assisted individual identification of sperm whale flukes. In P.S. Hammond, S.A. Mizroch, and G.P. Donovan (eds), *Individual Recognition of Cetaceans: Use of Photo-Identification and Other Techniques to Estimate Population Parameters*, pp. 71–77. Report of the International Whaling Commission, Special Issue 12, Cambridge, UK.

Whitehead, H. (1990b). Mark–recapture estimates with emigration and reimmigration. *Biometrics*, **46**, 473–479.

Whitehead, H. (1996). Variation in the feeding success of sperm whales: temporal scale, spatial scale and relationship to migrations. *Journal of Animal Ecology*, **65**, 429–438.

Whitehead, H. (1997). Analyzing animal social structure. *Animal Behaviour*, **53**, 1053–1067.

Whitehead, H. (2000). Density-dependent habitat selection and the modeling of sperm whale (*Physeter macrocephalus*) exploitation. *Canadian Journal of Fisheries and Aquatic Science*, **57**, 223–230.

Whitehead, H. (2003). *Sperm Whales: Social Evolution in the Ocean*. University of Chicago Press, Chicago, IL.

Whitehead, H. (2007). Selection of models of lagged identification rates and lagged association rates using AIC and QAIC. *Communications in Statistics—Simulation and Computation*, **36**, 1233–1246.

Whitehead, H. (2008a). Precision and power in the analysis of social structure using associations. *Animal Behaviour*, **75**, 1093–1099.

Whitehead, H. (2008b). *Analyzing Animal Societies: Quantitative Methods for Vertebrate Social Analysis*. University of Chicago Press, Chicago, IL.

Whitehead, H. and Dufault, S. (1999). Techniques for analyzing vertebrate social structure using identified individuals: review and recommendations. *Advances in the Study of Behavior*, **28**, 33–74.

Whitehead, H., Christal, J., and Dufault, S. (1997a). Past and distant whaling and the rapid decline of sperm whales off the Galapagos Islands. *Conservation Biology*, **11**, 1387–1396.

Whitehead, H., Gowans, S., Faucher, A., and McCarrey, S.W. (1997b). Population analysis of northern bottlenose whales in the Gully, Nova Scotia. *Marine Mammal Science*, **13**, 173–185.

Whitehead, H., Christal, J., and Tyack, P.L. (2000). Studying cetacean social structure in space and time: innovative techniques. In J. Mann, R.C. Connor, P.L. Tyack, and H. Whitehead (eds), *Ceatacean Societies*, pp. 65–87. University of Chicago Press, Chicago, IL.

WHO (1992). *International Statistical Classification of Diseases and Related Health Problems* (10th Revision). World Health Organization, Geneva.

Wikelski, M., Kays, R.W., Kasdin, N.J., Thorup, K., Smith, J.A., and Swenson, G.W. (2007). Going wild: what a global small-animal tracking system could do for experimental biologists. *Journal of Experimental Biology*, **210**, 181–186.

Wilkinson, I.S., Duignan, P.J., and Childerhouse, S.C. (2001). An evaluation of hot iron branding as a permanent marking method in New Zealand sea lions (*Phocarctos hookeri*). Abstracts of the *14th Biennial Conference on the Biology of Marine Mammals*, November 28–December 3, 2001, Vancouver, BC.

Williams, B.K., Nichols, J.D., and Conroy, M.J. (2001). *Analysis and Management of Animal Populations*. Academic Press, San Diego, CA.

Williams, R., Hedley, S.L., and Hammond, P.S. (2006a). Modeling distribution and abundance of antarctic baleen whales using ships of opportunity. *Ecology and Society*, **11**, 1. [online: URL: http://www.ecologyandsociety.org/vol11/iss1/art1/]

Williams, R., Lusseau, D., and Hammond, P. (2006b). Estimating relative energetic costs of human disturbance to killer whales (*Orcinus orca*). *Biological Conservation*, **133**, 301–311.

Williams, R., Leaper, R., Zerbini, A.N., and Hammond, P.S. (2007). Methods for investigating measurement error in cetacean line-transect surveys. *Journal of the Marine Biological Association of the United Kingdom*, **87**, 313–320.

Williams, T.M. (1989). Swimming by sea otters: adaptations for low energetic cost locomotion. *Journal of Comparative Physiology—A*, **164**, 815–824.

Williams, T.M., Kooyman, G.L., and Croll, D.A. (1991). The effect of submergence on heart rate and oxygen consumption of swimming seals and sea lions. *Journal of Comparative Physiology—B*, **160**, 637–644.

Williams, T.M., Friedl, W.A., and Haun, J.E. (1993). The physiology of bottlenose dolphins (*Tursiops truncatus*): Heart rate, metabolic rate and plasma lactate concentration during exercise. *Journal of Experimental Biology*, **179**, 31–46.

Williams, T.M., Davis, R.W., Fuiman, L.A., Francis, J., LeBoeuf, B.J., Horning, M., Calambokidis, J., and Croll, D.A. (2000). Sink or swim: Strategies for cost-efficient diving by marine mammals. *Science*, **288**, 133–136.

Williams, T.M., Fuiman, L.A., Horning, M., and Davis, R.W. (2004a). The cost of foraging by a marine predator, the Weddell seal *Leptonychotes weddellii*: pricing by the stroke. *Journal of Experimental Biology*, **207**, 973–982.

Williams, T.M., Estes, J.A., Doak, D.F., and Springer, A.M. (2004b). Killer appetites:

assessing the role of predators in ecological communities. *Ecology*, **85**, 3373–3384.

Williams, T.M., Marshal, G., Frank, L., and Davis, R.W. (2008). Living on fast food: Assessing the energetics and survival of big hungry carnivores hunting on land and at sea. In *Molecules to Migration: The Pressures of Life—Proceedings of the 4th CPB meeting in Africa*, Mara 2008, pp. 409–416. Medimond, Italy.

Wilson, B., Hammond, P.S., and Thompson, P.M. (1999a). Estimating size and assessing trends in a coastal bottlenose dolphin population. *Ecological Applications*, **9**, 288–300.

Wilson, B., Arnold, H., Bearzi, G., Fortuna, C. M., Gaspar, R., Ingram, S., Liret, C., Pribanic, S., Read, A.J., Ridoux, V., Schneider, K., Urian, K.W., Wells, R.S., Wood, C., Thompson, P.M., and Hammond, P.S. (1999b). Epidermal diseases in bottlenose dolphins: impacts of natural and anthropogenic factors. *Proceedings of the Royal Society of London, B*, **266**, 1077–1083.

Wilson, B., Grellier, K., Hammond, P.S., Brown, G., and Thompson, P.M. (2000). Changing occurrence of epidermal lesions in wild bottlenose dolphins. *Marine Ecology Progress Series*, **205**, 283–290.

Wilson, E.B. Jr (1991). *An Introduction to Scientific Research*. Courier Dover Publications, N. Chelmsford, MA.

Wilson, E.O. (1975). *Sociobiology: The New Synthesis*. Belknap Press, Cambridge, MA.

Wilson, R.P. and McMahon, C.R. (2006). Measuring devices on wild animals: what constitutes acceptable practice? *Frontiers in Ecology and the Environment*, **4**, 147–154.

Wilson, R.P., Putz, K., Charrassin, J.B., and Lage, J. (1995). Artifacts arising from sampling interval in dive depth studies of marine endotherms. *Polar Biology*, **15**, 575–581.

Wilson, R.P., White, C.R., Quintana, F., Halsey, J.G., Liebsh, N., Martin, G.R., and Butler, P. J. (2006). Moving towards acceleration for estimates of activity-specific metabolic rate in free-living animals: the case of the cormorant. *Journal of Animal Ecology*, **75**, 1018–1090.

Wilson, R.P., Liebsch, N., Davies, I.M., Quintana, F., Weimerskirch, H., Storch, S., Lucke, K., Siebert, U., Zankl, S., Mueller, G.,

Zimmer, I., Scolaro, A., Campagna, C., Plotz, J., Bornemann, H., Teilmann, J., and McMahon, C.R. (2007). All at sea with animal tracks; methodological and analytical solutions for the resolution of movement. *Deep-Sea Research Part II—Topical Studies in Oceanography*, **54**, 193–210.

Winn, H.E., Goodyear, J.D., Kenney, R.D., and Petricig, R.O. (1995). Dive patterns of tagged right whales in the Great South Channel. *Continental Shelf Research*, **15**, 593–611.

Winship, A.J. and Trites, A.W. (2003). Prey consumption of Steller sea lions (*Eumetopias jubatus*) off Alaska: how much prey do they require? *Fishery Bulletin*, **101**, 147–167.

Winship, A.J., Trites, A.W., and Calkins, D.G. (2001). Growth in body size of the Steller sea lion. *Journal of Mammalogy*, **82**, 500–519.

Wood, S.N. (2006). *Generalised Additive Models: An Introduction with R*. Chapman and Hall. Boca Raton, FL.

Worthington, J.W., Allen, P.J., Pomeroy, P.P., Twiss, S.D., and Amos, W. (1999). Where have all the fathers gone? An extensive microsatellite analysis of paternity in the grey seal (*Halichoerus grypus*). *Molecular Ecology*, **8**, 1417–1429.

Worthy, G.A.J. and Lavigne, D.M. (1983). Energetics of fasting and subsequent growth in weaned harp seal pups. *Canadian Journal of Zoology*, **61**, 447–456.

Worthy, G.A.J., Innes, S., Braune, B.M., and Stewart, R.E.A. (1987). Rapid acclimation of cetaceans to an open-system respirometer. In A.D. Huntley, D.P. Costa, G.A.J. Worthy, and M.A. Castellini (eds), *Approaches to Marine Mammal Energetics*, pp. 115–126. Allen Press, Lawrence, Kansas.

Worthy, G.A.J., Morris, P.A., Costa, D.P., and Le Boeuf, B.J. (1992). Moult energetics of the northern elephant seal (*Mirounga angustirostris*). *Journal of Zoology, London*, **227**, 257–265.

Wray, S., Cresswell, W.J., and Rogers, D. (1992a). Dirichlet tessellations: a new, nonparametric approach to home range analysis. In I.M. Priede and S.M. Swift (eds), *Wildlife Telemetry: Remote Monitoring and Tracking of*

Animals, pp. 247–255. Ellis Horwood, New York.

Wray, S., Cresswell, W.J., White, P.C.L., and Harris, S. (1992b). What, if anything, is a core area? An analysis of the problems of describing internal range configurations. In I.M. Priede and S.M. Swift (eds), *Wildlife Telemetry: Remote Monitoring and Tracking of Animals*, pp. 521–537. Ellis Horwood, New York.

Wright, B.E., Riemer, S.D., Brown, R.F., Ougzin, A.M., and Bucklin, K.A. (2007). Assessment of harbor seal predation on adult salmonids in a Pacific Northwest estuary, *Ecological Applications*, **17**, 338–351.

Würsig, B. (1982). Radio-tracking dusky porpoises in the South Atlantic. *FAO Fisheries*, Series 5, **4**, 145–160.

Würsig, B. and Jefferson, T.A. (1990). Methods of photo-identification for small cetaceans. In P.S. Hammond, S.A. Mizroch, and G.P. Donovan (eds), *Individual Recognition of Cetaceans: Use of Photo-Identification and Other Techniques to Estimate Population Parameters*, pp. 43–55. Report of the International Whaling Commission, Special Issue 12, Cambridge, UK.

Würsig, B. and Würsig, M. (1977). The photographic determination of group size, composition, and stability of coastal porpoises (*Tursiops truncatus*). *Science*, **198**, 755–756.

Yang, J., Chen, L., Sun, L., Yu, J., and Jin, Q. (2008). VFDB 2008 release: an enhanced web-based resource for comparative pathogenomics. *Nucleic Acids Research*, **36**, D539–542.

Ydenberg, R.C. and Clark, C.W. (1989). Aerobiosis and anaerobiosis during diving by Western Grebes—an optimal foraging approach. *Journal of Theoretical Biology*, **139**, 437–447.

Yeates, L.C., Williams, T.M., and Fink, T.L. (2007). Diving and foraging energetics of the smallest marine mammal, the sea otter (*Enhydra lutris*). *Journal of Experimental Biology*, **210**, 1960–1970.

Yen, P.P.W., Sydeman, W.J., and Hyrenbach, K.D. (2004). Marine bird and cetacean associations with bathymetric habitats and shallow water topographies: Implications for trophic transfer and conservation. *Journal of Marine Systems*, **50**, 79–99.

Yodzis, P. (2001). Must top predators be culled for the sake of fisheries? *Trends in Ecology and Evolution*, **16**, 78–84.

Yonezaki, S., Kiyota, M., Koido, T., and Takemura, A. (2005). Effects of squid beak size on the route of egestion in northern fur seals (*Callorhinus ursinus*): Results from captive feeding trials. *Marine Mammal Science*, **21**, 567–573.

York, A.E. and Kozloff, P. (1987). On estimating the number of fur seal pups born on St. Paul Island, 1980–86. *Fishery Bulletin, US*, **85**, 367–375.

York, A.E., Thomason, J.R., Sinclair, E.H., and Hobson, K.A. (2008). Stable-carbon and nitrogen isotope values in teeth of Steller sea lions: Age of weaning and the impact of the 1975–76 regime shift in the North Pacific Ocean. *Canadian Journal of Zoology*, **86**, 33–44.

Young, N.M. (2001). The conservation of marine mammals using a multi-party approach: An evaluation of the take reduction team process. *Ocean and Coastal Law Journal*, **6**, 293–346.

Zar, J.H. (1999). *Biostatistical Analysis*. Prentice Hall, Upper Saddle River, NJ.

Zeppelin, T.K., Tollit, D.J., Call, K.A., Orchard, T.J., and Gudmundson, C.J. (2004). Sizes of walleye pollock and Atka mackerel consumed by the western stock of Steller sea lions in Alaska from 1998–2000. *Fishery Bulletin*, **102**, 509–521.

Zimmer, W.M.X., Johnson, M.P., Madsen, P.T., and Tyack, P.L. (2005a). Echolocation clicks of free-ranging Cuvier's beaked whales (*Ziphius cavirostris*). *Journal of the Acoustical Society of America*, **117**, 3919–3927.

Zimmer, W.M.X., Tyack, P.L., Johnson, M.P., and Madsen, P.T. (2005b). Three-dimensional beam pattern of regular sperm whale clicks confirms bent-horn hypothesis. *Journal of the Acoustical Society of America*, **117**, 1473–1485.

Zuoshuang Xiang, Yuying Tian, and Yongqun He. (2007). PHIDIAS: a pathogen–host interaction data integration and analysis system. *Genome Biology*, **8**, R150. http://www.phidias.us/ [Accessed 3rd March, 2010.]

Zurlo, J., Rudacille, D., and Goldberg, A.M. (1996). The Three Rs: The Way Forward. *Environmental Health Perspectives*, **104**, 878–880.

Index

absorption 166–167, 169, 175, 183–184
abundance 25, 42–67, 119, 128, 135, 222, 340, 342, 355–357
  trends 66
acceleration 229, 246, 247, 251, 254, 255
accelerometer 229–230, 246, 246, 248–249, 253
acoustic 266, 269
  alarms, stimuli 351, 357
  sampling and survey 43, 54, 58, 72, 96, 244
  tags – *see* tags
acoustic tracking or telemetry 227, 231–232, 248
AD Model builder 95
adrenal glands 153
aerial survey or observation 45, 53, 57, 58, 61–63, 66, 224, 243
aerobic (oxidative) metabolism 166–169
aerobic scope for activity (AS) 168
age and age estimation 13, 33, 48–49, 72, 81, 98, 107–118, 121, 123–124, 126, 285, 289, 311
age at sexual maturity (ASM) 122
age at physical maturity (APM) 123
age–length key 107
age-specific 121, 122, 126, 137
age structure 114, 120, 122, 126, 127
  stable age distribution 127, 138
  stationary 127
aggression 352
aircraft 44, 55, 57, 59, 66
air rifle 225–226
Akaike's information criterion (AIC) 133, 142–143
Akaike weights 133
Alborán Sea 64
Aleutian Islands 299, 308
Allee effect 140, 141
allele 310, 314–316, 319–322
allometric reconstruction 198, 200–201, 205–206
allometry 168, 178–179, 189
allozyme 312, 322
altimeter 62
altitude 70
American coast 308
American Society of Mammalogists 101–104
amino acids 115
ammonia vapour 112
ammonium oxylate or hydroxide 111
anaerobic 261–262
  metabolism 166–167, 169

anaesthetics and anaesthesia 13, 21, 29–30, 32, 33–35, 38–39, 109, 224
analgesics and analgesia 13, 33
anatomy 10
animal welfare 1–15, 16, 33, 35, 38
anisotropy 85
Antarctic 34, 66, 115, 170, 329, 332
Antarctic circumpolar current 290
antenna 232
anthropogenic 119, 158, 164
antibiotic 109
antibody 146, 149
anus 103, 105
apex predators 243
apnoea/breath-holding 167
apparent survival 130
approximations 138
aquatic v
Arctic 32, 40
area restricted search (ARS) 258
Argos 224, 228, 231–239, 249, 256
arithmetic mean 84
arthropods 146, 152
ascertainment 322
aspartic acid 115
Atlantic cod 284
Atlantic Ocean 328 – *see also* North Atlantic
ATP 166–168
attachment 224–227– *see also* tagging
attenuation 232
auditory meatus 102, 114
auditory range 232
Australia 66, 308
autocorrelation 56, 70, 78–79, 84–85, 88–89, 95, 238
autocovariance 89
availability bias 55–57
avoidance 56
axillary girth 100, 104 – *see also* body girth

back-propagation 91
bacteria 40, 145–146, 152
Bahamas 248
Balcomb, K. 295
bald eagle 300
baleen 114, 116, 315
baleen whale – *see* mysticete

Baltic Sea 328
bandwidth 84, 232
BAPS 318
Basque whalers – *see* whalers
bathymetry 234
battery 223, 228–229
Bay of Fundy 37
Bayesian statistics 45, 94, 138, 142, 257, 262, 317–319, 322
  mixed effects 142
  random effects 142
bearing 246
Beaufort Sea 302–304
Beck, B. 285
behaviour 9, 70, 107, 222, 233, 237, 307
  diving 249, 260
  haul-out 44, 70, 75, 222, 224–227, 233, 238
  of individuals 68
  resting 69
  social 68
  traits 145–146, 150, 161
behavioural
  associations 257–258, 264, 266, 269–278
  bouts 250–253, 261
  data 264, 266
  ecology v, 235, 291–293
  events 264, 267
  indicators 38
  interactions 83
  modelling 98
  patterns 223
  response 5, 9
  sampling 266–267, 275
  states 81, 257–258, 264, 267
  survey 266, 268
  terminology 281–282
Bering Sea 325, 334
best practice 2, 4, 7–8, 12–15
bias – *see* statistical bias
Bigg, M. 295
binning 76
binoculars 62
biocompatible 226
biodegradable materials 229
biodiversity 68, 96
bioelectrical impedance analysis (BIA) 182
biofouling 223
bioinformatics 156
biological diversity 340–341, 359
biomass reconstruction 198, 203–204, 206–208
bionomic equilibrium 326, 327
biopsy 50, 60, 125, 144, 154–155, 157
bio-sonar 244
biotoxin 146–147, 154

bimetallic 226
binocular microscope 110
bird 8
Bird Island, South Georgia 290, 292
birth and birth rate 47, 50, 109, 113, 119, 121, 124, 136
bleach marking 23
blood
  cell counts and chemistry 21, 38, 148–151
  chylomicrons 211–212
  leucocytes 150, 156
  sampling and collection 13
blowhole 102–103
blubber/adipose tissue 100–101, 104, 107, 116, 125, 141, 147, 153–155, 157, 165, 170, 179, 181, 211–212, 215, 217, 225–226
  biopsy 212, 217
  depth/volume 181, 212
body
  carbohydrate content 179
  composition 179–183, 252–253
  condition 98, 229, 100–101
  energy content 101, 180–181
  fat content 229, 179–183, 187–188
  fat-free mass (FFM) 179–182
  girth 98, 100–107
  lean body mass 179–182
  length 98, 100–107, 116
  mass or weight 99–107, 117, 165, 167–168, 178–181, 189
  protein content 179–181, 187–188
  shape 98
  size v, 98–118, 122, 168, 172–173, 227
  temperature – *see* thermoregulation
  volume 100
  water pool – *see* body water content
  water content 173–176, 179–182, 187
bootstrapping 54, 123, 138, 205, 208
bone 107, 114, 315
Boness, D. 285
boto – *see* odontocetes; Amazon river dolphin
bottom time 251
bow 40
bow-riding 226
Boyd and McCann index 100
Bradycardia 167, 177
branding 284–285
  freeze 20, 21–22
  hot-iron 4, 20, 131
breathing – *see* respiratory system
breathing hole 227
breeding 68, 83, 288, 307–308, 310
Britain 45
British Columbia 295

Brownian motion 85
buoyancy 229–230, 251–253
by-catch 42–43, 84, 144, 194, 308, 321, 342, 351, 353–359

C ++ 94
calibration 99
calorimetry 168–173, 179–180, 185
camera 229, 245, 351
   animal-borne 245, 246, 253, 256
   blimp-cam 244
   video 66, 229, 245, 246
Canadian 333
Canadian Council on Animal Care 9
CANOCO (software) 94
canonical correspondence analysis 90
CAPTURE (software) 52
capture and release vi, 11, 13, 16, 28–40, 99, 106, 127, 144, 152–154, 158, 160, 224–226, 284, 286, 293
capture histories 43, 46
   'band recovery' models 127
   carrying capacity (K) 139–140
   encounter history 128–129
   'known fate' models 127
   'live recapture' models 127
   probability of capture 128, 132, 134
capture probability 48, 50
capture–recapture 25, 28, 43, 45–53, 73–74, 81, 124, 127, 134
carapaces 199, 201, 205
carbohydrate(s) 166–167
   body carbohydrate content – see body composition
   catabolism 169, 188, 210
   in milk – see milk proximate composition
carbon dioxide (CO_2) 166, 168, 170–171
   rate of CO_2 production (VCO_2) 169–172, 174–175
carcass analysis 179–180
cardio-pulmonary status 150–151
cardiovascular system 33, 38
CART (software) 94
Cartesian system 70
Catabolism 166–170, 188
catch limits 15
catch-per-unit-effort 43
CATS 316
causal agents, factors and links 146–148, 158–159, 164
CCAMLR 291, 292
cell phone – see GSM
cellular respiration 166–168
cementum 109, 113

censoring 77
census 43, 124
central place foraging 256
centre of activity 84
cephalopods 195, 198–205, 209
cephalopod beaks 192, 195, 198–205
cetacean research vii
cetaceans 6, 11, 12, 22, 24–26, 35–36, 38–41, 43–46, 49, 53–58, 60, 62, 65–66, 84, 101–102, 106–107, 110, 115, 120, 122, 124–126, 132, 141, 165, 168, 170, 174, 178, 180, 183, 185, 194, 196, 204, 211–212, 215, 218, 221, 224–225, 233, 244, 248, 260, 283, 294, 297, 308–309, 316, 318, 340, 344, 346, 358 – see also odontocetes; mysticetes
CeTAP 297
Chapman's estimator 51–52
chemical restraint – see anaesthesia
chlorodifluoromethane 22
chlorophyll 58
cholesterol 114
chorionic gonadotrophin 125
chromatography 115
circulatory system – see cardiovascular system
CITES 308
classification tree 89–90
climate prediction and change 118, 119, 230
closing mode 63
cluster polygon 83
clustering 318
coast, coastal, and coastline 70, 233
code of conduct 8
co-dominance 316, 322
coefficient of variation 113
cohort 145, 152, 159, 163
collinearity 77
collisions 297, 298
colony 95
Committee on the Status of Endangered Wildlife in Canada (COSEWIC) 348, 358
companions 273–274
compass 248
computation 92–93
computer literature 91
computer power 77
computer software 94–95
concave polygon 83
conception 121
condition – see body condition
conductivity 246, 249
condylobasal length 104
conflicts v
confounding factor(s) 145, 147

conservation v, viii, 5, 6, 11–12, 14–15, 42, 68, 96, 118, 144, 155, 163, 306–315, 317, 325, 329, 334, 340–359
    assessment 342–346, 348–349, 358–359
    direct intervention 352–353
    implementation 342–343
    least concern 344–345
    problem detection 342–343, 354, 358
    response 342–343, 352, 357
    solution 342–343, 352
    status 341–342, 344–353, 355
contaminant 146–147, 153–155, 157–158, 160, 219, 296
controlled exposure experiments 5, 7
Cormack-Jolly-Seber estimation (CJS) 127–133, 134
Corpus albicans or luteum 122, 125, 126
correlated random walk 235, 238, 257
correspondence analysis 90–91
corticosteroid 38
counting and counts 44–45, 86, 88, 312
covariates – see statistical covariates
cranium 104
crossbow 225–226
cross-sectional sampling 120, 122, 123, 126, 127
crowding board 32
crustaceans 195–196, 199–201, 205, 209
cryostat 111
cryptic 76
CTD 230
cue-counting 43, 53, 57–58, 62, 64, 72
culling or killing 6, 11, 19
culture 1, 14, 279–280
currency 259
cyanoacrylate glue 224
cytochrome 313

data
    acoustic 58, 66
    analysis 63–65, 223
    ancillary 60
    availability viii
    catch 66
    coding 223
    compression 223
    effort 62
    environmental 60
    ephemeris 228
    gaps 233
    interpretation v, 7
    layers 81
    location 227–228
    management 16
    outliers 63
    path 235

    quality 13, 307
    sharing 10, 233
    spatial 69, 72–75, 79–96
    stratification 65
    transmission 223
    visual 62, 96
    visualization 223
database viii, 83, 235
data logger 243, 246, 249 – see also tags
day length 227
dead reckoning 228
death and dead specimens 16, 39, 43, 66, 107, 110, 114, 144, 145, 147, 152–153, 160, 224, 349, 351–353
deaths and death rate 47, 50, 113
decalcification 110–111
decision analysis and rules 14–15
delphinids – see odontocetes
demographics and demography 119–121, 123, 139, 308–309, 320, 342–343, 349–351, 354
density-dependence 122, 139–140, 141, 284
density estimation or modelling 54, 58–60, 69, 76, 80, 84
density gradient 61
density surface modelling 43, 55, 58–59, 65–66
density, water 230
dental elevator or extractor 110
dental glue 224
dental layers – see growth layer groups
dentine 112–113
dentition 98–99, 102, 107–114
dependant young 13
depth 68, 70, 90, 227–228, 233
detection
    function or probability 61, 63–64, 72, 74, 81–82
    range 227
deuterium (^2H_2O) 173, 175–176, 180–181
deviance 77
diagnosis, diagnostic techniques 147–149, 152, 155, 160
diamond saw 110
diazepam 33, 38
diel 250
diet vi, 68, 191–221, 285, 294, 302
    faecal (scat) analysis 191–192, 195–199, 203–204, 206–210, 221
    fatty acid analysis – see fatty acid(s)
    gastric lavage/enema collections 193, 195–196
    gastrointestinal analysis 191–196, 198, 203–208
    molecular identification 191–192, 208–210
    regurgitate analysis 192, 195–198, 204–205, 221
    skeletal/hard structure analysis – see skeletal/hard structure(s)
    soft tissue analysis 194, 196–199, 219–221
    stable isotope analysis – see stable isotopes

dietary marker 184
difference or differential equations 136
digestion 166–169, 172–173, 177, 184, 189, 192, 198, 201, 203–205, 208–211
digestion 204–205
digital photography 25–28, 66
dimethyl ether 22
diploid 311, 322
discounting 79
discriminant analysis 89–90
disease 40, 68, 144–164, 287, 294, 296, 340, 346, 349
  effect 148–149, 154, 157–160
  prevalence 145, 148–149, 154–155, 158
dispersal 310, 320, 322
Distance (software) 54, 58, 61, 64–66, 94
distance sampling 7, 53–65
Distinct Population Segment 246, 309, 356
distribution 75, 79, 84, 87, 91, 96, 222
disturbance 2, 7, 13, 16, 28, 42, 193, 196, 224
diurnal 233
diving and dive response 34, 82, 167, 170, 172–173, 177, 179, 226, 246
  bins 249
  classification or function 249–251
  depth and duration 249, 250, 260
  drift 182–183, 229, 252
  shape 246–249, 250, 251, 252, 254, 260
DNA 146, 155–156, 307–324, 356
  analyses of diet 191–192, 194–197, 199, 209–210, 211
  barcode 313
  fingerprinting 312, 322
  primers 192, 209
  pyrosequencing 210
dolphins – see odontocetes
dominance 277, 322
domoic acid 148
Doppler shift 228
dorsal fin 103, 225–226
dorsal ridge 226
double-counting 56, 64–65
doubly labelled water (DLW) 173–177, 180, 187–188
doxapram hydrochloride 38
DPX 113
drag 223–225
drugs and drugging – see also anaesthesia; analgesia; sedation
  dose 34
  inhalation 33–34
  injection 34
  legality 34
D-tag 225, 230, 235, 254, 255, 269
dugong – see sirenian

duty cycle 223
dye marking 23
dynamic programming models 262

earplug 114
ECG 229
echolocation 230, 244, 255
ecological
  fallacy vii
  fluctuations/climate 302, 304, 305
ecology v, 42, 45, 68, 74, 80, 83, 91, 95, 100, 321
economics 325–326, 333–335
ecosystem 96, 189, 308, 340–342, 351, 353
  management 291, 339
ecotourism 333–334
ecotype 116, 296
EDTA 111
effective population size 310, 322
eigenvalue-eigenvector pairs 138
electronics 222–223
electrophoresis 314–315, 322
Ellis, G. 295
emigration 47, 50, 53, 119, 135, 342
enamel 113
encounter rate 60, 63
Endangered Species Act (ESA) 308–309, 346–348, 358
endangered species or populations v, 7, 11, 12, 15, 95, 286, 340, 342, 344–348, 350, 352–355, 359
  critically endangered 344–346, 359
endocrine status 150–151, 154
energy (E) or energetic 166
  acquisition 165–167
  apparent digestible energy (DE) 166, 176, 183–184
  balance analysis 183–184 – see also nutrient/energy balance
  budget 230
  consumption 233, 260
  equivalents 179–180, 187
  expenditure (EE) 169, 171–172, 174, 176–178, 223, 229, 101
  faecal energy (FE) 166, 183
  gross energy (GE) 166, 183, 185–188
  gross requirements (GER) 189–190
  heat 165–170
  metabolizable energy (ME) 166–167, 183, 189
  net energy (NE) 166–167, 183
  storage 166–169, 180, 188
  swimming 229
  total body energy (TBE) – see body composition
  urinary energy (UE) 166, 183
  use/output 165, 183–185, 187, 189

energetics 165–190
  individual 166–189
  maintenance 165, 167–168
  population 189–190
engineering 86
entanglement 297–298, 352–353, 355, 359
environment 72, 222
  effects of research on 8, 13, 22, 145, 147–149,
    154, 157–158, 160–162, 164
environmental
  change 118
  covariates of factors 82–90, 96, 119
  envelopes 87
  gradients 90–91
  niche factor analysis 90
  optima 90
  selection function 76
  space 70–71, 75, 86–87, 89
  variables and data 72, 77–78, 80–82, 89–91,
    246, 249
  variability 139, 141
Environmental Protection Agency (EPA) 160
Environment Protection and Biodiversity
    Conservation Act (EPBC Act) 348
eosin 112
epidemic 146
epidemiology 144–146, 148, 149, 152–159, 164
  sample collection 152–155
  visual assessment 144, 152, 155
epididymus 122
epithelial cells 114
epizootiology 145
epoxy 224
ESRI (software) 94
etching 110
ethanol 110
ethical
  asymmetry 7, 12
  boundaries of right and wrong 3, 15
  consideration 144, 163, 350
  customs 1
  debate 8
  gaps 2
  guidelines 2, 4–9, 14
  principles 2, 3
  responsibility 350
  review 3, 5, 7, 12, 16
  standards 2
  values 1, 4, 8, 14, 341
ethics vi, vii, viii, 1–15, 16, 224, 340 – see also ethical
ethics committees 3, 8
ethnafoam 229
European Cetacean Society 8

euthanize 15
evolution 300–301, 309, 321
exon 316, 323
exothermic 224
experimental design vii, viii, 1, 7, 10, 13, 44, 50,
    60–61, 64, 66, 72–74, 131, 135, 306–307,
    350–252
expert knowledge 87
exploitation 6, 15 – see also harvesting
exposure 145–149, 153–155, 157–164, 351
  response 145–150, 153, 155, 157, 158, 160–164,
    351, 358
extinction v, 341, 343–349, 352–353, 356
extrapolation 70, 71
extrinsic 119, 256
eyeballing 83
eye lenses 115, 199–201

faecal sampling 125
faeces – see diet, faecal (scat) analysis
Fastloc 228, 239
fat(s) 116, 166–167, 179, 182
  body fat content– see total body fat
  catabolism 169, 188, 210
  deposition 187–188, 211
  in milk – see milk proximate composition
  storage 167, 210–212, 214–215
fatty acid(s) (FAs) 115–116, 191, 193–194, 210–216,
    221, 285, 294, 304
  analysis 211–214
  blood chylomicrons 211–212
  blubber/adipose tissue 193, 210–212, 215
  calibration coefficients (CCs) 193, 215
  milk 211–212
  prey 213
  quantitative FA signature analysis (QFASA) 193,
    211, 214–217
  signatures 191, 193, 210–216
  storage 210–212, 214–215
fecundity 120–121, 123, 135, 137, 154, 163, 284,
    291, 342–343, 349
feeding 243
  detection of 229
  energy acquisition 252
  grounds 311
  meal size 229, 253
  patch 260
  strategies 307
  success 245, 246, 252, 253, 255
first passage time (FPT) 258
fish 194–197, 199–201, 203–206, 208, 209,
    243
fish scales 196, 199, 209

fisheries 6, 31, 42, 308–309, 340, 342, 351, 353–359
  closure 357–358
  observers 342, 354–358
  restrictions 351, 355, 357
fishing effort 84
fishing gear 312
fishing weir 35
flipper dislocation 39
flipper length 101–105
flukes 103, 106
Follow 265, 266, 267
  Group 267, 268
  Individual 267
food intake 167, 169, 176, 183–184, 187, 189
food limitation 353
food web(s) 191, 211, 217–220
foraging 69, 95, 243, 245, 246–252, 256
  bouts 239 – *see also* diving
  decisions 250, 251, 259, 262
  dives 249–251
  ecology 285, 287, 291
  tactics 256
  trips 252, 261
formalin 110, 114
formic acid 110
Fortran 94
functional response 292
fungi 145–146, 152
fur clipping 23
fractal analysis 258
freeze-branding – *see* branding
freon 22
F-test 77
functional response 258–295
fungi 40
fur 224

GAM 65, 86, 89, 238
GAMM 89
gape 102
gas chromatography 213–214
gas exchange – *see* respirometry
gene 316
GENECLUST 318
GENELAND 318
gene flow 307–311, 321, 323
gene tree 314, 323
Generalized Additive Models – *see* GAM
generation time 310–311, 323
genetic vi, 145–146, 155, 157, 298, 306–324, 340, 346, 349, 356
  analysis 154, 157, 268, 277, 279
  diversity 320, 340

drift 314, 323
markers 312–317
material 146, 157
relatedness 98
genetical identification 50
genital 105
genome 311, 316, 323
genomics 155–157
genotype 141, 314–315, 323
Geobugs (software) 94
geographic range 344 – *see also* home range
geographical region 108
geographical space 70, 72, 75, 79, 83, 86
geolocation 227–228
geophysical variables or analysis 82, 85
geostatistics 85–86, 94
GeoZui4D 235
gestation 120
gillnet 35, 37, 351, 353, 355–357, 359
girth – *see* body girth
GIS 61, 66, 80–81, 83
glare 64
glaucous winged gulls 300
Global Mammal Assessment 346
Global Positioning System – *see* GPS
Global Telecommunications System 230
GLM 86, 88–89
GLMM 88
glue 224
glycolysis and respiration 166–168
Gompertz model 117
GPS 228, 233, 235, 239, 256
GRASP (software) 94
GRASS (software) 94
Greenland 66, 312
gross birth rate (GBR) 121
gross pregnancy rate (GPR) 121
gross reproductive rate 124
group size and composition 54, 60, 63
growth 70, 98–118, 165, 169, 183, 188–189
growth layer groups 98, 108–114
GSM 231, 233
Gulf of Maine 312
Gulf of Mexico 310
Gulf of St. Lawrence 332

habitat and habitat use or maps 2, 42, 71, 75, 80–93, 235–236, 241, 252, 340, 346, 358–359
habitat modelling 69, 82
habitat preference 71, 76, 80, 83, 85–87, 91–93
haematoxylin 112
haematology, haematological status 148–151
handshaking 234

haploid 311, 313, 323
haplotype 312, 319–320, 323
harm or harmful effects 2, 8, 11–12, 16
harmonic mean 84
harpoon 329
harvest or harvesting 6, 43, 308, 325–326, 328, 334, 340–341, 344
    commercial sealing 325, 329, 335
    commercial whaling 329, 331, 335
    control rules 335–338
haul-out – *see* behaviour
hazards 144, 160–162, 164
heading 246
health status and assessment 40, 144–164
    reference ranges 152
    risk 145–146, 148, 149, 157, 160–163
hearing – *see* auditory
heart rate (f_H) 38, 176–178, 246
heat increment of feeding – *see* specific dynamic action
helminths 146
Hensel 104
hepatic function/status 150–151, 154
herbivore 243
herring 37
heterozygote 316
heuristic analysis 236–237, 241
Hinde's framework 263–264, 269
hippocampal atrophy 148
histogram 271
histologist or histology 111, 115
home range 68–69, 81, 83–84, 87, 321
homodont dentition 109
homoplasy 314, 323
homozygote 316, 323
hoop net 29, 32, 35
hormone(s) 125
hot-iron branding – *see* branding
HPLC 115
Hudson Bay 303
hukilau 35, 37
human
    health and safety 3, 40–41, 144–145, 152, 154, 156–160, 163–164
    impacts or threats 340–343, 349, 352, 358
    society 1, 3
humanity vi, 6
hunters and hunting v, 43, 341
hurdle approaches to modelling 76
hydrochloric acid 110
hydrodynamics 223, 226
hydrogen isotopes 173–174, 180–183, 187
hydrogen peroxide hair dye 23
hydrophone 58, 66, 248, 254, 255

hypercubes 87
hypersaline solution 110
hypothesis generation and testing viii, 11, 86–87, 95, 229, 317–321, 351

ice and ice conditions 227, 302, 303, 304, 305
IDRISI 94
illuminance 246
imaging, images, and imagery 10, 26, 66, 81, 85
immigration 47, 50, 119, 135, 342
immobilization 33–35, 99
immune status and response, immunology 150, 151, 152, 154, 157, 161
immunoassay 125, 208
implantation 19–20, 121, 225
incidence – *see* risk
incisor 109–110
inclinometer 62
independent contrasts 279
individual-based models (IBM) 140
individual effects 95
infection and infectious 17, 19, 145–152, 154–155, 157
inference 70, 80, 100, 235, 307 – *see also* statistical inference
ingestion 229
inheritance 307
injury 19, 20, 23, 25, 35–36, 40
International Whaling Commission (IWC) 329, 330–331, 335
interpolation 71–72, 82, 85–86
intrinsic factors 68, 119, 256
intron 316, 323
instrumentation v, 9, 11
International Federation of Clinical Chemistry 152
International Statistical Classification of Diseases (ICD) 152
International Union for the Conservation of Nature (IUCN) 42, 308–309, 344–346, 348, 352, 358
invasive techniques 11–12
inverse square law 232
Iridium 231, 233
isoflurane 21, 34
isomers 115
isotopes 115–116
isotope dilution 173–176, 180–183, 187, 252
isovaleric acid 116
Iverson, S.J. 285

jackknife 54
jaw 103, 110–111
Jolly-Seber 52 – *see also* Cormack
juveniles 137

Kalman filter 239–240
Kelp 300
Keratin 114
kernel smoothing 81, 84–85
kidney 153
killing – *see* culling
kinship 279
kriging 81–82, 85–86
krill 245, 290

laceration 39
lactation 121, 184–189, 211, 215
lactic acid (lactate) 166–167, 169
landscape genetics 319
latitude 68, 70, 84, 86, 227–228
Lefkovitch matrix 137
legislation and legal v, vii, 3, 13, 14–15, 22, 34
length – *see* body length
lesions 17, 23, 25, 150, 152, 154, 158, 160
Leslie matrix 137
Lévy flights 237, 257
life cycle and life history v, 25, 40, 44, 69–70,
       98, 138, 224, 235–236, 283–284, 289,
       293, 295, 306
lifetime reproductive output 120
light levels 230
lineage and lineage sorting 308, 313, 318, 323
line transect 53–65, 72–75, 79, 81
linear regression 88
lipid(s) 100–101, 179, 181, 183–187, 210, 212–215,
       219–220 – *see also* fat(s) and fatty acid(s)
lithium thionyl chloride 223
liver 153
LMD index 100
Location Class 228
location determination 223, 227–228
locus 316, 323
logistic equation 123
log-normal distribution 52
longitude 68, 70, 84, 86
longitudinal sampling 120, 122, 126
long-term studies 109, 123–124, 283–305
lunar cycles 113

MacAskie, I. 295
magnetometer 230
maintenance
       costs/metabolism 165, 167–168
mammalogy v
mammary 103
MAMVIZ (software) 94, 233
management vii, viii, 6–7, 14, 42, 66, 144, 160,
       162–163, 309, 318–320, 325, 329, 332, 334, 338,
       340, 344, 349, 351, 354, 356–357

Manifold 94
MANOVA 87
Mansfield, A. 285
Mantel test 87
MapInfo (software) 94
Maple (software) 94
maps and mapping 75, 235
marginal value theorem 256
marine biology or ecology v, 83
Marine Mammal Protection Act (MMPA) 337, 354
marine mammalogy v–vi
Marine Mammal Science 8, 65, 95
marine top-predators 68
MARK (software) 52–53, 66, 94, 130–131
marking 16–28, 47–48, 50–51, 120, 123, 128,
       130–31 – *see also* tags and tagging
mark-recapture – *see* capture–recapture
mass – *see* body mass or weight
mass spectrometry 116
maternal 307
mating systems 277
MathCad 94
Mathematica (software) 94
mating 307
Matlab (software) 94
maximum likelihood 130, 320
maximum sustainable yield (MSY) 326–328
MCMC 240–241
McMurdo Station 288, 289
Medicine 144, 149, 155, 160
melanocytes 20, 22
melon 102
membranes 111
meta-analysis viii, 144
metabolic
       basal metabolic rate (BMR) 166–169, 178–179,
              189–190
       costs 165–167, 169–170, 172–174, 178, 189
       diving metabolic rate 172–173, 177
       field metabolic rate (FMR) 169, 174, 177–179,
              190, 229
       hood 171
       maximal metabolic rate (MMR) 168–169
       rate (MR) 167–170, 172, 174, 176–179, 183–184,
              188–190, 229
       requirements 165, 169, 184, 189–190
       resting metabolic rate (RMR) 167–169, 170,
              172–173, 178–179, 183
       standard metabolic rate (SDR) 166–169 – *see also*
              BMR
       water production (MWP) 176, 187–188
metabolites 146, 155, 167
metapopulation 138, 288
meteorological variables 82 – *see also* weather

methodology vi–viii
methylene blue 112
microbiology 153, 154, 157
microsatellite 314–315
microscope 112, 114
microtome 110–111
migration 44–45, 69, 226, 235, 307, 311, 324
milk 147, 169, 176, 183–189, 211–212, 219
minimum convex polygon 83
minimum heat loss method 170
mitochondrial DNA sequence 311, 313–314
mitochondrial genome 307–314
MjM (software) 94
mobile phone – *see* GSM
model
    autoregressive – *see* autocorrelation
    deterministic 135, 138
    dynamic 96
    empirical 70–71, 80, 95
    fitting 71, 264–276, 271, 278
    Gaussian 239
    habitat 80
    hazard rate 82
    hierarchical 95
    individual-based 96
    linear 88
    phenomenological 70
    predictive 116
    mechanistic 71, 80, 95–96
    metapopulation 138
    mis-specification 95
    mixed effects 79, 88, 142
    multivariate 86
    observation 239
    optimal foraging 259–262
    overfitting 77, 89, 95
    process 238
    random effects 142
    regression 71, 76, 78, 85–86, 88, 96
    selection 66, 71, 77, 79, 89, 95, 142–143
    state-space 81, 96, 142, 238–241, 257, 262
    stochastic 135, 139, 261–262
modelling vi, 10 – *see also* statistics
molar 109
molecular ecology 312
monitoring 342, 343, 351, 353–355, 357–358
monophyly 313, 324
Monte Carlo 138
morality 3
moratorium 331
morbidity 144, 152
mortality 134, 144, 146, 147, 152, 160–161, 163,
    340, 342–343, 349–351, 354, 356, 358
morphology 308–309

morphometrics 98–118
Mote Marine Laboratory 292
movement 79, 85, 227–229, 235–239, 256, 306, 318
multi-collinearity 90
multiparous 147, 163
multivariate regression tree 89
multivariate statistics 86
muscle 116
muscle temperature 246
museum 98, 315
mustelid
    northern sea otter 348, 351
    sea mink 341
    sea otter 109, 165, 170, 185, 291–301, 334, 346
    southern (California) sea otter 165, 348–349, 351,
        353
mutation 314–315
mysticete 43, 66, 103, 114–115, 172, 211, 358
    blue whale 25, 46, 329, 348
    bowhead whale 45, 115–116, 330, 348
    Bryde's whale 330
    fin whale 329–330, 348
    grey whale 45, 116, 172, 330
    humpback whale 25, 28, 45–48, 51, 66, 310–312,
        330, 348
    minke whale 56, 66, 115, 259, 330
    North Atlantic right whale 120, 155, 230,
        297–299, 348–349, 351, 353, 355, 359
    North Pacific right whale 308, 348
    right whales 25
    sei whale 348

natality 121 – *see also* birth rate
National Institute of Standards and Technology
    (NIST) 147
National Research Council (NRC) 160
navigation 227
Navsys 228
near infrared spectrometry (NIRS) 210
necropsy 144, 153, 155
necrosis 226
negative binomial 76
neonatal line 109
network analyses 274–276, 278
    assortative mixing 275
    binary 275
    measures 275
neural network 91, 96
neurological status 150–151, 154
neurones 91
Newfoundland 332
New Management Procedure (NMP) 331
New Zealand 227
niche factor analysis 90

niche partitioning 68
nitric acid 111
noise 232, 294
noose pole 29
non-infectious 146
non-parametric 81, 85, 89, 96, 114, 237
normal distribution – *see* parametric
North Atlantic 57, 310, 312
North Pacific Fur Seal Convention 332
North Pacific Ocean 325
North Sea 328, 339
Northwest Atlantic 332
North-western Hawaiian Islands 286
nuclear genome 307–314
nuclear tests 116
nucleotides 314
Nunavut 304
nutrient/energy balance 169–170, 179, 183–184
nutritional status/causes 145–146, 150, 151, 154,
      158, 161
nyanzol D 23

OBIS SEAMAP 234
observation error 81
odonbenidae – *see* walrus
odontocetes 35, 39–40, 109, 115, 123, 124, 153, 172,
      211–213, 230
   Amazon river dolphin 23, 37
   Australian snubfin dolphin 308
   baiji (Yangtze river dolphin) 95, 341, 352
   beaked whale 109, 255
   beluga 36–37, 66, 116, 318, 344, 346
   bottlenose dolphin 22, 23, 28, 37, 38–40, 46, 51,
      64, 107, 109, 129, 141, 152, 156–158, 160, 163,
      177, 181, 218, 227, 267, 278, 292–294, 333
   bottlenose whales 212
   common dolphin 56, 99
   Cuvier's beaked whale 255
   Delphinus sp 317–318
   dolphins and dephinids 11, 25, 35–40, 48,
      110–111, 226, 244
   dusky dolphin 35
   false killer whale 23, 35
   franciscana 109, 294
   harbour porpoise 37, 56, 57, 109, 112, 225, 339,
      355–358
   Hawaiian spinner dolphin 36, 37, 109
   Hector's dolphin 46, 274
   humpback dolphin 46
   Irrawaddy dolphin 308
   killer whale 25, 35, 116, 129, 137, 161, 171–172,
      196, 209, 226, 243, 273, 295–297, 301, 333–334,
      346, 348, 352
   long-finned pilot whale 278

melon-headed whale 226
Mesoplodon 246, 255
narwal 66, 109, 225
Neophocoena 255
northern bottlenose whale 46
open ocean 37
Pacific white-sided dolphin 23, 35
pan-tropical spotted dolphin 35
Perrin's beaked whale 308
pilot whale 35, 225
phocoenids 109
porpoise 35, 99, 111
Risso's dolphin 23
river dolphin 70
rough-toothed dolphin 23
short-beaked common dolphin 23
short-finned pilot whale 23
sperm whale 28, 46, 58, 66, 103, 109–110, 225,
      230, 244, 254, 255, 256, 266, 271, 272, 274, 328,
      330, 348
spinner dolphin 23, 39
spotted dolphin 308
Stenella 317
Tursiops 317
vaquita 95, 165, 325, 359
white-beaked dolphin 56
white-sided dolphin 56
Ziphius 255
oceanography 230–240, 242, 290, 292
oestradiol 125
offspring size 285
ontogeny 241
open access 330
optimal foraging 259–262
organ systems 149, 150–151, 154
osteology 99
otariid or eared seal 17–18, 20, 29–30, 32, 75, 181–182,
      196, 204, 210–211, 217, 252, 256, 307, 346
   Antarctic fur seal 30, 114, 180, 182, 195, 203, 245,
      256, 259, 261
   Australian fur seal 31, 46
   Australian sea lion 22, 23, 204
   California sea lion 20, 22, 30–31, 148, 177, 203, 253
   Cape fur seal 109, 330
   fur seals 20, 29, 329
   Guadalupe fur seal 348
   Japanese sea lion 341
   New Zealand fur seal 183
   New Zealand sea lion 20
   northern fur seal 23, 29, 109, 330, 332
   sea lion 29, 30
   Steller sea lion 20, 21, 27, 30, 117, 120, 170, 203,
      209, 215, 219, 229, 253, 308–309, 325, 348, 359
   Subantarctic fur seal 123, 134

otoliths 194, 197–201, 203–205
ovarian corpora 101
overexploitation 331, 332, 335
ovulation, rate 121
oxidative metabolism – *see* aerobic metabolism
oxygen (O_2) 166–168, 171
  debt 166–167
  rate of O_2 consumption ($\dot{V}O_2$) 168–177
  ^{18}O 173–174
  $\dot{V}O_{2max}$ 168–169 – *see also* MMR
  $\dot{V}O_{2rest}$ 168
oxytetracyclin 115
oxytocin 185

pack-ice 66
pain and distress 2, 9, 13, 17, 21, 33
paint marking 23
PARTITION 318
parturition 121
patch 251, 258, 260–262
paraffin 111
parametric 76, 84, 86–88
parasite 155, 195, 284
parentage 315
Pascal 94
passing mode 63
passive acoustics 244, 248–249
passive acoustic monitoring (PAM) 230
path 246, 257–258
pathogen 39, 40, 146, 153, 157, 161, 163, 349, 359
pathogenomics 156, 157
pathology 148, 153, 154, 157
pattern cell 26–28
pattern recognition and matching 25, 26–27
PCBs in blubber 141
PCR 209–210, 313–314, 324
pectoral fin 99
peeled convex polygon 83
peer review 3, 7, 9
pelage marks or patterns 25–28, 47, 225
pelagic 321
penguin 177
pens 199–200
perception bias 55, 56
periodic bone 115
permitting – *see* research permitting
perturbation analysis 138
  elasticity 138
  sensitivity 138, 142
Petersen estimator 46, 51, 66
phenol 111
phocid or true seals 17–18, 20, 21, 31–32, 34, 75, 178–182, 211, 217, 346

bearded seal 109, 269
Biakal seal 230
Caribbean monk seal 165, 341, 348
crabeater seal 34
elephant seal 46, 253
grey seal 20, 21, 24, 25, 27, 32, 45- 46, 49, 123, 134, 135, 142, 177, 180, 182, 186, 190, 197, 203, 211, 225, 240, 254, 257, 259, 284–286, 339
harbour seal 20, 27, 32, 44, 137, 143, 152, 177, 182, 196, 203–204, 227, 253, 334
harp seal 109, 180, 253, 329, 330, 332–333, 337
Hawaiian monk seal 123, 129, 161, 165, 286–288, 348, 352–353
hooded seal 330, 337
leopard seal 25, 34, 46
monk seal 95
northern elephant seal 20, 22, 32, 123, 138, 182–183, 196, 247, 251, 254
ringed seal 20, 302, 304
southern elephant seal 20, 21, 32, 34, 138, 182–183, 230
Weddell seal 32, 34, 46, 170, 246, 248, 288–290
photogrammetry 26, 99
photo-identification 25–28, 43, 46, 49–51, 74, 153, 154, 155, 160, 265, 268, 287–288, 293, 295, 297, 306
phylogeographic signal 318, 324
phylogenetics and phylogeny 313–318, 321, 324
physiological response 5
physiology v, 107, 222, 241, 246
pilot study 60
pinniped(s) 6, 8, 17–18, 20, 22, 24, 25–26, 29, 32–35, 40–46, 49, 70, 99, 101, 104, 106–110, 117, 120, 122, 124–125, 131, 135, 153, 165, 168, 170, 178, 180–183, 185, 195–196, 199, 203–204, 209, 212, 215, 221, 224, 238, 284, 288, 304, 306, 340, 344, 358 – *see also* otariid; phocid
pinniped research vii
point sampling or transects 53, 58, 72
Poisson process or distribution 72, 75
polar bear 40, 71, 109, 165, 180, 182, 185, 302–305, 344, 346
polar orbiting satellite 228
polar regions 230
polar seals 227
policy, policy makers vii, 1, 341–343, 346–347, 359
pollutant, pollution 147, 154, 160–161
polychlorinated biphenyls (PCBs) 141, 163
polygon methods 83
polymerase chain reaction – *see* PCR
polymorphic and polymorphism 314–316, 324
polynomial 86, 88, 237
polynya 304
polytope envelopes 87

population
  assessment 285, 287, 289, 291, 293, 297, 302, 303, 342–349, 355–356, 358–359
  biology 107
  breeding 307
  decline 341–342, 344, 349, 352–353, 359
  definition 43
  density 42
  dynamics 42, 68, 96, 122, 138, 283, 284, 299, 332, 349, 352
  energetic – see energetics
  extinction risk 141
  genetics vi
  growth rate 135–137, 138, 139, 140
  health 144–146, 148–149, 152–154, 156, 158–161, 163–164
  matrix models 121, 137–138, 139, 140
  models and modelling vi, 120–122, 135–143, 342–343, 349–350, 352, 354, 357, 359
  process or parameters 222, 224, 227
  recovery 341, 343–344, 351
  requirements 189–190
  risk, at-risk 341–350, 353, 356
  size 45, 50, 66, 119, 139, 189–190, 310, 344, 354
  status/trend 341–342, 344–350, 352–353, 355
  structure 310–311, 315–316, 321
  studies 229
  usage of space 72–75
  viability analysis (PVA) 139, 141–142, 349–350, 356–357
populations vii, 69–75, 80, 83, 95
post-canine 104, 110
Potential Biological Removal (PBR) 336–337, 354–358
power analysis – see statistical power
precaution 5
precautionary principle or approach 335
precision 60, 65, 67
predation 260, 299, 300, 301, 346, 352
preference 82, 87–88
pregnancy and pregnant 39, 121–125
presence/absence 76, 85, 88–91
prey and prey distribution 68, 77, 82, 191–221, 227, 309
  abundance 233
  cephalopods 195, 198–205, 209
  consumption 154, 157, 184, 189–190, 191–193, 198, 200, 203, 206–209, 211, 213–215, 218, 221, 252
  crustaceans 195–196, 199–201, 205, 209
  distribution and abundance 245, 250, 252, 259
  DNA – see DNA analyses of diet
  encounters 245

energy content 198, 206
fatty acids – see fatty acid(s)
field 229–230
fish 194–197, 199–201, 203–206, 208, 209
  quality 252, 261
  remains 191–221 – see also diet
  stable isotopes – see stable isotopes
  types 191, 195, 201, 206–209
primer 316, 324
primiparity or primiparous 122, 147, 163
principal components analysis 77, 86, 90–91
principal coordinates analysis 273
professional societies 8, 14
progesterone 125
projection matrices 137
protection v
protein(s) 111, 151, 155, 166–167, 179, 182
  body protein content – see body composition
  catabolism 188, 210
  deposition 187–188
  electrophoresis 208–209
  in milk – see milk proximate composition
protozoa 146, 152
pseudo-absence 72, 76, 87
pseudoequilibrium 71, 75, 96
pseudo-range 228
public consultation, interest, and trust 3, 7, 11–12, 14–15
pulp cavity 110–111, 113
pupping 70
purse seine net 35, 37

QAIC 276

R (software) 52, 65, 94
radio frequency tags – see tags and tagging
radio telemetry – see telemetry
raft trap 30
RAMAS 142
Random Forest (software) 94
range 68, 233 – see also home range
range estimation 69, 80
Ranges (software) 94
recapture 74, 81 – see also capture–recapture
reciprocal monophyly 318, 324
recoveries, from dead animals 134
recruitment 285
redundancy analysis 87
Red List 344–346, 348, 358
refinement, reduction, replacement 9–10, 13, 15
refuges 68
regression – see model
regression tree 89–90
regulation 346, 350, 357–358

relatedness 307
relationship measures 264, 269, 273
  dyadic 270
  interaction rates 269
  matrices 270
remote sensing 81, 222
renal function 150–151, 154
reproduction 70, 120, 124, 127, 163, 165, 184,
    189, 242, 285, 288, 294, 297, 298, 302,
    303, 304
reproductive 127
  cycle 121, 122
  failure 163
  history 120
  isolation 318
  output 122
  rate 25, 40, 119, 120, 121, 122, 125, 307
  senescence 123
  status 13, 147, 150–151, 154–155, 163
research
  alternatives to 6, 10
  constraints 15, 16
  costs and benefits of 3–5, 7, 9–10
  ethics 1–15
  financial costs 4, 7, 13
  for the public good 7
  funding and investment viii, 6
  logistics 13
  permitting 3, 7, 16, 59, 350
  publication 4, 7
resource or resource management viii, 72, 325,
    333–334
resource selection functions 76
respiratory quotient (RQ) 169–172, 175
respiratory system and rate 33, 38
respirometry (gas exchange) 170–176
response variable 72, 76–77
restraining – see capture
restraint board 29
restriction endonuclease 315, 324
reticule 62
Revised Management Procedure (RMP) 331, 335,
    337, 338
Richard's model 117
rifle 50, 225
rigor mortis 107
RINEX files 228
risk
  cumulative risk 145, 148–149
  odds ratio 145, 149, 160
  rate 145, 148–149
  factors and assessment 146, 148, 160–164
  management 144, 160–164
  ratio 149

assessment or analysis 6–7, 11–12, 16, 39, 320
  avoidance 68, 260
rivers 70
RNA 146, 156
rookery 79
rostrum 99
roto-radio 226
Russia 308

Sable Island 284, 286
salinity 230
sample bias 108
sample size 10, 79, 224, 310, 320–320, 351,
    354
sampling effort and methods 72–74, 83, 90
Sarasota Bay 40, 107, 141, 292, 294
SAS (software) 94
satellite relay data loggers (SRDLs) 230, 249,
    257
satellite tag – see tags
satellite telemetry – see telemetry
Satellite Tracking and Analysis Tool 235
scale 76, 250, 252, 256
scatter plot 83
scientific societies 8
scientific uncertainty – see statistical uncertainty
seabirds 253
sea otter – see mustelid
sea star 300
sea state 44
sea surface temperature 58, 90, 227
sea urchin 300
seals 224, 233, 262 – see also otariids; phocids;
    pinnipeds
seal finger 41
sealworm parasite 284
search strategies/tactics 252, 256, 257
seasonality 75, 79
seawater 232
sedation 33, 38
sein net 35–36, 40
selectivity 126
senescence 123
sensors 223, 228–229
serology 146, 148, 154
set-up latency 233
sex, effects of 33, 49, 72, 81, 100, 116, 284,
    314
sex ratio 45, 310, 352
sexual maturity 117
sexual selection 291
ship 53–65, 72
ship strike 312, 340, 351, 359
shore 72

sightings and sighting probability 62, 74, 120, 123–124, 132, 134
signal-noise ratio 232, 237
silicone 224
single nucleotide polymorphism 315
singly labelled water (SLW) 176, 180–181, 187–188
sirenian 43, 53, 101, 105, 109
    dugong 36, 54, 66, 105, 109, 225, 228, 243, 244, 351
    manatee 114–115, 129, 139, 165, 174, 225, 340, 344, 346, 358
    Steller's sea cow 301, 341
    West Indian manatee 348–349
SIS (software) 95
skeletal/hard structure(s) 192–194, 196, 198–202, 203–206, 210, 221
    allometric reconstruction 198, 200–201, 205–206
    alternative structures 199–201
    biomass reconstruction 198, 203–204, 206–208
    carapaces 199, 201, 205
    cephalopod beaks 192, 195, 198–205
    digestion correction factors (DCFs) 204–205
    fish scales 196, 199, 209
    lenses 199–201
    minimum number of individuals (MNI) 200, 203
    numerical correction factors (NCF) 192, 198, 203–204
    numerical importance (N_i) 206
    occurrence indices 197–198, 206–209
    otoliths 194, 197–201, 203–205
    pens 199–200
    quantification 206–208
    recovery rates 203–204, 206
    reference material 192–193, 198–199, 205
    statoliths 199
skin (epidermis) 153, 154, 155, 226
    disease (histopathology) 150, 151, 152, 153, 158, 160
skull 305, 308, 101
Smirnov index 100
Smith, T.G. 302
smoothing 72, 82, 85–86, 237–283
snout 105–106
social behaviour or groups 9, 13, 68, 76, 78, 95, 116
social structure 263–265, 268, 269, 273, 280, 288–293
    evolution 278–279
    fission–fusion society 265
    mother–pup relationships 265
social systems 265, 273, 275, 276
Society for Conservation Biology 8
Society for Marine Mammalogy 8
sociogram 272
software 91–93, 278–279

solar power 223
sonogram 126
SOSUS 248
sound exposure 351, 358
sound production 38, 58, 222, 269
Southern Ocean 290–291, 329
space usage 72–76, 80–93
spatial analysis 69, 72–74, 80–93
    autocorrelation 84–85
    coordinates 81
    coverage 82
    density 69, 79
    dependence 85
    distribution 68, 72, 83
    ecology 80, 95
    extrapolation 80
    grid 79
    modelling 58–59, 72
    prediction 89, 91
    scale 74, 250, 252, 256
Species at Risk Act (SARA) 348
species concept 308
specific dynamic action (SDA) 165–167, 169
species identification 63
speed 236
    filter 239–240
spermatogenesis 122
spline 237, 241
S-plus (software) 94
squeeze cage 30
squid 243
stable isotopes/ratios 191, 193, 216–221
    analyses 219–221
    baleen, nail, teeth, whiskers 217–219
    blood 217–218, 220
    blubber 217
    mass spectrometry 216, 220
    milk 218–219
    muscle/internal organs 217, 220
    skin 217–218
    teeth 217–219
    trophic modelling 218–219
staining 110–112, 114
stakeholder 343, 352, 354, 355, 359
standard error 77, 79
standard length 101 – see also body length
staphlycoccus 40
state-variable 122
Statistica (software) 94
statistical
    advice 15
    analysis vii
    bias 44, 60, 72, 81, 85, 95, 99–100, 307
    cluster analysis 273

statistical (*cont.*)
  confidence limits 52, 190
  correlation 83
  covariates 55, 58–59, 64–65, 70, 76–77, 80, 82–88
  error (Type I and Type II) 11–12, 15, 27, 72, 95, 100, 319, 190
  goodness of fit 80
  inference 66, 72, 77, 86, 224
  likelihood 79
  measures 271
  methods 28
  model 43, 51–52, 58, 64–66, 95, 189–190, 342–343, 349–350, 352, 354, 357–359
  multidimensional scaling 273
  packages 90 – *see also* software
  power vii, 5, 10, 16, 224, 306, 310–320, 351
  properties 90
  sampling design 354
  significance 11
  uncertainty 5, 7, 11, 14, 85, 87, 91, 133, 135, 138–139, 142, 190, 237
  variance or variability 7, 10, 44, 52, 56, 60, 67, 77
statistician 67
statistics v, 79, 319
statoliths 199
sterilize 110
steroid 125
stochasticity 72–73, 86
stock 309
stomach temperature 229, 246, 253, 254
strandings 15, 72, 144, 147, 149, 152, 153, 154, 160, 194, 217, 321, 342, 354
stratification 61, 81, 321
stress and trauma 2, 17, 28, 39
stretcher net 32
strip transects 53, 72
stroking rate (*f*S) 176–178
STRUCTURE 318
study area 81, 87
study design – *see* experimental design
suckling behaviour 189
suction cup 225–227
sudden death 39
surface interval or time 246, 249
surfacing 57, 82, 224, 233
surgical procedures 11, 20
survey design – *see* experimental design
survey effort 72
survival or survival rate 25, 40, 45, 70, 119–120, 126–135, 154, 160, 161, 163, 222, 284–285, 291, 307, 349, 352–353

survival models 134, 137, 139, 142
sustainable harvest or exploitation 228, 326, 333
swell 64
swimming
  movement 233
  speed 228, 243, 246, 247–248, 251, 255
  strokes 230
Systat (software) 94
systematics 315, 324

tags and tagging 11, 17–19, 37, 43, 75, 223–242, 289, 290–295
  accelerometer 178, 183
  acoustic 227, 231,351
  Allflex tags 17
  animal-borne video 178, 191, 193
  archival 183, 233
  attachment 223–227
  Dalton roto tags 17, 24, 25
  tesign 222–223
  Discovery 19
  D-tag 225, 230, 235, 254, 255, 269
  double-tagging 48, 131, 229
  effects on animal 223
  failure 229
  geolocation 227–228
  hat tags 49
  implantation 226
  loss rates and retention 19, 25, 48
  Monel tags 17
  Peterson disk 25
  PIT and RFID 19–20, 131
  radio frequency 19–20, 224, 288
  radio-release 224
  satellite 183, 225, 226
  stomach temperature 229
  telemetry 177, 222–242
tail 104, 106
tail grab 35
tangle net 32
taxonomy 98, 306, 308–309, 317–319
TDR or time-depth recorder 222, 246, 249, 256
teeth – *see* dentition
telemetry 44, 50, 58, 72–74, 78–82, 96, 222–242, 248, 253
temperature 70, 112, 223, 229, 230
temperature regulation 165, 169, 172, 177
temperature–salinity 232
terrestrial 236, 306, 309
  mammals and birds 66
TESS 318
tessellation 83
testes 101

tether 226
thermal neutral zone (TNZ) 168–169
thermoplastic 226
thermoregulation 33, 38, 165, 169, 172, 177, 226
threatened species or populations 4, 12, 340, 342–348, 350, 352, 355, 359
  Red List 344–346, 348, 358
thresholding 80
throw net 32
tiletamine 33–34
time-depth recorder – *see* TDR
timed release 226
toluidine blue 112
tooth decalcification 110–112
tooth sectioning 110–112
toothed whales – *see* odontocetes
total body energy (TBE) 180–181
total body fat (TBF) 179–183, 187–188
total body protein (TBP) 179–181, 187–188
total body water (TBW) 173–176, 179–182, 187
toxicogenomics 157
toxin 145, 146, 147, 153, 154, 155, 157, 161, 163
tracking vi, 57, 64, 72 – *see also* telemetry
TrackTag 228
trade-offs 74, 100
training 13
tranquilizers 13
transcriptomes 316
transect 72–75, 79–82, 96 – *see also* line transect
Transect (software) 94
transition probabilities 134
translocation 352–353
transmission latency 233
transponder 74
traps and trapping 29–39, 74
trap-happy and trap-shy 48, 52
trauma 145–147, 152
trend-fitting 86
trigonometric functions 86
trinitrotoluene 223
tritium (^3H_2O) 173, 176, 180–181
trophic
  cascade 300
  interactions 191, 214, 216
  level 193, 210, 218–219, 221
  modelling 218–219
  tracer/marker 210, 217
truncated cone 101
tuna 37
turtle 236, 258
tusk 105–106, 109
tympanic bulla 115–116

ultrasonic measurement 107, 124, 126, 154, 181–182, 252
  transrectal and transvaginal 126
ultraviolet light 115
umbilicus 105–106
unbalanced sampling 70
uncertainty 335, 337, 339 – *see also* statistical uncertainty
  Estimation error 336
  Implementation error 335, 336
  Model error 335, 336
  Observation error 335, 336
US Fish and Wildlife Service (USFWS) 346
US National Marine Fisheries Service (NMFS) 346, 349, 355–358
usage estimation – *see* spatial analysis
usage maps 84, 96
uterine horn 124

vaccination 146
variance inflation factors 77, 79
variogram 85
vasculature 226
velocity 228 – *see also* swimming speed
veterinary care 13, 33–34, 38–39
VHF – *see* telemetry
video – *see* camera
video range finding 62, 66
viral and virus 145–146, 152, 157
visual data 62, 72
vital rates 119, 120, 127, 129, 135, 141
vocalization 243, 249 – *see also* sound production
von Bertalanffy model 117
VORTEX 142
vulnerable species or populations 345–346, 348, 358

walrus 22, 40, 109, 173, 225, 304, 346
Washington state 295
water (H_2O) 166
  body water pool, total body water (TBW) 173–176, 180–182, 187–188
  deuterated water (^2H_2O) 173, 175–176, 180–181
  doubly labelled water (DLW) – *see* deuterium (^2H_2O); tritium (^3H_2O); water turnover
  singly labelled water (SLW) – *see* deuterium (^2H_2O); tritium (^3H_2O); water turnover
  tritiated water (^3H_2O) 173, 176, 180–181
water temperature 246
water turnover 169, 173–176, 187–189
weaning 121
weather 75, 81
weight 81 – *see also* body weight

welfare 341, 342
whalers 328, 329
whales 75, 224 – *see also* odontocetes; mysticetes
whaling – *see* harvesting
whistles 294
wildlife 66
Winbugs (software) 95, 240

World Meteorological Organization 230

xiphosternal blubber 100

zolazepam 34
zooplankton 243
zygomatic arch 104